Proceedings of the 12th Italian Conference

Sensors and Microsystems

Proceedings of the 12th Italian Conference

Sensors and Microsystems

Napoli, Italy 12 – 14 February 2007

Editors

G Di Francia
ENEA CR Portici, Italy

P Maddalena
University of Napoli Federico II, Italy

I Rendina
CNR-IMM, Italy

C Di Natale
University of Rome "Tor Vergata", Italy

A D'Amico
University of Rome "Tor Vergata", Italy

World Scientific

NEW JERSEY • LONDON • SINGAPORE • BEIJING • SHANGHAI • HONG KONG • TAIPEI • CHENNAI

Published by

World Scientific Publishing Co. Pte. Ltd.
5 Toh Tuck Link, Singapore 596224
USA office: 27 Warren Street, Suite 401-402, Hackensack, NJ 07601
UK office: 57 Shelton Street, Covent Garden, London WC2H 9HE

British Library Cataloguing-in-Publication Data
A catalogue record for this book is available from the British Library.

SENSORS AND MICROSYSTEMS
Proceedings of the 12th Italian Conference

ISBN-13 978-981-283-358-7
ISBN-10 981-283-358-7

Printed in Singapore by World Scientific Printers

FOREWORD

The present volume contains the proceedings of the 12th Italian Conference on Sensors and Microsystems, organized by AISEM (Associazione Italiana Sensori e Microsistemi) and held in the town of Napoli on February 12-14 2007.

This 12th edition has been hosted in the Federico II University Congress Centre, just in front of the Megaride islet, where in the 8th century B.C., almost 3000 years ago, Neapolis (or Partenope) was founded by the Greeks, and where an impressive castle built by the Normans in the 12th century, named Castel dell'Ovo, continues to dominate the Gulf, defying time.

ENEA, Ente per le Nuove Tecnologie, l'Energia e l'Ambiente, Università di Napoli "Federico II" and CNR, Consiglio Nazionale delle Ricerche, have been the organizers of this scientific event.

The Conference has gathered the continuously increasing italian community working on sensors and microsystems. The present proceedings contains about 80 of the regular contributions that at the Conference were organized in 9 sessions: Chemical sensors, Physical Sensors, Microsystems, Biosensors, Optical Sensors and Microsystems, Device Fabrication and Assembly, Nanosensors, Array processing and networks.

During this Conference edition, for the first time in Italy, a showroom could be organized where practically all the electronic noses and tongues produced or investigated in Italy have been presented. Many thanks are due to the several groups that have contributed towards the success of this event.

Deep thanks are also due to Lina Sarro, of the Delft University (NL) and to prof. Renè Maury, of the University of Naples "l'Orientale", for the very interesting plenary lectures.

Arnaldo D'Amico
Girolamo Di Francia
Corrado Di Natale
Pasqualino Maddalena
Ivo Rendina

CONFERENCE SPONSOR

ENEA
Ente per le Nuove Tecnologie,
l'Energia e l'Ambiente

UNIVERSITÀ' DEGLI STUDI DI NAPOLI

CNR
Istituto di Microelettronica
e Microsistemi

ALFATEST

ASSING

DEMA

DKSH - MARKET INTELLIGENCE

FEI COMPANY - FEI ITALIA

G. GAMBETTI KENOLOGIA

HAMAMATSU ITALIA

IONVAC PROCESS

ITECO ENGINEERING

KENOSISTEC

LOT ORIEL GROUP - EUROPE

2M STRUMENTI

MAROTTA ADVANCED TECHNOLOGIES

PI - PHYSIK INSTRUMENTE

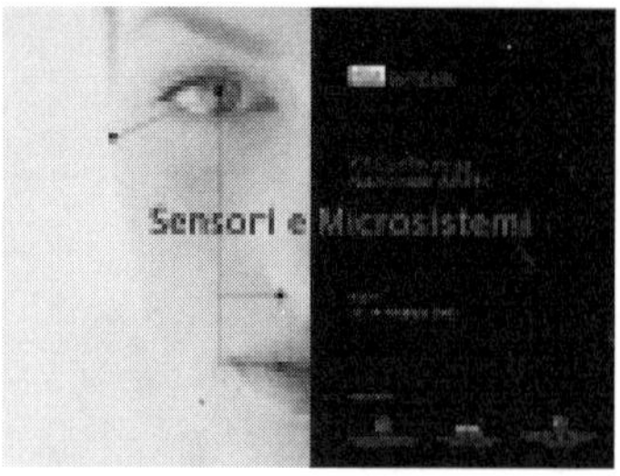

Organization

Girolamo Di Francia - *General Chairman*
ENEA - FIM MATNANO Portici

Pasqualino Maddalena - *Co-Chair*
Università di Napoli "Federico II"

Ivo Rendina - *Co-Chair*
CNR - IMM Napoli

Steering Committee

A. D'Amico - *Presidente AISEM*
Università di Roma "Tor Vergata"

L. Campanella
Università di Roma "La Sapienza"

P. De Gasperis
CNR-IMM Roma

C. Mari
Università di Milano

G. Martinelli
Università di Ferrara

U. Mastromatteo
ST Microelect. - Castelletto (MI)

A.G. Mignani
CNR-IFAC Firenze

M. Prudenziati
Università di Modena

G. Sberveglieri
Università di Brescia

P. Siciliano
CNR-IMM Lecce

G. Soncini
Università di Trento

Scientific Committee

M.C. Carotta
Università di Ferrara

P. Dario
Scuola Superiore S. Anna Pisa

F. Davide
Telecom Italia Roma

A. Diligenti
Università di Pisa

C. Di Natale
Università di Roma "Tor Vergata"

L. Dori
CNR-IMM-LAMEL Bologna

G. Faglia
Università di Brescia

C. Malvicino
CRFiat Orbassano (To)

G. Martinelli
Università di Ferrara

M. Mascini
Università di Firenze

N. Minnaja
Polo Navacchio SpA Navacchio - Cascina (PI)

B. Morten
Università di Modena

G. Palleschi
Università di Roma "Tor Vergata"

F. Villa
ST Microelctr. Castelletto (Mi)

M. Zen
ITC-IRST Trento

Local Organization AISEM 2007

S. De Vito
ENEA
Web site manager
http://aisem2007.portici.enea.it

Local Scientific Committee

R. Bernini
UniNa

L. De Stefano
CNR-IMM

S. De Vito
ENEA

M. Iodice
CNR-IMM

V. La Ferrara
ENEA

S. Lettieri
UniNa

D. Ninno
UniNa

G. Coppola
CNR-IMM

L. Quercia
ENEA

Organizing Secretariat

B. Alfano
ENEA

A. Ambrosio
UniNa

A. Castaldo
ENEA

A. Citarella
ENEA

E. Massera
ENEA

I. Nasti
ENEA

I. Rea
CNR-IMM

L. Rotiroti
CNR-IMM

A. Setaro
UniNa

V. Striano
CNR-IMM

T. Polichetti
ENEA

A. Del Mauro
ENEA

M. Gigliotti
CNR-IMM

D. Ascione
ENEA

Contents

Foreword v

Biosensors

Fabrication and characterization of the sensing element for glucose biosensor applications 3
V. Aiello, M. Fichera, F. Giannazzo, S. Libertino, A. Scandurra, M. Reins, F. Sinatra

Original tyrosinase organic phase enzyme electrode for the kinetic study of artificial rancidification of extra virgin olive oil 14
L. Campanella, A. Nuccilli, M. Tomassetti, S. Vecchio

Two new immunosensors and a biosensor for buffalo milk 19
L. Campanella, E. Martini, M. Tomassetti

Biosensors as new technologies for gene-doping investigation 24
T. Rusanova, D. Dell'Atti, S. Tombelli, M. Minunni, M. Mascini, P. Bogani, M. Buiatti

Graphite electrochemical sensors for the evaluation of PAHs-DNA adducts 31
M. Del Carlo, M. Di Marcello, M. Mascini, D. Compagnone

Antigen-antibody interaction on the gold surface modified by Langmuir-Shaeffer technologies with poly-pyrrole-derivatives monitored by μLibra 37
L. Schiavo, A. Scarpa, S. Greco

A novel technique for the direct detection of DNA hybridization 44
A. Savchenko, B. Snopok, M.G. Manera, J. Spadavecchia, P. Siciliano, R. Rella

Invited: Nanostructured-based sensors for analytical applications 49
F. Valentini, G. Palleschi, V. Biagiotti, M.L. Terranova, E. Tamburri

Screening of biomimetic receptors by means of high-density colorimetric microarray 57
M. Mascini, G. Guilbault, M. Del Carlo, M. Sergi, D. Compagnone

Performances of the immunogravimetric sensor μLibra 3.1 64
M. Passamano, S. Greco

LIVING PARAMETERS MONITORING

A study of indium oxide sensors for diabetes biomarker detection in the human breath 73
G. Neri, A. Bonavita, G. Micali, S. Ipsale, E. Callone, G. Carturan

Cytotoxicity of single-wall nanotubes on cultured human lymphocytes 80
O. Zeni, R. Bernini, M. Sarti, M.R. Scarfì, R.Palumbo, L. Zeni

Nanomaterials toxicity: An in-vitro investigation 86
G. Rametta, V. La Ferrara, G. Di Francia

Analysis of volatiles in the headspace of breast using a QMB based gas sensor array for breast cancer study: First evidences 93
A. D'Amico, C. Di Natale, M. Santonico, G. Pennazza, G. Mantini, M. Bernabei, E. Martinelli, R. Paolesse, S. Cabassi, A.G. Aronica, A. Calugi

GAS SENSORS

Pushing the limit of the silicon technology by using porous silicon: A CMOS gas sensing chip 103
G. Barillaro, P. Bruschi, F. Pieri, L.M. Strambini

Vapor sensor using thin film bulk acoustic resonator coated by carbon nanotubes-based nanocomposite layer 110
M. Penza, P. Aversa, G. Cassano, E. Serra, D. Suriano, W. Wlodarski, M. Benetti, D. Cannatà, F. Di Pietrantonio, E. Verona

Evaporation rate determination for water and alcohols in bubblers 116
A. Orsini, A. Bearzotti

All organic humidity sensors based on conjugated polymers and a tetracyanoquinodimethane salt 122
A. Arena, N. Donato, G. Saitta, G. Neri, G. Micali, G. Pioggia

Selective chemical sensors for NO_{2}- detection, using carbon nanotube/polymer composite nanowires 129
F. Valentini, V. Biagiotti, G. Palleschi, J. Wang

On the fabrication process of polymer-composites based sensors 137
A. De Girolamo Del Mauro, A. Citarella, E. Massera, L. Quercia, G. Di Francia

Development of QMB sensors based on iron porphyrins for carbon monoxide detection: A feasibility study 145
E. Mazzone, M. Mastroianni, C. Di Natale, R. Paolesse, M.I. Pistelli, F.Sintoni, A. D'Amico

Studies on chiral self-organization of amphiphilic porphyrin derivatives. Comparison between morphology in solution and in solid state 151
D. Monti, M. Stefanelli, M. Venanzi, M. Carbone, R. Paolesse C. Di Natale, A. D'Amico, S. Turchini, M. Girasole, G. Pompeo

Production and characterization of new Fe(TPP)Cl porphyrin films with improved optical gas sensing capabilities 157
M. Tonezzer, A. Quaranta, G. Della Mea, G. Maggioni, R. Milan, S. Carturan

Invited: Nanowires of semiconducting metal-oxides and their gas-sensing properties 162
C. Baratto, E. Comini, M. Ferroni, G. Faglia, A. Vomiero, G. Sberveglieri

Nanostructured conjugated polymers applied to sensors 172
I. Venditti, M.V. Russo, A. Bearzotti, A. Macagnano

Metal functionalised carbon nanotubes thin films gas chemiresistors 177
M. Penza, G. Cassano, R. Rossi, M. Alvisi, M.A. Signore, A. Rizzo, Th. Dikonimos, N. Lisi, E. Salernitano, E. Serra, R. Giorgi

Resistive λ-sensors based on Fe-$SrTiO_3$ nanopowders 185
G. Neri, A. Bonavita, G. Micali, G. Rizzo, R. Licheri, R. Orrù, G. Cao, D. Marzorati, E. Merlone Borla

Hydrogen sensor based on Pd nanowires 190
B. Alfano, V. La Ferrara, E. Massera, I. Nasti, G. Di Francia

Chemical sensors based on carbon nanotubes: Comparison between single and bundles of ropes 196
V. La Ferrara, B. Alfano, I. Nasti, E. Massera, G. Di Francia

Liquid Phase Sensors

Fiber optic sensors based on particle layers of tin dioxide for chemical detection in water and in air environments — 211
M. Consales, M. Pisco, P. Pilla, A. Cusano, A. Cutolo, A. Buosciolo, M. Giordano, R. Viter, V. Smyntyna

Synthesis and characterization of a polypyrrole nanowire modified electrodes for amperometric detection of ammonia in drinking water — 218
V. Biagiotti, F. Valentini, D. Moscone, G. Palleschi

Azulene based guest-host polymeric sensors — 223
A. Castaldo, L. Quercia, G. Di Francia

Optoelectronic nanosensors based on carbon nanotubes nanocomposites for the detection of environmental pollutants in air and water environment — 229
M. Consales, A. Crescitelli, A. Cutol, A. Cusano, S. Campopiano, M. Penza, P. Aversa, M. Giordano

Chemical Sensor Arrays and Networks

A multichannel quartz crystal microbalance for volatile organic compound analysis — 239
S. Pantalei, E. Zampetti, A. Macagnano, E. Proietti, C. Di Natale, A. D'Amico

Development of a new portable microsystem for wine quality analysis — 245
D.S. Presicce, L. Francioso, P. Siciliano, A. Adami, L. Lorenzelli, M. Malfatti, V. Guarnieri, M. Zen

Poly-pyrrole derivatives used as colorimetric sensors for volatiles detection — 251
F. Olimpico, A. Scarpa, O. Catapano, L. Fachechi, S. Greco

Invited: Analysis of NH_3/DMA/TMA mixtures by a multisensor miniaturised gas chromatographic system — 256
S. Capone, M. Zuppa, L. Francioso, I. Elmi, S. Zampolli, G.C. Cardinali, P. Siciliano

A gas microsensor array as new method to analyse the presence of unburned fuel in engine oil — 263
S. Capone, M. Zuppa, D.S. Presicce, F. Casino, L. Francioso, P. Siciliano

Enabling distributed VOC sensing applications: Toward TinyNose, a polymeric wireless e-nose 270
S. De Vito, E. Massera, G. Burrasca, A. Di Girolamo Del Mauro, D. Della Sala, G. Di Francia

Polypyrrole-derivatives sensor for traditional Italian cheeses discrimination by Libra Nose 278
A. Scarpa, L. Tortora, S. Greco

Neural calibration of portable multisensor device for urban atmospheric pollution measurement 283
S. De Vito, G. Di Francia, L. Martinotto

Optimization of support vector machines for quantitative e-noses 291
D. Esposito, S. De Vito, E. Massera, G. Di Francia, F. Tortorella

MICROFABRICATION AND MICROSYSTEMS

Experimental study of wetting phenomena in porous silicon by Raman scattering 303
M.A. Ferrara, L. Sirleto, G. Messina, M.G. Donato, S. Santangelo, I. Rendina

Invited: High flow rate permeation membrane on porous silicon for hydrogen filtering devices 310
R. Aina, U. Mastromatteo, F. Belloni, V. Nassisi, M. Renna, A. Romano

RF-MEMS coplanar shunt switches based on SU-8 technology 321
A. Lucibello, E. Proietti, S. Catoni, R. Marcelli, L. Frenguelli, G. Bartolucci

Phase shifters based on RF-MEMS coplanar shunt switches 329
D. Pochesci, S. Catoni, R. Marcelli, G. Bartolucci, F. Giacomozzi, B. Margesin

MEMS accelerometer calibration at low frequencies 337
F. Lo Castro, G. Brambilla, P. Verardi, A. D'Amico

Porous silicon membranes for drug delivery 343
I. De Santo, F. Causa, P. Netti, V. La Ferrara, I. Nasti, G. Di Francia

Silicon based transdermal drug delivery system 350
I. Nasti, V. La Ferrara, G. Rametta, G. Di Francia

Semiconducting nanoparticles in polymer films: Synthesis, characterizations, applications 356
T. Di Luccio, D. Carbone, M. Pentimalli, E. Piscopiello

Packaging methods for integrated thermal gas flow sensors 363
P. Bruschi, M. Dei, M. Schipani, M. Piotto

Electrical detection of cell adhesion in a single-cell electroporation biochip 370
A. De Toni, G. Cellere, M. Borgo, E. Zanoni, L. Santoni, L. Bandiera, L. Lorenzelli

Characterization of a silicon integrated micro-flow cytometer 377
R. Bernini, F. Brescia, M.R. Scarfì, R.Palumbo, E. De Nuccio, A. Minardo, L. Zeni, P.M. Sarro

Laser oxidation micropatterning of a porous silicon based biosensor for multianalytes microarrays 382
L. De Stefano, L. Rotiroti, I. Rea, E. De Tommasi, M.A. Nigro, F.G. Della Corte, I. Rendina

Feasibility of direct carbon nanotubes growth for sensing applications 388
T. Polichetti, O. Calò, P. Delli Veneri, T. Di Luccio, E. Massera, I. Nasti, P. Vacca, G. Di Francia

OPTICAL SENSORS AND MICROSYSTEMS

Invited: Metal-cladding leaky waveguides for chemical and biochemical sensing applications 401
R. Bernini, M. Tonezzer, G. Maggioni, S. Carturan, A. Quaranta, G. Della Mea, F. Mottola, A. Minardo, L. Zeni

Structured fiber Bragg gratings sensors: Perspectives and challenges 413
D. Paladino, M. Pisco, A. Cutolo, A. Cusano, A. Iadicicco, S. Campopiano, M. Giordano

CLASS: An innovative laser flow cytometer for the simultaneous measurement of size, refractive index, depolarization and fluorescence of cells 418
L. Fiorani, A. Palucci, V. Spizzichino

Optical probe for the turbine inlet temperature measurement in gas turbine plants 426
I. Gianinoni, E. Golinelli, U. Perini

Hollow-core optical fibers integrated with single walled carbon nanotubes as VOCs sensors 430
M. Pisco, M. Consales, A. Cutolo, A. Cusano, M. Giordano, P. Aversa, M. Penza, S. Campopiano

Multi-spectral extinction based optical system for the characterisation of particles and gases in thermoelectric power plants exhausts 435
E. Golinelli, S. Musazzi, U. Perini

Optical device for integrity assessment of thermal barrier coatings 440
L. De Maria, C. Rinaldi

Silicon resonant cavity enhanced photodetector based on the internal photoemission effect 448
M. Casalin, L. Sirleto, L. Moretti, F. Della Corte, I. Rendina

Electric and optical sensing in NIR environmental monitoring 458
M. Medugno

Multi-spectral infrared system for toxic gas detection 466
C. Corsi, N. Liberatore, S. Mengali, A. Mercuri, R.Viola, D. Zintu, M. Severi, G. Cardinali, I. Elmi, M. Passini

Characterization of a piezoelectric cap using a fiber Bragg grating 479
S. Rao, F.G. Della Corte

Optical fluorescence enhancers for trace detection of M1 Aflatoxin 483
C. Cucci, A.G. Mignani, C. Dall'Asta, G. Galaverna, A. Dossena, R. Marchelli

Magnetic field sensor at different pre-stress level 489
C. Ambrosino, D. Davino, C. Visone, A. Cutolo, A. Cusano, S. Campopiano, M. Giordano

The hyper-spectral optical signature of extra virgin olive oil 494
L. Ciaccheri, A.G. Mignani, H. Thienpont, H. Ottevaere, A. Cimato, C. Attilio

Physical Sensors

Fast scintillation readout by multi-pixel photon counting 503
R. Scafè , F. Pisacane, P.G. Gabrielli, G. Alonge, D. Della Sala, M. Salmi, R. Salmi, R. Pani, S. Salvatori, F. Zoccoli, G. Conte, F. De Notaristefani

Built-in strain measurements in porous silicon by Raman scattering 508
M.A. Ferrara, L. Sirleto, G. Messina, M.G. Donato, S. Santangelo, I. Rendina

Diamond detectors for X-ray beam monitoring 515
G. Conte, M. Girolami, S. Salvatori, R. Scafè, F. Pisacane, D. Della Sala

A finite element 2-dimensional model for the prediction of the frequency response of thermal gas velocity detectors 520
P. Bruschi, M. Schipani, N. Bacci, M. Piotto

Development of a multisensor layout for robots 526
M. Santoro, C. Moriconi

Optical strain sensor based on polymeric diffraction gratings 532
V. La Ferrara, I. Nasti, E. Massera, G. Di Francia

Systems and Electronic Interfaces

Uncalibrated high-dynamic range resistive sensor front-end with parallel capacitance estimation 539
A. De Marcellis, G. Ferri, V. Stornelli, A. Depari, A. Flammini

A 77 Hz lock-in amplifier for sensor applications 545
G. Ferri, A. De Marcellis, V. Stornelli, A. D'Amico, C. Di Natale, C. Falconi, E. Martinelli

CCII-based oscillator for sensor interface 551
V. Stornelli, G. Ferri, A. De Marcellis

Integrated wireless temperature sensor with on-chip antenna 556
F. Aquilino, M. Merenda, F. G. Della Corte

BIOSENSORS

FABRICATION AND CHARACTERIZATION OF THE SENSING ELEMENT FOR GLUCOSE BIOSENSOR APPLICATIONS

V. AIELLO

Università degli Studi di Catania, Dipartimento di Chimica Biologica, Chimica Medica e Biologia Molecolare, Viale A.Doria, 6, 95125 Catania, Italy

CNR – IMM unità Catania, Stradale Primosole 50, 95121 Catania, Italy

M. FICHERA, F. GIANNAZZO, S. LIBERTINO

CNR – IMM unità Catania, Stradale Primosole 50, 95121 Catania, Italy

A. SCANDURRA

Laboratorio Superfici e Interfasi (SUPERLAB), Consorzio Catania Ricerche, Stradale Primosole 50, 95121 Catania, Italy

M. REINS

Università degli Studi di Catania, Dipartimento di Chimica Biologica, Chimica Medica e Biologia Molecolare, Viale A.Doria, 6, 95125 Catania, Italy

F. SINATRA,

Università degli Studi di Catania, Dipartimento di Scienze Biomediche, Via S. Sofia, 87, 95100 Catania, Italy

Aims of this work were the optimization of a protocol for the immobilization of a biological molecule on an inorganic platform, the manufacture of the sensing element of a biosensor and its full characterization. We used the Glucose Oxidase (GOD) as biological molecule to immobilize and Si-based surfaces as inorganic platform. To define and optimize the best protocol, non biological techniques such as Atomic force Microscopy (AFM), X-ray Photoelectron Spectroscopy (XPS) and Energy Dispersive X-ray (EDX) coupled to Scanning electron Microscopy (SEM), were used. AFM and XPS measurements were employed to study the surface coverage by linker molecule. Finally, EDX measurement allowed us to provide a direct measurement of the enzyme presence into the sample cross section. Through these techniques, normally used in the microelectronic field, we demonstrated that the immobilization protocol allows one to obtain an uniform layer of linker molecule and its effectiveness.

1. Introduction

Recently, research in the area of biosensors has grown rapidly due to the need of miniaturization, mass production, accurate analysis and prompt measurement felt by the users community [1]. A biosensor is an analytical device combining a biological molecule and a physical transducer. The biological molecule acts as sensing element (recognition element), while the transducer produces an electrical or optical signal output proportional to the analyte concentration.

We used the Glucose Oxidase (GOD) extracted from the *Aspergillus niger*, as recognition element. It is an enzyme and catalyzes the β-D-glucose oxidation in δ-gluconolactone and the molecular oxygen reduction in hydrogen peroxide. GOD is a dimeric protein with a molecular weight of 160 KDa (80 KDa per monomer) and dimension of 7nm x 5.5nm x 8nm. Each monomer binds a FAD cofactor, which acts as a redox carrier in catalysis, with a noncovalent bond [2-3]. The polypeptide chain of each monomer has 538 amino acid residues [4]; three of these amino acid residues are cysteins: two are involved in disulfide bonds, while the third is a free thiol group [5]. This enzyme is used to monitor the glucose concentration in the blood [6]. For this reason, GOD is used for the fabrication of macro and, tentatively, micro glucose biosensors. A miniaturized glucose sensor could have immediate applications to diabetes monitoring.

We tested and optimized a protocol to immobilize the GOD on bulk SiO_2 and, subsequently, on porous silicon dioxide ($PSiO_2$). The choice of Si-based materials as inorganic platforms is due to the fact that Si has a mature and low cost technology and offers the possibility to integrate microelectronic devices. Moreover, the device dimension could be shrunk and the surface micromachined in order to fabricate new structures [7]. Therefore the Si-based materials are considered the best candidates for the next generations of biosensors.

The use of $PSiO_2$ is evaluated since it is biocompatible, can be easily manufactured using ULSI (Ultra Large Scale Integration) technology, has a wide surface/volume ratio and allows device performances 350 times higher than non-porous surface fabricated devices [7].

In this work, we applied a covalent immobilization protocol consisting of four step: 1) oxide activation, 2) silanization, 3) linker molecule (we used glutaraldehyde) deposition and 4) GOD coupling.

Surface measurement, such as Atomic force Microscopy (AFM) and X-ray Photoelectron Spectroscopy (XPS), demonstrated that the functionalized samples are uniformly covered with the organic layer and the enzyme is immobilized on their surface. Moreover, Energy Dispersive X-ray (EDX) measurements coupled to scanning electron microscopy (SEM) performed on the $PSiO_2$ cross section show that it is possible to immobilize the enzyme into $PSiO_2$ with good results. In fact, SEM-EDX allowed us to make a direct analysis of the enzyme presence into the sample.

2. Materials and Methods

The enzyme immobilization protocol used in this work is summarized in four steps and it is described in detail elsewhere [8]. In the first step the samples were immersed in a solution (SSC) of NH_3:H_2O_2:H_2O (1:1:10). After, they were treated with vapors of 3-aminopropyltriethoxysilane (APTES). The linker deposition was carried out using glutaraldehyde (GA) 2.5% in 0.1 M phosphate buffer solution (PBS), pH 6.5. Finally, the samples were immersed in a GOD solution 2 mg/ml in 0.1 M PBS, pH 6.5, overnight at room temperature (RT).
3-aminopropyltriethoxysilane (APTES), glutaraldehyde (GA) solution (grade II), glucose oxidase (GOD, type X-S, *Aspergillus niger*), di-sodium hydrogen phosphate anhydrous (Na_2HPO_4), sodium di-hydrogen phosphate (NaH_2PO_4), hydrogen peroxide at 30% (H_2O_2) were purchased from Sigma Chemical Co., St Louis, USA. The other chemicals used were purchased by Carlo Erba Reagenti (Italy). The water used is deionized, milliQ, having 18 MΩ resistivity.
To optimize the immobilization protocol, we prepared a set of samples. In particular, some samples were stopped after the various steps of the protocol (SSC; SSC+APTES; SSC+APTES+GA); others underwent the full protocol, while the references were treated only with water.
Si wafers, 6 inches, underwent an oxidation process at 950°C for 30 min in O_2 ambient and a thin oxide layer (~8 nm, as measured by ellipsometry) was grown on the Si surface. The $PSiO_2$ provided by Dr. L. La Magna of STMicroelectronics, was obtained following the protocol reported elsewhere [8].
AFM measurements were performed in air at room temperature using a XE 150 by PSIA scanning probe microscope operated in no contact mode. The scanned area was 1μm×1μm. The surface root mean square (RMS) roughness measured was defined as the standard deviation of the heights (Z values) in a XYZ three dimensional AFM map.
XPS measurements were carried out using a Kratos AXIS-HS spectrometer. The Mg K$\alpha_{1,2}$ of 1253.6 eV was used at the conditions of 10 mA and 15 keV with a

pass energy of 40 eV. During the analysis the residual pressure in the chamber was on the order of 10^{-7} Pa.

SEM measurements were carried out with a LEO 1550 equipped with a Oxford 7426 EDX using the INCA software program. The samples were cleaved in order to observe their cross section.

3. Results and Discussion

AFM measurements were carried out to detect the surface modification of planar SiO_2 samples during each step of the immobilization protocol. These analysis show an evolution of the surface modification due to the immobilization steps. The sample that underwent only the fist step (only the SSC solution) (fig. 1A) exhibited an RMS of 0.11 nm, similar to the reference sample RMS (not shown). This result showed that the sample surface is not modified by the oxide activation step, as expected. After the other protocol steps, we observed and increase in the RMS value up to 0.61 nm, measured on the sample that underwent the full process of immobilization (fig. 1B).

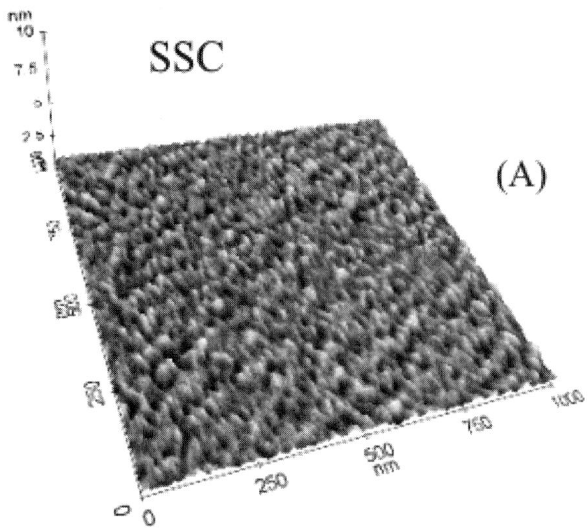

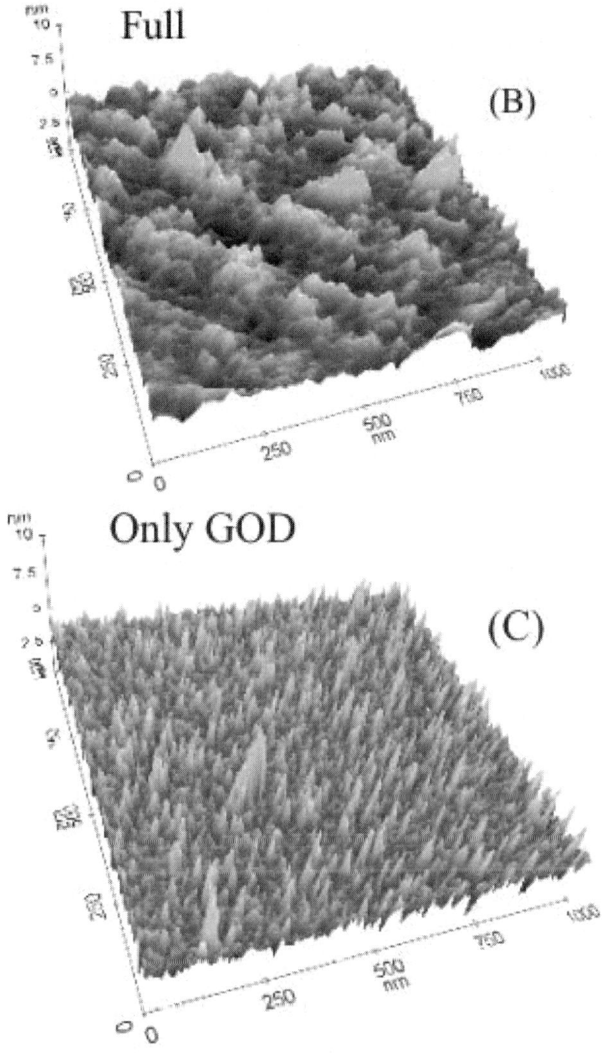

Figure 1:Three-dimensional AFM images of: (A) reference, (B) fully processed, (C) only GOD samples.

When GOD was deposited directly on the sample unfunctionalized surface, an RMS of 0.64 nm (fig. 1C), was measured by AFM. This value is similar to the one of the fully processed sample. Nevertheless, the two samples showed a different surface morphology. In fact, high and broad features were detected in the fully processed sample, while smaller and sharp features were observed in

only GOD sample. Moreover, the peaks heights of the only GOD sample were around 5 nm, in agreement with the enzyme dimension [9].
AFM measurements allowed us to detect the surface modification after the functionalization, but the study of the deposited chemical species was carried out by XPS measurements. This is a powerful technique, able to provide chemical bonding information and molecular composition of the materials lying on the sample surface or in the first layers beneath it [10-11]. The C1s, N1s and Si2p spectra of the all set of samples were carried out, but only the most meaningful data are shown in this paper (the full data are presented in Ref. 9). The C1s spectrum of the fully processed sample, shown at the bottom of fig. 2, has three peaks, at 285 eV, 286.3 eV and 288.3 eV. The peak at 285 eV, that was detected also in the reference sample (fig.2, top), is due to C-C and C-H bonds. On the other hand, the peaks at 286.3 eV and 288.3 eV are typical of R-CH_2*-NH-(CO)-, and R-CH_2-NH-(C*O)- chemical groups respectively, hence clearly due to the GOD presence.

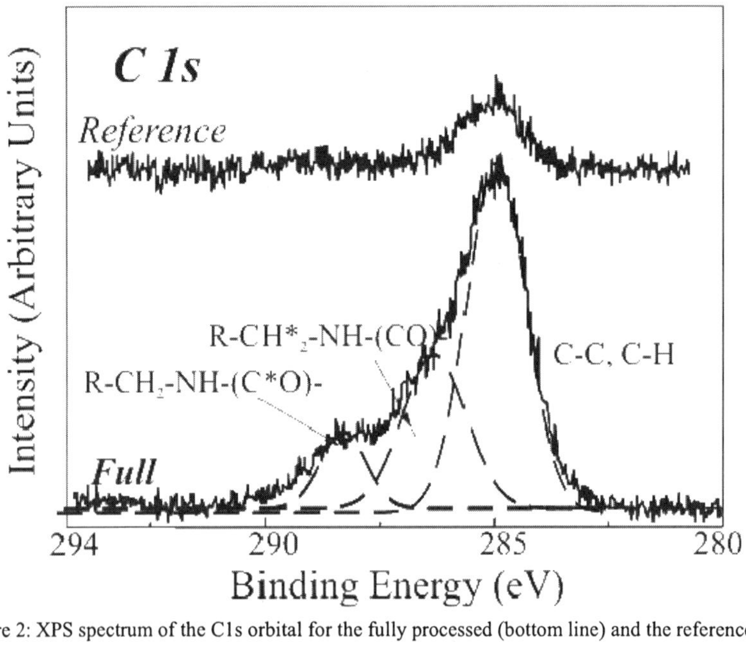

Figure 2: XPS spectrum of the C1s orbital for the fully processed (bottom line) and the reference (top line) samples.

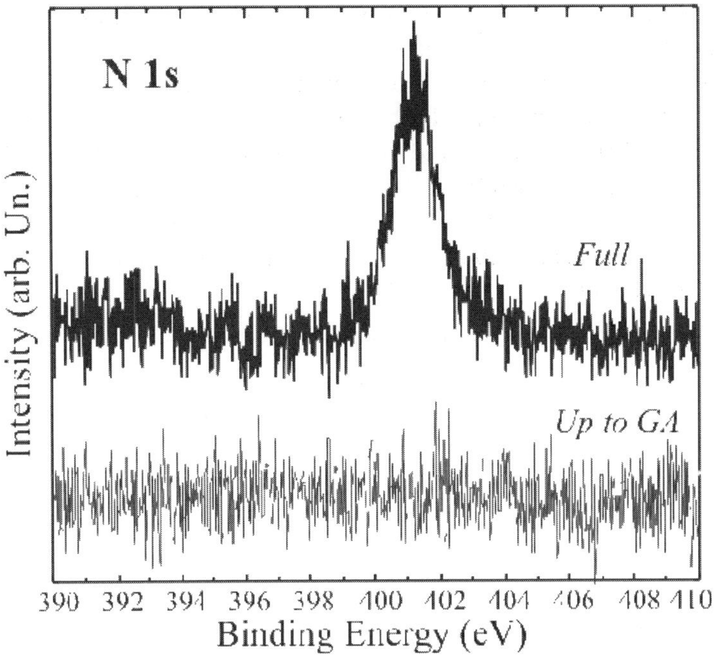

Figure 3: XPS spectrum of N1s orbital for the fully processed (top line) and the up to GA (bottom line) samples.

Moreover, the spectrum reported in fig.3 shows the N1s peak at 401 eV and indicated that N is detectable only in the sample that underwent the whole process. The other samples, stopped after the various steps did not exhibit this signal.

Finally, the uniformity of the organic layer deposited on the oxide surface up to the linker molecule deposition, was tested by XPS. We monitored the Si2p signal, as reported in fig. 4. In particular, we detect the Si2p signal of: reference (top line), up to GA (SSC+APTES+GA; second line from the top) and fully processed samples (bottom line). The reference sample (top line) exhibited two peaks with binding energies at 99.7 eV and 104 eV assigned to Si° and SiO_2 respectively. We monitored the Si° peak after each immobilization step to determine the uniformity of the extra layer produced on the SiO_2 surface during its functionalization. The film deposited on the Si substrate, given by both the SiO_2 and the organic layers (APTES+GA+GOx) had a thickness such to completely shield the substrate Si2p signal, as clearly observed from the spectrum of the fully processed sample of fig. 4 (bottom spectrum). The observation of the up to GA sample (second spectrum from the top of the same figure) clearly showed that such signal was fully shield already after GA deposition (SiO_2+APTES+GA). It is due to a perfect shield of the substrate operated by a uniform layer of organic material. Therefore, we demonstrated that the immobilization protocol let us to obtain an uniform layer of linker molecule. It should be mentioned that the sample that underwent only the deposition of GOD, still exhibited the substrate signal. This result demonstrated that when the surface was not properly functionalized a non uniform layer was obtained. This last result provided the final confirmation of the efficiency of our immobilization protocol [9].

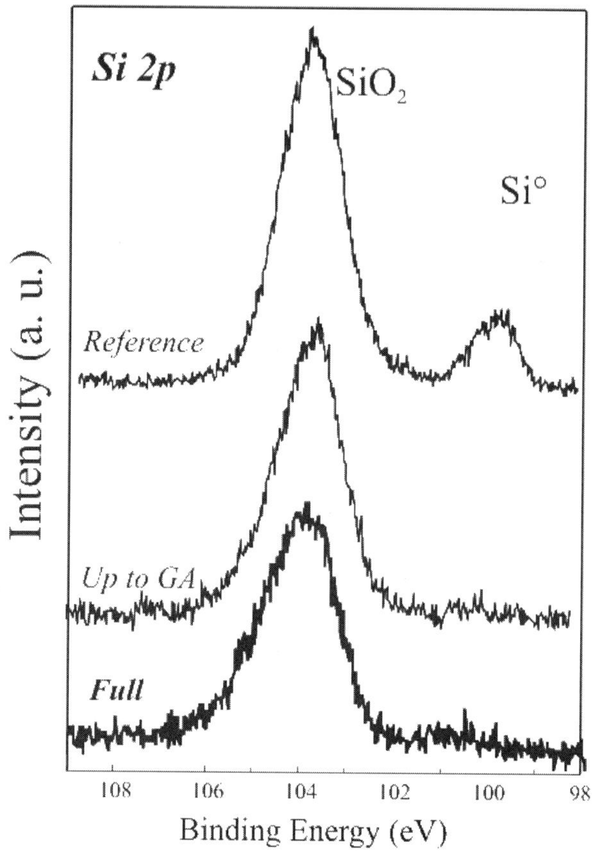

Figure 4: XPS spectrum of Si2p orbital for the fully processed (bottom line) and up to GA (second line from the top) anf reference (top line) samples.

Finally, to enhance the surface area, we applied the same immobilization procedure to a $PSiO_2$ substrate and carried out SEM-EDX measurements. The samples underwent the same immobilization procedure already characterized, were cleaved just before the measurements to image the sample cross section. The samples were only cleaved in order to reduce at minimum the artefacts due to sample preparation. In fig. 5A the SEM image of the $PSiO_2$ layer cross-section is shown, the porous layer thickness was about 3 μm. The EDX spectrum of the fully process sample (fig. 5 B) showed three peaks at 0.34 keV, 0.51 keV and 1.73 keV corresponding to the K_α emission of C, O and Si atoms, respectively. Moreover, also the N atom peak was detected. It was not observed in the reference sample (not shown), while, in this sample the C peak was still present but with a lower intensity. These measurements provided an experimental direct evidence of the GOD immobilization in the porous matrix [12].

A

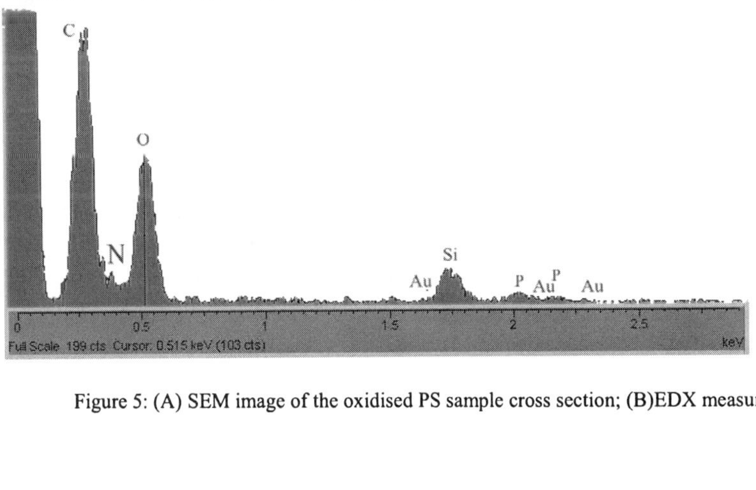

B

Figure 5: (A) SEM image of the oxidised PS sample cross section; (B)EDX measurements

Conclusion

We optimized the immobilization protocol of GOD on Si-based surfaces, using techniques typical of microelectronic devices characterization. In particular, we monitored the protocol on bulk SiO_2 by AFM and XPS techniques. AFM measurements allowed us to study the samples surface topography and its modifications after the various steps of immobilization. XPS measurements permit us to detecte the GOD presence.. In fact, the fully processed sample showed the peak at 285 eV, typical of C-C and C-H bonds, and two more peaks at 286.3 eV and 288.3 eV, typical of R-CH_2*-NH-(CO)-, and R-CH_2-NH-(C*O)- chemical groups, respectively. This result and the N presence demonstrated that there was enzyme immobilized on the sample. Moreover, XPS analysis allowed us to test the uniformity of the organic layer deposited on the oxide surface up to the linker molecule deposition. In fact,. the Si2p orbital showed two well distinct peaks having binding energies of 99.7 eV and 104 eV, assigned to Si° and SiO_2 respectively. These signals were detected in all the samples that underwent the first three immobilization protocol steps. The sample that underwent the overall protocol exhibited only the peak at 104 eV, since the organic film deposited on the Si substrate had a thickness such to completely shield the Si° signal. Moreover, the sample that underwent only GOD deposition, still showed such signal. This was a clear demonstration that the immobilization protocol allowed us to obtain an uniform layer of linker molecule, hence its efficiency.

Finally, we used the protocol to functionalize the Porous Silicon dioxide and demonstrate directly the enzyme presence into the sample cross-section by Energy Dispersive X-ray (EDX) coupled to Scanning electron Microscopy (SEM).

Acknowledgments

The authors acknowledge Dr.L. La Magna of the Catania site of STMicroelectronics for providing the oxidized porous Si samples, Dr. V. Raineri for the useful discussions about AFM measurements, Mr. A. La Mantia and Mr. M. Torrisi of the failure analysis lab. of the Catania site of STMicroelectronics for the SEM–EDX measurements, A. Spada and N. Parasole of CNR – IMM and G.F. Indelli of SUPERLAB for the expert technical assistance. This work was partially founded by the regional project "POR Sicilia 2000-2006, Mis. 3.15" and by Univ. of Catania within the project

"Piano ICT per l'Eccellenza del Settore Hi-Tech nel Territorio Catanese (ICT-E1)".

References

1. S. Bharathi, M. Nogami. Analyst **126** (2001) 1919
2. Q.H. Gibson, B.E.P. Swoboda, V. Massey. J. Biol. Chem. **239** (1964) 3934.
3. G. Zoldák, A Zubrik, A. Musatov, M. Stupák, E. Sedlák. J. Biol. Chem. **279** (2004) 47601.
4. K.R. Frederick, J Tung, R.S. Emerick, F.R. Masiarz, V. Massey. J. Biol. Chem. **265** (1990) 3793.
5. V.R. Sarath Babu, M.A. Karanth, M.S. Thakur. Bios. & Bioel. 19 (2004) 1337.
6. J.G. Wagner, D.W. Schmidtke, C.P. Quinn, T.F. Fleming, B. Bernacky, A. Heller. PNAS **95** (1998) 6379.
7. M. Fichera, S. Libertino, G. D'Arrigo. Proc. SPIE Int. Soc. Opt. Eng. 5119 (2003), 149.
8. S. Libertino, Fichera, V. Aiello, G. Statello, P. Fiorenza, F. Sinatra. Microelect. Eng.**84/3** (2007) 468.
9. S. Libertino, F. Giannazzo, V. Aiello, A. Scandurra, F. Sinatra, M. Renis, M. Fichera. submitted for publication on Langmuir.
10. D. Briggs, J. T. Grant. Eds. IM Publications, Surface Spectra Ltd., Chichester, 2003.
11. D. Briggs, M.P. Seah. Eds. John Wiley & Sons, Chichester, 1990; Vol. 1 Second Edition.
12. S. Libertino, V. Aiello, P. Fiorenza, M. Fichera, A. Scandurra, F. Sinatra. submitted for publication IEEE.

ORIGINAL TYROSINASE ORGANIC PHASE ENZYME ELECTRODE FOR THE KINETIC STUDY OF ARTIFICIAL RANCIDIFICATION OF EXTRA VIRGIN OLIVE OIL

LUIGI CAMPANELLA[a], ADRIANO NUCCILLI[a], MAURO TOMASSETTI[a], STEFANO VECCHIO[b]

[a]*Dipartimento di Chimica, Università "La Sapienza", Piazzale Aldo Moro, 5, 00185 Roma (RM).*

[b] *Dipartimento di Ingegneria Chimica, Università "La Sapienza", Via del Castro Laurenziano, 7, 00161 Roma.*

Abstract

There is no need to point out the great importance of extra virgin olive oil both as a foodstuff and from the agroindustrial point of view. As for all fats, the proper conservation of the product is essential also for extra virgin olive oil. Stability to self-oxidation, which leads to product rancidification, is thus very important and has been the subject of many studies, also by our group. In the present research, attention was focused on the artificial isothermal rancidification process at 5 different temperatures ranging from 98 to 180°C using a tyrosinase biosensor operating in n-hexane. This made it possible to calculate all the principal kinetic parameters of the process under examination.

1. Introduction

There are several potential advantages in carrying out enzymatic reactions in organic media instead of in aqueous solution, and consequently in fabricating enzyme sensors working in organic solvents (OPEEs): efficient catalysis may be achieved with substrates or real matrices that are poorly soluble in water, for instance, oils and fats; undesirable side reactions in organic media as well as substrate and product inhibition can be reduced; the thermal stability of the enzyme is enhanced; hard immobilization is often unnecessary as enzymes are insoluble in organic solvents; microbial contamination is eliminated and recovery and reuse of the enzyme are easier [1]. One of the most useful OPEEs developed in recent times is the tyrosinase OPEE, which has proved very useful in polyphenols determination in extra virgin olive oil [2] working in organic media. One of the most important problems concerning extra virgin olive oil is its conservation and therefore the kinetics of the rancidification process. The focus of the present study was consequently the artificial rancidification process produced in extra virgin olive oil heated in an oxidizing environment for the purpose of constructing a kinetic model of oxidizing thermal breakdown of the polyphenols contained in it. To this end a series of isothermal oxidative breakdowns of extra virgin olive oil were carried out at different temperatures (98, 120, 140, 160, 180°C), respecting the main indications referring to the study of rancidification of the oil, characteristic of the procedure developed by AOM [3].

2. Methods

2.1. *Polyphenols Determination*

In all the tests performed the variation of polyphenol concentration of the extra virgin olive oil over time at different temperatures was determined using a tyrosinase biosensor (OPEE) operating directly in n-hexane. Thanks to this original biosensor, constructed entirely of Teflon and which allows the measures to be performed while immersed in n-hexane, a solvent in which the extra-virgin olive oil samples are highly soluble, it was possible to construct experimental curves describing how polyphenol concentration varied, that is the concentration of the principal natural antioxidants contained in the extra-virgin olive oil, as a function of time at 5 different temperatures ranging from 98 to 180°C. The "model-fitting" method [4] was applied to acquire the kinetic constant values at various temperatures (including the activation energy value) and to identify the equation that best fits the experimental curve representing the trend of the entire degradation process. Lastly, interesting observations were made concerning the half-life of polyphenol concentration reduction at different temperatures ranging from 98 to 180°C. Essentially an exhaustive kinetic model of the rancidification process was obtained on the basis of measures that it was possible to perform using the special tyrosinase OPEE developed by us [2].

2.2. *"Model Fitting" Method*

The "model-fitting" method [4] is designed to determine the kinetic parameters of a chemical process. This is achieved by selecting from among the more likely, mathematically defined, kinetic models the one(s) that best fit(s) the experimental data obtained. The aim is thus to decide which of the more common $f(\alpha)$ functions reported in the literature best represent the trend of the experimental data referring to the degradation process, considering that the integral function $g(\alpha)$ must be a linear function of temperature, i.e.:

$$g(\alpha) = \int d(\alpha)/ f(\alpha) = kt$$

then calculating from the experimental data obtained using the sensor the values of the degree of conversion α for the entire extra virgin olive oil degradation process at the various set temperatures (Table 1). Consideration was then given

to all the principal f(α) functions reported in the literature [4], together with the corresponding g(α) functions, where α is the degree of conversion. In applying the method it was necessary to construct the regression curves performing a graphic interpolation of the experimental data. Then, plotting the values of g(α) as a function of time t_α for each of the equations corresponding to the different kinetic models considered at the different degradation temperatures, the trends were obtained with different degrees of linearity according to the greater or lesser correspondence between the trend of the experimental data (Table 1) and the kinetic model used to represent them. The values of the slopes of the straight lines best fitting these trends provide the value of the kinetic constant (k) at the different isothermal temperatures considered.
The evaluation of which of all the kinetic models examined is the one that best fits the kinetic thermal degradation trend of the polyphenols in extra virgin olive oil was performed by determining the coefficient of correlation r^2. Furthermore, by introducing Arrhenius equation it was then possible to compute the value of the activation energy.

3. Results

Table 2 shows the selected equation that best fits the isothermal degradation process of polyphenols in extra virgin olive oil. The value of the computed activation energy as well as those of the kinetic constants at the 5 different isothermal temperatures considered are also shown in Table 2.

4. Conclusion

The tyrosinase OPEE (organic phase enzyme electrode) used in n-hexane proved quite valid in monitoring the oxidative degradation of the polyphenols contained in extra virgin olive oil. The experimental values obtained throughout the experiment allowed the principal kinetic parameters to be computed, including the kinetic constant values at temperatures ranging from 98 to 180°C. It is very interesting to observe that at 180°C the kinetic constant is about twenty times greater than at 98°C. This indicates the decisive role of the temperature in the rancidification of the oil and therefore the great importance of temperature when the oil is used for example to cook food, in order to determine the maximum period over which the oil may be used.

Table 1. Values of the conversion degree α (α=(Ci –Ct)/ Ci) for the thermal degradation of polyphenol pool obtained by isothermal measurements, at different temperatures, for extravirgin olive oil.

T (°C)	*t* (min)	*α*	*t* (min)	*α*	*t* (min)	*α*	*t* (min)	*α*
98°C	0	0.00	60	0.60	1620	0.81	3120	0.91
	10	0.02	120	0.62	1800	0.86	3360	0.94
	20	0.03	480	0.76	1980	0.87	3600	0.97
	30	0.30	1380	0.77	2820	0.88	3840	1.00
120°C	0	0.00	20	0.49	150	0.74	570	0.90
	3	0.07	30	0.55	210	0.76	720	0.91
	6	0.09	60	0.56	330	0.80	840	0.93
	10	0.11	90	0.86	450	0.88	960	0.96
	15	0.27						
140°C	0	0.00	20	0.47	150	0.74	570	0.94
	3	0.17	30	0.53	210	0.79	720	0.98
	6	0.20	60	0.55	330	0.90	840	1.00
	10	0.23	90	0.71	450	0.93		
	15	0.41						
160°C	0	0.00	10	0.32	30	0.53	210	0.85
	3	0.02	15	0.34	60	0.67	330	0.91
	6	0.32	20	0.36	90	0.74	450	0.97
180°C	0	0.00	10	0.67	30	0.80	120	0.91
	3	0.22	15	0.70	60	0.87	210	0.95
	6	0.65	20	0.72	90	0.89		

Table 2. Calculated values of principal kinetic parameters concerning the isothermal polyphenol degradation of extra virgin olive oil.

Best kinetic f (α) function	$(1-\alpha)^2$				
Best kinetic g (α) function	$1/(1-\alpha)-1$				
Activation Energy (kJ mol^{-1})	36 ± 4				
Isothermal degradation temperature (°C)	98	120	140	160	180
Kinetic constant (s^{-1})	$0.7\ 10^{-4}$	$2.5\ 10^{-4}$	$3.2\ 10^{-4}$	$6.1\ 10^{-4}$	$12.7\ 10^{-4}$

References

1. L. Campanella, G. Favero, M.P. Sammartino, M. Tomassetti, J. Mol. Cat. B, Enzym., 7, 1997, pp. 101-113
2. L. Campanella, G. Favero, M. Pastorino, M. Tomassetti, Biosensor & Bioelectronics, **14**, 1999, pp. 179-186
3. V. Baran, M. Colonna, M. Di Toro, V. Greco, Phys. Rev. Lett., **86**, 4492 (2001). AOCS Official and Tentative Method: Cd 12-57. Reapproved 1997 Fat Stability, Active Oxygen Method
4. F. Rodante, S. Vecchio, M. Tomassetti, J. Pharm. Biomed. Anal., **29**, 2002, pp. 1031-1043

TWO NEW IMMUNOSENSORS AND A BIOSENSOR FOR BUFFALO MILK

LUIGI CAMPANELLA, ELISABETTA MARTINI, MAURO TOMASSETTI

University of Rome "La Sapienza", Department of Chemistry, P.le A. Moro 5, 00185 Rome, Italy

The increasing commercial interest in exploiting the therapeutic value of Lactoferrin has stimulated the need for reliable assays for its determination in milk at the endogenous level. In this study we developed and characterized an immunosensor for the determination of antibacterial protein (Lactoferrin), with the aim of suggesting this procedure for routine control of an important foodstuff product such as buffalo milk. At the same time, using also another new electrochemical immunosensor for the measurement of immunoglobulin G, it proved possible to determine also immunoglobulin G in the same buffalo milk samples. Briefly, milk contains several protective proteins, including Lactoferrin and immunoglobulin G, that can contribute to the preservation of milk. In the present paper we successfully studied two new immunosensor devices for the analysis of both of these important proteins.

1. Introduction

Milk contains various protective proteins, including lactoferrin and immunoglobulin G, which can contribute to the preservation of the milk itself [1-3] and give it its interesting characteristics as a foodstuff. In the present study we developed two different immunosensors for the determination of antibacterial proteins (lactoferrin and immunoglobulin G) in buffalo milk, with the aim of proposing these devices for the routine control of an important foodstuff product such as bovine milk. To this end we developed two different kinds of immunosensors: one for the analysis of immunoglobulin G, the novelty of which is the use of an enzyme sensor as detector; the other a new immunosensor for the quantification of lactoferrin. Thorough-going research was carried out in the development of the latter immunosensor. Therefore the increased commercial interest in exploiting the therapeutic value of lactoferrin has actually stimulated the need for reliable assays for its determination in milk at the endogenous level [4, 5]. Furthermore, a competitive procedure was used for lactoferrin determination in which the antibody (anti-lactoferrin) was conjugated with horseradish peroxidase enzyme using a biotinylation process.

2. Methods

The biotinylation of lactoferrin and the subsequently used competition procedure for the measurement by the lactoferrin immunosensor is illustrated in figure 1.

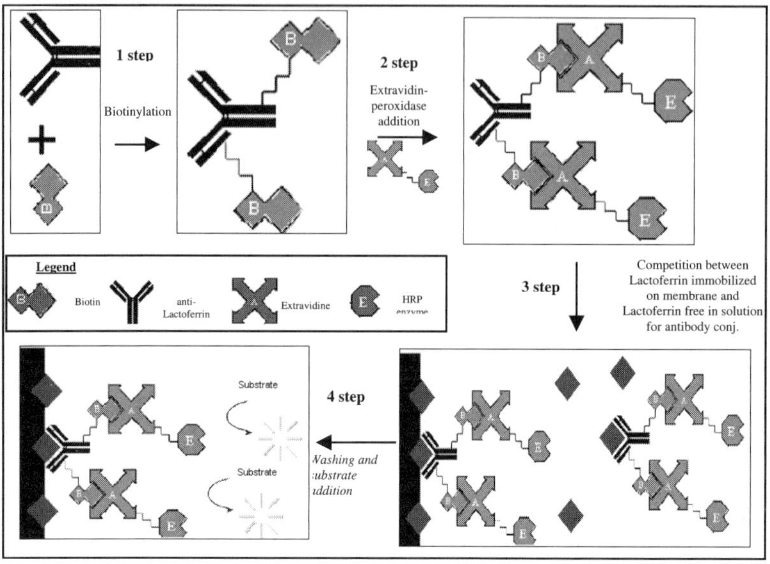

Figure 1. Biotinylation of the antibody (anti-lactoferrin) and "competition" scheme of measurement.

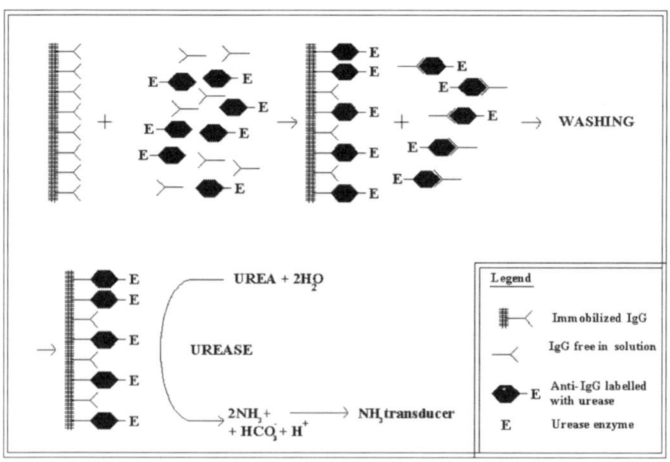

Figure 2. Determination of IgG by immunobiosensor using tyrosinase enzyme electrode as detector and Clark electrode as transducer.

However, the competition procedure for the measurement by the IgG immunosensor using an enzyme sensor as detector is illustrated in figure 2. Lastly, also the antioxidant activity of the milk samples was measured by means of a superoxide dismutase (SOD) biosensor recently developed by us [7]. This was done in order to determine the superoxide radical obtained by coupling a transducer (an amperometric hydrogen peroxide electrode) with the superoxide dismutase enzyme immobilized on the electrode. The superoxide radical is produced in aqueous solution by xanthine, which changes to uric acid during the oxidation reaction catalyzed by the xanthine oxidase enzyme. The disproportion reaction of the superoxide radical in the presence of the SOD enzyme produces oxygen and hydrogen peroxide. The latter is detected by the H2O2 amperometric sensor.

3. Results

After optimizing the "competitive" measurement procedures, the lactoferrin immunosensor was used for the determination of lactoferrin in buffalo milk on three different lactation days (see Table 1).

Table 1. Analytical characterization of immunosensor method for both Lactoferrin and IgG determination, using competitive procedures.

Methods	**Determination of Lactoferrin by means of immunosensor. Test geometry: competition between Lactoferrin Biotin-Avidin-peroxidase conjugated and Lactoferrin free in solution for Anti-Lactoferrin immobilized in membrane**	**Determination of anti-IgG by means of immuno-biosensor, which uses as detector a tyrosinase enzyme sensor. Geometry of the test: competition for the IgG, immobilized on the membrane, between the anti-IgG conjugated with the alkaline phosphatase and not conjugated, both free in solution.**
Transducer employed	H_2O_2 electrode	Clark electrode
Regression equation (**Y=** a.u., **X=** mol l^{-1}) Level of confidence $(1-\alpha) = 0.95$;	$Y = 35.0 (\pm1.2) \log X - 68.5 (\pm6.0)$ $(n - \nu) = 8$; $(t = 2.31)$	$Y = -0.86 (\pm0.07) \log X + 1.33 (\pm0.3)$ $(n - v) = 7$; $(t = 2.36)$
Linear range (mol l^{-1})	**7.0×10^{-8}- 1.0×10^{-5}**	**$(0.26 - 13) \times 10^{-10}$**
Correlation coefficient	0.9891	0.9831
Repeatability of the measurement (as pooled SD%)	5.7	4.2
Lower detection limit (LOD) (mol l^{-1})	**3.5×10^{-8}**	**1.3×10^{-11}**

On the same days, using the electrochemical immunosensor for the measurement of immunoglobulin G and with the help of the particular construction geometry of this extremely selective "immunobiosensor" [6], it was possible to determine also immunoglobulin G in the same buffalo milk samples.
Lastly the antioxidant capacity of these milk samples was also measured using a superoxide dismutase (SOD) biosensor. Experimental lactoferrin and immunoglobulin G concentration values and those of the antioxidant capacity determined at three different lactation days are shown in Table 2.

Table 2. Antioxidant capacity (RAC), Lactoferrin and immunoglobulin G values determined in buffalo milk on three different lactation days.

Lactation days	RAC % (RAC units) n= 3; RSD% ≤5	Lactoferrin ($mg\ l^{-1}$) n= 3; RSD% ≤5	Immunoglobulin G (μg^{-1}) n= 3; RSD% ≤5
52	45	244	50
80	18	235	150
134	28	83	1696

4. Conclusion

The lactoferrin level in buffalo milk was successfully determined by the optimized immunosensor. In addition, the IgG immunobiosensor was suitably applied to detect the immunoglobulin G level in the same milk. In the absence of any currently accepted reference methods for direct measurement of these proteins, the two immunosensors described herein may be used to make good this shortcoming. The sensitivity, the specificity and the linear range of the two corresponding methods are satisfactory. The limits of detection of these methods are respectively of the order of 10^{-8} and 10^{-10} mol l^{-1}. Lastly, the antioxidant capacity was also measured using a biosensor based on the SOD enzyme sensor, which proved suitable also for the analysis of milk sample.

References

1. Ensminger AH, Esminger M. K. J. et al. Food for Health: A Nutrition Encyclopedia. Clovis, California: Pegus Press; 1986.
2. Hajjar IM, Grim CE, Kotchen TA. J Clin. Hypertens., 5 (2) (2003), 122-6.
3. Wood, R. The Whole Foods Encyclopedia. New York, NY: Prentice-Hall Press; 1988.
4. Britigan BE, Serody JS, Cohen MS. Adv. Exp. Med. Biol. 357 (1994), 143-156.

5. Adamik B, Zimecki M, Wlaszczyk A, et al., Arch. Immunol. Ther. Exp. (Warcz). 46 (1998), 169-176.
6. Campanella L., Martini E., Tomassetti M.: Proceedings of the 11th Italian Conference, AISEM 2006, Sensors and Microsystems, Word Scientific Publishing Co, Singapore, (in press).
7. Campanella L., Favero G., Tomassetti M.; Analytical Lett. 32 (13) (1999), 2559-2581.

BIOSENSORS AS NEW TECHNOLOGIES FOR GENE-DOPING INVESTIGATION

T. RUSANOVA, D. DELL'ATTI, S. TOMBELLI, M. MINUNNI,*M. MASCINI

Universita degli Studi di Firenze, Dipartimento di Chimica, Via della Lastruccia 3 Sesto Fiorentino, 50019, Italy

P. BOGANI, M. BUIATTI

Universita degli Studi di Firenze, Dipartimento di Biologia Animale e Genetica Firenze, 50127, Italy

This work reports on the development of a piezoelectric DNA biosensor for the identification of enhanced Green Fluorescence Protein gene often used as a marker of gene delivery. The oligonucleotide sequence of 20 bases selected from the starting DNA has been immobilized on quartz crystal surface via biotin-streptavidin interaction. The detection of the eGFP gene was examined and optimized and the analytical parameters of the system were studied. The developed sensor was applied to the analysis of samples extracted and amplified by polymerase chain reaction.

1. Introduction

The use of gene transfer methods for athletic enhancement is inevitable [1]. To prepare for such a possibility, it will be necessary to develop more efficient and more effective methods for detection of the foreign genetic information and/or the vector used to deliver the transgene [2]. Gene doping is defined by the World Anti-Doping Agency (WADA) as "the non-therapeutic use of cells, genes, genetic elements, or of the modulation of gene expression, having the capacity to improve athletic performance". Gene doping is a process in which DNA of performance-relevant genes is introduced into the cells of athletes. This transgenic DNA leads to the increased production of the performance-enhancing substances within the body. This is made possible by using suitable gene shuttles that help to integrate the transgenic DNA into the human genome or into the cell plasma. There are different types of vectors for gene transfer, however viral vectors (retroviruses, adenoviruses and etc.) are the most efficient. So it would be useful to trace the presence of foreign DNA into the athlete body for detection

* Corresponding author: minunni@unifi.it

of illegal case of gene therapy. For this purpose DNA-based biosensors are a good candidate [3,4]. Among them piezoelectric sensors offer the possibility of monitoring the hybridisation reaction in real time, without the use of any label and could be applied to gene-doping detection [5,6].

In this work a piezoelectric sensor has been applied to the detection of eGFP gene used as a marker of gene transfer. The green fluorescent protein (GFP) is a protein (238 amino acids) from the jellyfish *Aequorea victoria* that fluoresces green. The enhanced green fluorescent protein gene (eGFP) has been optimized for brighter fluorescence and higher expression in mammalian cells (excitation maximum = 488 nm, emission maximum = 507 nm) and it is frequently used as a reporter of expression in cell and molecular biology [7].

2. Experimental

2.1. *Chemicals*

11-mercaptoundecanol, 1-ethyl-3-(3-dimethylaminopropyl)carbodiimide, and streptavidin were purchased from Sigma (Milan, Italy), Dextran 500 from Amersham Biosciences (Uppsala, Sweden), (+)/–epichlorohydrin and *N*-hydroxysucciminide from Fluka (Milan, Italy). Ethanol and all the reagents for the buffers were purchased from Merck (Italy).

Two different buffers were used for probe immobilisation (NaCl 300 mM, Na_2HPO_4 20 mM, EDTA 0.1 mM, pH 7.4) and for target hybridisation (NaCl 150 mM, Na_2HPO_4 20 mM, EDTA 0.1 mM, pH 7.4). All the other reagents were purchased from Merck (Italy). Oligonucleotides were purchased from MWG Biotech (Milan, Italy).

The base sequences of the probe (20-mer) selected from the starting DNA (Figure 1) and of the complementary target are given below:
Biotinylated probe eGFP: 5'- biotin - ACG GCA TCA AGG TGA ACT TC - 3'
Complementary target eGFP: 5'- GAA GTT CAC CTT GAT GCC GT - 3'

2.2. *Apparatus*

9.5 MHz AT-Cut quartz crystals (14 mm) with gold evaporated (42.6 mm^2 area) on both sides were purchased by International Crystal Manufacturing (USA). The measurements were conducted in a methacrylate cell where only one side of the crystal was in contact with the solution. The quartz crystal analyzer used for the measurements was the QCMagic analyser by Elbatech (Marciana, Livorno, Italy).

The concentration of the DNA was determined with fluorescence by using Picogreen® dye (TD-700 Fluorometer, Turner Biosystem from Analytical Control, Milan, Italy).

2.3. *Immobilisation procedure*

Before the immobilisation of the probe, the crystals were washed in a boiling solution of H_2O_2 (30%), NH_3 (30%) and milliQ water in a 1:1:5 ratio for ten minutes and then rinsed with milliQ water.

The biotinylated probe was immobilised via biotin-streptavidin binding on the gold sensor surface previously modified with thiol/dextran/streptavidin. The details of the immobilisation procedure are reported in literature [8].

2.4. *Hybridisation detection using synthetic oligonucleotides*

Once the probe was immobilised on the gold surface, the hybridisation reaction with synthetic oligonucleotides in hybridisation buffer solution was conducted by adding to the sensor cell 100 μL of the oligonucleotide solution at different concentrations in the range 0.050-1 μM. The reaction was monitored for 10 minutes. After each cycle of hybridisation, the single-stranded probe on the crystal surface was regenerated by 1 min treatment with 1 mM HCl.

2.5. *DNA sample preparation*

The plasmid DNA pEGFP-C1 containing *eGFP gene* was from Clontech (USA). "One Shot TOP10 Chemically Competent" cells from *E. coli* (Invitrogen) were used for transformation of the plasmid pEGFP-C1. Thank to the *eGFP* marker gene, the transformed cells were recognized by irradiating with UV light and recording the fluorescence. For plasmid DNA extraction from *E. coli* NucleoSpin®Plasmid kit (Macherey Nagel, M-Medical, Firenze) was used.

The DNA fragment of 219 bp containing the target sequence was amplified by using the sense and antisense primers (Figure 1). The PCR conditions were as follows: denaturation at 95°C for 3 minutes, then 40 cycles: 1 minute at 95°C, 30 seconds at 56°C for primers' annealing and finally an extension phase of 5 minutes at 72°C, then cooling to 4°C. All PCR experiments were conducted with Termocycler MJ-Research Ptc-200 (Peltier Thermal Cycler) DNA Engine. Screening of the PCR products was performed by gel electrophoresis and visualised through a UV transilluminator.

a)

...atcgacttcaaggaggacggcaacatcctggggcacaagctggagtacaactaca
acagccacaacgtc**tatatcatggccgacaagca**gaagaacggcatcaaggtgaact
tcaagatccgccacaacatcgaggacggcagcgtgcagctcgccgaccactaccagc
agaacacccccatcggcgacggccccgtgctgctgcccgacaaccactacctgagca
cccagtccgccctgagcaaagaccccaacgagaagcgcgatc**acatggtcctgctgg**
agttcgtgaccgccgccgggatcactctcggcatggacgagctgtacaagtaaagcg
gccgcgactctagaattccaactgagcgccggtcgctaccattaccaa...

b)

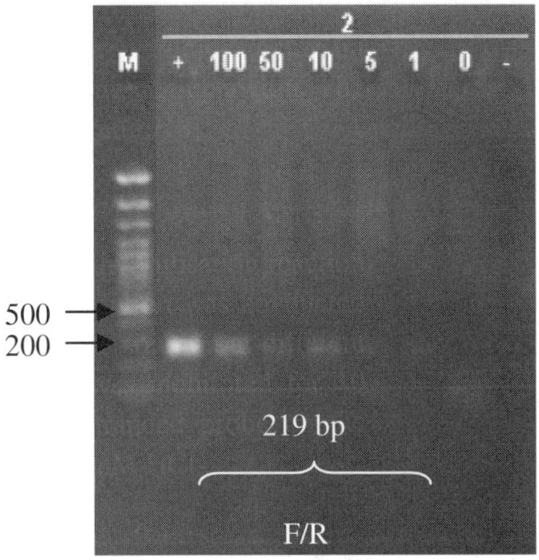

Figure 1. (a) Sequence of bases of the fragment amplified from the eGFP gene. The sense (F) and antisense (R) primers used for the amplification are reported in bold. The immobilised probe eGFP is underlined. (b) Gel electrophoresis of the PCR amplification conducted as reported in Section 2.5, "+" plasmid, "-" blank PCR, 100-1: plasmid genome; 0: human genome.

2.6. *PCR samples detection*

After the optimization of the sensor with synthetic oligonucleotides, PCR samples were tested. The samples were first thermally denatured (5 min at 95 °C and 1 min cooling in ice), and then left in contact with the probe for 20 min.

3. Results and Discussion

3.1. *Hybridisation detection using synthetic oligonucleotides*

The sensor was first optimized with synthetic oligonucleotides and the main analytical parameters were studied, e.g. specificity, sensitivity, reproducibility. The experiments were performed with different concentrations of the target oligonucleotides (0.05–1.00 μM), complementary to the probe immobilised on the crystal. One hundred microliters were added to the crystal surface modified with biotinylated probe. The reaction was monitored for 10 min, the solution was then removed and the surface washed with the same hybridisation buffer to eliminate the unbound oligonucleotide. The analytical datum was given by the difference between the signal recorded before the hybridisation (baseline) and after the washing, once the hybridisation has occurred; both values are taken when the crystal is in contact with the same solution (Figure 2). The sensor allows detection of the target in a concentration range of 50 nM to 250 nM (Figure 3). For the concentrations higher than 250 nM plateau has been observed. A reproducibility (relative standard deviation, R.S.D. %) of 9.3 % was found for the system. The specificity of the interaction was tested by using a 20-mer non-complementary oligonucleotides (1 μM). The signal resulting from this interaction was negligible (<3 Hz), evidencing the specificity of the system.

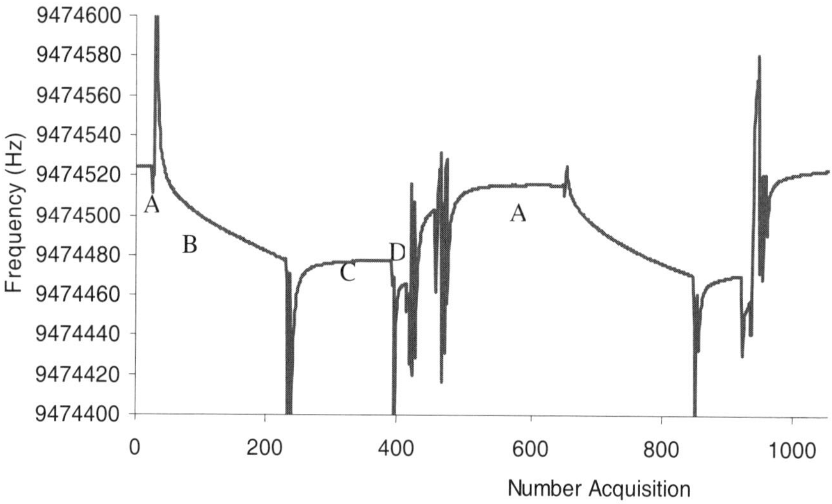

Figure 2. Hybridisation cycles: A - Hybridisation buffer, B - target, C - Hybridisation buffer, D - Regenerating solution HCl 1mM.

3.2. *Hybridisation detection using PCR-amplified samples*

After the study with synthetic oligonucleotides, the system has been applied to the analysis of DNA samples from plasmid DNA transformed with *E.coli* and amplified by polymerase chain reaction. The amplified DNA fragment, internal to the eGFP gene, was 219 bp in length and contained the complementary sequence to the immobilised probe eGFP. These samples were pre-treated by denaturation (5 min at 95 °C and 1 min cooling in ice), and then left in contact with the probe for 20 min. The signals were detected for all the tested samples (table 1).

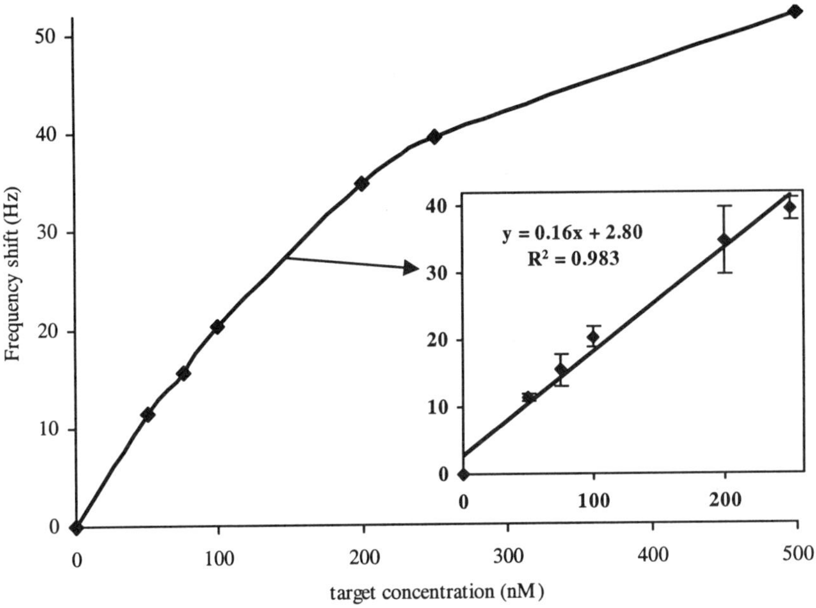

Figure 3. Calibration curve obtained with synthetic oligonucleotide on the crystal modified with the biotinylated probe eGFP.

Table 1. The results obtained for PCR-amplified samples of DNA

Sample N.	Conc., ppm	ΔF (Hz)
1	9.8	-7
2	9.7	-6
3	24.0	-11*
4	21.2	-12

* The measurement was performed in triplicate and the average value is reported with a CV% = 9.0.

4. Conclusion

The development of a piezoelectric biosensor for the detection of the eGFP gene, used as a marker for gene transfer, is described. The analytical parameters of the sensor were studied. Moreover, the developed sensor was applied to the analysis of DNA samples amplified by polymerase chain reaction.

These preliminary findings demonstrate the potential applicability of DNA-based sensing to detect a marker such as eGFP, eventually present in vectors used in gene transfection. This is a first approach for gene doping control, which might become a problem as early as the next summer Olympic Games in 2008 in Beijing.

Acknowledgments

The Italian Ministry of Health is acknowledged for funding of the project "Metodi bioanalitici basati su biosensori a DNA per le nuove frontiere del doping: l'individuazione di geni e proteine esogene"

References

1. H.J. Haisma, O. de Hon, P. Sollie, I. Vorstenbosch. Gene Doping. Netherlands centre for doping affairs. Netherlands, 2004. 38 p.
2. H.M.E. Azzazy, M.M.H. Mansour, R.H. Christenson, *Clinical Biochemistry*. **38,** 959 (2005).
3. M. Minunni, S. Tombelli, M. Mascini, A. Bilia, M.C. Bergonzi, F.F. Vincieri, *Talanta*. **65,** 578 (2005).
4. T. Jiang, M. Minunni, P. Wilson, J. Zhang, A.P.F. Turner, M. Mascini, *Biosensors and Bioelectronics*. **20,** 1939 (2005).
5. S. Tombelli, M. Minunni, A. Santucci, M.M. Spiriti, M. Mascini, *Talanta*. **68,** 806 (2006).
6. D. Atti, S. Tombelli, M. Minunni, M. Mascini, *Biosensors and Bioelectronics*. **21,** 1876 (2006).
7. B.P. Cormack, R.H. Valdivia and S. Falkow, *Gene*. **173,** 33 (1996).
8. S. Tombelli, M. Minunni, M. Mascini, *Methods*. **37,** 48 (2005).

GRAPHITE ELECTROCHEMICAL SENSORS FOR THE EVALUATION OF PAHs-DNA ADDUCTS

MICHELE DEL CARLO, MANUELA DI MARCELLO, MARCELLO MASCINI, DARIO COMPAGNONE

Dipartimento Di Scienze Degli Alimenti, Università Degli Studi Di Teramo, Via Carlo R. Lerici 1, 64023 Mosciano Stazione, Teramo

**Corresponding author: mdelcarlo@unite.it*

The aim of this experimentation was the evaluation of benzo(a)pyrene-DNA adducts (BaP-DNA adducts) formation using graphite screen printed electrodes (SPE) and chronopotentiometry. Graphite electrochemical sensors appeared a useful tool for the realization of a rapid method of analysis for the detection of BaP-DNA adducts. To test the performance of these devices the electrochemical response produced by genomic DNA sequences adsorbed on the electrochemical transducer was characterized. A concentration dependent oxidation peak was observed at +1.050 mV vs Ag/AgCl reference electrode, this peak is due to the guanine oxidation. Moreover the response of the immobilized DNA to BaP, and other polycyclic aromatic hydrocarbons (PAHs), was characterized. Depending on the hydrophobicity of the target molecule either a decrease (>3 aromatic rings) or an increase (≤3 aromatic rings) of the guanine oxidation peak was observed. A combination of UV light and oxidising conditions (0.4 μM H_2O_2) were used to obtain oxygenated species such as diol epoxide. Under these experimental conditions the formation of oxidised BaP ((ox)BaP) derivatives was followed by UV spectra monitoring the appearance of an absorbance maximum at λ=261 nm. The formation of (ox)BaP-DNA adducts was evaluated by chronopotentiometry using screen printed electrodes.

1. Introduction

Polycyclic aromatic hydrocarbons (PAHs) are a class of ubiquitous contaminants characterised by multiple aromatic rings, including those that incorporate atoms such as sulfur and nitrogen. Many of the PAHs tend to bio-accumulate, mainly in lipids. They are mainly toxic in kidney and liver where they may act as mutagenic and carcinogenic factors. These compounds along with their metabolic oxidation products have been determined in foods of animal and vegetal origin as well as environmental samples (industrial and municipal wastewater, effluents, rainwater, and drinking water). There are convincing evidences suggesting that the real hazards of PAHs depends on their oxidised products (1). PAHs metabolic oxidised products may be produced by intracellular singlet oxygen and other reactive oxygen species (ROS) that cause oxidative damage in biological systems. Also a photo induced oxidation, has

been described. The photo-oxidised products, often exert stronger bioactivity than the parent compounds (2).
Hence the ecological or human health threat is deriving from degradation of a parent compound and the fate of the oxidation by-products is ignored. In our opinion bioassay may complement chemical analysis to give an integrated measure of toxicity. Toxicity bioassays can be used not only to evaluate efficiency of oxidation processes but also to study the fate of the parent compounds as well as the by-products formed during oxidation. Traditionally, toxicity measurement employing aquatic organisms require long exposure times and large sample volume (3). Other toxicity tests are based on micro-organisms which are rapid, cost effective and reproducible and they are becoming widely applied. Luminescent bacteria have been found to be particularly useful in evaluating toxicant impacts. Other methods can be devised as an integration of biological molecules and chemical/physical transducers obtaining the so called biosensors. Among these DNA biosensors appears promising in order to detect the eventually occurring binding between DNA and BaP and/or its by-products (ox)BaP.
The aim of this experimentation was the evaluation of graphite screen printed electrodes (SPE) for the detection of BaP-DNA adducts formation. Chronopotentiometry was used as detection strategy to monitor the adduct formation. To achieve this goal, the specific aims of this experimentation were: i) to develop an experimental procedure to detect guanine oxidation electrochemical signal, ii) to develop an experimental protocol to oxidize BaP to produce reactive species able to form DNA adducts, iii) to evaluate the BaP-DNA adduct formation.

2. Experimental

Reagents and instrumentation

Genomic salmon testis ssDNA was purchased from Sigma-Aldrich (Italy). Graphite screen printed electrodes were obtained from EcoBioServices & Research (Florence, Italy). Acetonitrile, dichloromethane, hexane, sodium acetate and potassium chloride were purchased from Merck. PAHs standards were obtained from Sigma (Milano, Italy). All chemicals and reagents used were of analytical grade. Spectrophotometric determination were made with a Lambda Bio20, Perkin Elmer (Monza, Italy). Electrochemical measurements were performed with an Autolab electrochemical analysis system with a GPES 4.5 software package (Ecochemie, Utrecht, Holland), in connection with a VA-Stand 663 (Metrohm, Milano, Italy).

Electrochemical DNA analysis

The electrode surface was pre-treated by applying a potential of + 1.6 V for 2 min and +1.8 V for 1 min. The single stranded salmon testis DNA was immobilised at fixed potential (+ 0.5 V vs Ag screen-printed pseudo-reference electrode for 120 s) onto the screen-printed electrode surface. During the immobilisation step, the strip was immersed in acetate buffer solution containing 60 μg/ml of single stranded salmon testis DNA. Then a cleaning step was performed by immersion of the sensor strip in a clean acetate buffer solution, at open-circuit condition. The chronopotentiometry scan was carried out to evaluate the oxidation of guanine residues on the electrode surface. Potential range was 0.3-1.3 V and the oxidising current was 2 μA. The height of the guanine oxidation peak (around + 1.0 V vs. Ag screen-printed pseudo-reference electrode) was measured.

Evaluation of PAHs interaction with immobilised ssDNA

The PAHs interaction with ssDNA was evaluated by the change of the peak height of the anodic signal produced by guanine oxidation. The DNA modification was estimated via the percentage of response decrease (% R), which is the ratio of the guanine peak height after the interaction with the sample, and the guanine peak height after the interaction with buffer solution.

Photo-oxidation of B(a)P

Hydrogen peroxide-assisted BaP degradation was studied by adding 0.4 μM H_2O_2 to the solution. All experiments were conducted in duplicate. The appearance of oxidation products was followed by spectrophotometric analysis in the UV range. In a first experimentation the photo-oxidised BaP solution was used to directly asses the possibility of (ox)BaP-DNA adduct formation in solution. In a successive experimentation, the oxidation products were extracted in hexane, then solution was dried under nitrogen stream, and re-dissolved in a reduced volume of acetonitrile. The purified solution was then used for (ox)BaP DNA adduct formation.

Evaluation of oxidised BaP interaction with immobilised ssDNA

To test the formation of BaP-DNA adducts the above described electrochemical protocol was used. The photo-oxidised solution was used to obtain BaP-DNA. The interaction between oxidised BaP and DNA was evaluated by changes of the electrochemical signal of guanine. After ssDNA immobilisation an incubation step at open circuit was carried out.

3. Results and Discussion

The electrochemical signal obtained by chronoamperometric analysis of ssDNA adsorbed onto the electrode surface was used to obtain a dose response curve in the concentration range 5-80 μg/ml. The guanine base oxidation peak was observed at +1.050 mV. In figure 1 an example of the guanine oxidation peak using different ssDNA concentration is reported. The peak height resulted concentration dependent.

The variation of this peak was used in successive experiments to evaluate the interaction of different PAHs on the immobilised ssDNA. In figure 2 the dependence of the relative biosensors response on the hydrophobicity of the investigated compound is reported. The more polar compound caused an increase of the guanine oxidation peak, whereas the more hydrophobic PAHs produce a diminishment of the response with respect to the blank analysis. A possible explanation of this behaviour is that the small (≤ 3 aromatic rings) less hydrophobic compounds (fluorene and phenantrene) may destabilise the single DNA helix analysis determining an enhancement of guanine base availability for the electrochemical oxidation process. On the contrary, the PAHs with > 3 aromatic rings (benzo(a)pyrene and benzo(a)antracene), that is with higher hydrophobicity, determine a stabilization of the single helix and a diminishment of the oxidation rate of guanine bases.

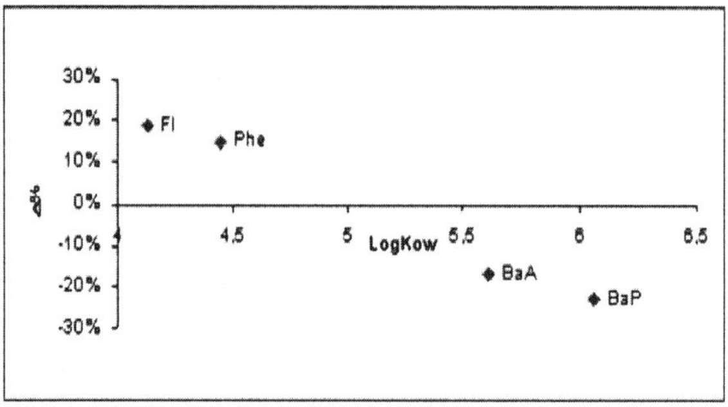

Figure 1: Chronopotentiogram of ssDNA

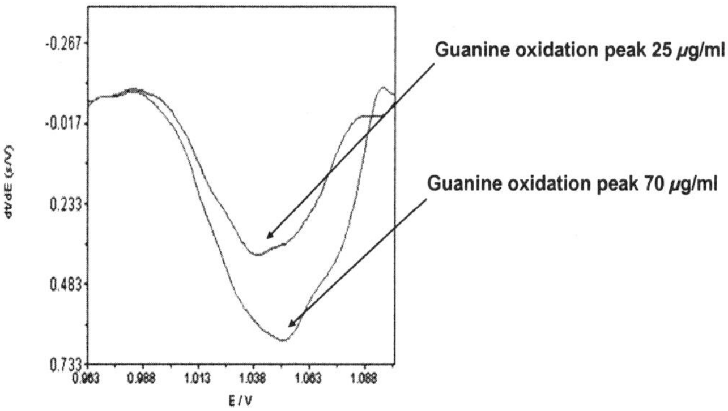

Figure 2: Effect of PAHs hydrophobicity on the guanine oxidation peak with respect to blank measurements

In order to investigate the (ox)BaP-DNA adduct formation a photo-oxidation protocol for BaP has been carried out. The appearance of an absorbance peak at

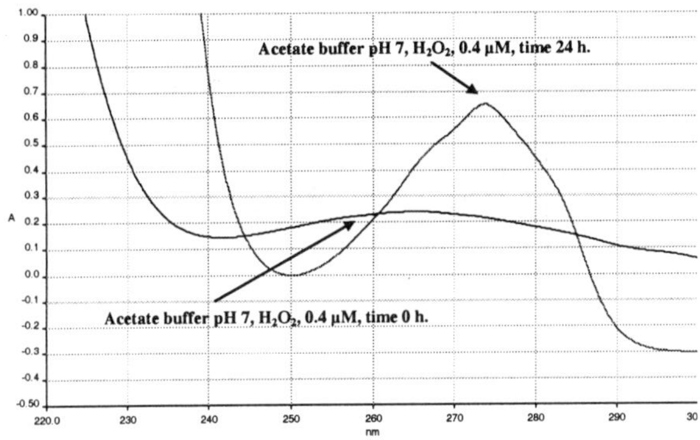

Figure 3: UV spectra of BaP, before and after photo-oxidation process.

λ=264 nm confirmed the hydroxylation of BaP aromatic rings (Figure 3), the reaction was followed over a 24 hours period.
In that time the complete disappearance of BaP was observed, while an increase of the absorbance at the observed λ was appreciated. The reaction products where sampled at diverse reaction time, 2-4-8-24 hours using liquid-liquid extraction and used in further studies including their characterisation by LC-MS and FTIR analysis (data not shown). Moreover these solutions where used to evaluate (ox)BaP-DNA adduct formation in dedicated experiments.
A first attempt to asses the formation of (ox)BaP-DNA was performed by addition solution during the UV/H_2O_2 driven photo-oxidation process. In that case a significant decrease of the guanine oxidation peak was obtained after two hours of photo-oxidation process with respect to all the control solutions (Figure 4). Particularly the decrease of guanine oxidation peak was significantly lower than that obtained under the same conditions in absence of BaP (control 2) and in absence of H_2O_2 (control 3). Control 1 was ssDNA in absence of both BaP and H_2O_2.
Using longer incubation time no definitive data were obtained. This could be due to the prolonged exposure of the ssDNA to oxidative conditions that caused a generalised destabilisation of the ssDNA structure.
Finally, the reaction occurring between ssDNA and the products of the photo-oxidation reaction, previously obtained by liquid-liquid extraction, was assessed. A diminishment of the guanine oxidation peak was also observed. Further studies are needed for a thorough evaluation of the (ox)BaP-DNA adducts electrochemistry.

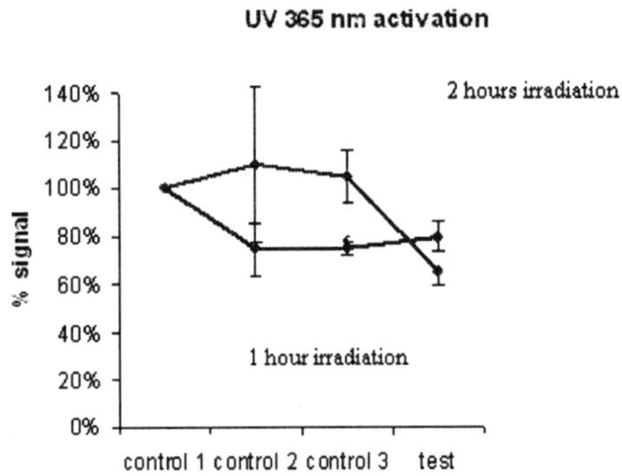

Figure 4: evaluation of (ox)BaP-DNA adduct formation

4. Conclusions

Graphite screen printed electrodes appeared a useful analytical device for the detection of PAHs with immobilised ssDNA. The electrochemical protocol enabled the detection of the reaction occurring between PAHs and ssDNA.

References

1. H. Shemer, K. G. Linden, Water Research 41 (2007) 853
2. J. Sabate´, J.M. Bayona, A.M. Solanas, 44 (2001) Chemosphere 119
3. S. Parvez, C.Venkataraman, S. Mukherji 32 (2006) Environ. Intern. 265

ANTIGEN-ANTIBODY INTERACTION ON THE GOLD SURFACE MODIFIED BY LANGMUIR-SHAEFFER TECHNOLOGIES WITH POLY-PYRROLE-DERIVATIVES MONITORED BY μLIBRA

LUIGI SCHIAVO*, ANTONIO SCARPA & SIMONA GRECO

Biological Division Technobiochip S.c.a.r.l, Via Provinciale per Pianura, 5 (Loc. San Martino), 80078, Pozzuoli (Na), Italy

Here we report, for the first time, the use of poly[ferrocene]-1H-pyrrole, a patented poly-pyrrole derivative synthesized at Technobiochip, to create, using Langmuir-Shaeffer thin film deposition method, an interfacial chemistry on the crystal quartz golden surface to firmly stabilize antibody molecules. Coated quartzes were used to perform a classical antigen-antibody binding reaction on Technobiochip's μLibra 3.1 (Quartz Crystal Microbalance). Our results demonstrate that the antibody binding on coated quartzes was extremely increased (70%) over that shown by bare quartzes. In addition, the antigen-antibody interaction was also significantly improved (30%).

1. Introduction

From the discovery of DNA, genomic has provided many information about gene sequences and regulations and has thirsted for finding a link between genetic map and disease. However, it is easy to understand that links between gene products and pathologies are very difficult to be established through genomic technologies only, therefore proteomic approach has allowed to move attention from genes to proteins that. Several methods and analytical techniques are used for studying proteins. Among these, biosensors are very attractive because allow the real-time analysis of reactions without labeling requirements and provide quantitative information on the rate and equilibrium binding constants. Biosensor methodology appears to be a promising technique for investigating specific protein/ligand interactions and its applications are expanding rapidly [1]. To address this problem and in line with the post-genomic era, Technobiochip company invests many of its human and economic resources in studying the development of biosensors that could be used in bio-

* Address Correspondence to: Dr. Luigi Schiavo, BSc, PhD, Senior Researcher Biological Division Technobiochip S.c.a.r.l, Via Provinciale per Pianura, 5 (Loc. San Martino), 80078, Pozzuoli (Na), Italy. E-mail: l.schiavo@technobiochip.com - Phone: +39 081 5264315 – Fax: +39 081 5265116

medical field. In particular, at Technobiochip is under study the development of more and more sensitive and specific immuno-microgravimetric sensors based on the quartz crystal microbalance (QCM) μLibra 3.1. In brief, a QCM consists of a thin quartz crystal sandwiched between a pair of golden electrodes. Due to the piezoelectric properties of quartz, it is possible to make resonant the crystal by applying an AC voltage across its electrodes. The resonant frequency of the crystal depends on the total oscillating mass. When a mass is adsorbed to the sensor crystal the frequency decreases and the oscillation frequency shift is proportional to the mass adsorbed onto the surface as stated by the Sauerbrey relation [2]. Technobiochip's *μLibra 3.1* is a two-channels high-resolution Quartz Crystal Microbalance (QCM) specifically designed to work both in air and in liquid phase. It is a very low noise microbalance system based on quartz crystal resonators that uses 10 MHz AT-cut quartzes with gold electrodes on chromium layer. Each quartz crystal is housed in a low volume, flow-through cell (25 μl), so a very small amount of sample in needed. The sensitivity of μLibra is 4.4 ng for a 10 MHz quartz crystal. From a biological point of view, a biomolecule can bind the golden surface by passive adsorption only. However, sometimes, both a stronger binding and a complete surface coating are required. In order to increase the linking of the anti-IgG antibody to the gold surface we attempted the coating of the electrode surface using patented poly-pyrrole-derivatives synthesized at Technobiochip.

2. Material and Methods

2.1 Reagents

Both Mouse Polyclonal anti-IgG Antibody (Ab) and Mouse IgG (Ag) were purchased from SIGMA (Milan, Italy). The quartz crystals, 10 MHz AT-cuts, gold electrodes (Figure 1a), were obtained from International Crystal Manufacturing Company (Ohio, USA).

2.2 Technobiochip μLibra 3.1

μLibra 3.1 is a Quartz Crystal Microbalance (QCM) produced by Technobiochip Scarl (Pozzuoli, Italy). As showed in Figure 1b, the instrument (EC_{06} Certified) is composed by a Main Unit and a Cell Base Unit that contains two oscillators and equipped with two low-volume flow-through cells for real-time measurements. μLibra output data are elaborated by the LibraVIEW software. More technical informations and specifications of the instrument are available consulting Technobiochip website (**www.technobiochip.com**).

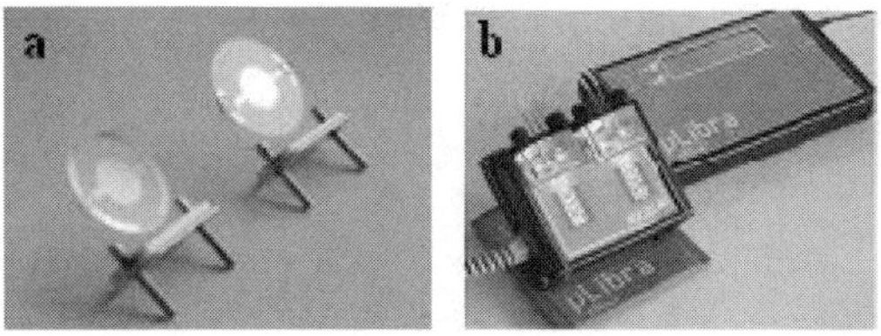

Figure 1. (a) μLibra 3.1 quartz crystal resonator. (b) μLibra 3.1 equipment with two flow cells chambers inserted.

2.3 Chemical synthesis of poly-pyrrole derivatives and quartzes functionalization

Equimolar pyrrole and aldehyde solution were incubated in a BF_3-saturated environment for poly-pyrrole polymers catalysis [3,4,5], following the reaction showed in Figure 2.

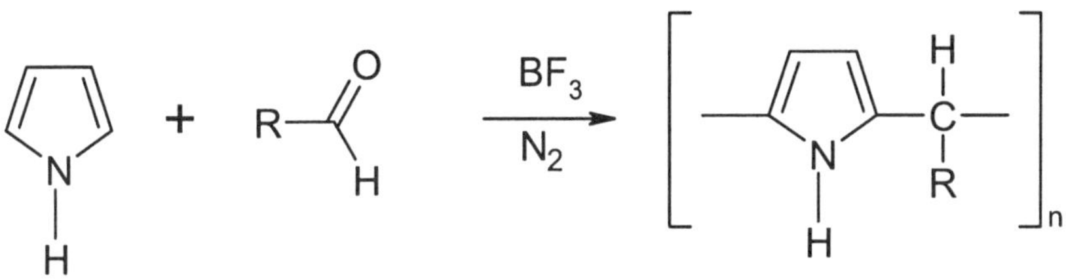

Figure 2. Chemistry of a poly-pyrrole polymers linear synthesis.

By using similar synthesis reactions, five different patented poly-pyrrole-derivative polymers have been obtained at Technobiochip (Figure 3). After quartzes cleaning with piranha solution (NH_3–H_2O_2 1:1 v/v), 20 layers of five different polypyrrole-derivatives synthesized were deposed onto the gold quartzes surface by LS technique using KSV LB-5000 instrument.
The Langmuir-Shaeffer method for the deposition of thin films is an easy and efficient way to deposit ordered layers of molecules on solid substrates. In this method, the substrate is aligned almost parallel to the air-water interface and is lowered to touch the compressed monolayer, until the latter adheres to the surface. Finally, the polymer-modified quartzes were used to perform a typical antigen-antibody binding reaction using Technobiochip μLibra, based on QCM technology.

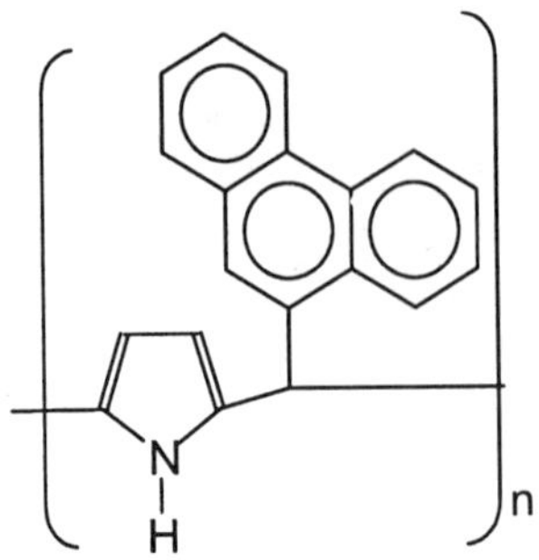

Poly[2-(-9 phenantrhryl-ylmethyl)]-1H-pyrrole

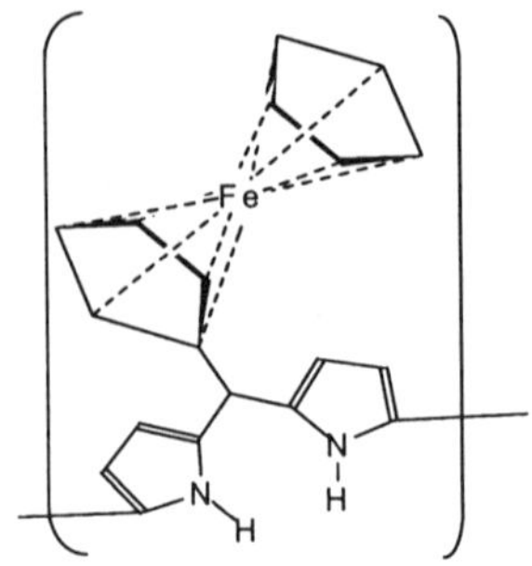

Poly[ferrocene]-1H-pyrrole

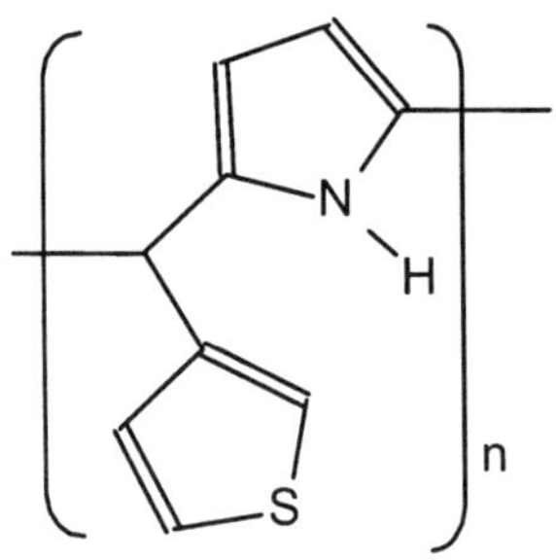

Poly[2-(thien-2-methyl)]-1H-pyrrole

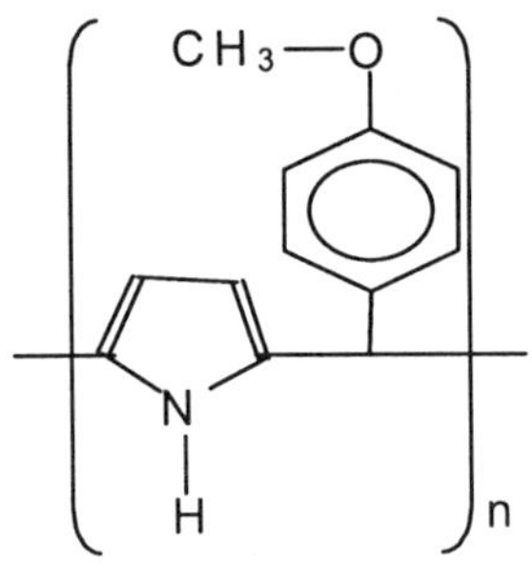

Poly[2-4(methoxybenzyl)-1H-pyrrole

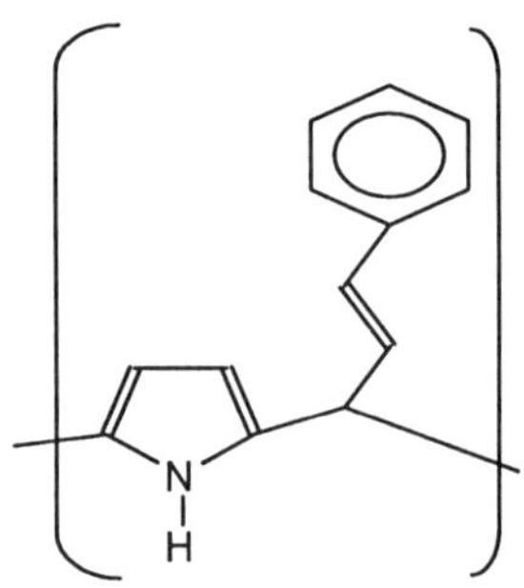

Poly{2-[2-(2E)-3-phenylprop-2-enyl]-1H-pyrrole}

Figure 3. Chemical structures of used poly-pyrrole-derivatives.

3. Results and Discussion

As shown in Figure 4, Anti-Mouse IgG is able to bind both the bare and the coated quartz surface by passive adsorption. Using five different patented poly-pyrrole-derivatives synthesized at Technobiochip, we demonstrate for the first time that quartzes treated with Poly[ferrocene]-1H-pyrrole show a 70% stronger capacity to bind the antibody over that shown by bare quartzes; moreover, the Anti-Mouse IgG and Mouse IgG interaction was improved (Figure 4).

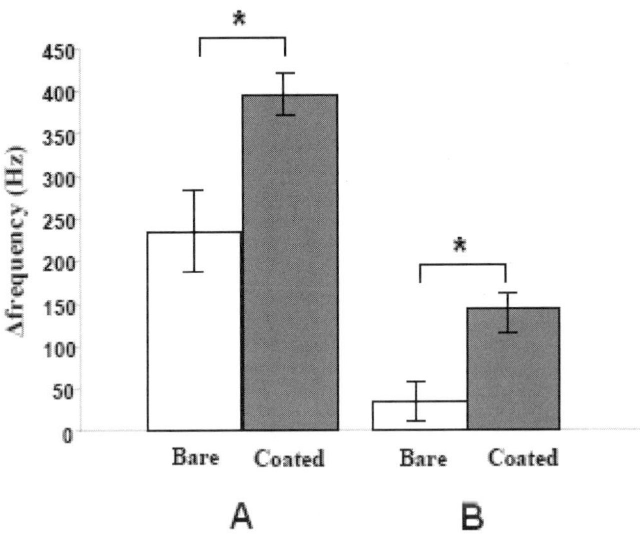

Figure 4: Measured quartz crystal frequency variations after antibody binding (A) or antigen-antibody binding (B) on bare (clear columns) or poly[ferrocene]-1H-pyrrole coated quartzes (dark columns). All the experiments were performed in triplicate. Data are means ± SD (*$P<0.05$ by Student's *t*-test).

On the contrary, quartzes treated with Poly[2-(-9phenantrhryl-ylmethyl)-1H-pyrrole, Poly{2-[2-(2E)-3-phenylprop-2-enyl]-1H-pyrrole, Poly[2-(thien-2-methyl)]-1H-pyrrole and Poly[2-4(methoxybenzyl)-1H-pyrrole are not able to enhance the binding capacity of the immunocomplex (Figure 5).
The experimental evidence that only the Poly[ferrocene]-1H-pyrrole is able to enhance the binding capacity of the antibody on the crystal quartz surface let us to suppose that the interaction between the antibody and the polymer does not involve the pyrrolic group, present in all the used poly-pyrrole derivatives, but

the Ferrocene one. To valorize our hypothesis, we are currently studying by Fourier Transformate InfraRed (FTIR) the nature of that interaction.

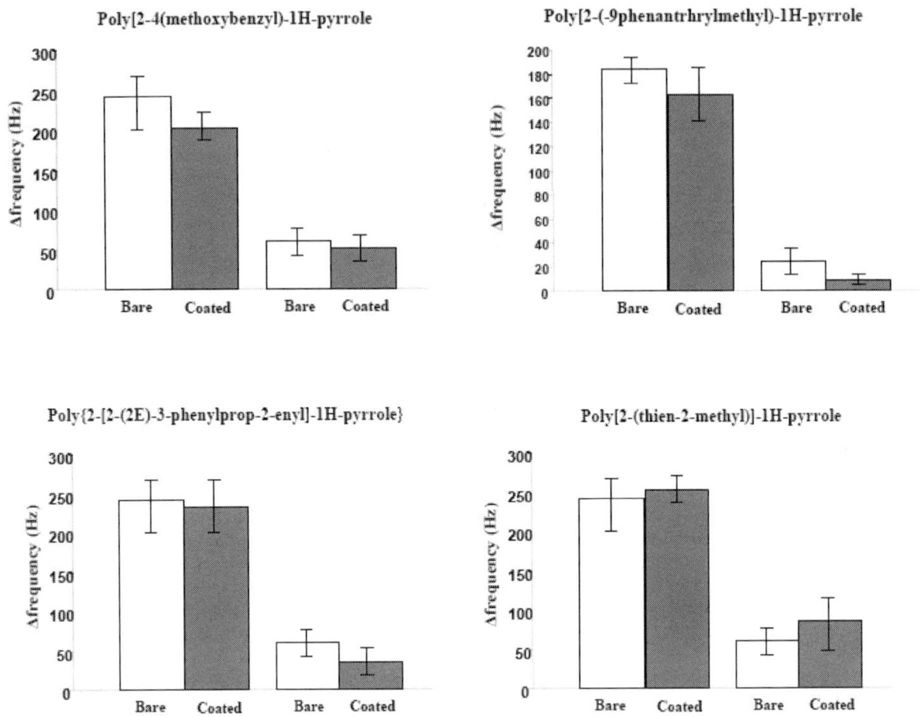

Figure 5: Measured quartz crystal frequency variations after antibody binding (A) or antigen-antibody binding (B) on bare (clear columns) or poly[-pyrrole polymers-coated quartzes (dark columns). All the experiments were performed in triplicate. Data are means ± SD

4. Conclusions and Future Perspectives

These preliminary data indicate for the first time that Technobiochip's patented poly[ferrocene]-1H-pyrrole may be used to enhance the binding capacity of a specific antibody on crystal quartzes surface, allowing the creation of a specific biosensor able to successively recognize a specific antigen. Consequently, other patented polypyrrole derivatives synthesized at our company are also being tested currently.

References

1. L. Murphy, *Current opinion in chemical biology.* **10**, 177 (2006).
2. G. Sauerbrey, *Arch. Elektrotech.* **18**, 617 (1964).
3. A.D. Alder, F.R. Longo, F. Campus and J.J. Kim, *Inorg. Nucl. Chem.* **32**, 2443 (1970).
4. J.S. Lindsey, H.C. Hsu and I.C. Schreiman, *Tetraherdon Lett.* **27**, 4969 (1986).
5. J.S. Schreiman, I.C. Hsu, H.C Kearney, P.C. Marguerettaz, *J. Org. Chem.* **52**, 827 (1987).

A NOVEL TECHNIQUE FOR THE DIRECT DETECTION OF DNA HYBRIDIZATION

ANDREY SAVCHENKO, BORIS SNOPOK

V.Lashkaryov Institute of Semiconductor Physics, National Academy of Sciences, Prospect Nauki, 41, Kyiv, 03028, Ukraine

MARIA GRAZIA MANERA, JOLANDA SPADAVECCHIA, PIETRO SICILIANO, ROBERTO RELLA

Insituto per la Microelettronica ed i Microsistemi, via perMonteroni, Lecce, 73100, Italy

We propose a novel technique to detect DNA hybridization. The technique involves measuring scattered light under surface plasmon resonance (SPR) conditions. We have shown that the maximum scattering angle correlates with the traditionally employed reflection minimum. Measuring of scattered light under SPR conditions has all the functional advantages of the SPR technique. In addition, the proposed technique simplifies device design, increases the dynamic range of analysis, and integrates data with those from surface-plasmon field-enhanced fluorescence spectroscopy. We demonstrate the technique by showing direct registration of 20 bases oligonucleotide hybridization. Also we have investigated the amplification effect of 20nm gold nanoparticles (GNP) on the oligonucleotide immobilization process.

1. Introduction

Sensor systems based on surface plasmon resonance (SPR) are widely used in the basic and applied sciences [1,2]. Here we propose a novel technique to detect DNA hybridization. The technique involves measuring scattered light under surface plasmon resonance conditions. We have shown that the maximum scattering angle correlates with the traditionally employed reflection minimum [4]. The maximum scattering angle carries information on processes occurring at or near the transducer surface. The technique is simple to use and is very efficient. In this work we demonstrate the technique by showing direct interaction between single strand 20 bases oligonucleotides.

2. Experimental Setup

The scanning SPR spectrometer "BioHelper 2.0", registering scattered light under SPR conditions, was developed at the V. Lashkaryov Institute of

Semiconductor Physics of the National Academy of Sciences of the Ukraine [5]. The SPR spectrometer uses open measurement architecture and employs a simple gold-covered glass slide (Figure 1). The optical part of the instrument includes a semicylindrical lens (1) and a replaceable glass plate (chip) with a layer of gold (3) that is in optical contact with the lens through an immersion liquid (polyphenyl ether, n = 1.61). A cell (4) is made of transparent material (plexiglass). The cell supplies material to the transducer in the flow-through mode. A laser beam formation unit (dashed line) is placed on the angle scanning system. The unit consists of a semiconductor laser (λ = 650 nm) and collimating optics (2). A concave cylindrical lens (that prevents the collimated laser beam from focusing on the convex surface of the main lens (1)) also is placed in the scanning system. A stepper motor maintains the scanning angle Θ at 23° (from a starting angle of 52°). The angular resolution is 0.035°. The gold nanoparticles with expected diameter of 20nm were produced by well-known procedure [6] by citrate reduction form $HAuCl_4$ solution. There two approaches were used to immobilize gold nanoparticles on the gold substrate. In the first case the self assembled monolayer of phenylene diisotiocyanate was formed on the gold surface with sequential immobilization of GNP. The second one was consist in immobilization on the gold surface a poly-l-lysine as intermediate layer for GNP immobilization on the gold surface of SPR chip.

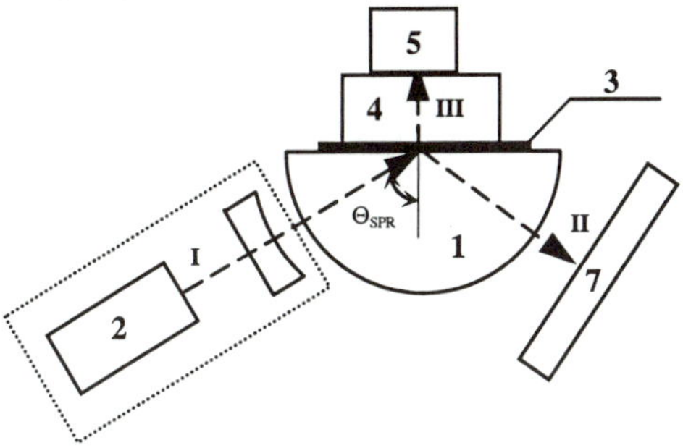

Fig.1 Experimental setup of the scanning SPR spectrometer "BioHelper".

3. Scattering Measurement Validation

The registering of reflected light in SPR condition is a well known technology for detection of biological interactions. In this work to receive the analytical signal we use both scattered and reflected light in SPR conditions.

To check linear dependency of the shifts in extremes of the angular dependences of scattering and reflection under SPR, two sets of experiments was carried out. These tests enabled us to obtain the dependencies for two limiting cases. We have investigated variation in bulk refractive index of the volume phase over the bare physical transducer and after formation of a thin homogeneous film on the transducer surface.

Typical measured reflectivity (R) and scattering (S) of a gold film in water as a function of the angle of incidence is shown on figure 2. The slight shift in extremes of the curves can be observed.

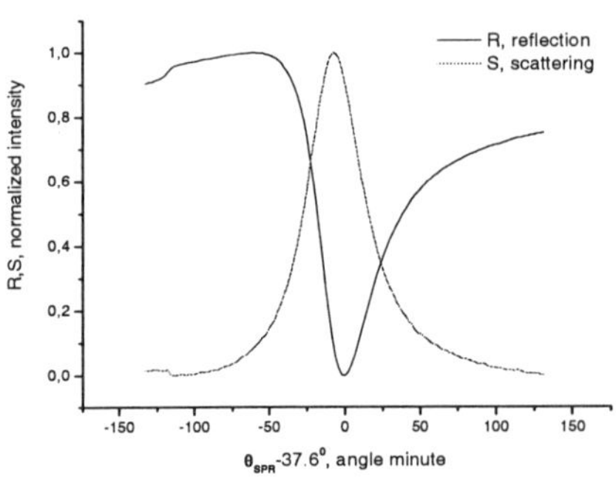

Fig. 2. Measured reflectivity (R) and scattering (S) of a gold film in water as a function of the angle of incidence.

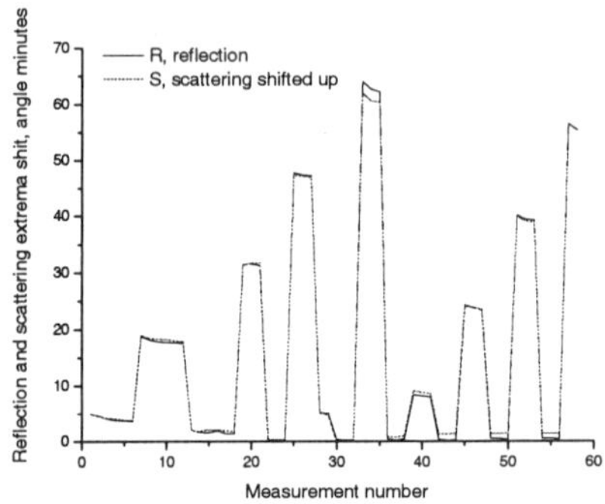

Fig. 3. The extreme shift in scattered and reflected light for different glycerol concentrations.

On fig. 3 the response in shifts of extreme both for scattered and reflected light for different glycerol concentration are shown. It was shown that the shifts in extremes of the angular dependences of scattering and reflection are mutually proportional to each other with a high degree of accuracy (the correlation coefficient is >0.99).

Thus, the data allow us to contend that the measurement of scattered light under SPR conditions may be efficiently used to obtain information on processes occurring at the interface.

4. Results and Discussion

To test the value of "BioHelper 2.0" for investigating DNA interactions, we have performed real-time measurements for immobilization of probe, hybridization and regeneration. The data on figure 4 shows kinetic curves for immobilization of thiolated oligonucleotide (probe) on the clean gold. The response that was observed is approximately 2.0 angle minutes in both scattering and reflection channels.

We have investigated the amplification effect of 20nm gold nanoparticles (GNP) on the oligonucleotide immobilization process. We found that the signal on probe immobilization can be greatly increased by using GNP. The data on figure 5 shows kinetic curves for immobilization of thiolated oligonucleotide on the 20nm gold nanoparticles that was previously immobilized on SPR chip though supporting layer of poly-l-lysine. The response observed is approximately 22 angle minutes for scattering channel and 29 angle minutes for

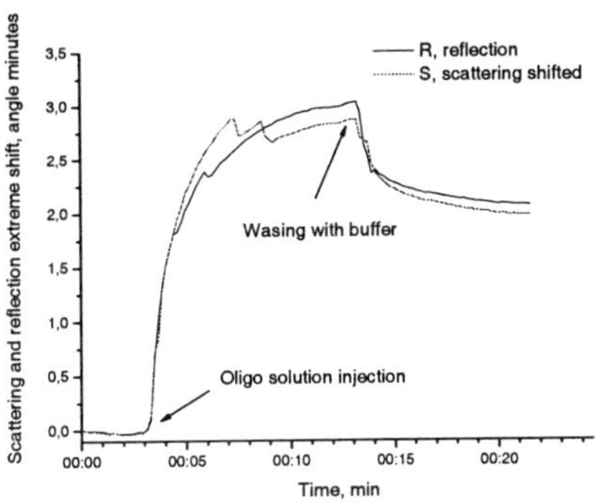

Fig. 4. The extremes shift in scattered and reflected light for 20 bases thiol-modified oligonucleotide immobilization on clean gold surface.

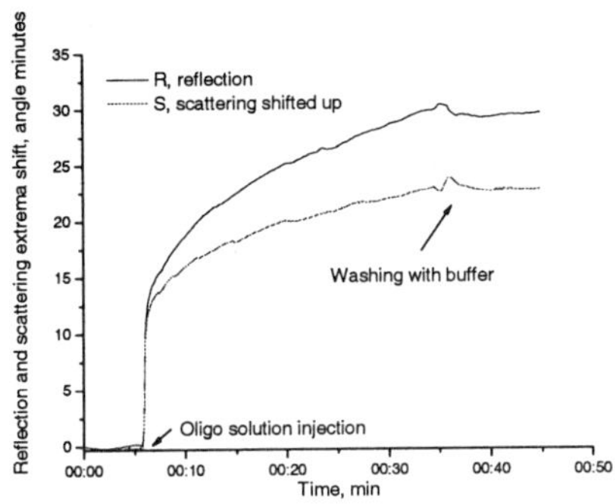

Fig. 5. The extremes shift in scattered and reflected light for 20 bases thiol-modified oligonucleotide immobilization on GNP surface.

reflection channel that more than 10 times bigger than on clean gold surface.

However, immobilization of probe on the gold nanoparticles that was immobilized through phenylene diisothiocyanate gives results that comparable with the clean gold (approx. 2.0 angle minutes in both channels, data not shown). This difference in amplification effect can be explained by different GNP packing density that was not investigated in this work.

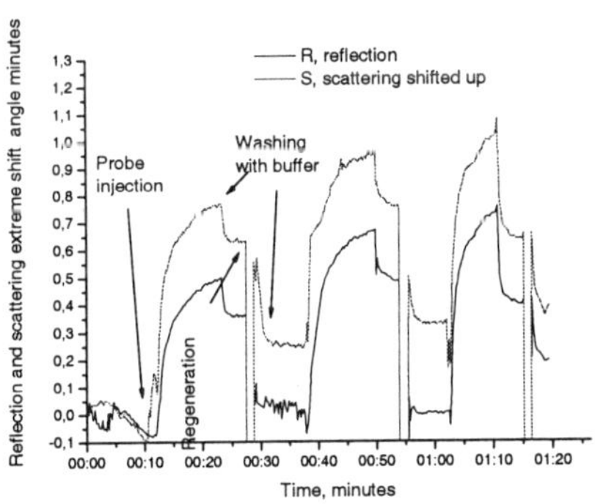

Fig. 6. The extremes shift in scattered and reflected light for repeatable 20 bases complementary oligonucleotide hybridization on 20 bases probe.

We have checked the ability of the system to register hybridization process. We used complementary 20 bases oligonucleotides to check the hybridization process and non complementary 20 bases oligonucleotides to check the specificity of reaction.

We have found that immobilization and hybridization process for as small molecules as 20 nucleotide bases can be registered by both procedures based on an analysis of scattered light and based on an analysis of reflected light. Also we have found that the chip can be regenerated and reused.

The data on figure 6 shows kinetic curves for hybridization of target, washing process and regeneration cycle that was repeated three times. The signal on immobilization of non complementary strands gives the signal that less than 5% of that complementary (data not shown).

5. Conclusions

Here we show that measurements on scattered light yield valuable information under SPR conditions in sensor applications. A scanning spectrometer records variations in both the volume refractive index and oligonucleotides layer formation on a metal surface and has all the functional advantages of the SPR technique. Using the proposed technique we detected kinetic of oligonucleotides immobilization and hybridization. We registered amplification effect of immobilization on 20nm gold nanoparticles.

The novel technique to measure scattered light intensity seems to be very promising. Considering that the design is simple, the technological and measurement reliability of the device optical system is very good. The proposed technique simplifies device design, increases the dynamic range of analysis, and integrates data with those from surface-plasmon field-enhanced fluorescence spectroscopy. All this will result in increased universality, reliability, and cheapness of analytical equipment, thus permitting development of a new generation of efficient and scalable instrumentation using SPR phenomena.

Acknowledgments

This work was supported by INTAS grant # 05-109-5077.

References

1. Rich, R. L.; Myszka, D. G. Drug Discovery Today: Technologies 2004, 1, 301-308.
2. Ramsden, J. J. Optical biosensors. J. Mol. Recognit. 1997, 10, 109-120.
3. Kim, S. J.; Gobi, K. V.; Harada, R.; Shankaran, D. R.; Miura, N. Sens. Actuators B: Chem. 2006, 115, 349-356.
4. Snopok, B.A., Yurchenko, M., Szekely, L., Klein, G., Kasuba, E. Anal Bioanal. Chem. 2006, 386, 2063–2073.
5. Grabar, K.C.; Freemant, G.R.; Hommer, M.B.; Natan M.J. Anal. Chem. 1995, 67, 735-743.

NANOSTRUCTURED-BASED SENSORS FOR ANALYTICAL APPLICATIONS

FEDERICA VALENTINI[†]

Chemistry Department, Tor Vergata University, via della Ricerca Scientifica 1, 00133 Rome, Italy

GIUSEPPE PALLESCHI[†], VANESSA BIAGIOTTI[†],

Chemistry Department, Tor Vergata University, via della Ricerca Scientifica 1, 00133 Rome, Italy

M. L. TERRANOVA[†,2], E. TAMBURRI[†,2],

MINASlab Chemistry Department, Tor Vergata University, via della Ricerca Scientifica 1, 00133 Rome, Italy

In this work a chemical sensor prototype had been electrochemically characterized using a Single-Walled Carbon Nanotubes (SWCNTs) modified gold microelectrode. Several molecular probes (such as ascorbic acid, uric acid, acetaminophen) were investigated by cyclic voltammetry. These compounds are useful to evaluate the selectivity of the electrochemical biosensor prototypes interesting for several analytical applications in environmental monitoring, food quality control and clinical diagnosis.

1. Introduction

Recently a wide range of nanomaterials had been synthesized and nanotechnologies are used to assemble chemical sensors, biosensors and immunosensors [1]. Among all the nanomaterials, carbon nanotubes (i.e.; SWCNTs and Multi-Walled Carbon Nanotubes-MWCNTs) have attracted a lot of attention because of their unique properties, such as their mechanical strength, the higher electrical conductivity, biocompatibility and the possibility to functionalize them with chemical groups. In this work we describe the assembling of selective electrochemical sensors using SWCNTs, having a high signal to noise ratio for the detection of some interesting molecules. The purpose of this work is to investigate the enhancement of the Signal/Noise ratio observed at SWCNT/Au modified microelectrodes. In particular, the SWCNTs functionalization (which exhibit several electrically charged chemical groups, present on their walls and at the caps end) seems to be responsible for the improved selectivity of the analytical response (probably due to the electrostatic interactions between the functional groups and the molecules present in the solution, for measurements). Finally, the miniaturization of these electrodes (Au microwire having radius of 125 μm) combined with the deposit of SWCNTs, by using the Electrophoretical Deposition Process –EPD [2] could represent an interesting and innovative way to fabricate sensitive and selective portable electrochemical devices. In addition, considering the higher surface area exhibited by the carbon nanotubes and their higher bioaffinity towards several biological molecules such as enzymes, antibodies and proteins, these nanomaterials could represent an ideal platform to immobilize biocatalysts for biosensor assembling.

2. Experimental

2.1. *Materials*

Single-Walled Carbon Nanotubes (CarboLex AP-Grade; diameter 12-15 Å) were purchased from Sigma-Aldrich (Steinheim, Germany). Potassium ferricyanide, hexaammineruthenium(III) chloride, ascorbic acid, uric acid, 4-acetaminophenol, potassium chloride, sodium chloride, potassium dihydrogen phosphate, and dipotassium hydrogen phosphate were purchased from Sigma (St. Louis, MO). CH_3CN, H_2SO_4, H_2O_2, and ethanol were from Fluka (St. Louis, MO). All the chemicals from commercial source were of analytical grade. Phosphate buffer (0.1 M, pH 7.0) was prepared with bidistilled deionised water using potassium dihydrogen phosphate, and dipotassium hydrogen phosphate. Only for the interference evaluation (i.e. ascorbic acid) a 0.1M acetate buffer solution, pH 5.4 was used for the experiments. Diluted solutions were prepared just before use. The insulating varnish was purchased from RS (Milan, Italy) and the gold microwire (radius: 125 μm) was from Goodfellow Cambridge Limited (England).

2.2. *Apparatus and Procedures*

For Au microelectrodes a cleaning "piranha" solution, prepared immediately before its use by mixing concentrated H_2SO_4 with H_2O_2, 30% in a ratio of 7:3 v/v, was used. Instead for GC bare electrode polishing, alumina powder (Al_2O_3, Buehler, Evanston, IL) particle size of 1, 0.3, and 0.05 μm was respectively applied on GC electrode surfaces. Then, the working electrodes were copiously rinsed with double distilled water, and finally sonicated in absolute ethanol for 10 min before use. All experiments were carried out at room temperature, working in a phosphate buffer solution (0.1M, pH 7.0). Initial cyclic voltammetry experiments were carried out over the range 5-100 mV/s, while a scan rate of 100 mV/s was eventually chosen to survey the behavior of the various electrodes being evaluated. For the electrochemical sensor assembly, firstly a SWCNT coating on Au microelectrodes (r= 125 μm) surface was performed by an electrophoretical deposition process (EPD), widely described in our previous work [3]. Cyclic Voltammetry, and Chronoamperometry experiments were carried out using an Autolab electrochemical system (Eco Chemie, Utrecht, The Netherlands) equipped with PGSTAT-12 and GPES software. The electrochemical cell was assembled with a conventional three-electrode system: working electrodes made of Au/SWCNTs, and GC bare electrode (Model AMEL GC/492/2; Milan, Italy), an Ag/AgCl/KCl was the reference electrode, and finally a Pt was used as counter electrode. Morphology of the deposits was studied using a Hitachi S-4000 Field-Emission Scanning Electron Microscopy (FEG-SEM).

3. Electrochemical study performed at SWCNT/Au microelectrode

Figure 1 shows the SEM micrograph of the SWCNT/Au microelectrode surface that appears completely covered by bundles of aligned nanotubes. The SWCNT deposit is like a densely packed layer with a rough surface.

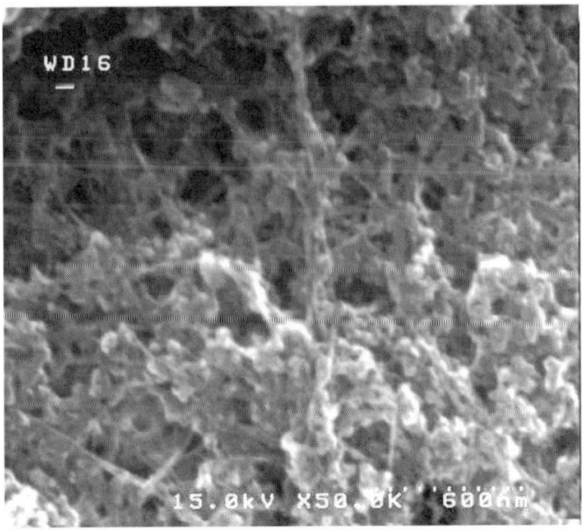

Figure 1.The FE-SEM micrograph of the SWCNT deposit on Au microelectrode

The electrochemical reactivity was investigated using 1mM potassium ferricyanide at SWCNT/Au modified microelectrodes, and Au bare microelectrodes (used here for comparison). At the same time, the electrochemical profile of 1mM potassium ferricyanide was also recorded on GC bare electrode which represents a reference because of the excellent electrochemical profiles previously observed with inorganic probes and also biological redox systems. This specific investigation provides convincing evidence that the electrochemistry obtained at SWCNT-modified Au microelectrodes is related only to the SWCNTs coating and not to the Au substrate. Figure 2A showed an increasing of the peak currents at SWCNT/Au microelectrodes (curve a) for 1mM $Fe(CN)_6^{3-}$compared to the superimposable voltammogram recorded on Au bare microelectrode (curve b) and the electrochemical profile recorded at GC bare electrode (Figure 2B). The increased values of the peak currents could be explained considering the higher electrochemical area (see Table 1, calculated by chronoamperometry, according to the literature [4]) and the higher electrical conductivity of SWCNTs.

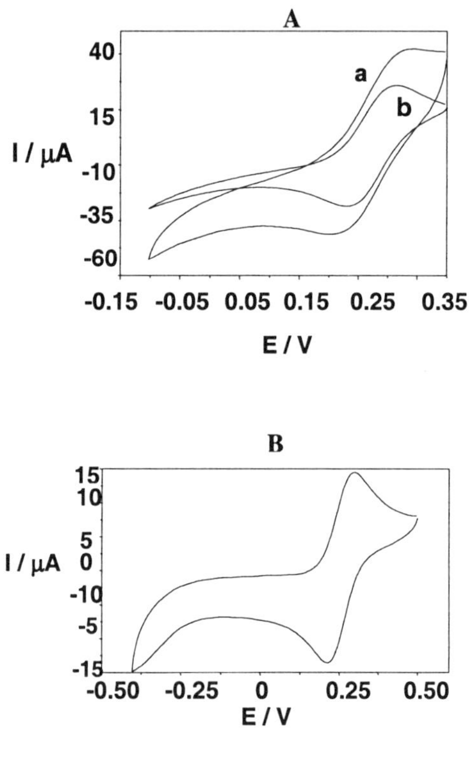

Figure 2. (A) Cyclic voltammograms for 10^{-3} M potassium ferricyanide recorded at SWCNT/Au microelectrode (a) and at Au bare-microelectrode (b). (B): Cyclic voltammograms for 10^{-3} M potassium ferricyanide recorded at GC bare electrode.

Table 1. Electrochemical area evaluated by Chronoamperometry, using Cottrell's equation.

Electrodes	Geometric Area cm^2	Electrochemical Area cm^2
Au[a] bare E	0.039±0.004	0.044±0.004
SWCNT/Au E	0.039±0.004	0.060±0.005
GC[b] bare E	0.031±0.003	0.040±0.003

Measurements of the potassium ferricyanide faradaic current as a function of the scan rate was also carried out and the analysis of the peak currents ($I_{pa,pc}$) vs. $v^{1/2}$ $(mV\ s^{-1})^{1/2}$ over the range studied (Figure 3) showed that the electron-transfer mechanism is controlled by a linear diffusion process. Considering that working at 100 mV/s on SWCNT/Au modified microelectrodes, the signal to noise ratio resulted higher (in 1mM potassium ferricyanide solution), this scan rate was selected for all the voltammetric measurements, because it represented a right compromise to obtain a high sensitivity, and a faster electron-transfer process.

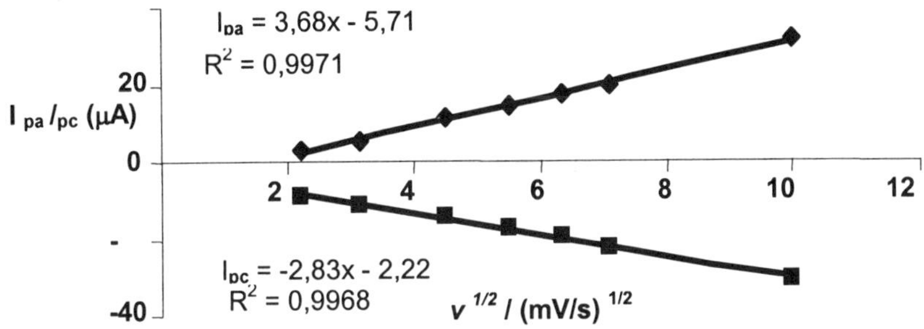

Figure 3. Plot of the catalytic current vs the square root of different scan rates: 5 – 100 mV/s.

In addition, to test the reproducibility of the SWCNT EPD deposition, six SWCNT/Au modified microelectrodes were assembled the same day. The reproducibility study was performed by cyclic voltammetry in 1mM potassium ferricyanide solution, working at 100 mV/s, in 0.1M phosphate buffer, pH 7.0. The inter-electrode reproducibility (i.e.; RSD%) appeared to be a little less than 1%.

3.1. *Selective response to several biomolecules at the functionalised SWCNTs modified Au microelectrodes*

The performances of SWCNT/Au based microelectrodes were also evaluated using acetaminophen, ascorbic acid, and uric acid (see Figure 4 A, C, E, respectively). These molecules were investigated because they represent the most common interferences in biosensor application fields. Figure 4A showed the irreversible oxidation peak for acetaminophen at SWCNT/Au microelectrode having higher current peak, and lower oxidation potential (potential shift of 151 mV) compared to the bare Au electrode. This confirmed that only in the presence of SWCNTs an evident electrocatalytic effect was observed toward the acetaminophen oxidation, having a quasi-reversible electrochemical profile, if compared to the irreversible trend detected on Au bare microelectrodes. This electrochemical profile was also compared with that recorded at GC bare electrode (Figure 4B). A higher oxidation potential (0.512 V) and a lower peak current (27.5 μA) detected at the GC bare electrode, confirmed that the best performances were observed at SWCNT/Au modified microelectrodes. Moreover, considering the responses of ascorbic acid and uric acid, no signals were obtained on Au/SWCNT modified microlectrodes (see Figure 4C, and E, respectively), compared to those recorded on the Au and GC bare electrodes (see Figure 4 D, F) working in the same experimental conditions. This could be related to the presence of an electrostatic barrier on the SWCNT/Au microelectrode surface, for the presence of several electron-donor OH groups [2,3], which repelled the negatively charged molecules (as ascorbate and urate in 0.1M acetate buffer solution, pH 5.0, and 0.1M phosphate buffer pH 7.0, respectively). The possible effect due to the presence of an electrostatic barrier, was confirmed using a neutral molecule, (as acetaminophen) for which no electrostatic repulsion was observed working with the same SWCNT-modified Au microelectrodes. In addition, using 0.1mM epinephrine, which resulted positively charged in 0.1M phosphate buffer solution, pH 7.0 (considering its pKa = 8.87,), a higher signal to noise ratio was observed at SWCNT/Au microelectrodes[5], compared to the Au, and GC bare electrodes (data not shown), confirming that the electrostatic barrier due to the presence of SWCNT-OH (the electron-donor groups) could attract molecular systems having opposite electrical charges[5].

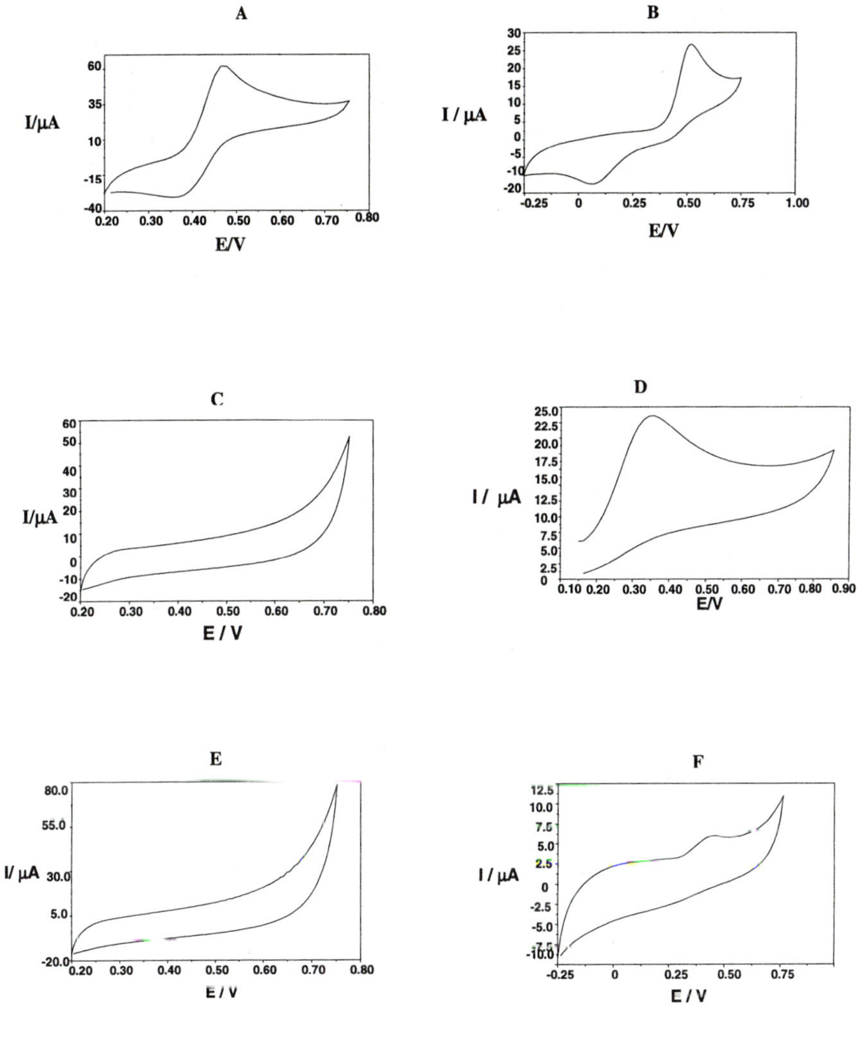

Figure 4 (A) Cyclic voltammograms for 0.1mM acetaminophen; (C) 0.1mM Ascorbic Acid; and (E) 0.1mM uric acid, recorded at SWCNT/Au microelectrodes (tip diameter: 250 μm; tip length: 0.5 cm). (B) Cyclic voltammograms for 0.1mM acetaminophen; (D) 0.1mM Ascorbic Acid; and (F) 0.1mM uric acid recorded at GC bare electrode .

4. Conclusions

In this work new electrochemical sensors have been proposed modifying the Au microwire surface with functionalized SWCNTs. The analytical response observed for the molecules tested resulted more sensitive and selective compared with that obtained with traditional GC electrodes. In addition the higher signal to noise ratio observed at SWCNT modified Au microwires combined with the miniaturization of the electrochemical transducers could open a new interesting way to assemble nanostructured biosensors and devices useful for analytical applications in environmental monitoring, food quality control and clinical diagnosis.

References

[1]. J. Wang, *The Analyst*, **130,** 421 (2005).

[2]. F. Valentini, S. Orlanducci, M. Letizia Terranova, G. Palleschi, *Sensors & Actuators: B. Chemical* **123**(1), 5-9 (2007)

[3]. F. Valentini, S. Orlanducci, M. L. Terranova, and G. Palleschi, *THERMEC 2006, Proceedings of the International Conference on Processing e Manufacturing of Advanced Materials*, July 4-8, 2006 Vancouver, Canada.

[4]. J. Wang, Analytical Electrochemistry, ***ISBN 1-56081-575-2***,1994 VCH Publishers, Inc.

[5]. F. Valentini, G. Palleschi, E. Lopez Morales, S. Orlanducci, E. Tamburri, M. L. Terranova, Electroanalysis, **19**(7-8), 859-869 (2007)

SCREENING OF BIOMIMETIC RECEPTORS BY MEANS OF HIGH-DENSITY COLORIMETRIC MICROARRAY

MARCELLO MASCINI[AB*], GEORGE GUILBAULT[B], MICHELE DEL CARLO[A], MANUEL SERGI[A] AND DARIO COMPAGNONE[A]

[a] *Università di Teramo, Dipartimento di Scienze degli Alimenti , 64023 Teramo, Italy*

[b] *Department of Chemistry, University College Cork (UCC), Cork, Ireland*

**Corresponding author: mmascini@unite.it*

The aim of this work was to optimise biomimetic-based strategies for possible pathogen detection. The use of new affinity ligands such as the oligopeptides computationally designed is particularly suitable for large-scale synthesis and overcomes the disadvantages of antibodies or enzymes which are often unstable and expensive. The pathogen prototype system chosen was *Listeria* bacterium because of the large amount of X ray and NMR structures associated to that pathogen available from the literature. 13 oligopeptides libraries (298 oligopeptides in total) mimicking the binding pocket of the *listeria* target, the protein cadherin, were designed. The contribution of individual peptide to binding was investigated using Hex a protein-ligand docking program. Four peptides different in length and binding energy were selected for experimental part. High-density colorimetric microarray was used to assess the effective binding and the selectivity of the receptors towards the target. The concentration of pathogen cells in solution ranged from 10^5 to 10^9 cells/ml. The raw signals were read using a conventional scanner and elaborated with an image software. The pixel intensity decrease of the signal was interpreted proportional to the probe-target binding. With gasket facilities, cross section data approach was applied detecting up to 512 entities simultaneously. In this way the main analytical parameters were investigated. This study was finalised to gain an understanding of how the parameters calculated by computational modelling could be helpful to select biomimetic receptors to bind the target in experimental conditions.

1. Introduction

The use of new affinity ligands such as oligopeptides computationally designed is particularly suitable for large-scale synthesis and overcomes the disadvantages of biological molecules which are often unstable and expensive. This latest class of potential receptors can be made mimicking the nature that use as bricks aminoacids to construct a large amount of variable and adaptable molecules for life (1). To test the efficiency of these new ligands the prototype system chosen was the pathogen *Listeria* bacterium because of the large amount of X ray and NMR structures associated to that pathogen available from literature. To enter and survive within the host cells, bacteria frequently exploit intracellular signal transduction mechanisms. Internalin is the listerial surface protein that mediates host cell specific internalization binding its host cell receptor cadherin. Information available on this complex, were used as a guideline to reduce the number of structures to be computationally tested (2). Starting from the crystallographic structure studied by Schubert and co-workers, we produced a series of libraries using as backbone the aminoacids from cadherin involved in the binding site. Studying the geometry conformation of the Internalin/Cadherin complex, 13 oligopeptides libraries (298 oligopeptides in total) mimicking the binding pocket were designed. The contribution of individual residues to binding was investigated using Hex a protein-ligand docking program. Hex is a program that treats conformers and protein structure as rigid during the majority of the docking process, exploring all possible positions of each ligand in the active site (3).

In experimental part high-density colorimetric microarray was used to assess the effective binding and the selectivity of the receptors towards the target.

Microarray technologies permit systematic approaches to biological discovery that have begun to have a profound impact on biological research, pharmacology, and medicine. The ability to obtain quantitative information about the complete transcription profile of cells promises to be an exceptionally powerful means to explore basic biology, diagnose disease, facilitate drug development, tailor therapeutics to specific pathologies, and generate databases with information about living processes (4).

In this way colorimetric arrays can be a good candidate as alternative of fluorescence detection for producing inexpensive analytical devices (5).

The advantages of colorimetric microarray-based assays are their sensitivity, repeatability, rapidity, simplicity, portability, and cost-effectiveness. They have applications in the clinical and medical in vitro diagnostic fields and even if they have to be fully validated they are believed to be equivalent to other immunoassays (6).

Four peptides different in length and binding energy were selected for experimental part. The experimental strategy, consisting of label direct capturing assay, produced raw signals that they could read using a conventional scanner and elaborated with an image software. The pixel intensity decrease of the signal was interpreted proportional to the probe-target binding. Four bacteria *Listeria monocytogenes*, *Listeria innocua*, *Listeria monocytogenes without internalin* (genetically modified), *Lactococcus lactis* were selected to check the specificity and the cross reactivity of the peptides.

2. Experimental

Reagents

All chemicals and reagents used were of analytical grade. The peptides were purchased from JPT Peptide Technologies GmbH (Germany) and were biotinylated. All reagents for buffers and Extravidin-alkaline phosphatase conjugate (extravidin-AP) were obtained from Sigma-Aldrich (Ireland). Blocker solution (BSA-like), developer solution (like paranitrophenylphosphate, pNPP) were supplied from the Miragene Inc. (US). The Blocker and Developer were undiluted.

The tris buffer used to make extravidin-AP solution was at the final concentration of 50mM with 1mM $MgCl_2$. The phosphate buffer used to make peptide and bacteria solutions was at the final concentration of 10 mM.

All bacterial cultures were grown in 10 ml of Brain Heart infusion broth (BHI). Overnight cultures were diluted 1:20 in fresh BHI and grown at 37°C with shaking at 200 rpm to an optical density 600 (OD600) of $5*10^9$ cell/ml. These cultures were harvested by centrifugation at 4,000 xg for 10 minutes. Cultures were washed with an equivalent volume of buffer with centrifugation / resuspension repeated additional two times (7).

Apparatus

The Zeta-Grip chips, gaskets and the FAST Frame were all provided by Miragene Inc. (US). In each of the 64 wells was allowed to spot up to 8 drops of 1μl, resulting in a cross-section data set of 512 different entities. The hard-scanner used was Epson 2400 Scanner coupled with the software Photoshop CS2 (Adobe Systems) for acquiring and inverting the image. The free "share ware" software UTHSCSA Image Tools 3.0 was used for quantification of spots after inverting and saving in tiff file the image scanned.

Colorimetric Microarray protocol

The array procedure consisting of 5 principal steps are reported below:

1. After placing the gaskets into the Fast Frame 1μl drop of the biotinylated peptides, at the 1mM concentration in phosphate buffer, were hand spotted onto the surface, put into cooled conditions (4-8°C) 60-80% humidity for 60 minutes and at 37°C for 60 minutes.
2. To each well 120 μl of Zeta-Blocker was added and was left at room temperature for 30 minutes.
3. The incubation wells were then emptied and the washings were disposed of appropriately. 120 μl of the different bacteria sample was added for 60 minutes to each well and incubated at 37°C.
4. The incubation wells were emptied and washed appropriately. 120 μl of Extravidin-AP at a concentration of 6 μg/ml n 50 mM tris buffer was added to each well and incubated at 37°C for 30 minutes.
5. After washing 120 μl of Zeta-Developer was incubated at room temperature for 10 minutes. The Zeta-Developer was discarded and a quick rinse of the

gasket with tap water was performed and then the surface was covered with water for 2 minutes to stop further colour development. The surface was left air-dry by positioning it upright as to allow excess water to run off all night.
The analytical signal was an average of 5 points randomly acquired within the spot, expressed as pixel intensity having arbitrary scale ranged from 0 to 255. To all analytical signals, the background, always less then 15, was subtracted. The statistical data calculation came from the average of three drops spotted in different wells.

3. Results and Discussion

In the table 1 the binding energies of the peptides selected for experimental part were reported. E total, E shape and E force refer to different values in energy. E shape determines how the target fits into the protein and E force is the measure or electronic interaction between the complexes. The sum of these two values gives E total. The four peptides came from different parts of the protein cadherin and were selected from the 298 possible peptides designed and tested by molecular modelling using as target internalin. The peptides were chosen different in structures and energies to gain an understanding of how the parameters calculated by computational modelling could be helpful to select biomimetic receptors to bind the target in experimental conditions. It should be noted that at the peptide end a Lysine was introduced to bind biotin.

Table 1. The binding energy of the four peptides used in experimental part versus Internalin.

ID	Peptide	E total	E shape	E force
A	NLDK	-276.2	-168.8	-107.4
B	VGVPE	-339.8	-181.3	-158.5
C	EQPPFPKN	-459.8	-215.2	-244.6
D	WRQIKSNKDK	-590.4	-319.1	-271.2

In experimental part the high density colorimetric microarray system was used to test simultaneously the response of the four peptides computationally selected with four bacteria *Listeria monocytogenes* (LM), *Listeria innocua* (LI), *Listeria monocytogenes without internalin* (genetically modified (LMGM), *Lactococcus lactis* (LL) selected to check the specificity and the cross reactivity of the peptides. The results obtained for this experimental part was reported in table 2. The peptides gave a concentration-dependent response towards the sample LM. Especially looking at the results from the peptide D the sample LM could be detected up to 10^7cell/ml. At this last concentration the other peptides suffered by cross reactivity effect. The solution of 10^5cell/ml was found to be too diluted. The relative standard deviation was all over within 10%.

Table 2. Pixel intensity signals obtained using the four bacteria in the range 10^5-10^9 cells/ml incubated in the wells spotted with the four peptides. The numbers in bold represented the scores (column) and the loads (row) used as labels in Figure 1.

Peptide		**A**	**B**	**C**	**D**
Sample		**1**	**2**	**3**	**4**
LM at 1E+09	**1**	12	10	28	44
LM at 1E+07	**2**	32	34	45	76
LM at 1E+05	**3**	55	46	63	143
LI at 1E+09	**4**	53	45	62	110
LI at 1E+07	**5**	68	55	57	126
LI at 1E+05	**6**	60	56	55	121
LM GM at 1E+09	**7**	40	46	45	123
LM GM at 1E+07	**8**	36	42	56	114
LM GM at 1E+05	**9**	41	38	67	108
LL at 1E+09	**10**	36	40	39	102
LL at 1E+07	**11**	41	48	46	112
LL at 1E+05	**12**	38	44	43	104
Buffer	**13**	53	56	60	145

In order to get a clearer view of this set of results a principal component analysis (PCA) was carried out in auto scaling mode. The results of this elaboration was reported in Figure 1. The three principal components were used obtaining an information up to 98.61% of the total. Looking at the contribution of the scores it was possible to clearly distinguish the *Listeria Monocytogenes* (point 1 highlighted with a circle) between the sample set.

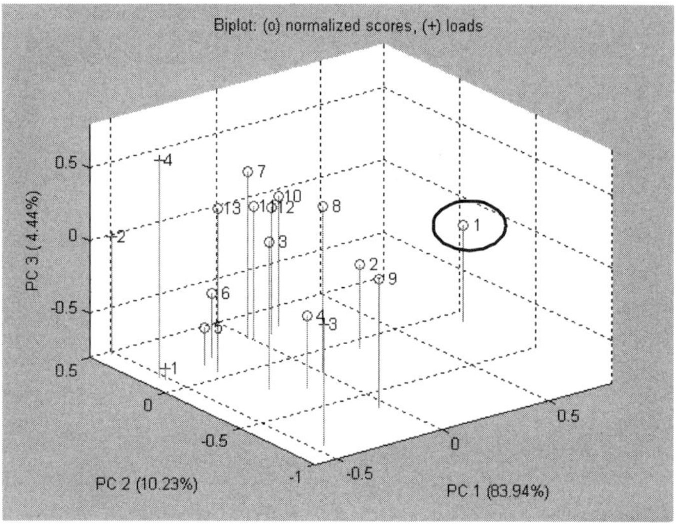

Figure 1. Biplot Auto-scaled. The principal component analysis of the matrix reported in table 1. The scores and loads labels are reported in Table 1.

4. Conclusions

The Development of a highly sensitive microarray-based system using colorimetric detection for the detection of *Listeria Monocytogenes* was demonstrated. The method of assembling the protocol and in particular the peptides choice using molecular modelling stategy was considered of paramount significance in obtaining system selectivity.

Acknowledgments

The authors would like to express their gratitude to Prof. Colin Hill from University College Cork for his contribution to the work. This research was supported by MEIF-CT-2005-011588 within the 6th European Community Framework Programme.

References

1. C. Falciani, L. Lozzi, A. Pini and L. Bracci Chemistry & Biology, 12 (2005), 417
2. W.-D. Schubert, C. Urbanke, T. Ziehm, V. Beier, M.P. Machner, E. Domann, J. Wehland, T. Chakraborty, D.W. Heinz, Cell 111 (2002) 825
3. http://www.csd.abdn.ac.uk/hex/
4. R. A. Young. Cell 102 (2000) 9
5. K. S. Suslick, N. A. Rakow, A. Sen, Tetrahedron, 60 (2004) 11133

6. S.J. Lebrun, W.N. Petchpud, A. Hui, C.S. McLaughlin, J. Immunol. Methods, 300 (2005) 24
7. S. Corr, C. Hill, C. Gahan, G.M Cormac, Microbial Pathogenesis 41 (2006) 241

PERFORMANCES OF THE IMMUNOGRAVIMETRIC SENSOR μLIBRA 3.1

M. PASSAMANO*, S. GRECO
Biological Division Technobiochip S.c.a.r.l., Via Provinciale per Pianura n° 5, 80078 Pozzuoli (Na), Italy

Here we present a study on the performances of *μLibra* v3.1, a new and more compact Quartz Crystal Microbalance (QCM) developed at Technobiochip. Using a classical antigen-antibody binding reaction, we tested the potentiality of the instrument (CE_{06} Certified) to be used as immunosensor valuating accuracy, sensitivity, response linearity and pH buffer effects, respectively. The experiments were performed immobilizing the antibody both on bare and coated quartzes using a self-assembled monolayer (SAM) of 16-mercaptohexadecanoic acid (MHDA). Obtained data demonstrate a good intra-assay accuracy between channels, linearity up to 40 μg/ml (plateau value) and 1 nmol/l as sensitivity. Nevertheless, measured and calculated number of antibody molecules is in agreement indicating a good reliability.

1. Introduction

The Quartz Crystal Microbalance (QCM), as a simple yet powerful technique, has been widely employed in chemical and biological sensing thanks to real-time response, label-free features and low cost. QCM immunosensors, that use highly specific antigen-antibody reactions on piezoelectric crystal microbalance, have been developed in a wide range of applications, e.g. microbial studies in food analysis, environmental monitoring and clinical diagnostics [1]. The potential of piezoelectric devices as biosensors is based on the relationship between the change in the oscillating frequency of a piezoelectric crystal and the

* Address Correspondence to: Dr. Myriam Passamano, BSc, MS, Researcher Biological Division Technobiochip Scarl, Via Provinciale per Pianura, 5 (Loc. San Martino), 80078, Pozzuoli (Na), Italy. E-mail: m.passamano@technobiochip.com - Phone: +39 081 5264315 – Fax: +39 081 5265116

mass deposited on its surface. The equation describing the frequency-to-mass relationship in the air phase has been given by the Sauerbrey equation [Eq. (1)].

$$\Delta F = \frac{-2.3x10^6 F_0^2 \Delta m}{A} \quad (1)$$

where ΔF is the change in frequency (Hz), F_o is the resonant frequency of the crystal (MHz), Δm is the mass deposited (g) and A is the area coated (cm^2). This equation is only valid for a rigid coating [2]. Taking into account the effect of viscosity and density of solution, modification of the equation has been made by Kanazawa and Gordon [3] to develop equivalent circuit model for the response of piezoelectric crystal in liquid phase also using a more elastic biomaterials. Technobiochip S.c.a.r.l. carries out activities in the design and production of *μLibra,* a quartz-based immuno-microgravimetric balance. Instruments are constantly being checked for quality performances. A new, more compact microbalance, *μLibra* v3.1, with EC certification, has been developed at Technobiochip laboratories and the goal of the present study was to evaluate the instrument's performances as immunosensor. A simple system model, mouse IgG as antigen and anti-mouse IgG as antibody, has been used to evaluate accuracy, sensitivity and linearity of the system. Moreover, we have investigated the pH buffer effects on antigen-antibody interaction and the influence of the self-assembled monolayer (SAM) of 16-mercaptohexadecanoic acid (MHDA) on antibody binding on quartz golden surface.

2. Materials and Methods

2.1. *Reagents*

Mouse Polyclonal IgG, goat Anti-mouse IgG and other reagents were purchased from Sigma (Milan, Italy). The quartz crystals, 10 MHz AT-cuts, gold electrodes on chromium layer, were obtained from International Crystal Manufacturing Company (Ohio, USA).

2.2. *Apparatus*

μLibra 3.1 is a Quartz Crystal Microbalance system based on quartz crystal resonators, produced by Technobiochip S.c.a.r.l. (Pozzuoli, Italy). The instrument is composed of a main unit and a cell base unit with two 10MHz oscillators that can support two measuring chambers (Figure 1). The two-channel acquisition system allows for single cell operation as well as working/reference measurements. The oscillators either work in air or in solution by two low-volume flow-through cells. The LibraVIEW software elaborates μLibra output data.

Figure 1. Quartz Crystal Microbalance μLibra 3.1.

2.3. *Experimental Protocol*

Quartzes, with or without functionalizations, were placed in a static flow-cell and the basal frequency oscillation (ΔF) was stabilized by buffer adding (Phosphate Buffered Saline, PBS, 10 mM pH 7.4). Then the antibody solution (Anti-Mouse IgG, SIGMA) was flown and the surface saturated with a BSA solution (100 μg/ml). After PBS washing, antigen solution (Mouse IgG, SIGMA) was added and ΔF of antigen-antibody interaction recorded. All the experiments were performed in replicates.

2.4. *Strategies of Antibodies Immobilization*

Two different procedures of antibodies immobilization were used on quartzes surface: physical adsorption of antibody or covalent binding on functionalized quartz with self-assembled monolayer (SAM) of 16-mercaptohexadecanoic acid (MHDA).

MHDA-SAM: Quartzes, after cleaning with piranha solution (H_2SO_4:H_2O_2 3:1 v/v), were dipped into 1 mM MHDA in ethanol for 12 hours at roomtemperature and then washed with distilled ultra pure water. The monolayer was successively activated with a mixture of 0.2 M EDC/0.05 M NHS for 30 minutes at room temperature.

3. Results and Discussion

Using a classical antigen-antibody binding reaction a study on the performances of the immunogravimetric sensor μLibra 3.1 was performed.

The reproducibility between working and reference channels was firstly investigated by using physical adsorption of antibody on golden quartz. The experiments were carried out in replicates in both chambers by using 100 μg/ml anti-IgG and 40 μg/ml IgG. *μLibra* 3.1 showed a good intra-assay accuracy and reproducibility between channels (data not shown). In order to evaluate the immunosensor's sensitivity, increasing antigen concentrations from 0.62 to 60 μg/ml were analyzed for binding to 100 μg/ml anti-IgG and the respective frequency variations have been recorded (Figure 2).

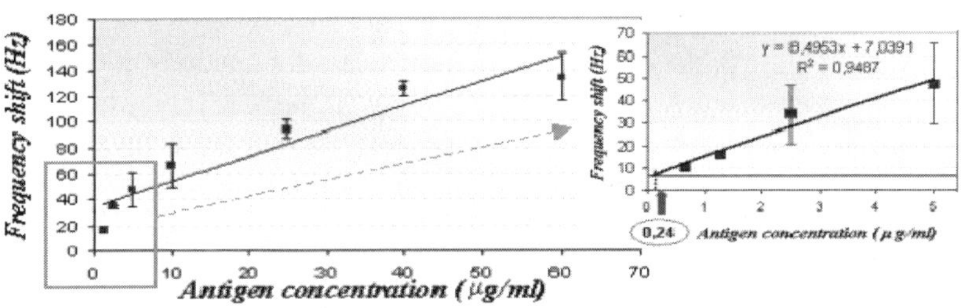

Figure 2. Calibration plot for mouse IgG. Red line represents the instrument noise. On the left an inset shows more in detail the response in the antigen concentrations range between 0 and 5 μg/ml.

The sensor's response showed a linearity until 40μg/ml (plateau value). The sensitivity was 0.24 μg/ml (1.0 nmol/l), corresponding to the lowest detectable antigen concentration that gave a signal (Δf) above the instrument noise (±2.5 Hz) (Figure 2). The resulting working range was 0.24-40 μg/ml. Furthermore, increasing antibody concentrations (5.0–100 μg/ml) were analyzed for binding to 5.0 and 0.62 μg/ml of antigen. The first antigen concentration represented an intermediate value between the detection limit and the saturation value while the second one was near to limit detection value (Figure 3).

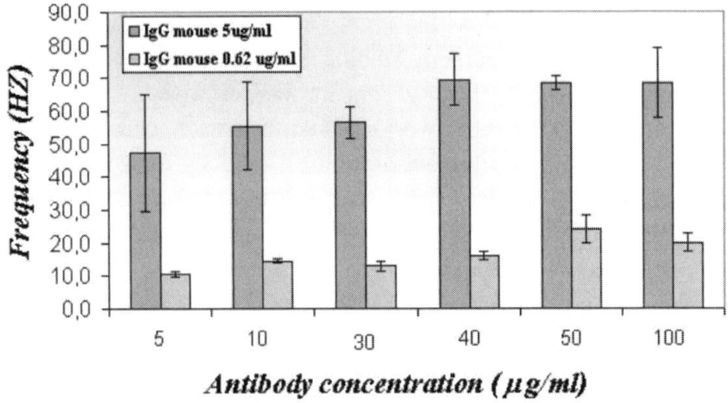

Figure 3. Signal saturation curve at 0.62 and 5.0 μg/ml antigen concentration.

As shown in Figure 3, the observed saturation values for antibody at 0.62 and 5 μg/ml antigen concentrations were 40 and 50 μg/ml, respectively. Moreover, at low antibody concentrations, higher frequency variations were obtained after antigen binding, probably due to the unmasking of antibody binding sites.

The pH influence on antigen-antibody interaction was also measured by using 10 mM PBS at different pH values (pH 6.0, 7.4 and 8.0). Higher and more stable responses were obtained at pH 7.4 (data not shown). Finally, we evaluated the influence of SAM of MHDA, a long-chain carboxylic acid-terminating alkanethiol, on antibody binding on quartz golden surface [4]. No significant differences between bare or SAM-coated quartzes were obtained (data not shown). Moreover, the calculated numbers of antibody molecules on the quartz surface were 9.3 and 10×10^{11} molecules, corresponding to 232 and 265 ng for SAM and bare quartzes, respectively. These values were in agreement with the theoretical value [5], so indicating a good sensor's reliability.

4. Conclusions

A comprehensive analysis of the immunogravimetric sensor *μLibra* v3.1 was performed. The sensor shows a good intra-assay accuracy between channels, linearity up to 40 μg/ml and 1 nmol/l as sensitivity. Nevertheless, measured and calculated numbers of antibody molecules are in agreement, so indicating a good sensor's realibility. In conclusion, *μLibra* 3.1 demonstrates to have realistic potentiality as immunosensor.

References

1. P. Skladal, *J. Braz. Chem.* **4**, 491 (2003).
2. G. Z. Sauerbrey, Z *Phys.* **155**, 206 (1959).
3. K. K. Kanazawa and J. G. Gordon, *Anal Chem.* **57**, 1770 (1985).
4 X. L Su and Y. Li, *Biosens. Bioelectron.* **19**, 563 (2004).
5. Y. Lee, E.K. Lee, Y. W. Cho, T. Matsui, I. Kang, M. H. Han, *Proteomics.* **3**, 2289 (2003).

LIVING PARAMETERS MONITORING

A STUDY OF INDIUM OXIDE SENSORS FOR DIABETES BIOMARKER DETECTION IN THE HUMAN BREATH

G. NERI, A. BONAVITA, G. MICALI, S. IPSALE

Dept. of Industrial Chemistry and Materials Engineering, University of Messina, C/da di Dio, Vill. S. Agata,98166 Messina, Italy

neri@ingegneria.unime.it

E. CALLONE, G. CARTURAN

Dept. of Materials Engineering and Industrial Technologies, University of Trento, 38050 Trento, Italy

Summary

Breath analysis allows monitoring several pathologies in a non invasive way. Notwithstanding semiconductor-based sensors show optimal features for clinical application (easy to manufacture, small in size and cheap) until now specific devices based on MOS sensors for breath analysis are lacking and sensitivity is still poor. In this study are reported preliminary data on the development of an acetone MOS sensor in order to analyze diabetic patient's breath. In_2O_3 nanopowders synthesized by two different sol-gel methods have been compared as sensing layer.

1. Introduction

Breath analysis is rapidly gaining ground as means to non invasively diagnose many diseases and monitor various aspects of the metabolism [1-4]. The bulk matrix of exhaled human breath is a mixture of nitrogen, oxygen, carbon dioxide, water and inert gases, the remaining small fraction consisting of a variety of VOCs (Volatile Organic Compounds) in very small concentrations. In the presence of a particular disease, some specific VOCs are found in anomalous concentration. For example, the breath acetone concentration is higher in diabetics than in healthy peoples and has been found to correlate with plasma-ketones and β-hydroxybutyrate concentration in venous blood [5-8]. The increase of ketone bodies in the blood results in ketoacidosis, a severe clinical conditions in diabetic peoples [8, 9]. Therefore, breath acetone is a suitable marker to monitor frequently and in a non invasive way diabetics at risk for ketoacidosis.

Recent advances in analytical instrumentation has lead to the availability of a new generation of real-time, portable breath test instruments that could, in time, become ordinary in medical screening. Sensor-based electronic devices are expected to play a main role in breath analyzers and have potential applications in the detection of VOCs in the human breath [10].

However, sensitivity of MOS sensor is still poor considering the very low level of VOCs of interest in the breath. In the case under investigation, normal individuals acetone breath concentration varies between 0.3 – 1.3 ppm. On the other hand, increased concentrations of acetone (in the range 1.7–3.7 ppm) is reported in the breath of patients suffering from diabetes. Higher levels are typical of the ketoacidosis state.

One way to increase the sensitivity of MOS sensors is to use the metal oxide sensing layer in the nanostructured form. As the final microstructural and functional characteristics of nanostructured metal oxide powders are highly influenced by the synthesis process used, great attention is devoted to this aspect. Therefore, we focused our study on In_2O_3 nanopowders prepared by a non-acqueous sol-gel method and a starch-aided hydrolytic sol-gel synthesis [11, 12].

2. Indium Oxide Nanopowders Synthesis

In_2O_3 nanopowders were synthesised via a nonaqueous sol-gel method (route 1) involving the solvothermal reaction of indium (III) isopropoxide with benzyl alcohol (Fig. 1). The dopant, Pt, has been dispersed on the surface of the semiconducting metal oxide by wetness impregnation, contacting the In_2O_3 nanopowders with the proper amount of an aqueous solution of H_2PtCl_6. The nanopowders have been finally dried and fired at controlled temperature (between 150 and 450 °C) in air.

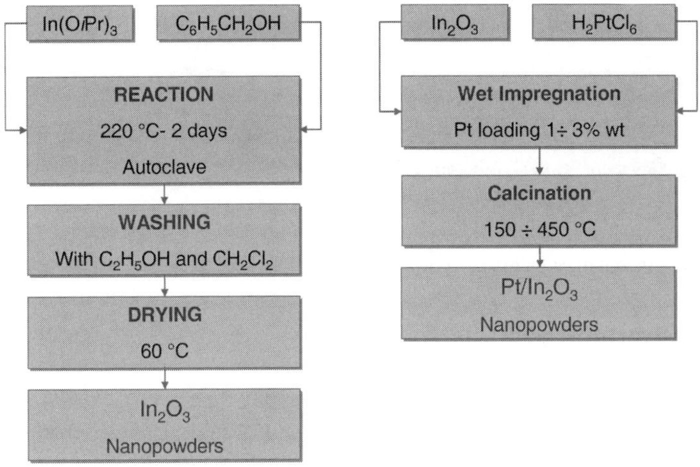

Fig.1. Synthesis of In_2O_3-based nanopowders by the non aqueous sol-gel method.

Synthesis of In_2O_3-based nanopowders by the starch-aided hydrolytic sol-gel synthesis (route 2) was carries as follows. Pure precursors ($SnCl_4$, $InCl_3$), diluted in ethanol absolute, were added by a doping funnel to starch gel in a three-necked flask under vigorous mechanical stirring, the pH being kept around 7.0 by simultaneous addition of 0.5 M NaOH at 100 °C. At room temperature, the starch component was further degraded by α-amylase in a few hours, with the formation of a colloidal suspension. Products were recovered after washing with doubly distilled H_2O to remove soluble by-products, and treatment with H_2O_2 (30% w/w in water) at 50 °C for 24 h. The pure oxides In_2O_3, were dried at 120 °C for 12 h and stored in dry conditions. Samples were heated in air up to 550 °C at a rate of 5 °C/min. A schematic view of the stabilization of metal oxide nanoparticles by this method is reported in Fig. 2. Subsequently the nanopowders are doped with 1 wt% Pt by wetness impregnation and treated at 400°C for 1 hour.

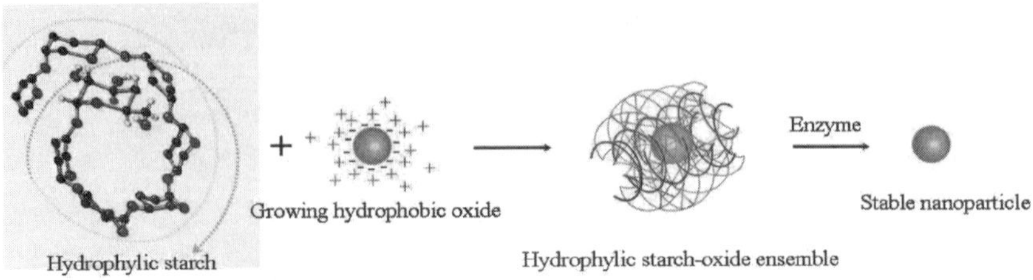

Fig. 2. Stabilization of nanopowders by the starch-aided hydrolytic sol-gel synthesis.

3. Microstructural Characterization

XRD and TEM characterization of the nanopowders were carried out by a Italstructure diffractometer mod. APD2000 and a Jeol JEM 2010 microscope. XRD show only the presence of the bixybite phase on both samples (Fig. 3).

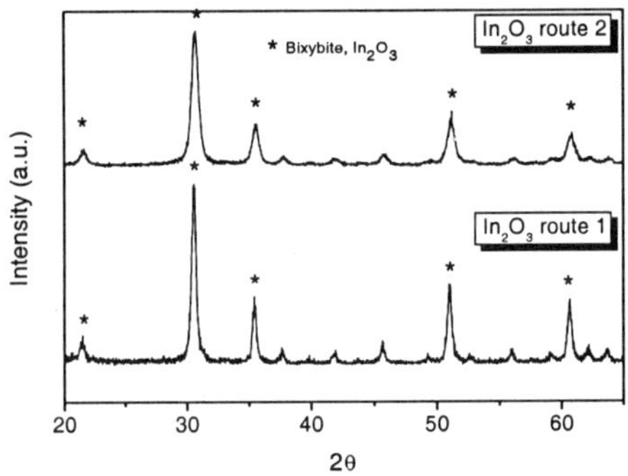

Fig. 3. XRD patterns forIn$_2$O$_3$ samples synthesized by route 1 and route 2.

The mean grain size, confirmed further by TEM observations, are 25 nm for the sample prepared by the non aqueous sol-gel synthesis (route 1) and 15 nm for the sample prepared by the starch-aided hydrolytic method (route 2).

4. Acetone Sensing Tests

Sensing tests were preformed on devices obtained by mixing In2O3 powders with water to create a paste which was successively screen-printed on standard alumina substrates (3x6 mm2) supplied with comb-like electrodes and a Pt-heater. The devices were dried and then treated in air at 350°C for one hour in order to stabilize the film texture and microstructure. Resistance measurements were carried out in a home made apparatus (see Fig. 4) by a Agilent 34970A data acquisition switch, maintaining the sensors under controlled stream of dry air and pulsing different acetone concentrations coming from a permeation tube (Fine permeation tubes).

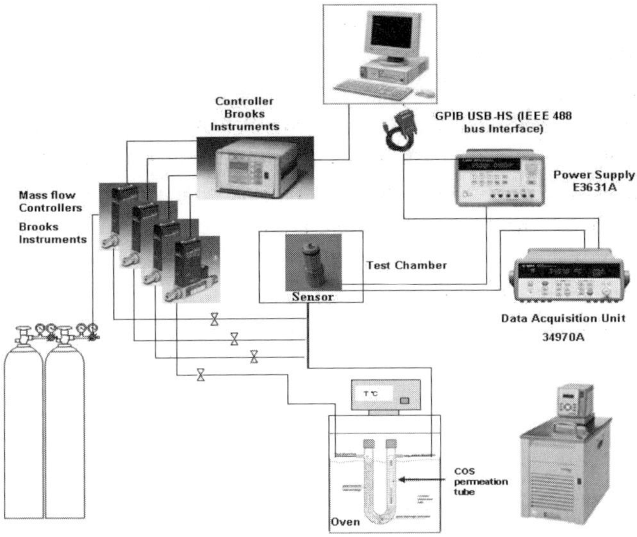

Fig. 4. Schematic view of the sensing test apparatus.

The acetone concentrations tested (1-10 ppm) cover the range from healthy to diabetic peoples. Acetone is a normal breath constituent and is responsible for the sweet odor of the breath of diabetic individuals. It is produced mainly from the spontaneous decarboxylation of acetoacetate and, to a lesser degree, by the enzymatic conversion of acetoacetate to acetone via the enzyme acetoacetate decarboxylase. Breath acetone level rises further during ketosis or ketoacidosis, a type of metabolic acidosis most common in untreated diabetes which is caused by high concentrations of ketone bodies, formed by the deamination of amino acids and the breakdown of fatty acids [9]. In ketoacidosis, the accumulation of keto acids is so severe that the pH of the blood is substantially decreased.

The current widely used assays of ketones in blood to monitor ketosis and ketoacidosis have some drawbacks. Frequent blood sampling is invasive, particularly in children. To have the opportunity to perform frequent assessments of ketone bodies concentrations friendly, breath acetone analysis is potentially useful.

Therefore we investigated the possibility to develop acetone MOS sensors with high sensitivity. We focused our attention on nanostructured In_2O_3 powders as sensing layer. The sensors were first tested to evaluate the best working conditions in order to obtain the higher sensitivity and faster response/recovery time. Sensing data indicate that samples prepared by means of route 1 are the most sensitive. Pt doping is further effective in enhancing the response of In_2O_3 sensors. Sensors response at different temperature towards 6 ppm of acetone indicate that the highest sensitivities are obtained operating the sensors at 200°C. In order to have acceptable response/recovery times it is however necessary to increase the operating temperature. The temperature of 250°C represents the best compromise between high sensitivity and fast response/recovery time (Figure 5a). The calibration curve is linear in the concentration range between 1 and 10 ppm (Figure 5b).

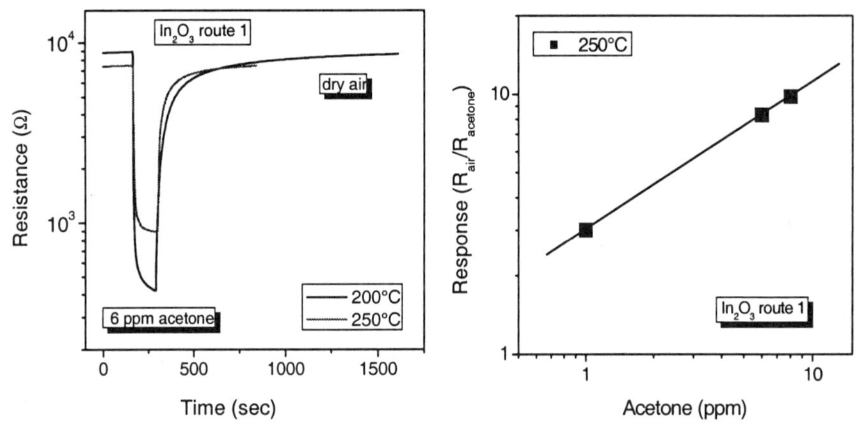

Fig. 5. (a) Transient response at 200°C and 250°C; (b) calibration curve at 250°C.

These preliminary results suggest that indium oxide nanopowders are promising sensing materials for acetone detection in medical applications (i.e. ketoacidosis in diabetics). Further studies are in progress in order to: i) evaluate the response to acetone in the presence of interfering gases (CO, CO_2, humidity, etc.) present in the breath; ii) investigate the implementation of these sensors in a sensors array.

On this basis, a portable, hand-held breath acetone analyzer will be developed for monitoring ketoacidosis in diabetic patients.

Acknowledgment

The financial support for this work by MIUR under the PRIN project (Contract number 2005039547) is gratefully acknowledged.

References

1. W. Miekisch, J.K. Schubert, G.F.E. Noeldge-Schomburg, "Diagnostic potential of breath analysis;focus on volatile organic compounds" *Clinica Chimica Acta* 347 (2004) 25.
2. Marczin N, Kharitonov SA, Yacoub MH, Barnes PJ (Eds), *Disease markers in exhaled breath*, Dekker Ed., New York, 2002.
3. F. Di Francesco, A. Ceccarini, M. G. Trivella, R. Fuoco, "Breath analysis: trends in techniques and clinical applications" *Microchemical Journal* 79 (2005) 405.
4. T.H. Risby, S.F. Solga, "Current status of clinical breath analysis" *Applied Physics B* 85 (2006) 421.
5. M. Phillips, R. N. Cataneo, T. Cheema, J. Greenberg, "Increased breath biomarkers of oxidative stress in diabetes mellitus" *Clinica Chimica Acta* 344 (2004) 189.
6. M. Phillips, J. Greenberg, "Detection of endogenous acetone in normal human breath" *J. Chrom.* 422 (1987) 235.
7. J. Peinado, F.J. López-Soriano, J. M. Argilés, "Gas chromatographic method for estimation of acetone and its metabolites in biological samples" *J. Chrom.* 415 (1987) 372.
8. N. Maksimovich, V. Vorotyntsev, N. Nikitina, O. Kaskevich, P. Karabun, A. Martynenko, "Adsorption semiconductor sensor for diabetic ketoacidosis diagnosis" *Sensors and Actuators B*, 35-36 (1996) 419.
9. O. E. Owen, V. E. Trapp, C. L. Skutches, M. A. Mozzoli, R. D. Hoeldtke, G. Boden, G. A. Reichard "Acetone metabolism during diabetic ketoacidosis" *Diabetes* 31 (1982) 242
10. S. V. Ryabtsev, A. V. Shaposhnick, A. N. Lukin, E. P. Domashevskaya, "Application of semiconductor gas sensors for medical diagnostics" *Sensors and Actuators B* 59 (1999) 26.
11. N. Pinna, G. Neri, M. Antonietti, M. Niederberger, "Nonaqueous synthesis of nanocrystalline semi-conducting metal oxides for gas sensing" *Angew. Chem. Int. Ed.*, 43 (2004) 4345.
12. E. Callone, G. Carturan, M. Ischia, A. Sicurelli, "Nanometric oxides from molecular precursors in the presence of starch: coatings of glass with these oxides in silica sols" *J. Mater. Res.*, 21 (2006) 1726.

CYTOTOXICITY OF SINGLE-WALL NANOTUBES ON CULTURED HUMAN LYMPHOCYTES

O. ZENI, R. BERNINI, M. SARTI, M.R. SCARFÌ
CNR-IREA, 80124 Napoli – Italy

R. PALUMBO
CNR-IBB, 80134 Napoli – Italy

L. ZENI
Seconda Università di Napoli-DII, 81031 Aversa – Italy

Carbon-based nanostructures are becoming increasingly studied essentially for their possible applications in electronics, optics and biology, such as imaging, sensing and drug delivery. Among the new carbon structures, the single-wall carbon nanotubes (SWNTs) have a prominent position due to their unique functions. Nevertheless, although the size of these new nanoparticles, with their high surface area and unusual surface chemistry and reactivity, poses unique problems for biological material, there is a paucity of information on their toxicological properties to assess potential human health risk [1]. Cytotoxicity is an important factor in understanding the mechanisms of action of such materials, moreover it is thought to play an important role in a number of pathological processes, including carcinogenesis and inflammation. At the moment, studies on cytotoxicity of CNTs are scarse and report conflicting results mainly due to the different dispersion methods of CNTs employed [2]. The aim of this study was to investigate the cytotoxicity of SWNTs on human peripheral blood lymphocytes from healthy donors, since they participate in the recognition of foreign material in the blood. Cell viability, cell growth and primary DNA damage were evaluated to gain and assess cytotoxicity information. A statistically significant decrease in cell growth was found in human lymphocytes treated with SWNTs at final concentration of 25 and 50 μg/ml. Such decrease was not associated to loss in cell viability or DNA damage but was demonstrated to be related to a decrease in metabolic activity as assessed by resazurin assay.

Methods

Lymphocytes isolation and culturing

Human peripheral blood lymphocytes (HPBLs) were obtained with informed consent from anonymized buffy coats of healthy donors. They were isolated through lymphoprep density gradient centrifugation, in accordance with the manufacturer's instructions. Mononuclear cells were seeded in RPMI 1640 medium supplemented with 15% heat-inactivated foetal bovine serum and 1% L-glutamine. Phytoemagglutinin (PHA, 1%) was added as mitogen to stimulate T-lymphocytes to enter the cell cycle [3]. HPBLs from 3 healthy donors were employed for all the biological targets examined, except for comet assay where 4 donors were employed. All the reagents employed for the experiments were of analytical grade.

Preparation of nanotubes

SWNTs were obtained in a 90% purity form from HeJi, Inc. (Hong Kong, China). They are water insoluble and were suspended in RPMI medium at a concentration of 500 μg/ml, sterilized by autoclavation and dispersed by 3h treatment in ultrasound bath prior to being administered to the cells.

Cytotoxicity evaluation

Cell viability and cell growth were investigated by means of the trypan blue exclusion method in PHA stimulated lymphocytes. Cells were seeded at a density of $5x10^5$ cells/ml and SWNTs concentrations of 0, 5, 10, 25 and 50 μg/ml were tested in duplicate by collecting cell aliquots after 24, 48 and 72 h from seeding. Moreover, SWNTs were also added 24 h after PHA stimulation. Cells were counted in a Burker hemocytometer, and cell viability was calculated as the ratio of live to dead cells, expressed as percentage.

Metabolic activity was evaluated by applying the resazurin reduction assay [4]. Following 24 and 48 h SWNTs treatments, cells were collected and tested with resazurin 10 μg/ml. The production of resorufin was analyzed with a fluorometer (Perkin-Elmer, LS50B) at an excitation wavelength of 530 nm and an emission wavelength of 590 nm. To quantify resorufin production, a standard curve of fluorescent product was assessed. The results were expressed as μg of resorufin produced/ml/min.

Primary DNA damage was evaluated by the alkaline comet assay on the entire leukocytes population after 6 h treatments with SWNTs at final concentrations of 1, 5 and 10 μg/ml. Positive controls were included in the experiments by treating leukocytes with hydrogen peroxide (HP) for 30 min at 50 μM final concentration. Following the treatments, cells were processed to make up microscope slides according to the protocol developed by *Singh et al.* [5], with minor modifications [6]. Following ethidium bromide staining, for each donor, images of 300 randomly selected nuclei (150 from each of two replicate slides) were analysed by a computerized Image Analysis System fitted with a fluorescence microscope. DNA migration was evaluated by calculating the tail length, the percentage of migrated DNA, and the tail moment.

Results and Discussion

When cell growth was evaluated with trypan blue by counting cells at 24, 48 and 72 h after PHA stimulation, a decrease in cell number in all the SWNTs treated groups with respect to control group was detected. The decrease resulted dose dependent and became statistically significant when treatments of 25 and 50 μg/ml were considered as reported in Figure 1. This behaviour was also detected when SWNTs treatments were performed after 24 h of PHA stimulation, as shown in Figure 2. The latter finding excludes that the observed effect could derive from an interference of CNTs with PHA. These findings were not associated to loss in cell viability that never decreased below 90% at any SWNTs dose tested (data not shown). On the contrary when metabolic activity was investigated, 24 and 48 h treatments resulted in a decrease of resorufin production for all the concentrations tested. In particular, the average decreases ranged from 15 to 27% and from 15 to 40% for 24 and 48 h treatments respectively, resulting statistically significant in all cases. Data are presented in Figure 3, where the amount of resorufin produced is reported as control percentage. Absence of cytotoxicity was also confirmed when primary DNA damage was evaluated on the entire population of human leukocytes. In fact no increase in DNA migration, was found in human leukocytes following 6 h treatment with SWNTs at the doses tested. On the contrary, a statistically significant increase was detected when HP treated samples were compared to control ones ($P<0.001$). Results are shown in Figure 4. Work is in progress to better understand the interaction between SWNTs and human white blood cells by investigating other cellular endpoints.

References

1. B.J. Panessa Warren, J.B Warren, S.S.Wong, J.A. Misewich, *J Phys: Condens. Matter,* **18**, 2185-2201 (2006).
2. S.K. Smart, A.I. Cassady, G.Q. Lu, D.J. Martin, *Carbon* **44**, 1034-1047 (2006).
3. Boyum, Scand. J. Clin. Lab. Invest. **97**, 31-50 (1968).
4. J. O'Brien, I. Wilson, T. Orton, F. Pognan, *Eur. J. Biochem.* **267**, 5421-5426 (2000).
5. N.P. Singh, M.T. McCoy, R. Tice, E.L. Schneider, *Exp. Cell. Res.* **175**, 184-187 (1988).
6. A. Sannino, M.L. Calabrese, G. d'Ambrosio, R. Massa, G. Petraglia, P. Mita, M. Sarti, M.R. Scarfì, *IEEE Transaction on Plasma Science* **34,** 1441-1448 (2006).

Figure 1. Cell growth data for PHA stimulated lymphocytes treated with SWNTs. Data are presented as mean ± SD obtained from 3 healthy donors. Significant difference for treated groups *vs* control group; *P<0.05; *** P<0.001; two way ANOVA for repeated measures.

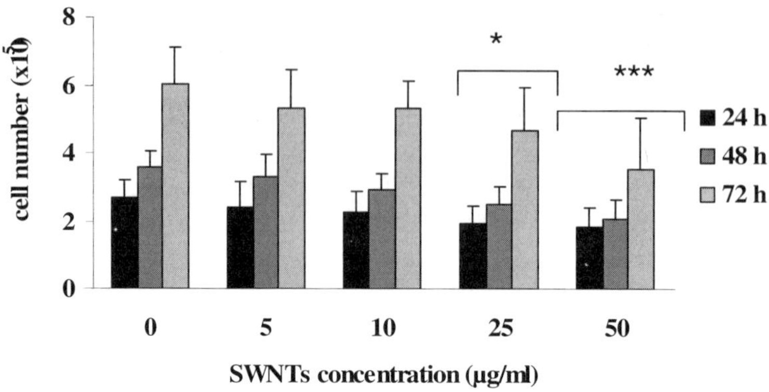

Figure 2. Cell growth data for PHA stimulated lymphocytes treated with SWNTs after 24 h of PHA stimulation. Data are presented as mean ± SD obtained from 3 healthy donors. Significant difference for treated groups *vs* control group; * P<0.05; ** P<0.01; two way ANOVA for repeated measures.

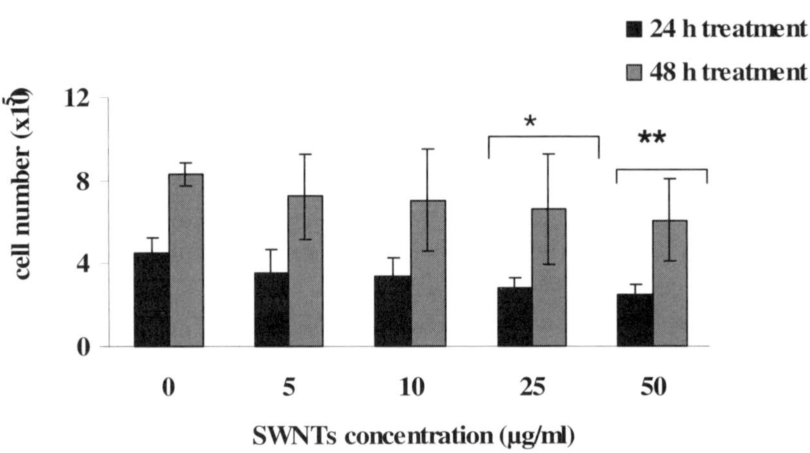

Figure 3. Concentration-response curve for metabolic activity of HPBL measured after 24 and 48 h SWNTs treatments. Data are presented as control percentage of resorufin production. Each data point represents the mean ± SD obtained from 3 donors. * P<0.05; ** P<0.01; *** P<0.001; one way ANOVA for repeated measures.

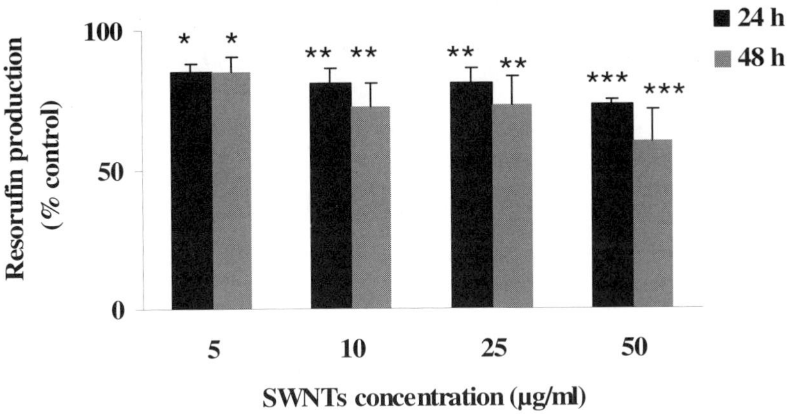

Figure 4. Percentage of migrated DNA (tail DNA %) in human leukocytes treated for 6 h with SWNTs. Data related to positive control (HP, 50 µM for 30 min) are also reported. Each data point represents the mean ± SD of 4 donors examined. *Significant (P<0.001) difference from negative control (0): two tailed paired Student *t* test.

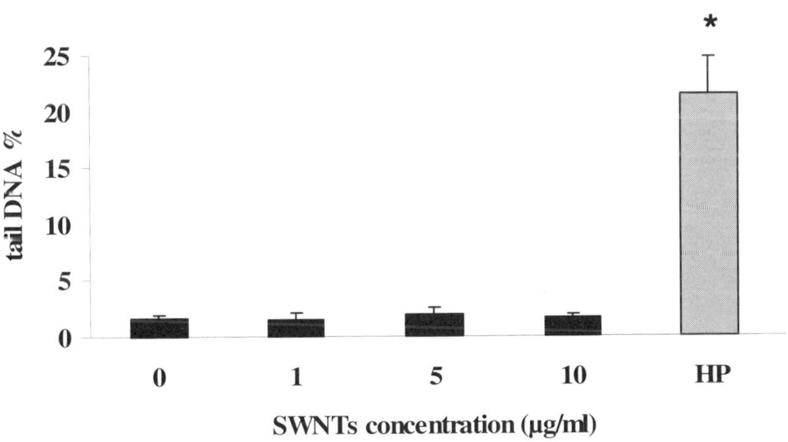

NANOMATERIALS TOXICITY: AN IN-VITRO INVESTIGATION

GABRIELLA RAMETTA*, VERA LA FERRARA AND GIROLAMO DI FRANCIA

ENEA Research center, Via Vecchio Macello, 80055 Portici (NA)

In this work we present a study about behavior in physiological solution of two kind of nanomaterials, as carbon black and porous silicon. Dissolution of these nanoparticles have been controlled resulting in CB aggregation and PS solubility. These results are useful to evaluated the effects of nanoscale materials into biological system for understanding the human health implications.

1. Introduction

The use of nanotechnology in consumer and industrial application will likely have a profound impact on a number of products from a variety of industrial sector. Potential health risk will dependent on the magnitude and nature of exposures to nanostructured particles, and on the release, dispersion, transformation and control of materials in the workplace. Several recent studies that evaluated the effects of nanoscale materials on biological system have provided very useful data for understanding the human health implications from exposure to these materials and have helped to identify priorities for future research. The increased reactivity of nanoscale materials that arises as a consequence of their larger surface area has created considerable interest in development of a better understanding of the effect of nanoscale materials on biological system. Nanoscale materials have been evaluated both in vitro and in vivo system to explore effect from dermal and inhalation exposure. Disaggregation, deagglomeration, and dissolution in biological fluids are important factors potentially contributing to a complete understanding of nanoparticle fate. Particle dissolution studies can be performed through the evaluation of particulate properties such as particle size or number, concentration. For toxicological interpretation, it may be relevant to try to simulate one compartment within the body or in the environment. The

---

* Corresponding author: tel. +390817723280 fax. +390817723344
email: gabriella.rametta@portici.enea.it

fluid used that may or may not contain additives intended to simulate a biological fluid, thus enabling measurements under well-controlled or ideal conditions.

In this work we present a study about behavior in physiological solution of nanostructured carbon black (CB) and porous silicon (PS). Dissolution of these nanoparticles have been controlled resulting in CB aggregation and PS solubility.

2. Experimental

We have used a commercial sodium chloride 0.9% solution to simulate a biological fluid. As nanoparticles we have used commercial carbon black and porous silicon, electrochemically fabricated, starting from n-type silicon substrate, <100> oriented, 1Ωcm resistivity, in HF:IPA=70:30 solution using a 40 mA/cm^2 current density for 10 minutes. CB and PS , before its dispersion in physiological solution, have been characterized by nitrogen sorption experiment by means of a surface area and pore analyzer.

Nanostructured materials have been then put in solution and placed in a rotating plate at room temperature to make an homogeneous solution. An aliquot has been periodically taken to control eventual changes in solution, due to dissolution phenomena. Solution has been filtered through a 0.2 μm syringe filter and introduced in a quartz cuvette to detect particles sizes by a Dynamic Light Scattering (DLS) technique, pH measurements and optical absorbance data by means of an optical fiber spectrometer.

3. Results and Discussion

3.1 *Carbon Black dissolution*

Carbon Black, with a surface area of 1500 m^2/g evaluated by nitrogen sorption experiment, has been dissolved in biological solution. By DLS, we have found that carbon black solution presents 100 nm average diameter particles and few particles in the range of 10 nm (fig.1). These data show the CB tends to aggregation and agglomeration in biological fluid.

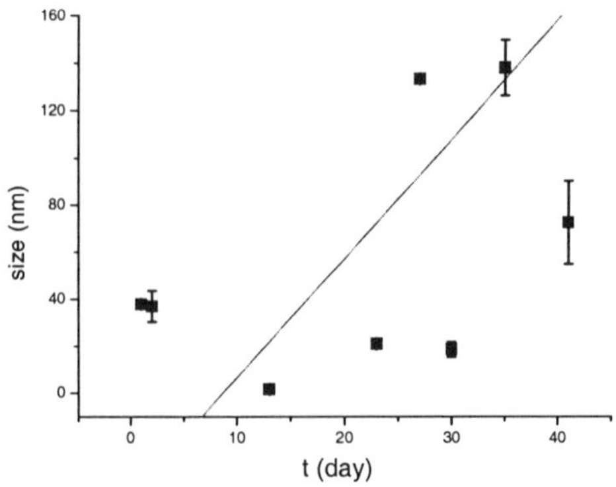

Fig. 1 The graphic shows the particle size measurement with DLS technique for the CB solution with the corresponding error bar.

Periodically we have measured pH of CB solution aliquot. For pH measurements we have found that starting CB solution shows a pH near 7, while aged solution has a pH around 5 (Fig. 2).

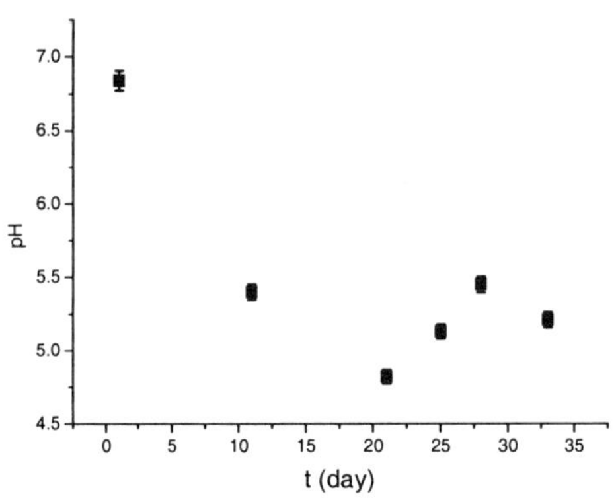

Fig. 2 CB solution pH measurements with the corresponding error bar.

To perform absorbance data we have used an optical fiber spectrometer observing a slightly change of absorbance signal during the observation time (Fig. 3).

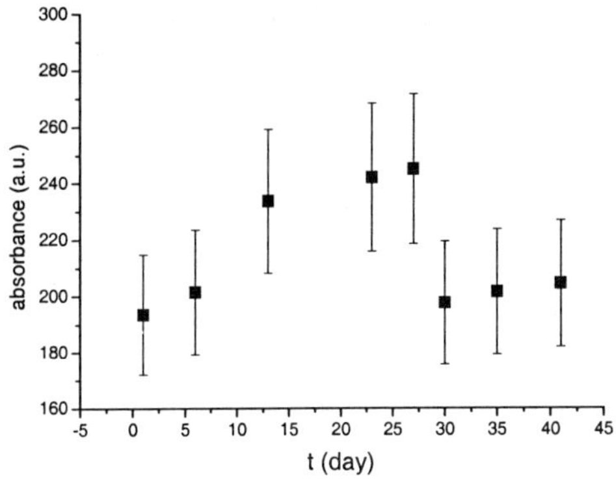

Fig. 3 The graphic show the optical absorbance data for CB solution, with the corresponding error bar.

All of these analysis show an aggregation during the experimental time of carbon black nanoparticles in solution.

3.2 *Porous Silicon dissolution*

Before PS dispersion in solution, nitrogen sorption experiment has been carried out to know nanoparticle shapes and dimensions (fig. 4) showing a surface area of 450 m^2/g and a pore diameter around 50 Å.

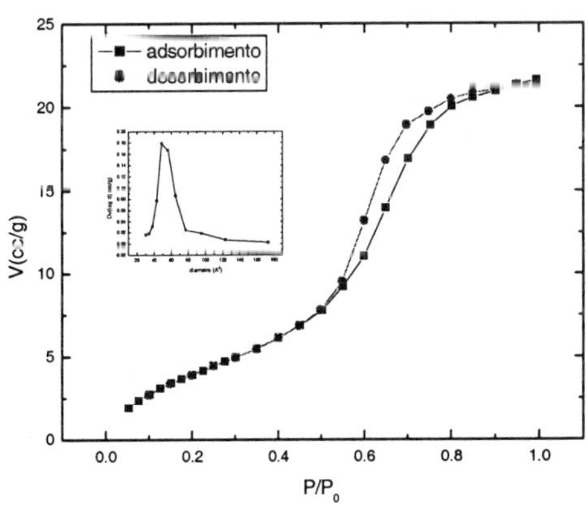

Fig. 4 isotherm and Pore distribution (in inset) of porous silicon

When we put PS in solution we have evaluated its solubility and consequently its decrease in nanoparticles size by DLS (Fig. 5)

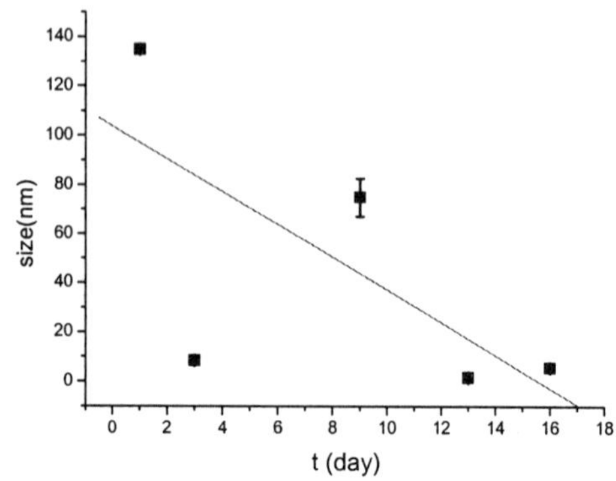

Fig. 5 The graphic shows the particle size measurement with DLS technique for the PS solution with the corresponding error bar.

Periodically, we have measured pH of the sample aliquot (Fig. 6). The pH value slightly decreases during the dissolution from its starting value of 4.8 to a value of 3.7 after 18 days of dissolution.

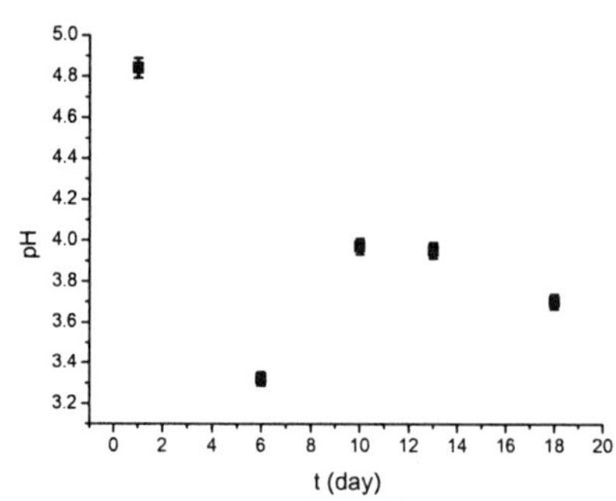

Fig. 6 Porous silicon solution pH measurement with corresponding error bar.

To perform optical absorbance data we have used an optical fiber spectrometer observing an increasing of absorbance signal during the observation time (Fig.7) that meaning concentration increase in solution.

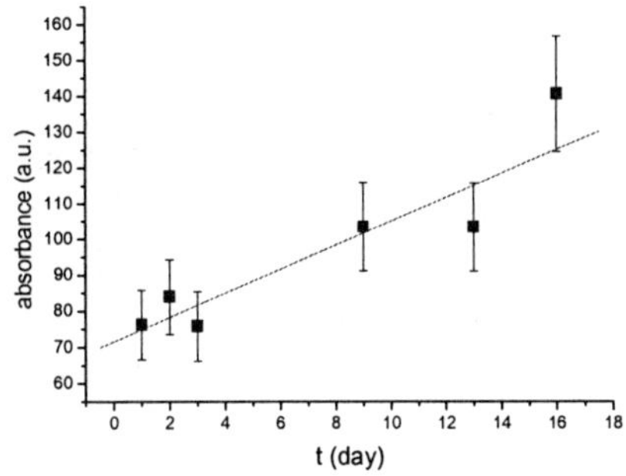

Fig. 7 The graphic shows absorbance data for PS nanostructured in solution and the corresponding error bar. Optical absorbance increases during time.

Another experimental confirmation of PS solubility has been given recording peculiar property of porous silicon: its photoluminescence (PL). HeCd laser with 441 nm line, a monochromator blazed at 500 nm and a CCD detector have been used to collect the spectra. In Fig. 8 PL before and after PS immersion in solution are reported. Comparing results it is possible to note a PL quenching after immersion in solution confirming its high solubility in biological environment.

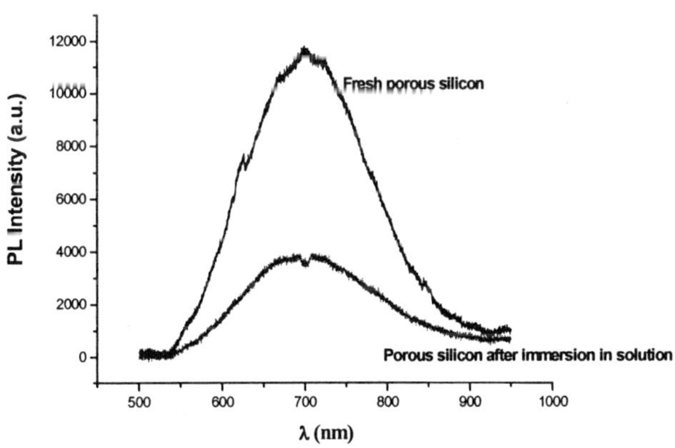

Fig. 8 Porous Silicon photoluminescence. The black and red curves show PL before and after immersion of the sample in solution.

4. Conclusions

A preliminary study about dissolution of two kind of nanostructured particles as carbon black and porous silicon, is reported. These result confirm that the carbon black particles result in aggregation during the experimental time while nanosized porous silicon particles show a decreasing of the size and an increase of the solubility. Work is in progress to better understand CB and PS in physiological solution and to correlate the behavior with toxicological properties of nanoscale materials.

References

1. K. Thomas and P. Sayre *Toxological Sciences* **87(2)**, 316-321 (2005)
2. M.P. Holsapple, W.H. Farland, T.D .Landry, N.A. Monteiro-Riviere, J.M. Carter, N.J. Walker, and K. Thomas *Toxological Sciences* **88 (1)**, 12-17 (2005)
3. P. Borm, F.C. Klaessig, T.D. Landry, B. Moudgil, J. Pauluhn, K. Thomas, R. Trottier, and Stewart Wood. *Toxological Sciences* **90(1)**, 23-32 (2006)
4. K. Thomas, P. Aguar, H. Kawasaki, J. Morris, J. Nakanishi, and N. Savage. *Toxological Sciences* **92(1)**, 23-32 (2006)

ANALYSIS OF VOLATILES IN THE HEADSPACE OF BREAST USING A QMB BASED GAS SENSOR ARRAY FOR BREAST CANCER STUDY: FIRST EVIDENCES

A. D'AMICO[1,2], C. DI NATALE[1], M. SANTONICO[1], G. PENNAZZA[1], G. MANTINI[1], M. BERNABEI[1], E.MARTINELLI[1], R. PAOLESSE[3], S. CABASSI[4], A.G. ARONICA[4], A. CALUGI[4]

1. Department of Electronic Engineering, University of Rome 'Tor Vergata',

Via del Politecnico 1, 00133 Roma; Italy

2. CNR-IDAC, Via del Fosso del Cavaliere 100, 00133 Roma; Italy.

3. Department of Chemical Science and Technology, University of Rome

'Tor Vergata', Via della Ricerca Scientifica, 00133 Roma; Italy

4. S. Eugenio Hospital; 00151 Roma, Italy

The purpose of this study was to test the ability of a QMB based gas sensor array in the analysis of the volatiles present in the headspace surrounding the breast. The goal of the experiment was to use this instrument as support of other diagnostic tools, for an early and non-invasive diagnosis of breast cancer.

The experiment involved a population of 30 women. 14 of them were affected by various forms of breast cancer, 13 were apparently healthy individuals considered as reference. The last three patients were 3 cases whose breast had been replaced by a prosthesis. The measurements were performed by the electronic nose of the University of Rome 'Tor Vergata'. The results obtained by a multivariate analysis of the data were very encouraging. The electronic nose showed 86% sensitivity, 93% specificity, 14% false negative, 7% false positive. These evidences suggest that the headspace surrounding the breast contains important information about breast healthstate, and the electronic nose is an efficient instrument to reveal and interpret these information.

Introduction

There are several breast cancer diagnostic tests available today. However an early diagnosis is the best strategy to defeat the disease. A preferable way to achieve an early diagnosis is a mass screening of the women population over 40 years.

The standard procedure to perform this screening is composed of a triple testing: clinical examination tests, imaging (traditionally x-ray mammography, ultrasounds, Magnetic Resonance Imaging (MRI), Positron Emission Tomography (PET)) and needle biopsy (fine needle or core biopsy). Each of these techniques presents some drawbacks in term of invasivity and effectiveness, and the false negative percentage of each single test is in the range from 10% to 18%. Moreover it is worth to remark that the combination of the different imaging techniques increase the sensitivity respect to cancer up to 91%-98%[1].

X-ray mammography is of course the most commonly used technology, but MRI is preferable for screening of high-risk cases because of its higher sensitivity, even if it is more expensive. Breast MRI is currently under investigation, but its application is of course very effective in some selected cases: highly suspicious findings, preoperative staging, women with lymphnode metastases and unknown primary cancer, monitoring of anti-cancer therapies [2].

Misclassified cancers are mainly very aggressive tumors that develops in one year period in women less than fifty years old, this supports the importance of a frequent monitoring strategy, which is achievable by mean of non-invasive and cheap diagnostic thecniques. In summary, none of the reported tests has the necessary features and accuracy to be the exclusive diagnostic method to perform an early diagnosis of breast cancer, except for their combination [3].

The effectiveness of 'imaging' tests suggests the use of a challenging technology which is non-invasive in principle: to perform an 'olfactive image' of the breast healthstate. VOCs emitted by human body are a powerful source of information on individual healthstate, and of course their examination does not need an invasive sampling procedure.

Recent studies by Phillips et al.[4,5] proposed a new non-invasive technique to predict breast cancer using volatile biomarkers in the breath. The breath test developed was similar to that used by Phillips in the research on lung cancer[6]. A fuzzy logic model was used to predict breast cancer, employing five different volatile organic compounds. Phillips achieved a sensitivity of 93.8% in the prediction set.

The authors of the present paper, have applied the electronic nose technology to the analysis of the headspace surrounding women breast. This technique allows to perform a measurement in situ instead of an indirect measure such as the case of breath analysis. The efficiency of this methodology is supported by other publications of the same group in the analysis of skin, sweat and breath [7,8,9], and also by promising researches about melanoma, bladder cancer and breast cancer performed by well-trained dogs [10,11,12].

Methods

A total number of 30 volunteers were recruited at the S.Eugenio Hospital in Rome. Fourteen of them were affected by various forms of breast cancer and were hospitalized waiting for a surgical treatment. Three volunteers were women whose breast had been replaced by a prosthesis. Thirteen volunteers have been recruited among the nurse and medical staff of the Hospital as references. These controls were not affected by any apparent disease, and were not taking any drugs. The cancer patients did not show of any different pathology. Each person participating to the experiment was informed about the nature of the measurement to obtain an informed consent.

The electronic nose developed at the University of Rome "Tor Vergata" was used in this experiment. It is based on an array made up of eight quartz microbalance (QMB) sensors, coated with different kinds of metalloporphyrins. The principle and application of QMB is based on the variations of the fundamental oscillating frequency of a thin quartz crystal, as a consequence of the adsorption of molecules from the gas phase.

The purpose of this experiment was the analysis of the Volatile Organic Compounds (VOCs) present in the headspace surrounding the breast, in order to obtain information about the breast healthstate. To perform such an analysis a sampling protocol was developed to convey the VOCs to the measure chamber of the electronic nose. Each subject was required to discontinue the use of deodorants or perfumes on the day of the measurement, not to influence the volatile compounds measured. No other attempts were made to control the VOCs emission.

The sampling protocol required each subject to maintain a sampler over the breast during the time of the measurement. Measures from both the breasts were taken into consideration in this analysis. An additional measure relating to skin thorax was performed in order to obtain a reference measure regarding individual odor.

The sampler was designed to match the anatomy of the breast and it was provided with two openings: the first used to convey reference air into the sampler and the second to collect breast headspace air, which was then conveyed to the electronic nose chamber. An online measurement was thus performed. The headspace air was flown at the constant speed of 0.5 ml/s into the sensor chamber. A stabilization of the sensor response occurs when the thermodynamic equilibrium between adsorbed and desorbed molecules is achieved.

The sensor responses, utilized for the data analysis, were chosen as the difference between the frequency variation measured during the cleaning and the measure phases. The analysis of the data was performed in the MATLAB environment, by means of the PLS Toolbox. A partial least square discriminant analysis (PLS-DA) was performed, considering two classes (breast cancer disease and reference group). In order to evaluate the identification performances of the method, a leave-one-out validation criterion has been adopted.

Results

As the aim of this study was to test the capability of electronic nose to correctly classify groups of subjects (in this case cancer patients and controls), a supervised technique has been used. The simplest possible choice was a discriminant analysis, where a linear model between sensor data and classes is built.

As in any supervised classification techniques, the classes has to be chosen a-priori. The natural choice for the samples in this experiment was to choose two classes including the patients with breast cancer and the control group. The prediction error was minimized using the leave-one-out cross validation method. The best fitting method included two latent variables. Data have been scaled by mean center (zero mean). Figure1 shows the score plot of the two latent variables. In the plot about 99% of the total variance of the data is represented. It can be observed a clear separation between the data related to the cancer patients and the control group.

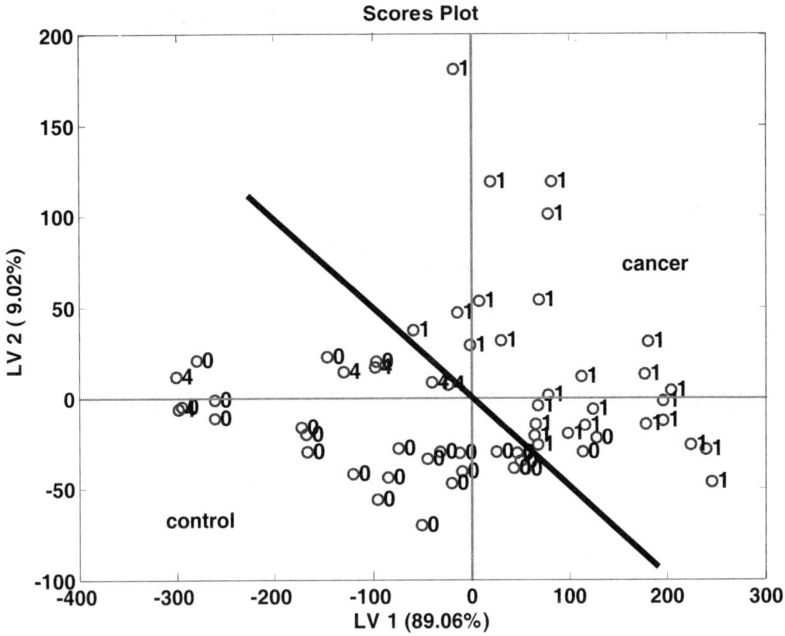

Figure1: scoreplot of the first two Latent Variables of the PLS-DA model built on e-nose data. (Labels: 0.controls; 1.breast cancer; 4.breast prosthesis)

Table 1 shows the confusion matrix of the PLS-DA method. The percentage of correct classification is about 93% for both the groups. The percentage of missed tumor identification is about 7%, of course less than 10-20% reported above for imaging diagnostic[1].

True/estimated	Control sample	Cancer disease
Control sample	30	2
Cancer disease	2	26

Table 1: Confusion matrix of the PLS-DA model built on e-nose data

It can also be observed from Fig.1 that the measures related to women with prosthesis (labelled as 4), even though belonging to the control group, as they do not manifest any disease, made up a distinctive cluster.

Summing up, the results obtained by a multivariate analysis of the data are very encouraging. The electronic nose showed 86% sensitivity, 93%

specificity, 14% false negative, 7% false positive, in the discrimination between the cancer patients and the control group.

Conclusions

First evidences shows that electronic nose could be a suitable instrument to be used as an adjunct device for the diagnosis of breast cancer. The electronic nose will never replace other consolidated diagnostic techniques, but it can help to reach an early diagnosis of cancer. The non invasivity of the measurement and the low cost of the instrument are other advantages of this technique.

Several authors reported studies on the composition of volatile compounds emitted by the body for other kinds of cancer, and in particular for lung cancer, for which breath composition analysis is the simplest and more obvious way to obtain useful information about helthstate. The composition of the VOCs expected to be found in the breath is a problem quite well studied, even at physiological level.

Instead, in the case of breast cancer, it is more difficult to have a-priori information about what to expect in a breast skin odour analysis. For this reason is also very difficult to give an analytical explanation for the good results obtained with the electronic nose in this preliminary study on breast cancer.

At this moment, the investigation here reported does not allow to infer any explanation for the discrimination obtained with the e-nose. A deeper analytical investigation with GC-MS equipment will be necessary to explain these results.

However the results here illustrated confirm that the headspace air surrounding the breast of women with breast cancer is different from that of healthy people.

References

1. Ciatto S, Houssami N, (2007) Breast imaging and needle biopsy in women with clinically evident breast cancer: does combined imaging change overall diagnostic sensitivity? The Breast (in press) DOI 10.1016/j.breast.2007.01.007
2. ehrens S, Laue H, Althaus M, Boehler T, Kuemmerlen B, Hahn HK, Peitgen HO, (2007) Computer assistance for MR based diagnosis of breast cancer: Present and future challenges. Computerized Medical Imaging and Graphics (in press) DOI10.1016/j.compmedimag.2007. 02.007
3. Ciatto S, Rosselli Del Turco M, Burke P, Visioli C, Paci E, and ZappaM. (2003) Comparison of standard and double reading and computer-aided detection (cad) of interval cancers at prior negative screening mammograms: blind review. British Journal of Cancer 89, 9: 1645-1649

4. Phillips M, Cataneo RN, Ditkoff BA, Fisher P, Greenberg J, Gunawardena R, Kwon CS, Rahbari-Oskoui F, Wong C, (2003) Volatile Markers of Breast Cancer in the Breath. The Breast Journal 9, 3: 184-191.
5. Phillips M, Cataneo RN, Ditkoff BA, Fisher P, Greenberg J, Gunawardena R, Kwon CS, Tietje O, Wong C, (2006) Prediction of breast cancer using volatile biomarkers in the breath. Breast Cancer Research and Treatment 99,1: 19-21.
6. Phillips M, Cataneo R, Cummin A, Gagliardi A, Gleeson K, Greenberg J, Maxfield R, (2003) Detection of lung cancer with volatile markers in the breath. Chest 123: 2115-2123.
7. Di Natale C, Macagnano A, Martinelli E, Paolesse R, D'Arcangelo G, Roscioni C, Finazzi-Agrò A, D'Amico A, Lung Cancer identification by the analysis of breath by means of an array of non-selective gas sensors, Biosensors and Bioelectronics, 18 (2003) 1209-1218.
8. Di Natale C, Paolesse R, D'Arcangelo G, Comandino P, Pennazza G, Martinelli E, Rullo S, Roscioni MC, Roscioni C, Finazzi-Agrò A, D'Amico, Identification of schizophrenic patients by examination of body odor using gas chromatography-mass spectrometry and cross selective gas sensor array. Medical Science Monitor 11 (2005) 366-375.
9. Di Natale C, Macagnano A, Paolesse R, Tarizzo E, Mantini A, D'Amico A, Human skin odor analysis by means of an electronic nose, Sensors and Actuators B 65 (2000) 216-219.
10. Nickel D, Manucy G, Walker D, Hall S, Walker J, Evidence for canine olfactory detection of melanoma, Applied Animal Behaviour Science, 89 (2004) 107-116.
11. Willis C, Church S, Guest C, Cook A, McCarthy N, Bransbury A, Church M, Church J, Olfactory detection of human bladder cancer dogs: proof of principle study, British Medical Journal, 329 (2004) 712-718, 2004.
12. McCulloch M, Jezierski T, Broffman M, Hubbard A, Turner K, Janecki T, Diagnostic accuracy of canine scent detection in early- and late-stage lung and breast cancer, Integrative Cancer Therapies 5 (2006) 1-10.

GAS SENSORS

Pushing the limit of the silicon technology by using porous silicon: a CMOS gas sensing chip

G. Barillaro, P. Bruschi, F. Pieri, L. M. Strambini
Dipartimento di Ingegneria dell'Informazione: Elettronica, Informatica, Telecomunicazioni, Università di Pisa, via G. Caruso 16, 56126 Pisa-Italy,
g.barillaro@iet.unipi.it

Summary

Air monitoring is a challenging objective both for city safeguard and human health. However, in order to get information on the real population exposure with the necessary spatial and temporal resolution a high density monitoring network would be needed. To this aim, small, reliable, low cost, integrated silicon-based sensors would be required. Among the different materials proposed in literature for gas sensor integration, porous silicon (PS) is one of the most promising because of its intrinsic compatibility with the silicon technology. In this work, the design and fabrication of a gas sensing chip, containing an array of PS gas sensors integrated along with several electronic basic blocks, by using an industrial CMOS process is reported, along with its electrical characterization. Finally, to demonstrate the functionality of the sensing chip, a simple readout electronic interface connected to one of the sensors is used to implement an integrated current-voltage converter.

1 Introduction

Nowadays, air pollution, mainly associated with NO_2 and particulate fine matter PM_{10} and $PM_{2.5}$, is a serious problem in many highly populated and industrialized areas. The monitoring of air quality is therefore crucial, especially in areas where pollution sources and the human population are dense. To prevent or minimize damages caused by atmospheric pollution to human health and to the environment, real-time *in situ* monitoring systems, able to rapidly and reliably detect and quantify pollution, are needed.

Unfortunately, today's standard monitoring systems are extremely expensive, time-consuming, and can seldom be used in real-time mode. For these reasons, at present only a limited number (in the order of tens) of fixed monitors are used in European cities, with the intent of reflecting the urban areas average exposure. Therefore, even if valuable for minimizing environment damages, their measurements may not reflect the actual exposure of the population. Hundreds or even thousands of fixed-site monitors, depending on the monitored area, would be required for health safety.

To achieve this goal small, reliable, low cost sensors able to detect the main air pollutants are needed. To this end, the development of miniaturized silicon-based sensors, compatible with well-established microelectronic technologies for integrated circuits (ICs) fabrication, could be one of the feasible solutions. The integration of such sensors along with the necessary readout electronic circuits offers an outstanding alternative for environmental monitoring, due to the possibility of integrated monitoring chip fabrication having light weight, small size and low cost.

In last years, a number of nanostructured materials, such as metal-oxides [1], porous silicon (PS) [2], carbon nanotubes [3] and others, have been considered for gas sensing application. Among the different materials proposed in literature, porous silicon is one of the most promising, specially because of its intrinsic compatibility with silicon technologies. In fact, the integration of PS-based sensors with the silicon technology, i.e. CMOS or CMOS-based processes used for today's ICs fabrication, could push the functionality of silicon chips toward novel applications and markets. However, nowadays, the compatibility of PS-based sensors with standard processes for ICs fabrication has been often overlooked.

Actually, in order to ensure the compatibility between PS-based devices and an IC industrial process some basic issues need to be satisfied: i) changes in a well established process flow are not easily accepted by microelectronic factories, so that the PS integration should leave the standard process flow unmodified; ii) in order to avoid PS damaging by standard technological steps (i.e. high energy implants, high temperature annealings, etc.) and contamination from the PS to the processing equipment, PS formation should not be an intermediate step of a standard process. Both the requirements can be satisfied if the PS formation is the last step of the industrial process.

To this aim, differently from PS sensors reported in literature, a new approach for the integration of PS-based gas sensors has been recently proposed, in which a PS layer, integrated along with a standard solid-state device (i.e. a diode, a FET, etc.), is exploited to modifie the electrical properties of the device itself upon gas adsorption in PS [4-6]. This approach allows the fabrication of PS layers after the integration of solid state devices, as the last step of the process, ensuring thus a full compatibility with industrial processes.

In this work, according to the aforementioned approach, a gas sensing chip, composed of an array of PS gas sensors integrated along with several electronic basic blocks, was designed and fabricated by using a standard CMOS process. Both the electronics and the sensors have been electrically characterized in dry air and/or in presence of NO_2 at hundreds ppb. Finally, in order to demonstrate the simultaneous functionality of electronics and sensors, a preliminary characterization of the chip in presence of hundreds ppb of NO_2 was performed by connecting the integrated devices to form a simple readout electronic interface, for instance a current-voltage converter, connected to one of the sensors.

2 Sensing chip fabrication

The fabrication sequence encompasses the PS formation as a post-processing option of a standard CMOS process. This allowed the fabrication of PS-based sensors and standard electronics on the same chip, with no modification of the process flow or equipment contamination, and with the addition of a single lithographic step to be applied at the end of the standard flow.

The fabricated chip (Fig. 1), with overall dimension 4 mm × 4 mm, contains: the electronics (upper part of the die), the PS sensors (centre of the die), and other structures not involved in this work.

The electronic area includes three operational amplifiers, an instrumentation amplifier, a reference voltage/temperature sensor circuit based on a band-gap architecture, and a ring of DMOS power transistors, arranged around the sensor area, to be used to set the chip temperature and perform purge operations of sensors.

Fig. 1 Optical photograph of the sensing chip: the upper part contains the electronics, while the central area contains an array 2x4 of PS sensors.

Here we will briefly describe only the operational amplifier and the band-gap circuit, since these cells have been used to build the simple interface described in the next section. For the op-amp cell, a compact topology offering low voltage – low power operation has been chosen [7]. The op-amp has been designed to yield a 130 dB static voltage gain, a 3 MHz gain-bandwidth product, a maximum input offset voltage of 3 mV, and an input voltage noise of 20 μV peak-to-peak in the frequency band 0.01–10 Hz. The class AB output stage delivers a 8 mA maximum current with only 2.15 mW quiescent power consumption. The band-gap circuit produces the reference voltage V_{REF} and a voltage V_{TEMP} proportional to the absolute temperature. The operating principle is that of "ΔV_{BE}" circuits

[8].The nominal reference voltage is 1.157 V with a total variation of 1 mV over the temperature interval –20 ÷ 100 °C. The area occupancy of the op-amp and of the band-gap cell is 360 × 230 μm^2 and 100 × 110 μm^2, respectively.

The sensing part of the chip is constituted of a 2 × 4 array of APSFET (Adsorption PS FET) sensors [4]. The APSFET is a FET-like device with a PS layer integrated between the drain and source terminals acting as a floating sensing gate. A sketch of the APSFET structure is given in Fig. 2 (left). Adsorption of molecules into the PS modifies the charge in the PS layer and/or at the PS/crystalline silicon interface, with the consequence of a modulation of the free charge carriers in the conduction channel between drain and source and, in turn, of the device current. The current variation both depends on the type and concentration of the sensed species.

Apart from the PS formation, the APSFETs were fabricated simultaneously with the other electronic components, by exploiting standard process steps. For instance: 1) the *p*-type substrate material was used for the active area, where the PS layer will be formed as the last step of the process; 2) the n^+ implant used for drain and source terminals fabrication of NMOS transistors was also exploited for drain and source contacts formation (interdigitated, in order to maximize the sensing area) of APSFETs.

The chip fabrication sequence can be summarized into three main steps: 1) design, of the electronic blocks and PS sensor structure with a standard IC design environment (CADENCE^{TM}); 2) fabrication, performed by using the 0.35 μm BCD6 (Bipolar + CMOS + DMOS) process of STMicroelectronics; 3) post-processing, consisting in a selective electrochemical silicon etching to produce the PS sensing layer in specific areas.
After the electronics fabrication has been completed, that is at the end of the BCD6 process flow, the die is covered by several passivation layers, in order to

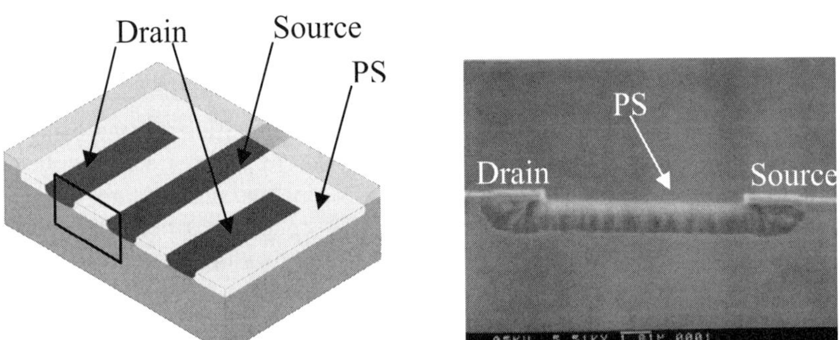

Fig. 2 Schematic view of an APSFET in BCD6 (left) and its SEM cross-section showing the PS layer between drain and source contacts (right).

protect the electronics itself. A few post-processing steps are then required to selectively remove the passivation and uncover the silicon surface in correspondence of the active area of sensors, for the next PS formation. A wet etching using a BHF solution was used for this purpose, after the definition of a 10 μm thick AZ4562 photoresist layer by means of a standard photolithographic step. The electrochemical etching for PS formation in the active areas of sensors was then selectively carried out throughout the photoresist-free spaces using an Ethanol:HF(48%):H_2O (2:1:1 vol.) solution. During this step, the electronics was protected by the thick photoresist layer. As this phase takes place in the dark, only the exposed p-type material is converted into PS, while the n-type drain and source contacts are un-etched.

Typical anodisation times and current densities (few seconds at about 20 mA/cm^2) give rise to PS layers a few hundreds nanometre thick. The samples are finally rinsed in acetone, to remove the photoresist layer, and in ethanol and pentane, to reduce cracking during the next PS drying step, performed in a N_2 atmosphere.

A SEM cross-section of an APSFET with the PS layer between drain and source is shown in Fig. 2 (right).

3 Sensing chip testing

In order to verify the correct operation of the electronics after the post-processing steps, an electrical comparative test was performed on critical electronic blocks before and after the PS formation. All tests showed that the performances of the electronics are not altered by the electrochemical etching. As an example, Fig. 3 (left) shows a comparison between the performances of an op-amp mounted in a buffer configuration, before and after the PS production. The output voltages V_{out_a} and V_{out_b}, before and after the PS formation, respectively, are actually a close replica of the triangular input voltage V_{in}.

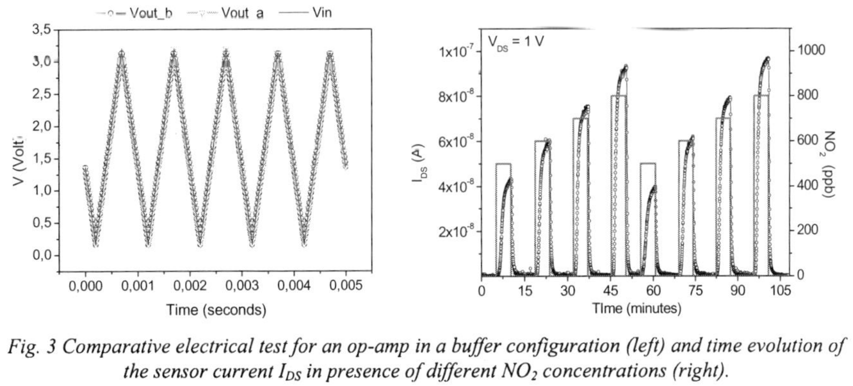

Fig. 3 Comparative electrical test for an op-amp in a buffer configuration (left) and time evolution of the sensor current I_{DS} in presence of different NO_2 concentrations (right).

Afterwards, the APSFET performances, in terms of sensitivity and response-time, were evaluated by means of an electrical test in presence of NO_2 at hundreds ppb.

The sensors were tested in a temperature-stabilized chamber at room temperature using NO_2 concentrations between 500 and 1000 ppb, and synthetic dry air as a carrier gas. The typical time evolution of the I_{DS} current of an APSFET, biased with $V_{DS} = 1$ V, for different NO_2 concentrations is shown in Fig. 3 (right). The response time, which is limited in our experimental set-up by the testing chamber volume, can be estimated in few minutes. The presence of NO_2 significantly increases the APSFET current: 1000 ppb of NO_2 produce variations higher than one order of magnitude, with respect to the baseline. Once the NO_2 is removed, the I_{DS} is restored to its initial value in dry air, and no significant drift is observed.

Finally, to test the simultaneous operation of electronics and sensors, so demonstrating the potentiality of the sensing chip, a simple interface was implemented (Fig. 4 (left)). One of the APSFETs was connected to an operational amplifier mounted in a current–voltage converter configuration and biased with a 3.3 V supply. The voltage reference V_{REF}, about 1.1 V, used to provide the APSFET drain voltage through the virtual ground, is generated with the above-mentioned band-gap circuit. Components included into the grey box of Fig. 4 (left) are all integrated on the chip, and the only external component is the feedback resistor R_F.

The time evolution of the converter output voltage V_{OUT} is shown in Fig. 4 (right) for different NO_2 concentrations. The output voltage closely follows the transient response of the sensor. The V_{OUT} signal increases as NO_2 is injected into the test chamber, as a consequence of the APSFET current increase. As soon as the NO_2 is removed, the output voltage is restored to its initial value, once again according to the APSFET behaviour.

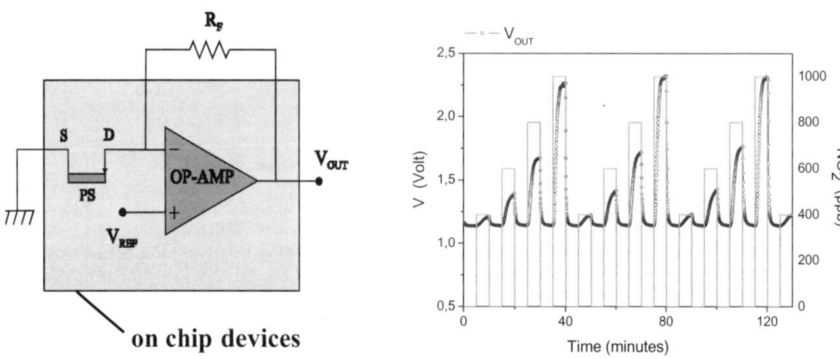

Fig. 4 On chip driving/readout electronic interface connected to an APSFET (left) and output voltage signal V_{OUT} in presence of different NO_2 concentrations (right).

4 Conclusions

In this paper, the integration of a sensing chip by using an industrial BCD6 technology containing PS-based gas sensors along with standard electronics was reported. The fabrication sequence, which includes the PS integration as a post-processing option of a standard microelectronic process, allows the fabrication of PS gas sensors along with solid-state devices. The comparative characterization of the electronics, before and after the PS formation, showed that its behaviour was not significantly affected by the anodisation process. Finally, after the sensors characterization in NO_2 at hundreds ppb was performed, a simple current-voltage converter by using an on-chip buffer connected to one of sensors was implemented as a demonstration of the chip functionality.

It is worth mentioning that the proposed method can be, in principle, extended to develop smart sensor chips with advanced on-board signal processing capabilities.

Acknowledgments

The authors wish to thank STMicroelectronics (Italy) for fabricating the chips in the framework of the University of Pisa–STMicroelectronics Joint R&D Center.

This work was partially supported by the Fondazione Cassa di Risparmio di Pisa.

References

[1] T. K. H. Starke, G. S. V. Coles, H. Ferkel, Sens. Actuators B, Chem., 85, (2002), 239.

[2] L. Valentini, I. Armentano, J. M. Kenny, C. Catalini, L. Lozzi, S. Santucci, Appl. Phys. Lett., 82, 6, (2003), 961.

[3] L. Boarino, C. Baratto, F. Geobaldo, G. Amato, E. Comini, A. M. Rossi, G. Faglia, G. Lérondel and G. Sberveglieri, Mater. Sci. and Eng. B, 69-70, (2000), 210.

[4] G. Barillaro, A. Nannini, F. Pieri, Sens. Actuators B, Chem., 93, (2003), 263.

[5] G. Barillaro, A. Diligenti, G. Marola, L. M. Strambini, Sens. Actuators B, Chem., 105, (2005), 278.

[6] G. Barillaro, A. Diligenti, L. M. Strambini, Phys. Stat. Sol. A, 5, (2007), 1399.

[7] R. Hogervost, J. P. Tero, R. G. H. Eschauzier, and J. H. Huijsing, IEEE J. Solid-State Circuits, 29, (1994), 1505.

[8] R. J. Baker, H. W. Li, and D. E. Boyce, CMOS – Circuit Design, Layout and Simulation (IEEE Press, New York, 1998), 474.

VAPOR SENSOR USING THIN FILM BULK ACOUSTIC RESONATOR COATED BY CARBON NANOTUBES-BASED NANOCOMPOSITE LAYER

M. PENZA, P. AVERSA, G. CASSANO, E. SERRA, D. SURIANO

ENEA, CR Brindisi, Department of Physical Technologies and New Materials, PO Box 51, Postal Office Br4, 72100 Brindisi, Italy

W. WLODARSKI

RMIT UNIVERSITY, School of Electrical and Computer Engineering, GPO Box 2476V, 3001 Melbourne, Australia

M. BENETTI, D. CANNATA', F. DI PIETRANTONIO, E. VERONA

CNR, Institute of Acoustics "O.M. Corbino" Via del Fosso del Cavaliere 100, 00133 Rome, Italy

A chemical sensor based on Thin-Film Bulk Acoustic Resonator (TFBAR) structure is designed, fabricated and functionally characterized. A novel acoustic gas microsensor is represented by TFBAR consisting of a piezoelectric AlN thin layer sandwiched between two metallic Al electrodes and excited by a microwave signal to build a vibrating membrane integrated onto silicon substrate and resonating at 1.045 GHz frequency. Nanocomposite film based on Single-Walled Carbon Nanotubes (SWCNTs) with nanostructured properties have been prepared by Langmuir-Blodgett (LB) material processing technique to functionalise the surface of TFBAR devices for gas detection. Based on the excellent integration compatibility with SWCNTs-based TFBAR sensor, the studied sensing device provides the potentiality in high-performance gas detection applications. The sensing characteristics towards vapors of acetone, ethylacetate, toluene, alcohols have been measured at room temperature with high sensitivity, fast response, good reversibility and repeatability of response.

1. Introduction

Mass-sensitive acoustic sensors have been largely used for gas [1,2] and liquid [3,4] detection and biosensing applications [5,6]. Surface acoustic wave (SAW) sensors and bulk acoustic wave (BAW) sensors are the most common acoustic sensors [7], which are highly sensitive miniaturized transducers involving different sensing mechanisms (e.g., mass density change, viscoelastic loading, acoustoelectric effect). A novel type of acoustic microsensor is represented by thin film bulk acoustic resonator (TFBAR) consisting of a piezoelectric thin layer sandwiched between two metallic electrodes and excited by a microwave signal to build a vibrating membrane resonating at 1-10 GHz frequencies. The piezoelectric layer transduces the applied microwave signal into acoustic standing waves confined between the two electrodes with the resonant frequency inversely depending on piezolayer thickness. Hence, TFBAR structures are very attractive as sensing devices [1-6] due to high Q-factor (200-1500), high mass sensitivity (1 ng/cm^2) and very small membrane sensing area (300-1000 μm^2). Compared to BAW resonator, TFBAR is much smaller in thickness (1-5 μm) of the vibrating membrane exhibiting considerably higher resonant frequency up to 10 GHz, consequently higher sensitivity and moderate signal-to-noise ratio. Compared to SAW device, TFBAR operates at the same possible resonant frequency range (1-10 GHz), but in a more limited space with a great potential for miniaturization and integration.

The carbon-derived nanostructures, in particular nanocomposite layers based on carbon nanotubes (CNTs), are widely perceived as very promising nanomaterials to develop high-performance chemical nanosensors. In fact, it has been demonstrated that the carbon nanotubes, at single-walled and multi-walled format, are very attractive for detecting vapour and gas molecules in nanoscale molecular sensors [8-11] with high sensitivity and fast response.

In this study, 28 nm-thick nanocomposite layers with filler of single-walled carbon nanotubes at 75 wt.% in an organic host-matrix of cadmium arachidate (CdA) have been deposited by Langmuir-Blodgett (LB) technique onto TFBAR sensors, implemented on (001) Si using a Si_3N_4 membrane covered by AlN piezolayer sandwiched by two Al electrodes, for organic vapour sensing of acetone, ethylacetate, toluene, m-xylene, alcohols, at room temperature.

2. Experimental Details

Figure 1 shows a schematic structure of a TFBAR device based on AlN/Si_3N_4 vibrating membrane and fabricated with a thickness of $Al/AlN/Al/Si_3N_4/Si$ = 0.1/1/0.1/1/380 μm, respectively. TFBAR device was implemented on (001) Si 380 μm thick substrate, coated on both sides with a layer of Si_3N_4 (1 μm). One of the substrate surfaces was sequentially covered with an Al bottom-electrode (100 nm thick), the sandwiched AlN piezoelectric film (1 μm) and the Al top-electrode (100 nm thick). The active area of the resonating membrane is 500 μm x 500 μm. AlN piezofilm and Al electrodes are deposited by PVD systems. The resonators were obtained by anisotropic etching of Si substrate in KOH solution at 80°C by windows opened in the Si_3N_4 layer at the back-side of the substrate.

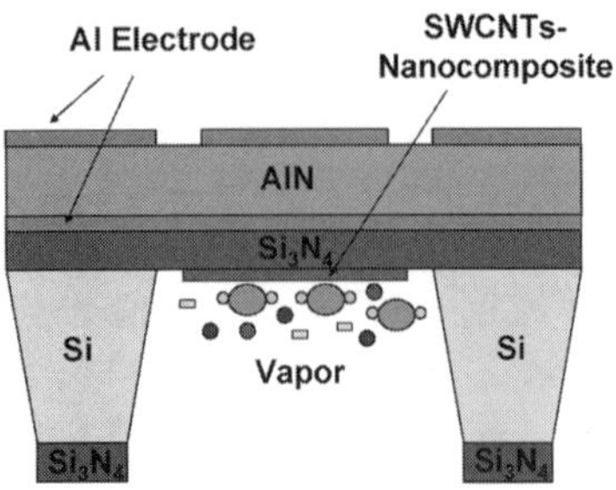

Figure 1

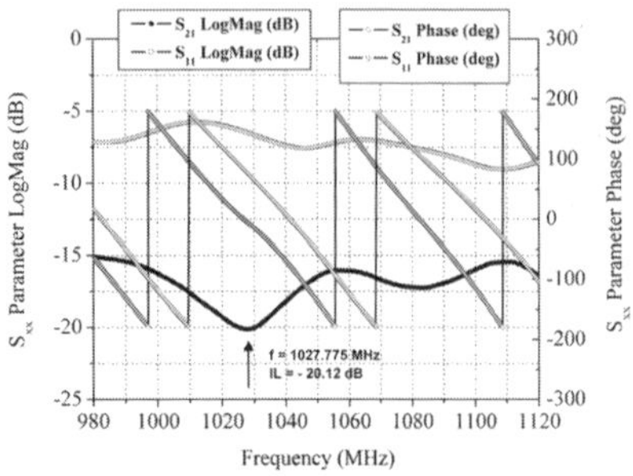

Figure 2

Figure 1. Cross-sectional view of the TFBAR sensor coated with SWCNTs-based nanocomposite LB film.

Figure 2. S_{21}, S_{11} parameters (LogMag and phase curves) of the TFBAR coated with 10 monolayers of SWCNTs-nanocomposite LB film.

The TFBAR membrane is coated by a LB film of SWCNTs-nanocomposite in a cadmium arachidate (CdA) host-matrix for enhanced vapor sensing [12]. The weight ratio of SWCNTs-filler in the CdA host-matrix is as high as 75 wt.%. The thickness of SWCNTs-nanocomposite film is of 10 LB monolayers that has a nominal thickness of 28 nm.

The S_{11} and S_{21} parameters of the TFBAR coated by 10 monolayers thick LB SWCNTs-nanocomposite have been measured by a network analyzer (Agilent 8753C) in both amplitude and phase formats, as reported in Figure 2. A resonance peak in the minimum point of S_{21} Log mag curve at 1027.775 MHz has been recorded. A frequency shift in the minimum point of S_{21} resonance of 17.6 MHz has been measured due to mass loading of the LB film onto TFBAR. A decrease in insertion loss due to layer deposition onto TFBAR is measured as about 4 dB as well.

The typical nanostructured morphology of the SWCNTs-based LB nanocomposite film has been observed by FEG-SEM with a randomly and densely distributed aggregation, as reported in Figure 3.

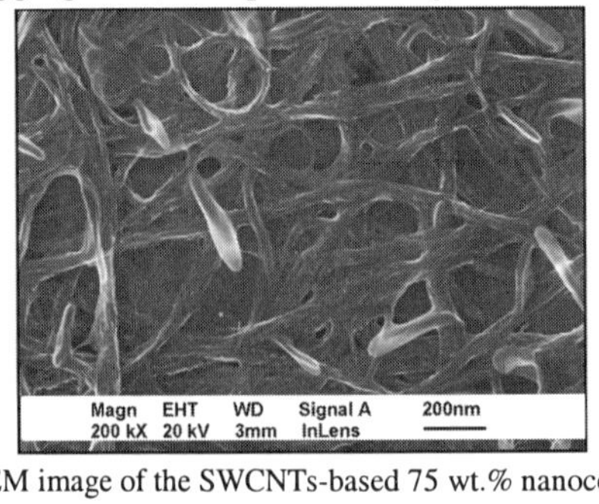

Figure 3. FEG-SEM image of the SWCNTs-based 75 wt.% nanocomposite LB film.

3. Results and Discussion

A typical room-temperature response to 500 ppm acetone using the TFBAR sensor coated by 10 monolayers LB SWCNTs-nanocomposite is shown in Figure 4. The frequency shift and insertion loss change upon vapor exposure of a 10-minute pulse have been measured relative to dry air by retrieving the minimum point at steady-state in the S_{21} transfer curve. Dry air is used as carrier gas. Another advantage of this TFBAR sensor is its ability to detect acetone vapors over a wide concentration range spanning from 100 to 500 ppm without saturating the sensor, as reported by calibration curves of frequency shift and insertion loss change in the Figure 5. Probably, this effect could be attributed to a high surface area of SWCNTs-nanocomposite by enhancing the gas molecules adsorption capacity. The TFBAR sensor configured as passive acoustic device with two independent output signals (insertion loss and minimum frequency) may enhance the useful information for pattern recognition applications.

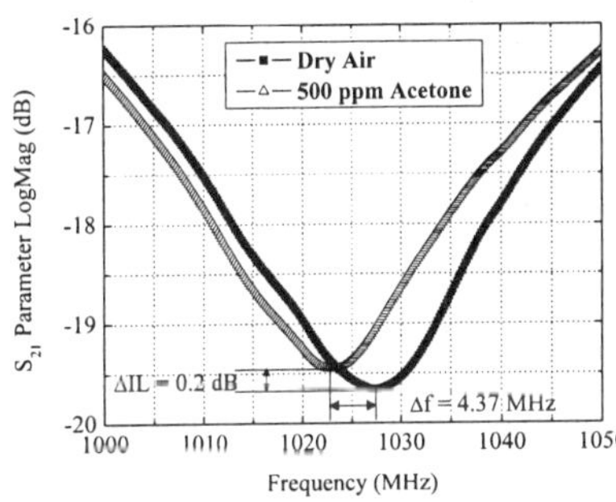

Figure 4

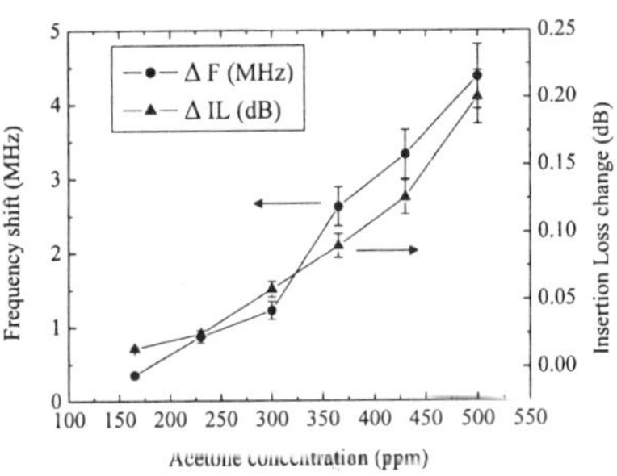

Figure 5

Figure 4. S_{21} parameter LogMag room-temperature response to 500 ppm acetone vapors relative to dry air using TFBAR sensor coated by a SWCNTs-based 75 wt.% nanocomposite 10 monolayers thick LB film.

Figure 5. Room-temperature calibration curves to acetone vapor of the frequency shift and insertion loss change by retrieving the minimum point in the transfer curve of S_{21} parameter using TFBAR sensor coated by 10 monolayers thick SWCNTs-based 75 wt.% nanocomposite LB film.

Additionally, the response of the TFBAR sensor shows good reversibility when the vapor under test is switched off with recovery gas of dry air; and acceptable repeatability is also measured upon period of three months.

As seen in Figure 6, the TFBAR sensor coated by SWCNTs-nanocomposite film is able to follow the acetone concentration steps with different resolution using the amplitude and phase output signal for continuous vapor monitoring. The TFBAR sensor results cross-sensitive to other tested organic vapors of ethylacetate, toluene, m-xylene and alcohols, at room temperature. A chemical pattern of the mean sensitivity is reported in the Figure 7. The highest gas mean sensitivity has been measured at room temperature for toluene, then ethylacetate, and finally acetone. The vapor sensitivity of TFBAR sensor can be controlled by varying the weight ratio and thickness of LB SWCNTs-nanocomposite.

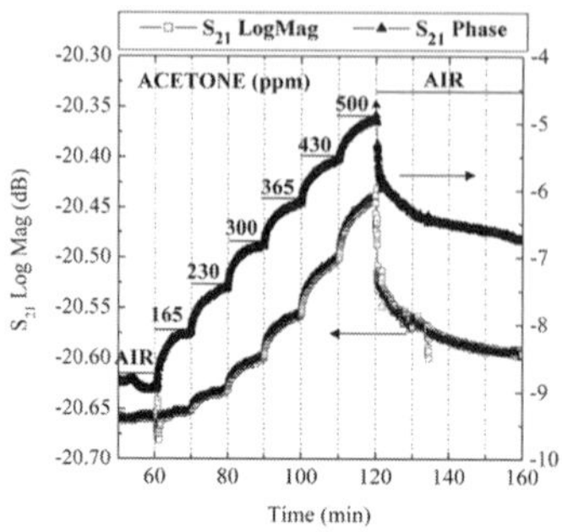

Figure 6

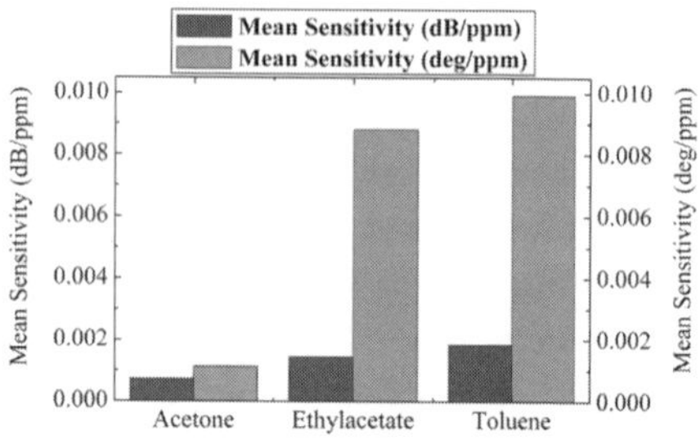

Figure 7

Figure 6. Room-temperature time response to step-pulses acetone vapors of S_{21} LogMag and Phase using TFBAR sensor coated by 10 monolayers SWCNTs-based 75 wt.% nanocomposite LB film.
Figure 7. Room-temperature mean sensitivity to vapors of acetone, ethylacetate and toluene using as output signal the phase and insertion loss change by retrieving the minimum point in the transfer curve of S_{21} parameter with the TFBAR sensor coated by 10 monolayers thick SWCNTs-based 75 wt.% nanocomposite LB film.

4. Conclusions

This study presents experimental results on TFBAR sensor coated by SWCNTs-nanocomposite LB film for vapor sensing applications, at room temperature. We have successfully designed, fabricated and tested a TFBAR sensor covered with LB film for detection of organic vapors (acetone, ethylacetate, toluene, alcohols) with sensitivity enough to detect their TLVs in workplace. The gas sensing device based on high-performance and high resonating frequency (1.045 GHz) was demonstrated. The sensing characteristics show the possibility as a miniaturized sensor in silicon-integrated sensor-systems capable of gas analysis. These TFBAR devices covered with highly-sensitive carbon nanostructured materials are becoming very promising sensing devices with advanced functions and novel properties for future applications.

References

1. M. Benetti, D. Cannatà, F. Di Pietrantonio, V. Foglietti, E. Verona, *Appl. Phys. Lett.* **87**, 173504 (2005).

2. M. Benetti, D. Cannatà, A. D'Amico, F. Di Pietrantonio, V. Foglietti, E. Verona, *Proceedings of 2004 IEEE Ultrasonics, Ferroelectrics, and Frequency Control Conference,* pp. 1581 (2004).
3. H. Zhang, E. S. Kim, *Journal of Microelectromechanical Systems* **14(4)**, 699 (2005).
4. M. Link, M. Schreiter, J. Weber, R. Primig, D. Pitzer, R. Gabl, *IEEE Trans. Ultrason., Ferroelect., Freq. Contr.* **53(2)**, 492 (2006).
5. J. Weber, W. M. Albers, J. Tuppurainen, M. Link, R. Gabl, W. Wersing, M. Schreiter, *Sens. Actuators A* **128**, 84 (2006).
6. S. Rey-Mermet, R. Lanz, P. Muralt, *Sens. Actuators B* **114**, 681 (2006).
7. D. S. Ballantine, R. M. White, S. J. Martin, A. J. Ricco, E. T. Zellers, G. C. Frye, H. Wohltjen, *Acoustic Wave Sensors*, Academic Press, San Diego, 1997.
8. J. Kong, N. R. Franklin, C. Zhou, M. G. Chapline, S. Peng, K. Cho, H. Dai, *Science* **287**, 622 (2000).
9. M. Penza, M. A. Tagliente, P. Aversa, G. Cassano, L. Capodieci, *Mat. Sci. & Eng. C* **26**, 1165 (2006).
10. M. Penza, G. Cassano, P. Aversa, F. Antolini, A. Cusano, A. Cutolo, M. Giordano, L. Nicolais, *Appl. Phys. Lett.* **85(12)**, 2379 (2004).
11. M. Penza, M. A. Tagliente, P. Aversa, M. Re, G. Cassano, *Nanotechnology* **18**, 185502 (2007).
12. M. Penza, M. A. Tagliente, P. Aversa, G. Cassano, *Chem. Phys. Lett.* **409**, 349 (2005).

EVAPORATION RATE DETERMINATION FOR WATER AND ALCOHOLS IN BUBBLERS

ANDREA ORSINI

Dept. of Chemical Science and Technology, University of Rome Tor Vergata, Via della Ricerca Scientifica 1, Rome, 00133, Italy

ANDREA BEARZOTTI

IMM - CNR, Rome division
Via del Fosso del Cavaliere, 100, Rome, 00133, Italy

In these experiments we evaluated the behavior of water and alcohols rate of evaporation by forced bubbling. Changing the quantity of nitrogen flux into the liquid or the temperature of the measure chamber, a freezer for cold temperatures or an oven for higher ones, it has been possible to analyze and compare the liquid mass evaporation rate with that calculated by Antoine Parameters formula. The results obtained are very similar to the theoretical ones except for Methanol that showed a 15% of higher evaporation rate. A special case is that of Decane, which Antoine Parameters graph presents a lack of values between forty and ninety degrees centigrade. The experiment showed an intermediate behavior between the two set of parameters.

1. Introduction

1.1. *Experiments Involving Bubblers*

In experiments involving vapour detection, bubblers are often used in order to carry in the measuring chamber the substances we are examining. It is of particular importance to exactly estimate the amount of analyte that reach the sensor under test. We must in fact consider that, beyond the free evaporation phenomenon , there is also the mechanical transport due to the gas flow that gives rise to such a kind of aerosol.

Another matter of fact is the minimum quantity (3–5 ml) of the organic solvent necessary to ensure a homogeneous bubbling and hence a constant analyte concentration in the gas phase, therefore it is useful with very expensive analytes to know exactly the liquid evaporation rate to determine the necessary level of this in the bubbler volume.

In the experiments involving this phenomena, the assumed behaviour was that of a gas – liquid saturation interface.

A first consideration about this problem can be found in the work of Göpel[4] to improve the reproducibility of gas concentration by volatizing method. It is mentioned the problem of aerosols formation, which leads to higher analyte concentrations in the gas phase than calculated and hence adjusted (deviation estimated in the range of 10%).

Our studies have been focused on alcohol vapours and in particular on ter-butanol, n-butanol, n-propanol, decane, methanol, ethanol and hexane and, for comparison, water with the final goal of finding a relationship between the alcohol concentration and the carrier gas flux in the test chamber.

Temperature and Nitrogen flux dependant measurements where performed to evaluate the liquids mass loss for different environmental situations.

1.2. *Theoretical Formulas*

Experimentally[1] the evaporation rate in function of time can be expressed with the following equation:

$$P = K \cdot \frac{\Delta M / \Delta t}{M_w \cdot \Phi} \cdot T \qquad (1)$$

where ΔM is the liquid mass loss during forced bubbling in grams, Δt is the time slot in seconds, P the saturation pressure measured in Pascal, M_w the molecular weight, T the temperature in Kelvin, Φ the value of the flux in the bubbler measured in Sccm and K a multiplying factor which value (about 0,5 x 10^9) is determined by measure units conversion.

The saturation vapour pressure of the vaporised compound depends on the temperature of the liquid contained in the bubbler and can be calculated using the Antoine equation[3]:

$$\log_{10} P = A - \frac{B}{T_{liquid} + C} \qquad (2)$$

Here T denominates the temperature of the thermostat in °C; A, B and C are constants and can be taken from the literature[3].

Varying the temperature and the analyte gas flow by using mass flow controllers, provides a wide range of analyte concentrations, therefore we can compare the results obtained with the two different formulas and verify how much is trustworthy the value determined with Antoine Equation.

1.3. *Experimental Setup*

The experiment was settled in two different environments: a Memmert Laboratory Oven in order to measure the behavior of the liquids in question at temperatures higher than RT and a freezer for the lowest ones.

Inside these climatic chambers there was the bubbler, filled with one of the aforementioned liquids, connected with nitrogen carrying tubes and placed on a precision balance (Sartorius BP221S) to measure the liquid mass in the vessel (see Fig. 1).

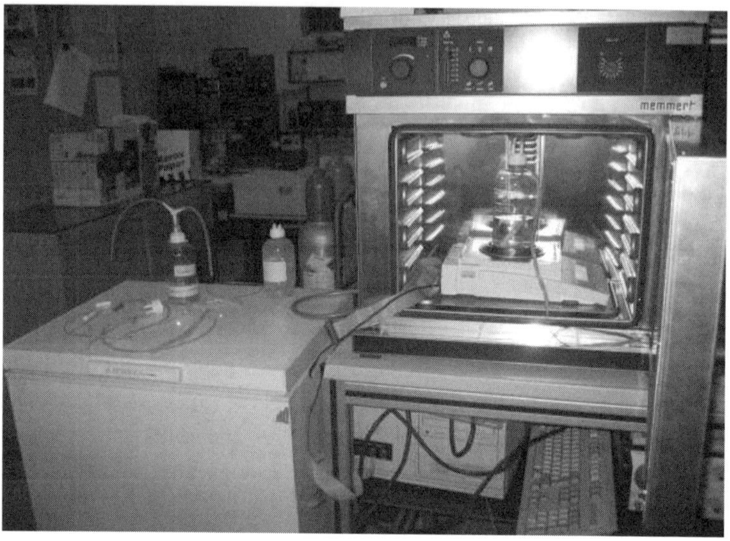

Figure 1. Experimantal instruments involved in the experiment and their arrangement in the laboratory.

The temperature sensor and the balance were connected with a serial port to a Pc for data acquisition (IE488 standard)

1.3.1. *Temperature Control*

The choice of a temperature sensor depends on the kind of appliance, considering ambient conditions, temperature ranges, precision of the measure.

In our experiment Temperature was determined with a Resistance Thermometer (TR) put inside the liquid container. We used a PT100, a passive element made by Platinum, which resistive value grows in linearity with temperature from a value equal to 100½ at 0°C. At this temperature the tolerance is about 0,03 °C.

1.3.2. *Nitrogen Flux Control*

One of the variable parameters in the experiment was the quantity of gas flux (Nitrogen) into the bubbler. The flux was settled by using mass flow controllers and it was ranging from 20 sccm to 200 sccm.

2. Results

Varying the temperature and the analyte gas flow by using mass flow controllers, provides a wide range of analyte concentrations in the bubbler headspace that corresponds with the quantity of mass loss of the liquid in the vessel. Two kind of parameters variation have been performed, one with gas flux and the other with temperature change.

2.1. *Liquid – Gas Phase Transition with Flux Variations*

These kind of measures show a proportional behavior of the evaporated mass versus the quantity of nitrogen flux. This is in good agreement with our forecasts and with the phenomena description as Gas saturation.

In fact with the same temperature we have the same pressure in the headspace of the liquid and, looking at formula (2), the ratio between the mass loss rate and the flux should keep constant.

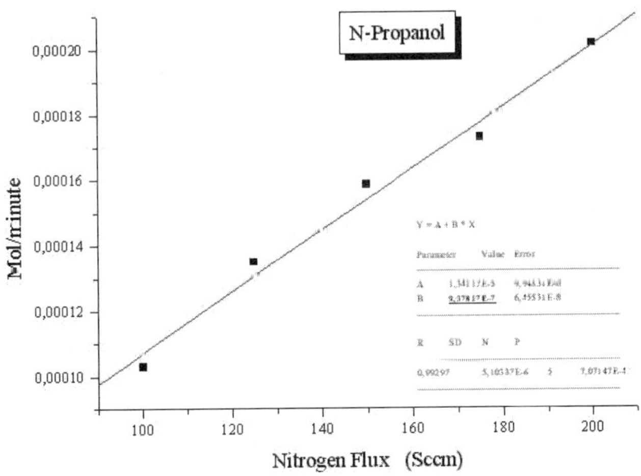

Figure 2: Graph showing the behavior of N-Propanol mass loss with increasing quantities of flux with constant room temperature.

Experiments conducted with different liquids at room temperature showed similar results. Moreover the same behavior has been observed for the liquids at forty degrees centigrade. Obviously the line tends to the Origin of the graph for zero flux of gas.

2.2. *Liquid – Gas Phase Transition Vs Temperature*

The behavior of alcohol evaporation rate at different fixed temperatures was analyzed with constant nitrogen flux. The data of interest was the slope of the liquid mass curve in time graph. The typical trend of the measured curve was an exponential decay at the beginning of flux through the liquid turning with time into a linear decrease. Taking the slope value for various temperatures we can build a graph for compare the measured value with that calculated with Antoine Parameters formula (2).

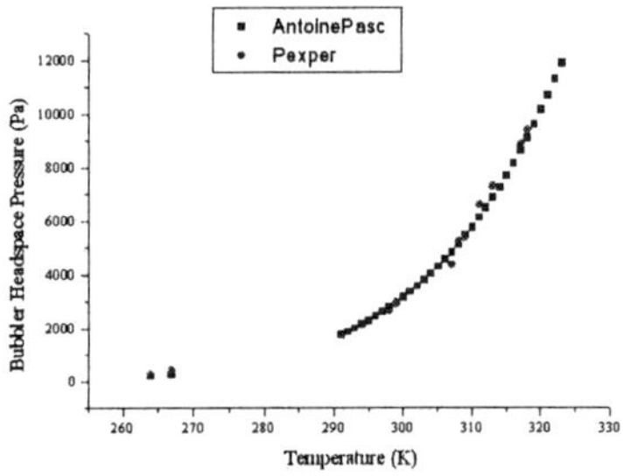

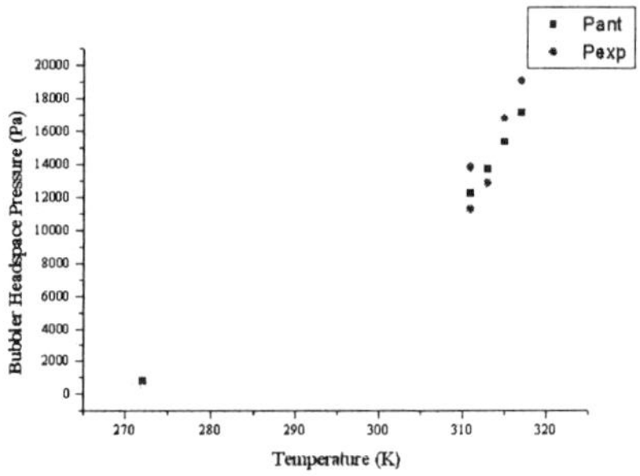

Figure 3: Graphs showing N-Propanol (left graph) and Ter-Butanol (right graph) experimental and theoretical evaporation rate. The data are reported as pressure in Pascal.

This was the approach for all the liquids under test except for Decane. In fact for this molecule there is a deficiency of Antoine Parameters at temperature slightly higher than forty degrees centigrade so it is not possible to do a real comparison between obtained results and foreseen evaporation rate. Using the 300 K Antoine parameters, the pressure (log scale) is 2,14 + 0,0215*T a bit less than experimental one 2,27 + 0,0245*T more similar to that obtained with higher temperature (about ninety degrees centigrade) Antoine Parameters by Williamham and Taylor[3].

3. Discussions

There is agreement between experimental data and the theoretical expression[2], although this one doesn't contemplate the mechanical transport in such bubblers, which leads to higher analyte concentrations in the gas phase than calculated. On the other hand a noticeable deviation was encountered only for Methanol. This one was the more volatile compound with an enthalpy of vaporization of 37,5 KJ/mol and it is more sensible to temperature variations.

Therefore the use of Antoine Parameters formula is a good approximation to forecast alcohol concentration in a measurement chamber

References

1. Zemansky & Dittman, 1981. *Heat and Thermodynamics*, McGraw Hill, New York.
2. Reif, F. 1965. *Fundamentals of Statistical and Thermal Physics*, McGraw Hill, New York.
3. Riddick, J. & Bunger, A. 1986. *Techniques of Chemistry, vol. II*, new edition, Wiley Interscience, New York.
4. K. Bodenhofer, A. Hierlemann, R. Schlunk and W. Gopel, *Sensors and Actuators B* **45**, 259-264V (1997).

ALL ORGANIC HUMIDITY SENSORS BASED ON CONJUGATED POLYMERS AND A TETRACYANOQUINODIMETHANE SALT

A. ARENA, N. DONATO, G. SAITTA

Dip. di Fisica della Materia e Tecnologie Fisiche Avanzate, Univ. di Messina

G. NERI, G. MICALI

Dip. di Chimica Industriale ed Ingegneria dei Materiali, Univ. di Messina

G. PIOGGIA

Centro Interdipartimentale di Ricerca "E. Piaggio", Univ. di Pisa

Summary

This work is about the development and the optical and electrical characterization of all organic heterojunctions consisting of solution deposited films of poly(3,4ethylendioxythiophene)/poly(styrenesulfonate) (PEDOT/PSS), a few micrometers thick, coated with an ion conducting material. This latter one is a complex obtained by the charge transfer reaction that takes place when acetonitrile solutions of the electron acceptor Tetracyanoquinodimethane (TCNQ) molecules are added to solutions containing the ion conducting sodium-polystyrenesulfonate (NaPSS). The principle of operation on which such devices are based on, relies on the possibility to change optical and charge conducting properties of conjugated polymers by electrically driven oxidization and reduction transitions.

1 Introduction

In the last few years conjugated polymers have been the subject of intense research activity mainly motivated by their high electron conductivity and their attracting optical properties, which find many applications in the development organic light emitting diodes, displays and photovoltaic cells. In addition to the well consolidated optoelectronic applications, recently there has been a renewed interest focused on the exploitation of the ion conducting features of conjugated polymers in electrochemical devices including supercapacitors [1], volatile memories [2], electrochromic windows [3] and even actuators emulating muscles [4]. The principle of operation on which such devices are based on, relies on the possibility to change optical and charge conducting properties of conjugated polymers by electrically driven oxidization and reduction transitions. In fact, according to a widely accepted qualitative model, the removal of electrons from the alternating single and double bonds of a conjugated polymer (oxidation), results in positive charges delocalized over the polymer chains. Charge balance is ensured by negatively charged ions that are incorporated into the polymer network to compensate the positive charge acquired in the oxidation. Oxidized polymers can be reduced as well, by reinserting electrons into the polymer matrix, which leave the backbone neutral or uncharged state. The state change of conjugated polymers due to redox reactions can produce several effects, e.g., variations in polymer optical transmittance, volume, porosity, and electrical transport features. These changes are related to the oxidation state of the polymer and are under electrochemical control: the neutral polymer, the reduced polymer, the oxidized polymer or any intermediate state can be reached by applying the appropriate potential. This paper is about the development and the optical and electrical characterization of all organic heterojunctions consisting of solution deposited films of poly(3,4ethylendioxythiophene)/poly(styrenesulfonate) (PEDOT/PSS), a few micrometers thick, coated with an ion conducting material. This latter one is a complex obtained by the charge transfer reaction that takes place when acetonitrile solutions of the electron acceptor Tetracyanoquinodimethane (TCNQ) molecules are added to solutions containing the ion conducting sodium-polystyrenesulfonate (NaPSS).

2 Optical characterization

Owing to the intimate relationship between the oxidation state of the materials under examination and their optical properties, optical absorption measurements carried out in the UV-VIS-NIR spectral range are of great aid in the attempt to give an interpretation of what happens at the interface between PEDOT-PSS and the TCNQ complex. It seems likely that owing to the interaction between the complex and the conjugated polymer, an insulating layer of TCNQ forms at the interface between the two materials. Several kinds of redox reaction could be involved in the mechanisms of the bilayer conduction. It can be hypothesised that, depending on the humidity condition, the TCNQ complex partially dissociates into Na+ and TCNQ- ions. In such a condition, a redox reaction involving the reduction of PEDOT from the highly conducting phase to the poor conducting one could take place at the negatively biased electrode:

$$PEDOT^{+}(PSS)^{-}+Na^{+}+e \rightarrow PEDOT^{0}+Na^{+}(PSS)^{-}.$$

Other kinds of reactions, for instance the formation of charge transfer complex between PEDOT and TCNQ:

$$PEDOT^{+}(PSS)^{-}+Na^{+}TCNQ^{-} \rightarrow (PEDOT)+(TCNQ)^{-}+ Na^{+}(PSS)^{-}.$$

could happen. Further, conduction electrons injected from the negatively biased electrode may reduce the high electron affinity $TCNQ^{0}$ molecules of the interfacial layer. The $TCNQ^{-1}$ molecules in their turn may recombine with the Sodium ions, at the expenses of $Na^{+1}PSS^{-1}$, yielding as a final result an increased conductivity, due to the oxidation of $PEDOT^{0}$ to $PEDOT^{+}$. In Figure 1 it can be seen the absorption spectrum of TCNQ (dashed) compared with the spectrum of NaPSS-TCNQ (solid). The bands positioned between 600 nm and 900 nm are ascribed to TCNQ anions. The inset compares the C≡N stretching peak of neutral TCNQ (dashed), with the cyano group stretching of NaPSS-TCNQ. The observed shift confirms that in the complex the TCNQ molecules are in the reduced state.

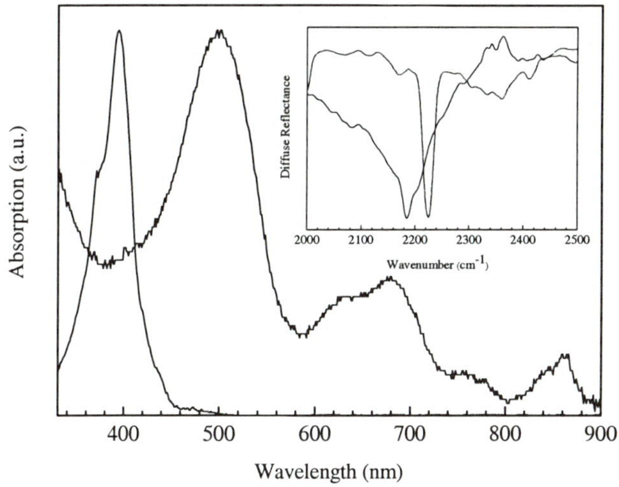

Figure 1 – Absorption spectrum of TCNQ (dashed) compared with the spectrum of NaPSS-TCNQ (solid).

3 Electrical characterization

From the electrical point of view, the behaviour of the heterojunctions consisting of PEDOT/PSS and NaPSS-TCNQ is found to remarkably depend on environmental conditions, in particular on the humidity. At room temperature and under average humidity condition, the bilayers, biased by triangular voltage sweep between -5V and +5V, show rectifying current-voltage characteristics, as the one shown in Fig. 2.

According to the experimental results, the current versus voltage exhibits a remarkable hysteresis, which persists after several hundreds cycles. Such a behavior, which is observed also in electrochemical supercapacitor, is likely to be due either to faradaic current, namely to redox transitions which take place at precise bias voltage, or to the formation of an immobile space charge region between the conducting polymer layers. The electrical behaviour previously described, in particular the relatively high current values and the rectification property are remarkably conditioned by the environment humidity. To show this, the electrical characterization has been carried out by inserting the bilayered devices into a measurement chamber equipped with an automatic control of the environmental conditions.

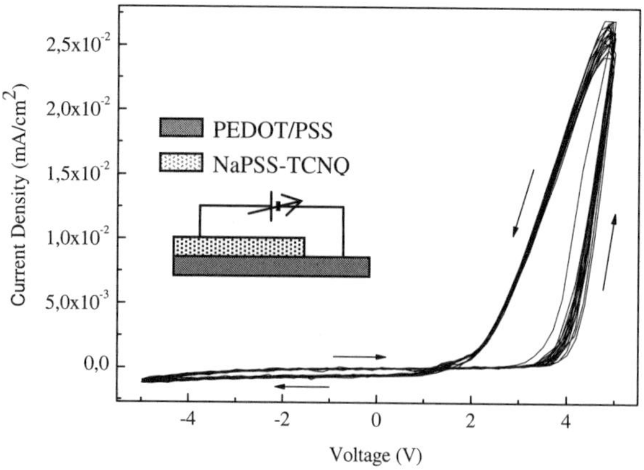

Figure 2 – Current measured by means of a Keithley 2400 sourcemeter through a PEDOT/PSS-NaPSSTCNQ bilayer biased by triangular voltage pulses (20 cycles).

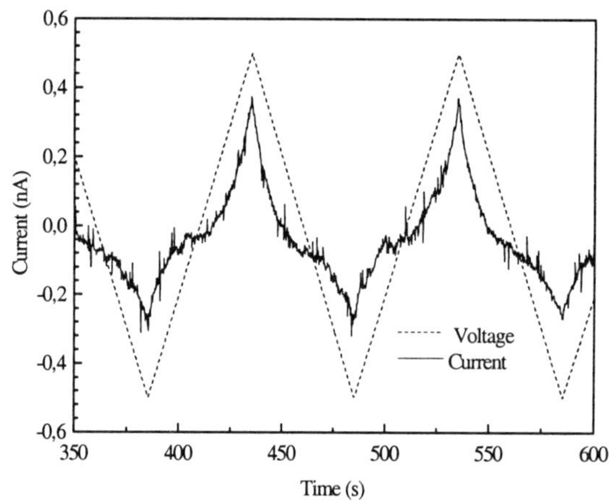

Figure 3 – Current measured in dry condition through a PEDOT/PSS-NaPSSTCNQ bilayer biased by a triangular voltage pulse between +5V and -5V (dashed line)

Here, in Figure 3-4 we show the measurement results carried out in the same bias voltage conditions of Fig. 2. The lower intensity curve is measured in dry conditions, the high intensity curve is measured at the humidity relative value of 50%.

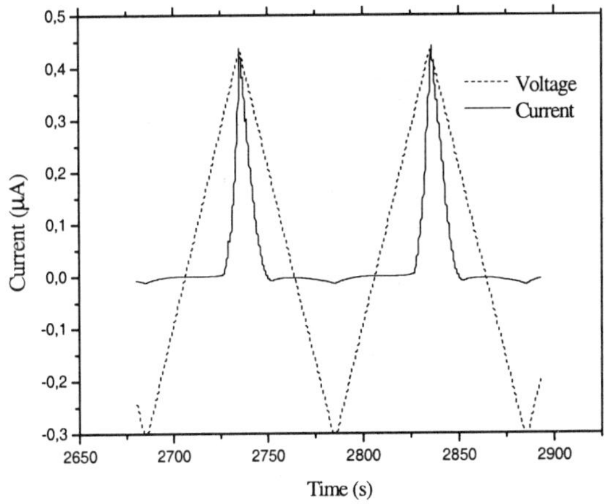

Figure 4 – Current measured at 50% relative humidity condition through a PEDOT/PSS-NaPSSTCNQ bilayer biased by a triangular voltage pulse between +5V and -5V (dashed line)

Further, the devices show humidity sensor functionality, as indicated by the ratio between the current in the 50% humidity condition and in the dry condition, which is found to reach values of the order of 10^3 at a bias voltage of 5 V.

4 Conclusions

In this paper we described the realization and characterization of humidity sensors based on conjugated polymer and NaPSS-TCNQ salt.

As it regards the humidity sensing functionality of the devices, such a behaviour has been observed elsewhere in devices that use doped PEDOT electrodes and electrolytic active layers, and is interpreted as the environmental dependent electrical conductivity of the ionic conducting transport layer. In the case examined here, under the hypothesis that the charge transfer complex of TCNQ behaves as an electronic conductor, a possible explanation of the experimentally observed dependence of current intensity upon humidity, could be that the higher is the humidity, the higher is the dissociation rate of the charge transfer complex into its charged components directly involved in the electrochemical reactions that are believed to control the current flow.

References

1. M.Z.-C. Hu, E.A. Payzant, H. Byers, J.Colloid Inter. Sci., 20 (2000) 222
2. Ben-Lin He, Ying-Ke Zhou, Bin dong, Hun-Lin Li, Material Science and Engineering A., 374(2004), 322
3. Swarup K. Majee, Himandri S. Majumdar, Alberto Bolognesi, Amlan J. Pal, Synthetic Metals, 156(2006), 828
4. S. Varis, M. Ak, G. Tanyeli, I. Mecidoglu Akhmedov, L. Toppare, European Polymer J., 42(2006), 2352
5. M.S. Cho, H. J. Seo, J. D. Nam, H. R. Choi, J. C. Koo, K. G. Song, Y. Lee, Sensors and Actuators B, 119(2006), 621

SELECTIVE CHEMICAL SENSORS FOR NO_2- DETECTION, USING CARBON NANOTUBE/POLYMER COMPOSITE NANOWIRES *

FEDERICA VALENTINI*, VANESSA BIAGIOTTI, GIUSEPPE PALLESCHI

Università degli Studi di Roma Tor Vergata, Dipartimento di Scienze e Tecnologie Chimiche, via della Ricerca Scientifica 1, 00133 Rome (Italy)

JOSEPH WANG

Department of Chemical and Material Engineering and Chemical and Biochemistry, Arizona State University, Tempe, Arizona 8528

*Corresponding author: **federica.valentini@uniroma2.it**

Bare Gold electrodes modified with multi-walled carbon nanotube (MWCNT)/poly(aniline) and MWCNT/poly(pyrrole) composite nanowires were assembled for sensitive and selective detection of nitrites. Several electroanalytical parameters were investigated for the template synthesis of composite nanowires, such as different monomers; different purification and functionalisation procedures of MWCNTs. The morphology of the nanowires was investigated by Atomic Force Microscopy (AFM), and Inverted Optical Microscopy. These new chemical sensors based on nanocomposite materials were fully characterized by Differential Pulse Voltammetry (DPV) working at an applied potential of +900 mV vs. Ag/AgCl. The best analytical performances were observed working at (MWCNT)/poly(aniline) composite nanowire based electrodes, in terms of linear range of concentration (1.00×10^{-6}-1.00×10^{-3} M), limit of detection (L.O.D. 1.70×10^{-6} M), sensitivity (31.85 A mol^{-1} L cm^{-2}), reproducibility (RSD % = 2), response time (8s), and higher selectivity towards common interfence. Finally, this nanostructured sensor prototype was tested in real samples of tap water, to which standard concentrations of nitrite were added, and subsequently evaluating the % of recovery ($\geq$ 90 %). Among the sensors tested, those prepared with poly(aniline)/MWCNT composite nanowires showed the best analytical performances.

1. Experimental

1.1 Synthesis of Polyaniline/MWCNT and Polypyrrole/MWCNT Composite Nanowires

The electrochemical-template preparation of the CNT-PANy and CNT-PPy composite nanowires was widely reported in literature [1].

The optimized conditions used in this work, concerns the electropolymerization which proceed by cycling the potential between -0.2 and 0.9 V vs Ag/AgCl reference electrode during the first cycle, and between -0.2 and 0.78 V vs Ag/AgCl during the following cycles at 0.05 V s^{-1} (for both polymer solutions). After the electropolymerization, the template-membrane was rinsed with water, air dried, and finally removed in order to collect the composite nanowires.

AFM samples were prepared by putting 2 μL of diluted nanowire suspension in water on a mica surface and allowing it to dry in air.

For the Inverted Optical Microscope imaging, 7 μL of diluted nanowire suspension in water were placed in a sealed well on a clean glass slide.

1.2 Electrode surface Preparation and Modification

Modified and unmodified gold disk electrodes were used as working electrodes. Electrode modification with nanowires was performed by adsorptive deposition of PANY/CNT and PPy/CNT nanowires. For this purpose, the electrode was immersed for 1 h under static conditions into a diluted PANY/MWCNT and PPY/MWCNT suspensions in water and finally dried at room temperature

1.3 Cyclic Voltammetry Procedure

Film's permeability to nitrite and interferent was evaluated by cyclic voltammetry in 0.1M acetate buffer, pH 4.0. The applied potential ranged from 0.2 to 1.1 V for all gold disk electrodes, at scan rate of 0.1 V s^{-1}. All measurements were performed at room temperature. The permeability study was carried out comparing the peak current for each one in 0.1M acetate working buffer, pH 4.0, at 0.1 V s^{-1} as scan rate. The nitrite concentration used for this study was 40 μM, while for all the interferent we used a concentration 1000 higher of 40 mM, as found in many kinds of real samples (i.e., AA in human body, as brain tissue, some biological fluids).

1.4 Differential Pulse Voltammetry Procedure.

DPV were performed with a pulse amplitude of 50 mV, a pulse width of 60 ms, a scan rate of 100 mV/s, a pulse interval of 200 ms, and a sampling time of 20 ms; E_i) 0.6 V and E_f) 1.2 V for nitrite calibration curves. All measurements were performed at room temperature, in 0.1M acetate working buffer, pH 4.0.

For recovery study in tap water samples (performed by the standard addition method), 0.5 M acetate buffer, pH 5.0, was used because the higher ionic strength value could minimize the pH changes during nitrite detection in real samples. 2mL fresh tap water samples were added to the electrochemical cell and diluted to 25 mL with the supporting electrolyte solution indicated above. All these measurements were performed under the optimised experimental conditions. The values of nitrite concentration were interpolated from the normal calibration curves (these last obtained in 0.5M acetate buffer, pH 5.0).

The recovery data (average of three determinations) was evaluated using the ratio between the expected value of nitrite concentration (after the standard additions) and the corresponding experimental value detected by DPV analysis. The precision parameter were also calculated as standard deviation of three determinations for each nitrite concentration value.

2. Results and discussion

2.1 MWCNTs functionalization

Fig. 1a shows not treated MWCNTs; Fig. 1b shows an image of shortened MWCNTs after the oxidizing treatment performed for 12h; Fig. 1c shows an image of shortened MWCNTs after the oxidizing treatment performed for 36h. They appears as dots.

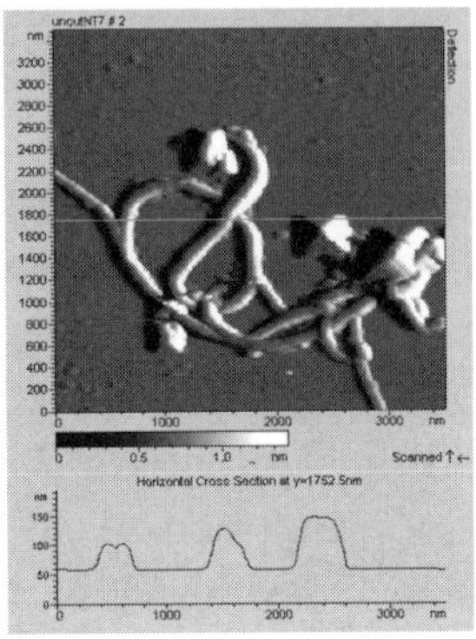

FIG 1a

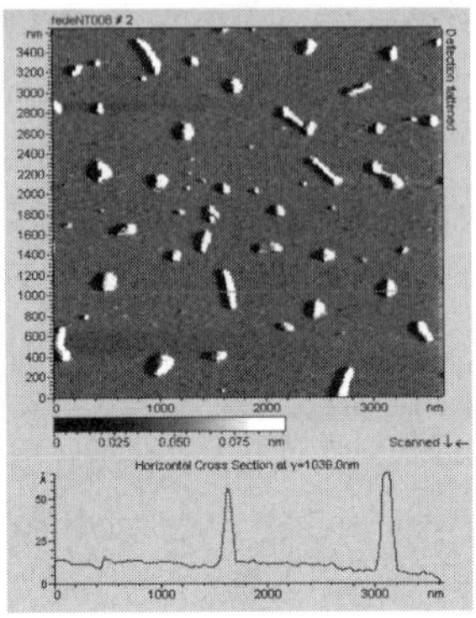

FIG 1b

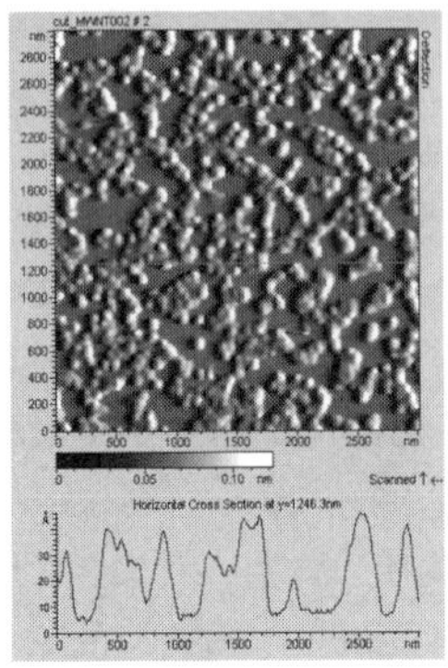

FIG 1c

MWCNTs treated in acidic mixture for 12h (1b) showed several oxygen groups (FT-IR data not shown) on their walls, so they were suitable to be entrapped into the polymeric chains during the electrochemical growth. In table 1 were summarized the typical fisical parameters of these nanomaterials [2].

Nanostructured material	Lenght (mm)	Diameter (nm)	Morphology (shape)
MWCNTs obtained by CVD synthesis	4.0	50	Typical cylindrical nanotubes structure
MWCNTs oxidised by H2SO4/HNO3 for 12h	0.3-0.4	5.0	Reduced cylindrical structure
MWCNTs oxidised by H2SO4/HNO3 a 36h	-	2.0-5.0	Quantum dots
Polyaniline nanowires	6.0	200	Nanowires
polyaniline nanowires bundle	10.0	500	Nanowires bundle

2.2 Morphological characterization

The morphological investigations of composite nanowires were performed by AFM and Inverted Optical Microscopy. In fact, Figure 2a shows a typical morphology of a single PANy/MWCNT nanowirc, which resulted 6.0 μm in length, and 200 nm in diameter (as the diameter of alumina template). In addition, Figure 2b reports a bundle of PANY/MWCNT nanowires. The shape and size of these PANy/MWCNT nanowires were also investigated by the use of Inverted Optical Microscope, and some typical images are shown in Figure 3a, and 3b.

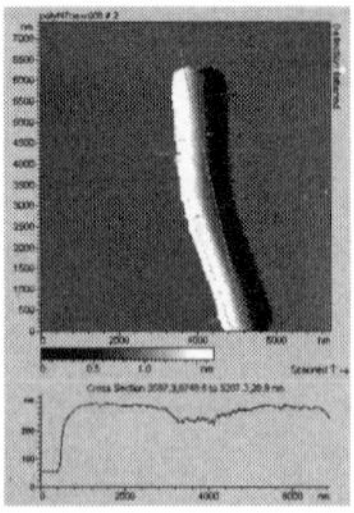
FIG 2a

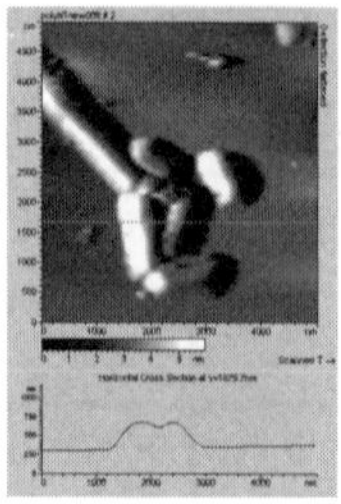
FIG 2b

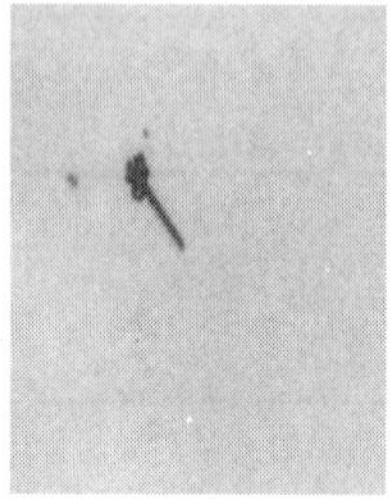
FIG 3[a]

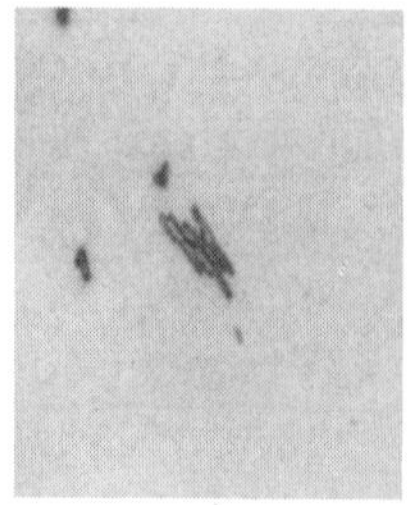
FIG 3b

2.3 Analytical study

Selectivity was studied towards common interferences as ascorbic acid, phenol, 2-chlorophenol, sodium sulphite
Permeability was evaluated by cyclic voltammetry using this expression:

$$P. \% = (I\ film / I\ bare) * 100$$

Where I bare is the peak current recorded at bare electrode and I film is the peak current at the modified electrode

Gold bare electrode modified by	PNO2-/ AA	PNO2-/ Phenol	PNO2-/ 2-chlorophenol	PNO2-/sodium sulfite
PANY/MWCNT by H2SO4/HNO3	2.86	2.57	3.33	1.40
PANY/MWCNT by KOH	1.52	1.55	1.38	1.14
PPY/MWCNT by H2SO4/HNO3	2.59	1.85	1.43	1.34
PPY/MWCNT by KOH	1.27	0.92	0.10	0.04

Better results in terms of perm-selectivty were obtained by PANy/MWCNT and PPy/MWCNT nanowires, using MWCNTs oxidized and shortened by a treatment in a mixture of concentrated nitric acid and sulphuric acid (3:1) under sonication for 12h.

Among these the best one was the PANy/MWCNT composite nanowire-based sensor. Especially, the presence of some chemical groups on MWCNT's walls seems to be responsible for the selectivity of the chemical sensors towards the common interferent, combined with the role of polymeric membranes to act as a diffusion barrier toward the electron transfer processes of several electroactive probes.

Calibration curves was obtained by DPV showing best results in terms of reproducibility and sensitivity for the gold-bare electrode modified by PANY/MWCNTs composite, which was used also to analyze real tap water samples.

Electrode	**Linear range**	**Equation of the linear regression**	**L.O.D. (3σ)**	**Sensitivity**	**RSD%**	**Time of response**
Gold bare electrode modified by	M		M	A mol^{-1} L cm^{-2}	(n = 15)	(s)
Polyaniline/MWCNTs	$1*10^{-6}$ – $1*10^{-3}$	y/mA =0.993x/mM + 0.193 R2 = 0.9998	$1.70*10^{-6}$	31.85	2.00	8
Polypyrrole/MWCNTs	$5*10^{-6}$ – $1*10^{-3}$	y/mA =0.875x/mM + 0.145 R^2 = 0.9989	$1.78*10^{-6}$	25.48	3.10	10

Real tap water sample analysis

Nitrites spiked	**Nitrites found before spiking**	**Expected value**	**Measured value**	**Recovery**	**RSD (n=3)**
(μM)	(μM)	(μM)	(μM)	%	%
20	0	20	17	85	1
50	0	50	45	90	1
200	0	200	191	96	2
500	0	500	490	98	1

3. Conclusion

In this work we demonstrate that the Polyaniline/MWCNTs and Polypyrrole/MWCNTs nanowire-based sensors led to very selective, sensitive, and reproducible probes for the detection of nitrite. The use of polymers combined with the oxidized and shortened nanotubes demonstrated a good perme-selectivity towards some common interferent, thanks to the presence of electrically charged groups which are able to attract and repel some molecules (for the electrostatic interactions). This property combined to the size exclusion performances (typical of the composite nanowire materials) could enhance the selectivity of these sensors but further investigations will be necessary in order to highlight this important aspect for the sensor performances. In addition, a low detection limit (according to Italian legal limit for nitrite in drinking water). The selective sensors described here offered also the possibility to measure nitrite in real samples minimizing the matrix effects.

3. Acknowledgments

This research was supported by Grants from FIRB N° RBNE01MBTC-002 Italian Project.
The authors wish to thank also the *Biodesign Institute at Arizona State University*, for technical assistance and support.

References

1. F. Valentini, A. Amine, S. Orlanducci, M. L. Terranova, G. Palleschi, Anal. Chem., 7, (2003) , 5413.
2. E.Tamburri, S. Orlanducci, M.L. Terranova, F. Valentini, G. Palleschi, A. Curulli, F. Brunetti, Carbon, 43(6) (2005) , 1213.

ON THE FABRICATION PROCESS OF POLYMER-COMPOSITES BASED SENSORS

A. DE GIROLAMO DEL MAURO,
A. CITARELLA, E. MASSERA, L. QUERCIA,
G. DI FRANCIA

Enea Research Centre of Portici, via Vecchio Macello,
80055 Portici, Naples, Italy

The results of investigations of the effects of two organic solvents of preparation, tetrahydrofuran (THF) and 1,1,1,3,3,3 Hexafluoro-2-propanol (HFIP), and of three different electrical geometry of devices on drop-coated poly (methyl-methacrylate)/ carbon black (PMMA/CB) composites gas sensors are presented. Using HFIP solvent, it is possible to obtain a good dispersion of the CB filler in the polymeric matrix but the thermodynamic responses of devices to acetone and ethanol vapours are independent from their morphology and from the geometry of devices. The highest sensor responses are to acetone vapor. This behavior could be probably attributed to the higher chemical affinity of less polar molecule as acetone towards PMMA. The filler dispersions, the current-voltage (I/V) characteristics and the stability of devices in the time were also studied and discussed.

1. Introduction

The conducting polymer-based composites have shown great potential in many applications such as gas sensors, pressure sensors, electrical shielding, thermistors and anti-static devices. Generally, these materials are composed of an insulating polymer matrix in which conductive fillers are homogeneously dispersed.

The versatility of polymer composites is of special interest above all to the gas sensor industry where arrays of these materials, called “electronic noses”, are used to identify and analyse gases and odours [1-3]. The response mechanism of polymer composite sensors have been described on the basis of percolation theory. When a conducting polymer composite gas sensor is exposed to vapour-phase analytes, it permeates into the polymer causing it to swell. This vapour-induced expansion of the polymer reduces the volume fraction of the conducting particles in the material and hence reduces the number of connected pathways for charge carriers. As a result, the conductivity of the composite decreases and the electrical resistance increases [4,5].

Several reports have studied PMMA/Carbon composites based gas sensors with the advantages of selective sensitivity and fast response [6,7]. Their sensing device performances to different gases, the reproducibility and the stability are closely related to the properties of carbon itself (particle size, surface area, surface conductivity) [6], to thin film morphology and to the dispersion of the carbon within the polymer [8-10].

Dong and co-workers have investigated the effects of processing conditions on electrical conductivity of the PMMA/CB composites and their electric responsivity against organic solvent vapors [11]. The experimental results showed that molecular weight of the polymer matrix, the carbon black content and the composite film thickness greatly influence the response behaviors of the composites in solvent vapors. The sensing performance of the composites is found to be related to the microstructure of the materials, which provides possibilities for further improve the overall properties of the composites by altering the processing parameters.

However, little studies of the influence of process of fabrication on the electrical characteristics of these composite gas sensors appear in the literature.

In this paper we discuss the process of fabrication of PMMA/CB composites and in particular the effect of solvent of preparation and of the three geometry of devices (D1, D2 or D3) on the final sensing elements performances.

Poly (methyl methacrylate) (PMMA) was chosen for our experimental because it is soluble in different organic solvents and it is often used in application requiring good resistance to weather or moisture.

2. Experimental section

The Carbon Black, used in preparing the composites, was carbon black pearls 2000, a furnace black material with 1500 m^2/g specific surface area, with 12nm average particle size and with 150g/l density. The solvents (tetrahydrofuran (THF), 1,1,1,3,3,3 hexafluoro-2-propanol (HFIP), acetone and ethanol) and the PMMA polymer were supplied from Sigma-Aldrich and were used as received.

2.1. *Preparation of PMMA/CB composite gas sensor*

The sensing polymer composites were prepared by dissolving the poly (methyl-methacrylate) (PMMA) (80mg) in THF and in HFIP and dispersing the filler (20mg), in the solution (0.4-0.5 wt% of polymer), by ultrasonic bath for 90 minute.

The devices have been fabricated by dropping the suspensions, by means of a microlitre pipette, onto a clean alumina (Al_2O_3) substrates (dimensions 5mm x

5mm). The electrical contacts geometry has been realized by a lift-off photolithographic process followed by e-beam Au evaporation [7]. The average volume of composite solution deposited was 3μl and the resulting film was allowed to dry in air.

A schematic structure of the multifingered substrates is shown in fig.1. Both the distance and the width of finger (l) are 0.1mm (D1), 0.2mm (D2) and 0.4 mm (D3).

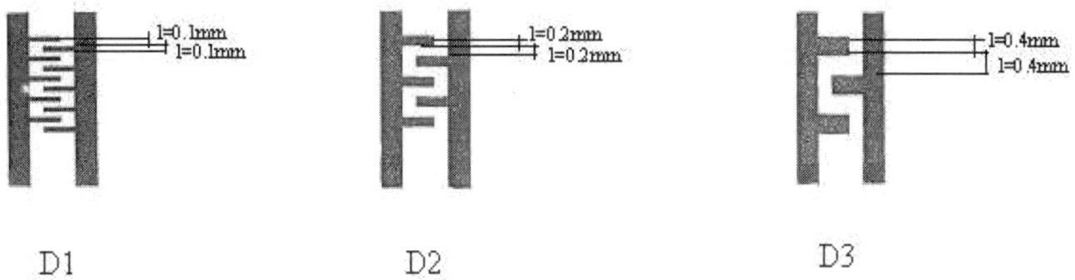

Fig.1: Schematic images of the three multifingered Al_2O_3 substrates (5mm x 5mm) fabricated by means of photolithographic technique.

In table 1 is reported the series of samples prepared.

Table 1. Condition used for the preparation of three type devices (Dx) and their base resistance (R_0).

Composite	Solvents preparation	Dx	R_0
PMMA/CB	THF	D1	2.0
	THF	D2	0.85
	THF	D3	4.58
	HFIP	D1	3.0
	HFIP	D2	5.0
	HFIP	D3	20.0

The single sensors were mounted on a commercial electronic TO8 case by soldering gold wires to the pins (fig.2).

Fig.2. Sensor device (5mm x 5mm) mounted on a commercial electronic case

2.2. *Measurement*

The electrical resistance of each device was measured using a Wavetek Meterman 35XP multimeter.
The particle size distribution of filler in suspensions and dispersions stability has been studied by Dynamic Laser Scattering (DLS) using a HPPS 3.1 (Malvern Instruments).
The gas sensing characteristics of the PMMA/CB composite were investigated in controlled atmosphere using a gas sensor characterization system (GSCS) described in previous paper [6]. The steel chamber with a volume of about 400cm^3 was placed in a climatic chamber to keep constant the temperature (20° C) and it was always under normal pressure during the measurements. An organic vapour diluted with nitrogen gas was passed in a chamber where the sensor was placed. The diluent was nitrogen gas that flowed through a condenser in the climatic chamber placed after a bubbler filled by the liquid of interest and kept at constant temperature bath (25°C). The gas concentration was controlled by the mass flow controller. The GSCS is interfaced with a computer for data acquisition, storage and analysis.
The devices were exposed to specific concentrations of acetone and ethanol at room temperature and pressure and were biased with an 0.1V constant voltage. Acetone and ethanol were chosen as gas analyte because prior experience have shown that easily measurable responses can be obtained with these gases.
The current-voltage characteristics of the samples were measured at room temperature using a Semi Automatic PA 200 SUSS Instruments.

3. Results and discussion

3.1. *Morphology of PMMA/CB composite devices*

Photographic images of PMMA/CB composites illustrate that the CB particles are quite homogeneously dispersed in the PMMA over alumina substrates when the HFIP is used as solvent of preparation of composite (see fig.3A).

(A)

(B)

Fig.3: Photos of the devices type D1 prepared with: A) HIFP; B) THF.

The better filler distribution in composites prepared with HFIP is also in good agreement with DLS analysis (tab.2) that present lower values of the size of filler in suspension and of the polydispersity index (PDI).

Table 2. Results of DLS analysis made on CB suspensions obtained adding filler in PMMA solution.

Solvent preparation	Average size of filler in suspension (nm)	Polydispersity index
THF	311	0.372
HFIP	260	0.171

Instead, using THF solvent, the CB particles are randomly dispersed within matrix and CB aggregation is observed (fig.3B). These CB aggregates increase the conductivity of devices and decrease their electrical resistance, as shown in fig. 4.

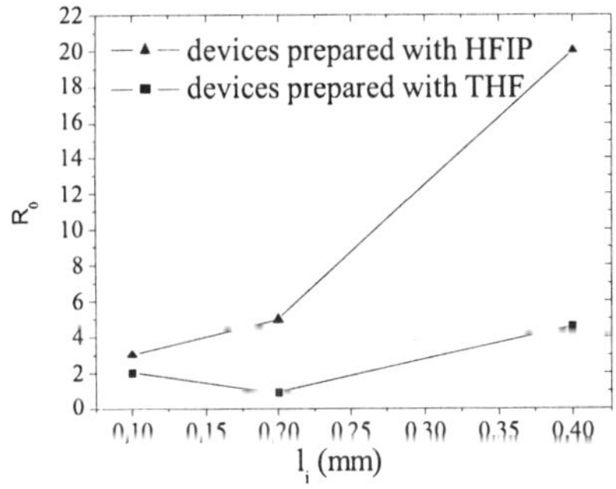

Fig.4. Variation of the base resistance (R_0) of PMMA/CB composites with finger length (l_i).

Due to the inhomogeneous nature of the composites prepared with THF, the electrical resistance doesn't depend on the geometry of the fingers, to difference of the devices prepared with HFIP.

3.2. *Current (I) – Voltage (V) Characteristics*

In fig.5 are illustrated the I/V characteristics of PMMA/CB nanocomposites type D1 prepared with THF and HFIP evidenced by curves 1 and 2, respectively. The I/V plots are linear, showing a positive slope, indicating ohmic behaviour.

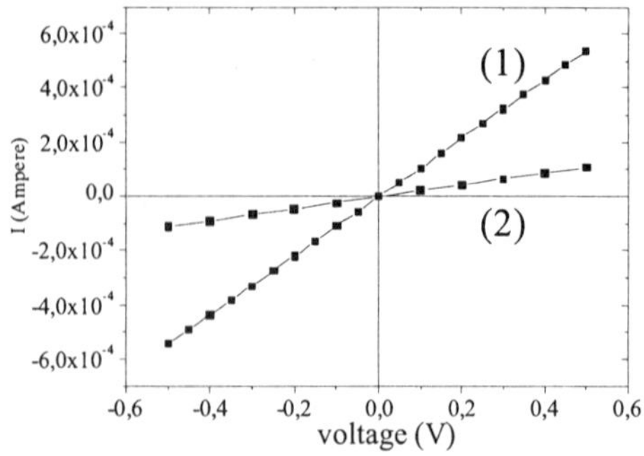

Fig.5. Current (I)-Voltage (V) characteristics of devices type D1 prepared with THF (1) and HFIP (2).

This variation of slope may be due to the difference of electrical resistance of two samples. The sample prepared with THF (fig.5, curve (1)) has a value of greater resistance because it is formed from large CB aggregates deposited on the device

3.3. *Device responses to acetone and ethanol vapors*

The fabricated PMMA/CB devices were exposed to acetone and ethanol vapors at room temperature and pressure and in the specific concentration range ppm. The responses of the devices to the two gas were represented by the current intensity change-defined as I_0-I/I_0 where I_0 is the sensor baseline current intensity and I is the sensor current intensity upon its exposure to solvent vapours.
Fig.6 shows the responses obtained for the polymer composite devices type D1, D2 and D3 prepared with THF and HFIP.

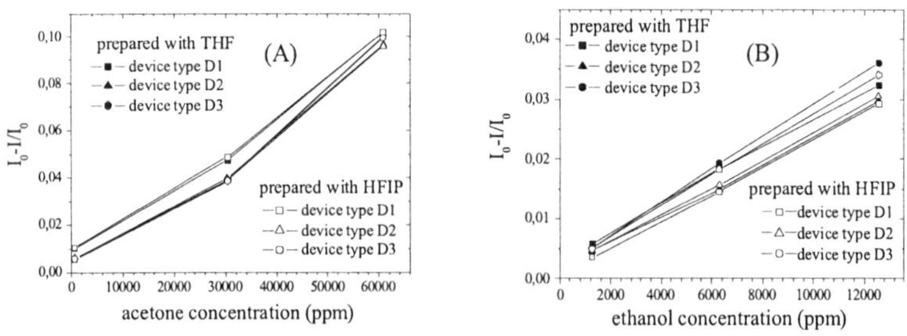

Fig.6. Change in current intensity of the PMMA/CB devices type D1, D2 and D3 prepared with THF and HFIP with increasing acetone (A) and ethanol (B) vapors.

It can be seen from fig.6 that the devices show an intensity change in the presence of the vapors and the sensors type D1, D2 and D3 prepared with THF and HFIP have linear and similar responses.

Also it is clearly apparent that the highest sensor response is to acetone vapor. This highest sensitivity can be probably attributed to the higher chemical affinity of less polar molecule as acetone towards PMMA.

Just as example, the responses to acetone vapor of the device type D1, prepared with HFIP, with time are shown in fig.7.

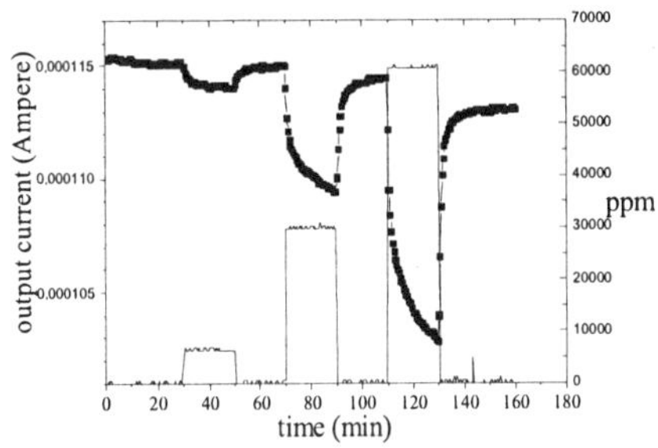

Fig.7: Responses of PMMA/CB sensors on exposure to acetone vapor.

3.4. *Stability of electrical properties.*

The resistance of the nanocomposites prepared with THF and HFIP was measured at increasing intervals of time up to about 90 days. Fig.8 shows the results obtained where it can be seen that the resistance of devices is stable for long periods of time.

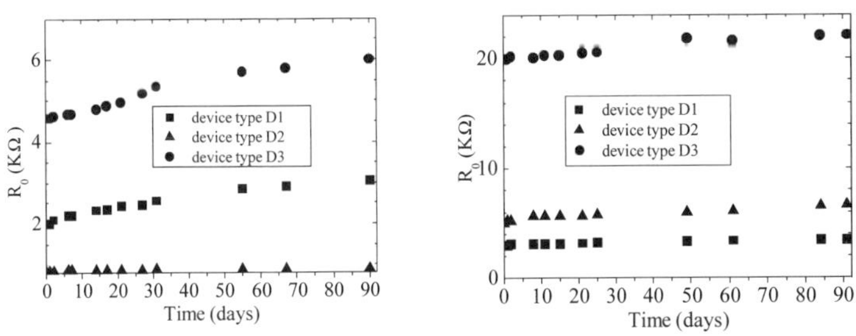

Fig.8: Variation of the conducting properties of devices prepared with: A) THF and B) HFIP in about 90 days.

In particular, the devices prepared with HFIP seem to be a little more stable with respect to those prepared with THF. This can be explained considering the volatility of the solvent of preparation. The HFIP, solvent lower boiling (bp=59° C), evaporates easily from the devices while the THF (bp=67°C) evaporates in the time.

4. Conclusion

We have prepared and characterized sensing devices to acetone and ethanol vapours based on composite films PMMA/CB, changing the solvent of preparation (THF and HFIP) and the devices geometry. The choice of solvent HFIP in the preparation of polymer/carbon black composite materials can significantly improve their morphology, as also confirmed by DLS analysis, but no their capabilities as gas sensor. In fact the thermodynamic responses of sensors are similar and linear. The same is also independent from the geometry of device

References

1. M. C. Lonergan, E. J. Severin., B.J. Doleman, S.A. Beaber, R.H. Grubbs, N.S. Lewis, *Chem Mater*, 8, 2298, (1996).
2. E.J. Severin, B. J. Doleman, *Analytical Chemistry, 72, 658,* (2000)
3. M. C. Burl, B. J. Doleman, A. Scaffer, N. S. Lewis, *Sensors and Actuators B*, 72, 149-159, (2001).
4. E.S. Tillman, B. J. Doleman, N. S. Lewis, *Sensors and Actuators B*, 96, 329, (2003).
5. X. M. Dong, R. W. Fu, Zhang, M. Z. Rong, *Carbon*, 42, 2551-2559, (2004).
6. L.Quercia, F.Loffredo, B.Alfano, V.La Ferrara, G.Di Francia, *Sensors and Actuators B* 100, 22, (2004).
7. B.Philip, J. K.Abraham, A. Chandrasekhar, V. K. Varadan, *Smart Mater. Struct.*, 12, 935, (2003).
8. F. Bueche, *J. Appl. Phys.*, 44, 532, (1973).
9. R. M. Scarisbrick, *J. Appl. D.*, 6, 2098, (1973)
10. L. Quercia, F.Loffredo, G. Di Francia, *Sensors and Actuators B*, 109, 153, (2005).
11. X. M. Dong, R. W.Fu, M. Q. Zhang., Z. P.Qin, B. Zhang, M. Z Rong., *Polymer Journal* 35, 1003, (2003).

DEVELOPMENT OF QMB SENSORS BASED ON IRON PORPHYRINS FOR CARBON MONOXIDE DETECTION: A FEASIBILITY STUDY

E. MAZZONE*, M. MASTROIANNI**, C. DI NATALE*, R. PAOLESSE**,
M.I. PISTELLI***,
F. SINTONI***, A. D'AMICO*
**Department of Electronic Engineering, University of Rome "Tor Vergata",*
Via del Politecnico 1, 00133 Rome, Italy
***Department of Chemistry, University of Rome "Tor Vergata",*
Via della Ricerca Scientifica 1, 00133 Rome, Italy
****Centro Sviluppo Materiali SpA,*
Via di Castel Romano 100, 00128 Rome, Italy

In the frame of a collaboration between the University of Rome "Tor Vergata" (Sensors and Microsystems group) and Centro Sviluppo Materiali SpA, a feasibilty study about the application of a transduction technique based on mass changes for carbon monoxide detection has been carried out. Sensors were developed using TSMR resonators as transduction elements and iron-porphyrins as chemically interactive materials. The results presented in this paper show that the Quartz Micro Balance (QMB) technology is promising for obtaining selective and highly sensitive CO sensors.

1. Introduction

Air quality monitoring in domestic and industrial environments focuses on toxic and flammable volatile compounds and among them CO is one of the most critical.

The importance of accurate carbon monoxide detection resides in its high toxicity even at low concentrations: CO binds haemoglobin which becomes no more available for oxygen transportation [1].

Effects of carbon monoxide poisoning range from headaches and dizziness to paresis, convulsions and unconsciousness until myocardial ischemia, atrial fibrillation, pneumonia, pulmonary aedema, acute renal failure and changes in perception of the visual and auditory systems [2].

Many different limits of exposure to carbon monoxide are recommended in various operating situations, the National Institute for Occupational Safety and Health (NIOSH, USA) has established a REL (*recommended exposure limit*) for CO of 35ppm (40mg/m^3) as an 8-hour time-weighted average.

Moreover CO is colourless and odourless, thus cannot be detected by the biological senses, while a concentration of 1% may cause death in few minutes.
For these reasons, reliability can be considered the crucial feature of a system designed for revealing the presence of such hazardous gas.
Existing systems for CO monitoring are based on the following working principles [3,4]:

- changes in ionic current in electrochemical cells;
- infrared absorption in optical sensor;
- conductivity changes in metal oxide based devices;
- catalytic combustion in pellistor sensors.

2. Sensors Development

The aim of this study was to investigate a new arrangement for sensing carbon monoxide with high sensitivity and resolution, using low cost devices at room temperature.
The working principle is based on the biological mechanism which is responsible for oxygen transportation within the blood: in normal conditions, haemoglobin reversibly binds oxygen molecules, but its affininity for CO is much higher (around 250 times).
This characteristic, which is responsible for carbon monoxide poisoning, has been used as transduction mechanism for the development of carbon monoxide sensors.
As the structure of haemoglobin contains four Fe(II)-porphyrins, the chemically interactive materials developed for this work were a set of different metalloporphyrins, whose sensitivity towards CO was previously reported. [5,6]
The compounds have been synthesized and characterized at the Chemistry Department of Tor Vergata, and their compositions are listed below:

1) $Fe(III)[T(3,5(OH)_2P)P](Cl)$
2) $Fe(III)[T(2,5(OH)_2P)P](Cl)$
3) $Fe(II)[T(2,5(OPiv)_2P)P](NMeIm)$

Porphyrins 1 and 2 present eight hydroxy groups on the meso phenyl substituents, in order to achieve a higher sensitivity for polar molecules and to increase the porosity of the sensing layer.
Figure 1 reports the structures of membranes 1) and 2).

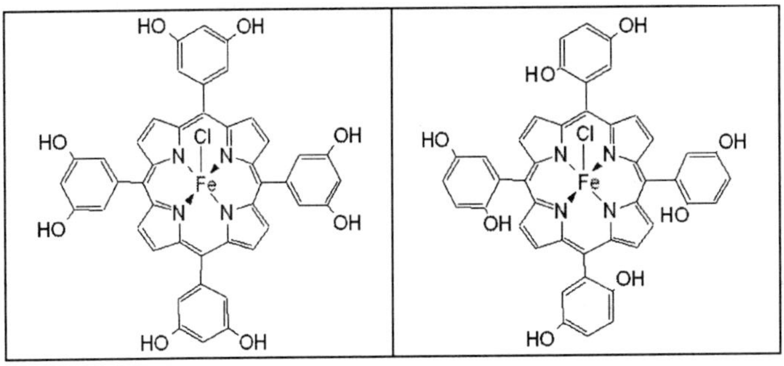

Figure 1: Structure of Fe(III)[T(3,5(OH)$_2$P)P](Cl) (left) and of Fe(III)[T(2,5(OH)$_2$P)P](Cl) (right).

For CIM 3, eight pivaloyl groups have been introduced in the molecular skeleton: this functionalization avoids the formation of the μ-oxo-dimer due to the oxidation of the metal centre Fe(II) to Fe(III), which is responsible for the loss of CO sensitivity. Stabilization of Fe(II) has been also obtained using the N-methylimidazole ligand, which reversibly binds CO [7].
The structure of Fe(II)[T(2,5(OPiv)$_2$P)P](NMeIm) is reported in Figure 2.
Sensitive materials have been deposed on TSMR devices by spray casting technique.

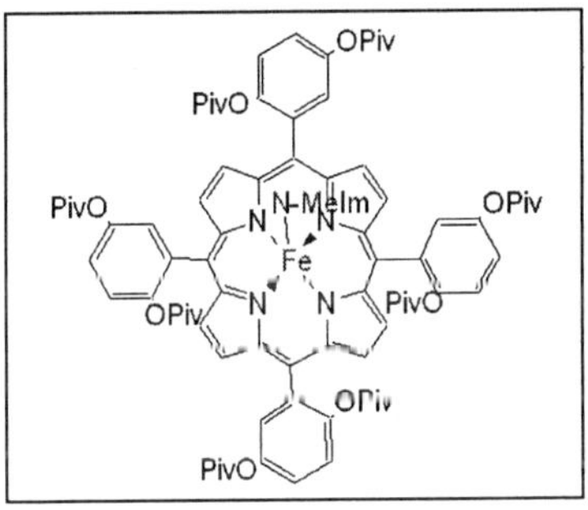

Figure 2: Structure of the Fe(II)[T(2,5(OPiv)$_2$P)P](NMeIm) porphyrin

3. Experimental

Sensors were mounted inside the chamber of an Electronic Nose: two devices for each CIM plus two empty resonators as references.
A Nafion filter has been used in order to obtain a controlled humidity level, to simulate a real working environment.

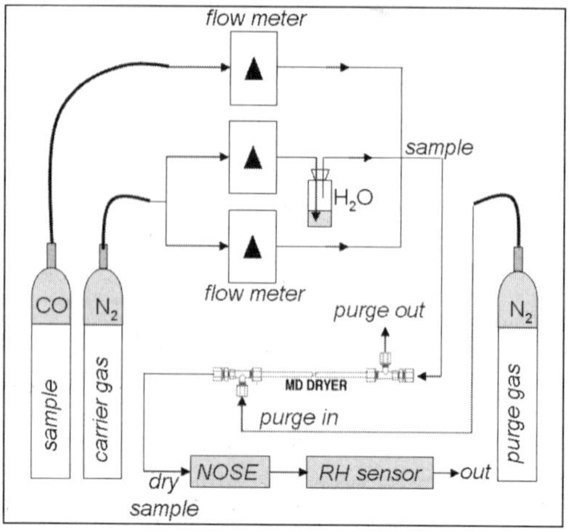

Figure 3: Schematic of the test bench assembled at the Chemistry Dept. of Tor Vergata University.

The measurement protocol was the following: 15 minutes of exposure to CO and 45 minutes of cleaning in nitrogen.
Several experimental sessions have been carried out:

1) Analysis of the performances of the Nafion filter;
2) CO measures without the Nafion filter;
3) CO measures with the Nafion filter;
4) CO, CO plus water vapour and water vapour measures with Nafion;
5) CO, CO plus water vapour and water vapour without Nafion.

Figure 4 reports a typical response curve, in this case from the devices covered with membrane 1).

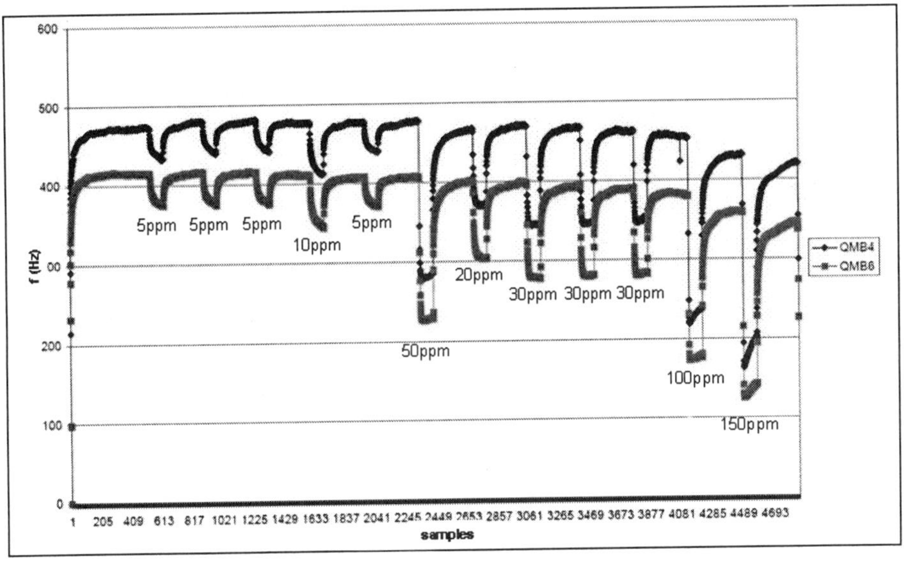

Figure 4: Response of Fe(III)[T(3,5(OH)$_2$P)P](Cl) to several CO concentrations.

Similar behaviours were shown by the other sensitive materials, demonstrating the applicability of such compounds for CO detection.

Measuremets at different concentration levels allowed to distinguish the sensitivity levels of each membrane: porphyrin 1) was the more sensitive, porphyrin 2) at an intermediate level, while porphyrin 3) showed the lowest sensitivity.

From the data, the following sensor parameters were estimated [8]:

- sensitivity = 6Hz/ppm @10ppm;
- resolution = 800 ppb.

Pratically, the limit of detection (LOD) was 5ppm, limited by the set-up, while the theoretical LOD is lower than 1ppm.

Moreover, tests performed at different humidity levels proved that this set-up, after calibration, may be used for CO detection in real samples, containing water vapour.

During the experimental phase, a critical effect was observed, consisting in a certain degree of membrane poisoning, mainly due to the measurement protocol. This phenomenon, if uncontrolled, can lead to a complete loss of sensitivity towards the target gas.

Further field trials demonstrated that at least two counteraction methods can be adopted in order to effectively reduce this effect: first of all the optimization of the protocol consisting in a shorter measurement cycle followed by a longer cleaning phase, then the use of a thermal treatment to enhance CO desorption. This last technique could eventually be applied using TSMR resonators with integrated heaters (these devices are already under development by our group).

Conclusions

Within this work, a new approach based on the application of QMB sensors for CO detection, has been proposed.
The transduction principle exploits the high affinity level of haemoglobin towards carbon monoxide, which is sometimes responsible for blood poisoning. On this assumption, several porphyrins, analogous to those which are part of the haemoglobin structure, have been synthesized and used as sensitive materials. Results demonstrated that the QMB technology, supported by proper poisoning avoidance strategies, is promising for developing fairly sensitive CO sensors.

References

1. C.R. Henry, D. Satran, B. Lindgren, C. Adkinson, C.I. Nicholson, T.D. Henry: *Myocardial Injury and Longterm Mortality Following Moderate to Severe Carbon Monoxide Poisoning*, **Journal of American Medical Association,** 2006, 295, 398-402.
2. A.L. Ilano, T.A. Raffin: *Management of carbon monoxide poisoning*, **Chest**, 1990, 97, 165-169.
3. N. Yamazoe, G. Sakai, K. Shimanoe: *Oxide semiconductor gas sensors,* **Catalysis Surveys from Asia**, 2003, 7(1), 63-75.
4. N. Barsan, U. Weimar: *Understanding the fundamental principles of metal oxide based gas sensors; the example of CO sensing with SnO2 sensors in the presence of humidity*, **J. Phys. Condens. Matter**, 2003, 15, R813-R83.
5. C. Di Natale, R. Paolesse, A. Macagnano, E. Martinelli, A. D'Amico: *Porphyrin Based Sensors for Carbon Monoxide Detection*, **Proceedings of Eurosensors XVI**, Prague (Czech Rep.), 16 -18 Sept. 2002.
6. C. Di Natale, R. Paolesse, A. Macagnano, A. D'Amico: *Development of porphyrins based sensors to measure the biological damage of carbon monoxide exposure*, **Proceedings of IEEE Sensors Conference 2003**, Toronto (Canada), Oct. 2003.
7. J.P. Collman, R.R. Gagne, C. Reed, T.R. Halbert, G. Lang, W.T. Robinson: *Picket fence Porphyrins, Synthetic models for oxygen binding hemoproteins,* **J. Am. Chem. Soc.**, 1975, *97(6)*, 1427-1439.
8. A. D'Amico, C. Di Natale: *A contribution on some basic definitions of sensor properties*, **IEEE Sensors Journal**, 7 (2001), 183-190.

STUDIES ON CHIRAL SELF-ORGANISATION OF AMPHIPHILIC PORPHYRIN DERIVATIVES. COMPARISON BETWEEN MORPHOLOGY IN SOLUTION AND IN SOLID STATE[1]

D. MONTI,* M. STEFANELLI, M. VENANZI, M. CARBONE, R. PAOLESSE

Dipartimento di Scienze e Tecnologie Chimiche, Università di Roma, "Tor Vergata", Via della Ricerca Scientifica 1, 00133 Rome, Italy

C. DI NATALE, A. D'AMICO

Dipartimento di Ingegneria Elettronica, Università di Roma "Tor Vergata", Via della Ricerca Scientifica 1, 00133 Rome, Italy

S. TURCHINI, M. GIRASOLE, G. POMPEO

ISM-CNR Area Ricerca Tor Vergata, Via del Fosso del Cavaliere, 00133 Rome, Italy

Spontaneous deposition of aggregates of a tetraphenylporphyrin derivative, possessing a cationic chiral appended functionality, straightforwardly occurs from aqueous solutions. The thickness of the layers depends on the initial macrocycle concentration. Combined UV-Vis, and CD spectroscopy studies and AFM morphological characterisation reveal that these layered films feature a peculiar fractal-type morphology. This aspect can be of great importance for the construction of solid state chemical sensors featuring interesting stereoselective properties.

1. Introduction

The quest to morphologically defined molecular assemblies onto a solid surface, with controlled size and shape, is a mandatory issue in material science [1] and for the development of molecule-based nano devices and electronics [2]. Porphyrin derivatives, and related congeners, are molecular tools of great interest in optical [3], and chemical sensors applications [4]. One of the most compelling characteristics influencing the properties of porphyrin films is, undoubtedly, the geometry and orientation of the deposited mesostructure. In particular, the realisation of supramolecular species featuring supramolecular

* This work is supported by FISR-SAIA 2001.

*Corresponding Author, E-Mail: monti@stc.uniroma2.it

chiraliy is of great, and almost yet unexploited, issue [5a]. In particular, the evolution of chiral porphyrin systems, at a supramolecular level, mainly relies i) on the aggregation of achiral porphyrin platforms on chiral templates, such as DNAs, RNA, and other natural or synthetic polymers [5b], ii) chiral ligand coordination [5c], and iii) spontaneous symmetry breaking upon self-association at air/water interface, or by directional stirring [5d,e]. The chirality of those above described films [5d,e], however, occurs in a random fashion, circular dichroism of opposite sign being obtained from different batches. We found that the presence of a cationic functionality on the periphery of a porphyrin macrocycle promote the formation of regular supramolecular structures by cationic-π drive self-recognition [6]. These architectures layer spontaneously onto hydrophobic surfaces to give well-defined mesoscopically oriented materials. These systems can be successfully employed in the detection of water pollutants, such as Hg or Pb salts [7]. By extending this protocol to chirally functionalised cationic derivatives we wish to obtained a molecular chiral synthon (i.e. molecular "chiron", $1H_2$, Fig. 1) for the construction of porphyrin mesostructures featuring supramolecular chirality [8]. In our case, the chirality of the layered porphyrin material should strictly follow that of the solution, being driven by the molecular information brought by the stereogenic centre. In this work we wish to present our new important results in the achievement, and the characterisation, of some solid-state supramolecular chiral assembly, by Atomic Force Microscopy (AFM).

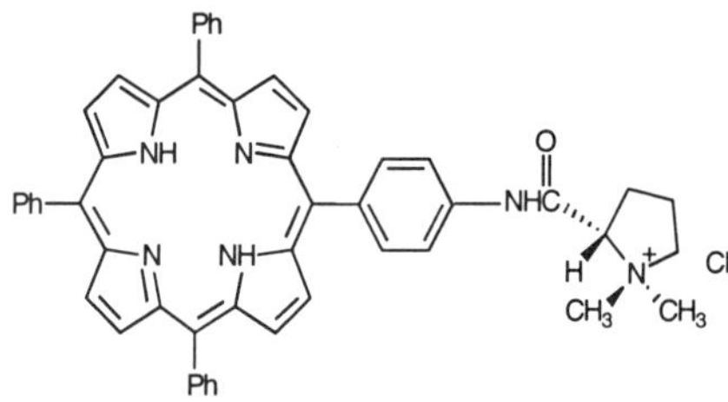

Fig. 1. Molecular structure of chiral porphyrin derivative $1H_2$ studied in the work.

2. Experimental

2.1 General. Porphyrin derivatives were prepared according to procedures previously reported [8]. UV-visible spectra were performed on a Perkin Elmer λ18 Spectrophotometer. CD spectra were performed on a JASCO J-600 spectropolarimeter.

2.2 *Preparation of samples*

Porphyrin aqueous solutions were prepared as follows. Proper aliquots of a porphyrin millimolar stock solution in ethanol (15÷150 μL), where added to a

1.0 mL of ethanol in an 8 mL glass vial. To this solution 3.0 mL of water were then added and the resulting solution vigorously shaken. A 3 mL portion was then transferred in a quartz couvette and the relative UV-Visible spectra acquired. This procedure ensures a 75:25 H_2O/EtOH (v:v) solvent composition, with a final porphyrin concentration spanning in the range of 1.5 to 9.0 x 10^{-5} M. Microscope glass slides (Forlab®; Carlo Erba, cut in 25 x 8 mm^2 pieces) were used as film substrates [4b]. Substrate depositions were achieved by dipping the glass slides into the appropriate volume of freshly prepared porphyrin solution ($2.0x10^{-6}$÷$5.0x10^{-5}$ M; H_2O/EtOH 9:1, v:v; 48 h; 35 °C). The slides were washed (H_2O/EtOH 9:1, v:v) and dried (N_2 stream, 40 °C). Reproducibility, in terms of layer thickness and homogeneity (UV-visible spectroscopy check), is within 5-10%. Molecular deposition on raw, untreated glass slides gave less reproducible results.

2.3. *AFM Studies.* The surface topography of porphyrin films was investigated in air by contact mode atomic force microscopy (AFM) using low tip force. The observations were performed by using an EXPLORER–VEECO system with a Si_3N_4 pyramidal tip having a curvature radius lower than 50 nm. For each sample, different images were recorded from different positions in order to check the lateral uniformity of the films. This allowed us to calculate the value of the root mean square of the surface average roughness, RMS, of the films.

The samples were prepared by deposition from a solution 5 μM of porphyrin on a chemically prepared H-terminated wafer of Si(100). Two different preparation procedures were followed: i) a 5 hours-long bath of the H-Si in the porphyrin solution followed by spin-coating, ii) a 5 hours-long bath of the H-Si in the porphyrin solution followed by a complete evaporation of the solvent. The two samples were probed by the AFM. The corresponding images are presented in section 3.

3. Results and Discussion

UV-visible spectra of **$1H_2$** (μM concentration) in H_2O/EtOH (75:25 v:v) solvent mixture showed the gradual formation of porphyrin J-type aggregates with time (Figure 2). Kinetic experiments of **$1H_2$** aggregation in H_2O/EtOH 75:25 (v:v), carried out by UV-Visible spectroscopy, gave interesting insights on the mode of

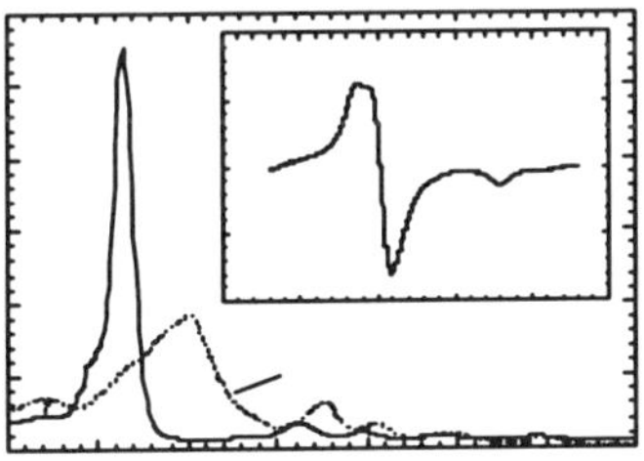

Fig. 2 UV-vis spectra of **$1H_2$** ($1.0x10^{-5}$M) in a) EtOH; b) H_2O/EtOH (75:25). Inset: Corresponding CD spectra.

aggregation. The extinction *vs* time profiles show an autocatalytic fractal-type behaviour, that can be excellently fitted by a non-conventional equation earlier proposed by Pasternack and others (Equation 1) [9]. Circular Dichroism (CD) spectroscopy reveals that the aggregation promotes the formation of large chiral suprastructures, steered by the presence of the appended chiral functionality (Fig. 2-Inset).

A concomitant molecular deposition spontaneously occurs with time, from the bulk solution. Dipping silanised microscope slides into an aqueous solution of **1H₂** ($2.x10^{-6} \div 7.5x10^{-5}$ M; H_2O/EtOH 75:25, v:v) resulted in the deposition of porphyrin aggregates, as evidenced by the typical yellow coloration of the glass surfaces. Complete deposition (UV-vis check of the Soret bands) is achieved within 24 hrs, depending on the concentration of the starting solutions. The extent of deposition depends on the initial concentration of the solution (Figure 3). The porphyrin films showed good mechanical stability. Intense rinsing and wipe-drying process causes no loss of material. Remarkably the layered porphyrin films features elements of chirality (Fig. 3-inset), showing coupled CD bands, indicating that the chirality of the supramolecular aggregates are retained during the deposition process. A plot of absorption values *vs* initial porphyrin concentration shows a linear dependence up to ca 2 x 10^{-5} M (Figure 3, Inset), likely indicating an uniform and regular deposition of layers. Atomic force microscopy technique was used in order to characterize the **1H₂** films.

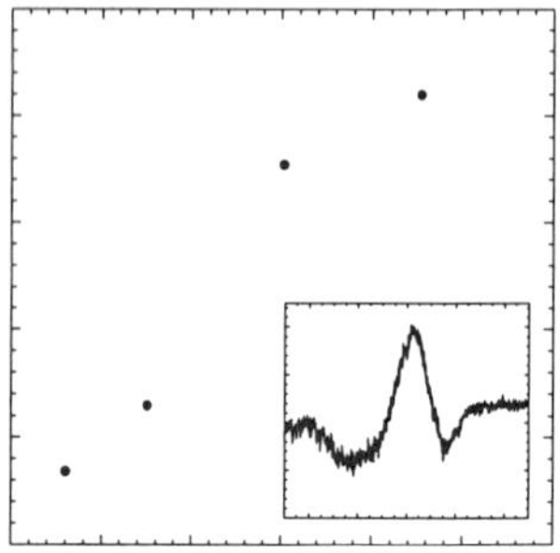

Fig. 3. **1H₂** OD (λ 431 nm) at different initial bulk concentrations. Inset: CD spectra of **1H₂** layer (5 x 10^{-5} M) on glass.

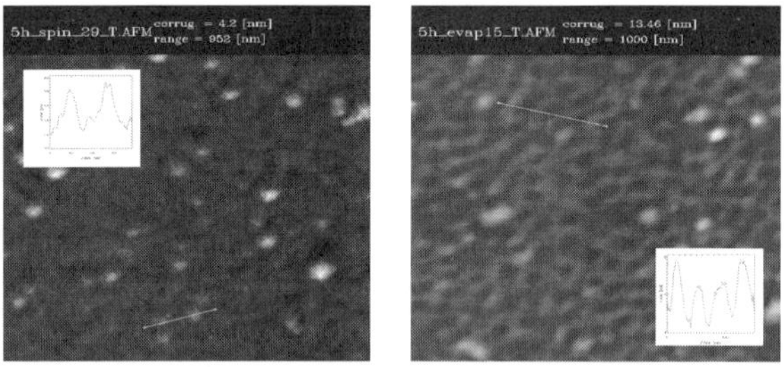

Fig. 4. AFM picture of layered **1H₂** (1.0x10-5 M) on H-Si(100) (a) after spin coating; (b) after evaporation.

Typical images are reported in Figure 4. Interestingly, the AFM picture showed peculiar morphology of the layered aggregates. The images show a regular corrugation of the surface, with alternating "humps" and "canals". The estimated RMS roughness value, measured into the "canals" is about 1.8 ± 0.1 nm. The typical distances between "canals" are 80 ± 10 nm. These values are compatible with the suprachiral organization of the aggregates in α-helix conformation, were 1.8 nm is the diameter of the helix. Furthermore, assuming the interplanar distance of porphyrin helix of 1 nm, we obtain an average aggregation of 80 porphyrin units. These values are common for the two preparations. These results are of importance in that is the first direct evidence of formation of "fractal-like" porphyrin aggregates, corroborating previous reported experimental evidences, mainly based on kinetic investigation [8a, 9]. Moreover, this morphology guarantees a high surface/volume ratio and consequently, a high number of absorption sites available for vapor-surface interaction, adding further values for the achievement of supramolecular systems for stereoselective sensor features [10].

4. Conclusion

In summary, the work presents a facile and straightforward way to obtain *chemosensitive* porphyrin film with ordered morphology. Application of the reported protocol to chiral porphyrin derivatives, for the achievement of enantioselective sensors is currently under investigation in our laboratories.

References

1. C.F.J. Faul, M. Antonietti, *Adv. Mater.* **2003**, *15*, 673.
2. C. Johachim, J.K. Ginzewski, A. Aviram, *Nature*, *408*, 541.
3. M.P. Debreczeny, W.A. Svcc, M.R. Wasielewski, *Science*, **1996**, 584.
4. a) Paolesse, R., Mandoj, F., Macagnano, A., Di Natale, C., in "*Encyclopedia of Nanoscience and Nanotechnology*". Vol 10. Nawa , H.S., Ed.; American Science Publ. **2003**. b) Paolesse, R.; Monti, D.; La Monica L.; Venanzi, M.; Froiio, A.; Nardis, S.; Di Natale, C.; Martinelli, E.; D'Amico, A. *Chem. Eur. J.* **2002**, *8*, 2476.
5. Supramolecular Chirality. *Topics in Current Chemistry*, 265. M. Crego-Calama and D.N. Reinhoudt Eds. Springer: Berlin, Heidelberg, New York. 2006. b) R.F. Pasternack, *Chirality*, **2003**, *15*, 329, and references therein. c) V.V. Borovkow, G.A. Hembury, and Y. Inoue, *Acc. Chem. Res.* **2004**, *37*, 449d) P. Chen, X. Ma, P. Duan, and M. Liu, *ChemPhysChem* **2006**, *7*, 2419. e) R. Rubires, J.-A. Farrera and J.M. Ribó, *Chem. Eur. J.* **2001**, *7*, 436.

4. D. Monti, M. Venanzi, M. Russo, G. Bussetti, C. Goletti, Montalti, N. Zaccheroni, L. Prodi, R. Rella, M.G. Manera, C. Di Natale and R. Paolesse, *New J. Chem.* **2004**, *28*, 1123.
5. L.S. Dolci, E. Marzocchi, M. Montalti, L. Prodi, D. Monti, C. Di Natale, A. D'Amico, R. Paolesse, *Biosens. Bioel.* **2006**, *22*, 399.
6. a) D. Monti, M. Venanzi, G. Mancini, C. Di Natale, R. Paolesse, *Chem. Commun.* **2005**, 2471. b) Monti, D.; Cantonetti, V.; Venanzi, M.; Ceccacci, F.; Bombelli, C.; Mancini, G. *J. Chem. Soc. Chem. Commun.* **2004**, 972.
7. R.F. Pasternack, E.J. Gibbs, D. Bruzewicz, D. Stezart, K.S. Engstrom, *J. Am. Chem. Soc.* **2002**, *124*, 3533.
8. R. Paolesse, D. Monti, L. La Monica, M. Venanzi, A. Froiio, S. Nardis, C. Di Natale, E. Martinelli, A. D'Amico, *Chem. Eur. J.* **2002**, *8*, 2476.

Production and characterization of new Fe(TPP)Cl porphyrin films with improved optical gas sensing capabilities

M. Tonezzer, A. Quaranta, G. Della Mea
Dipartimento di Ingegneria dei Materiali e Tecnologie Industriali, Università di Trento, via Mesiano 77, 38050 Povo(TN), Italy
tonezzer@lnl.infn.it

G. Maggioni, R. Milan, S. Carturan
INFN-LNL Viale dell'Università, 2- 35020 Legnaro, Italy

Summary

Thin porphyrin assemblies, widely used as sensing materials in different kinds of transducers, are usually produced through chemical deposition techniques. In this work novel 5,10,15,20 meso-tetraphenyl porphyrin (Fe(TPP)Cl) films have been deposited by means of a new physical technique named glow discharge induced sublimation (GDS). A preliminary morphological characterization has been performed by means of scanning electron microscopy (SEM): SEM images and highlight the great surface roughness and the high purity of GDS films with respect to films deposited by means of conventionally used procedures (e.g. spin coating (SPIN)) ones. Optical sensing measurements, performed in differently concentrated ethyl alcohol (EtOH) atmospheres, highlight that GDS samples yield higher response intensities than SPIN films, very fast responses and complete recovery.

1 Introduction

Porphyrins and their derivatives are a class of naturally occurring compounds having important biological representative including hemes, chlorophyll and Vitamins B12 among several others [1,2]. Porphyrin compounds have attracted much attention as promising materials because of their interesting properties in several research fields such as optoelectronic molecular semiconductors and non-linear optics: in particular, their physical and chemical properties make porphyrins promising sensing materials for monitoring different gaseous species, from highly reactive analytes, such as NO_x and HCl, to volatile organic compounds (VOCs) [3,4].

In order to be exploited as sensing materials, porphyrins compounds are usually deposited as thin solid films through chemical techniques (solvent casting, Langmuir-Blodgett, spin coating) [5]. Fewer efforts have been carried out to produce thin solid films by means of physical technique in spite of the fact that these techniques assure several advantages as great reproducibility, more uniformity and stricter control of the film thickness in comparison with standard chemical techniques. Moreover physical techniques, producing thin solid films without using any extraneous compound, allow to produce samples characterized by very high purity. The material purity is expected to play a very important role in gas sensing field because the traces of retained solvent can partially hinder the interaction between the film and the analyte molecules. Recent optical measurements demonstrated that vacuum evaporated thin porphyrin films exhibit higher responses towards analyte molecules than the chemically deposited ones [6]. This paper reports the employment of a new physical technique developed at LNL and named Glow-Discharge-induced Sublimation (GDS) for the deposition of thin iron chloride 5,10,15,20 *meso*-tetraphenyl porphyrin (Fe(TPP)Cl). GDS technique has recently showed to place as an attractive deposition method for the production of thin organic films for gas sensing applications allowing to produce samples characterized both by high purity and especially by extraordinarily large surface area to bulk ratios [7,8,9]. Thin Fe(TPP)Cl films were also deposited through the standard spin coating (SPIN) technique for comparison. In order to study and compare the gas sensing capabilities of the samples, optical measurements in presence of differently concentrated atmospheres of ethyl alcohol (EtOH) were performed.

Experimental

Fe(TPP)Cl compound is commercially available (Acros Organics) and was used directly without some further process. This material was deposited in the form of thin films onto clean silicon and quartz substrates by means of GDS technique: GDS technique is based on the use of a weakly ionized glow discharge produced in a standard radio frequency magnetron sputtering equipment: low energy ($E < 1$keV) He ions impinge on solid organic powder leading to the sublimation of the organic molecules and to their condensation onto the substrate. GDS set-up consisted of a vacuum chamber evacuated by a turbomolecular pump to a base pressure of 10^{-4} Pa. The glow discharge was sustained by a 1-in. cylindrical magnetron sputtering source connected to a radio frequency power generator (600 W, 13.56 MHz) through a matching box. The Fe(TPP)Cl powder was put on the surface of an aluminum target and placed on the sputtering source. The glow discharge feed gas was helium (99.9999 %), whose pressure inside the chamber was measured through a capacitance gauge: typical values of rf power, target DC self-bias, and working pressure were in the ranges $10 \div 20$ W, $-20 \div 300$ V, and 20.0 ± 0.1 Pa, respectively. Fe(TPP)Cl thin films were also deposited through spin-coating (0,1% wt of porphyrin powder in chloroform at 800 rpm for 30 sec) technique.

The surface morphology of the samples was investigated with a Philips XL-30 scanning electron microscope (SEM) and the optical absorbance spectra were acquired utilizing an experimental apparatus in which the measure chamber, placed inside a Spectrophotometer V-570 Jasco, is connected to two mass flow controllers suited to assure an accurate real-time control of the vapour concentrations into the chamber during the spectra collections. This apparatus allows to record both dynamic behaviours at a fixed absorbance wavelength and complete absorbance spectra in different vapour concentration atmospheres.

Results and discussion

Figure 1 shows SEM pictures of SPIN and GDS samples of Fe(TPP)Cl compounds. These images point out the different surface morphologies depending on the different deposition techniques. SPIN sample shows, in spite of the typical presence of craters formed by the solvent evaporation, a flat surface; on the contrary and GDS by flat morphology. On the contrary, GDS films shows a very rough surface. It is worth to note that the surface roughness represents a very important characteristic in gas sensing field because a large surface area improves the interaction between film and analyte molecules and therefore the sensing capability of the films.

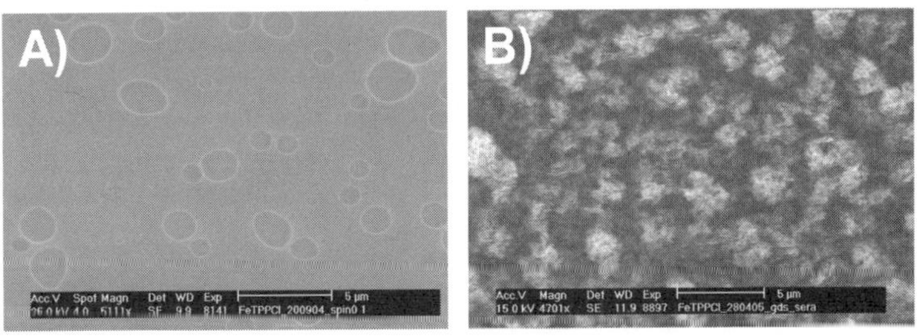

***Fig.1**: SEM images of SPIN (A) and GDS (B) films of Fe(TPP)Cl.*

The optical sensing capabilites of the SPIN and GDS films of Fe(TPP)Cl were investigated and compared by exposing the samples to differently concentrated ethanol-vapour-containing nitrogen fluxes. In particular, the evolution of the absorbance at 410 nm was monitored versus time: the optical responses of the samples were then calculated as the ratio between the absorbance variation $\Delta A = (A_t - A_0)$ and the signal noise N, where A_t and A_0 are the absorbance values during exposure to EtOH-containing and pure nitrogen atmospheres, respectively. Figure 2 depicts the $\Delta A/N$ evolution at 410 nm versus time, for the GDS and SPIN samples of Fe(TPP)Cl upon exposure to 2300 ppm EtOH. As can be seen, the behaviours of both the samples are characterized by a fast signal increase within a few seconds, followed by a slower increase until saturation values are reached. As the ETOH vapour stream is switched off, a dry nitrogen flux is activated and after a few hundreds of seconds the original signal strengths are restored.

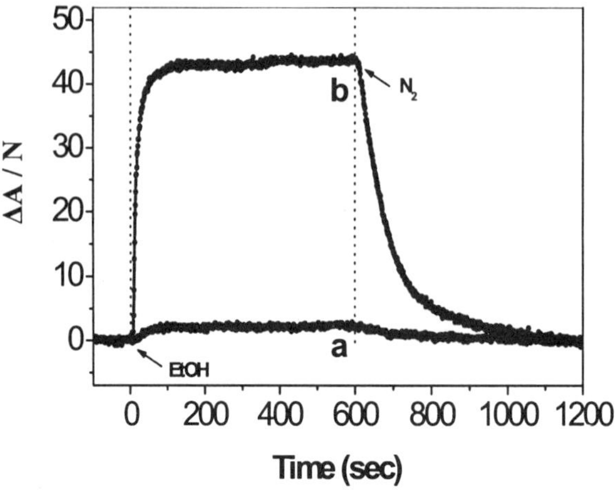

***Fig.2**: Single exposure/recovery cycle for a GDS film (b) of Fe(TPP)Cl in comparison to the responses of SPIN film (a) when exposed to 2300 ppm EtOH ($T = T_{AMB}$; $\lambda = 410$ nm)*

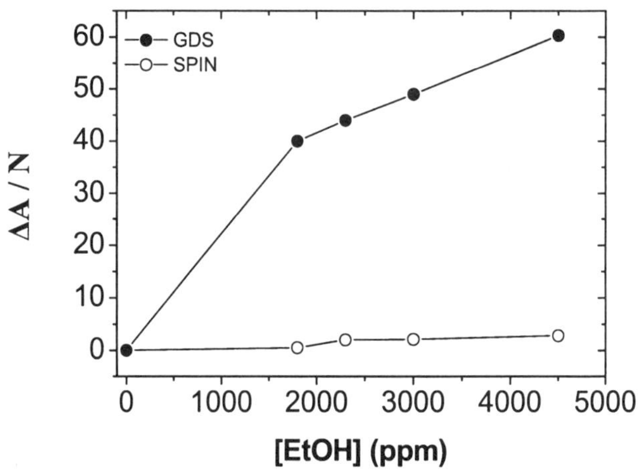

***Fig.3**: Calibration curves of SPIN and GDS films of Fe(TPP)Cl upon exposures to EtOH atmospheres over the range 1100 – 4500 ppm*

It is possible to point out that the optical behaviour of the GDS and SPIN films show a complete and fast recovery: taking into account that all the measurements have been performed at room temperature, these experimental results highlight the really interesting sensing capabilities of the two tested Fe(TPP)Cl samples.

The effect of the analyte concentration on the response of GDS and SPIN films of Fe(TPP)Cl samples was investigated with exposure-recovery cycles obtained over the ETOH 1100-4500 ppm range: in order to quantify the dependence of the response intensity on the concentration, the calibration curves of the samples are plotted in Fig. 3.

The comparison between the two curves shows that GDS samples are characterized by optical responses much more intense than SPIN ones at all the tested concentrations. Moreover, the more leaning slope of the calibration curve of GDS sample in comparison with that of SPIN one highlights the higher sensitivity of the GDS sample than the SPIN one.

From these preliminary tests it is possible to hold that GDS places itself as a very promising technique for the deposition of thin Fe(TPP)Cl solid film for gas sensing applications allowing to produce samples characterized by very fast responses, complete recovery and extraordinarily larger intensity and sensitivity in comparison with the samples deposited by means of the standard chemical deposition techniques.

Acknowledgements

This research was financially supported by the Fifth Commission of Istituto Nazionale di Fisica Nucleare (DEGIMON project).

References

[1] D. Dolphin (Ed.), *The Porphyrins*, VI(a), VII(b), Academic Press, New York, **1978**
[2] D. Dolphine (Ed.), *The porphyrins*, Vol. VII part B, Academic Press, New York, **1978**
[3] N.A. Rakow et al. - Nature, **406** (2000) 710
[4] O. Worsfold et al., *Colloids Surf. A* 198-200 (2002) 859-867
[5] A. D'Amico et al., Sens. Actuators B **65** (2000) 209
[6] M. Tonezzer et al., Sens. Actuators B **122** (2007) 620-626
[7] G. Maggioni et al., Chem. Mater., **17** (2005) 1895
[8] G. Maggioni et al., Surf. Coat. Technol., **200** (2005) 476
[9] M. Tonezzer et al., Sens. Actuators B **65** (2000) 613-619

NANOWIRES OF SEMICONDUCTING METAL-OXIDES AND THEIR GAS-SENSING PROPERTIES

C. BARATTO, E. COMINI, M. FERRONI, G. FAGLIA, A. VOMIERO, G. SBERVEGLIERI

Sensor Laboratory -CNR-INFM, Department of Physics and Chemistry for Materials Engineering, University of Brescia, ITALY

Summary

In this work, structural, morphological, electrical and optical properties of oxide nanowires, employed as active material for gas sensing devices, have been investigated. The reactive metal oxides were deposited by vapour phase condensation process from pure oxide powders on the polished alumina and silicon substrates. For the electrical characterization platinum was deposited as electrodes and a platinum meander was prepared to control the operating temperature. The samples were investigated by electron microscopy techniques. The sensing performance of the prepared devices has been investigated as a function of temperature and concentration of carbon monoxide and nitrogen dioxide.

1. Introduction

The prevention of the health risks due to the atmospheric pollution is one of the most important challenges in the countries with high industrial production.

Single metal oxides and their various combinations have shown particular sensitivity to gases such as carbon oxide and nitrogen dioxide. Bulk and surface properties of oxide film, such as band gap, electronic structure, position of Fermi level are determining its interaction with gases. Therefore, a contemplated choice of the metal oxide enables to obtain a sensing material where one of the above mentioned properties can be emphasized and employed for the gas monitoring.

Moreover, the necessity of direct pollution monitoring frequently requires that the devices works in critical conditions, such as high temperature and corrosive environment and metal oxide are robust materials that can be used in dangerous environments.

Quasi 1-dimensional nanostructures of semiconducting materials such as tin and zinc oxides are presently investigated in order to produce a novel class of sensing devices. These fascinating nanostructures may be produced in several unusual arrangements such as nanowires, nanorods, or nanocombs, resulting in great potential for fundamental study and application [1-5].

Quasi one-dimensional nanostructures are expected to show a variety of quantum confinement effects as two of their dimensions are comparable to the wavelengths of the electronic wave- function. The effect of reactive gases on the electrical conductance of semiconducting metal-oxides nanowires is the basic mechanism for the development of nanodevices with enhanced sensing performance.

The evaporation-condensation (EC) process, with Vapor-Phase (VP) or Vapor-Liquid-Phase (VLP) growth mechanism, is highly promising as deposition technique. Such a preparation method consists of thermally-driven evaporation of bulk metal oxides followed by condensation in controlled thermodynamic conditions. Un-catalyzed and size-controlled growth of nanowires, down to 10 nm in size, has been achieved by EC process.

Microstructural investigation of the nanowires and the analysis of the electrical properties is a key- feature for profound knowledge of growth process and for the exploitation of the unique properties of 1-D nanostructures. Electron microscopy techniques have been carried out in order to determine the degree of chemical homogeneity and crystalline arrangement.

The electrical and optical properties of nanowires have been investigated. Nanowires have been tested towards different gases, such as CO, NO_2. Photoluminescence (PL) spectroscopy was performed over a wide temperature and frequency range for the purpose of investigating the behavior of PL spectrum in presence of reactive gases.

2. Experimental

2.1. Nanowires preparation

The growth of nano-wires was carried out in a horizontal tube furnace by vapor transport process, oxide powder was preferred to the metal powder since it allows a better control of the evaporation process that happens at higher temperatures. The deposition conditions have been tailored in order to promote formation of 1D-nanostructures, through changing evaporation temperature, carrier gas and its flux.

Oxide powder was placed in an alumina tube inserted in a horizontal tube furnace, with temperature, pressure and evaporation time controlled. Evaporation took place by heating the source to 1370 °C. Mass transport was obtained using an Ar flow (75 sccm) at pressure of 1×10^4 Pa. Furnace heating from RT to 1370 °C took 1.5 hours. During furnace heating and cooling a reverse Ar gas flow (from the substrates to the powder) was applied, to avoid uncontrolled mass deposition under transient conditions. Once desired temperature was reached, deposition process lasted 5-30 minutes. Nanowires grew in a colder zone of the furnace, at a temperature between 400 °C and 500 °C on 2×2 mm^2 alumina substrates.

Table1. Deposition parameters.

Nanowires	Decomposition temperature (°C)	Substrate temperature (°C)	Duration (min)	Pressure (mbar)	catalyst
SnO_2	1370	450-500	30	100	Pt
ZnO	1370	250-400	30	100	Au

2.2. Electron microscopy

Scanning and transmission electron microscopy (SEM and TEM) has been carried out in order to determine the morphology, chemical homogeneity, and crystalline assembling. SEM analysis was carried out by field-emission LEO 1525 microscope, equipped with In-Lens detector for secondary-electrons imaging. The SEM was operated in the 2-3 kV accelerating voltage range to prevent the insulating substrate from electrostatic charging, thus allowing observation of uncoated specimens.

TEM investigation was carried out with a FEI Tecnai F20 microscope equipped with field emission source and capable of 0.17 nm point resolution. Nanowires were removed from the substrate and dispersed over a thin holey carbon grid.

2.3 Electrical gas testing

Alumina substrates, after the nanowires deposition, were equipped with platinum contacts on the top of the nanowires and a heating platinum meander on the backside, deposited by sputtering, in order to perform electrical measurements on ZnO nanowires as a function of the operating temperature.

Electrical characterization was carried out by volt-amperometric technique; the sensors were biased by 1 V and nanowires resistance was measured by a picoammeter. The flow-through technique was used to test the gas-sensing properties. A constant flux of synthetic air of 0.3 l/min was the gas carrier, into which the desired concentration of pollutants—dispersed in synthetic air—was mixed. All the measurements were executed in a temperature-stabilized sealed chamber at 20°C under controlled humidity.
For the gas sensing test, the operating temperature was varied between room temperature and 500 °C.

2.4. Optical gas testing

For optical characterization, devices were placed inside a test chamber kept at T=20°C, featuring a large quartz window granting easy access for optical measurements.

A constant flux of synthetic air equal to 0.3 l/min, mixed with the desired amount of gaseous species coming from certified bottles, flowed through the chamber at atmospheric pressure.

Photo luminescence (PL) measurements were performed using either a He–Cd laser at 325 nm or UV LED at 330 nm as light source. PL spectra were acquired perpendicular to the sample surface using a single spectrograph and a Peltier cooled CCD camera. Laser power impinging on the sample must be kept low ($20mW/mm^2$) to avoid PL signal drift.

3. Results and Discussion

3.1. Electron microscopy

Scanning and transmission electron microscopy (SEM and TEM) has been carried out in order to determine the degree of homogeneity and crystalline arrangement. High-resolution TEM imaging is useful for investigation of the termination of the nanowire lateral sides and apex. Electron diffraction (ED) and analysis of zero-order and higher-order Laue-zones allows precise determination of unit cell and space group.

Incoherent imaging techniques such as STEM with the High-Angle-Annular-Dark-field detector (STEM–HAADF) were used for the investigation of the shape of the nanowires and impurities and local variations in the composition (Z-contrast).

Figure 1. SnO_2 nanowires produced by the evaporation-condensation process.

As shown in Figure 1, SnO_2 nanowires prepared featured a very high aspect ratio as the length exceeds several microns and the width is about 50 nm. High resolution TEM and electron diffraction showed that the wire is single crystalline, with atomically-sharp termination of lateral sides. Measured Bragg reflections and the whole symmetry of the ED pattern agree with the cassiterite tetragonal SnO_2 phase (P42/mnm - SG 136). The nanowires grow along to the [100] direction. The length and flexibility allows nano-manipulation for removal and positioning over Si-based substrates for functional characterization.

As both composition and phase can be considered uniform for the crystalline SnO_2 nanowire; STEM-HAADF imaging directly visualizes variations in the projected thickness. From the line profile of the HAADF signal across the nanowires, a regular polyhedron-shaped section has been inferred, as it is expected for a wire with crystalline habit and crystal facets as lateral sides.

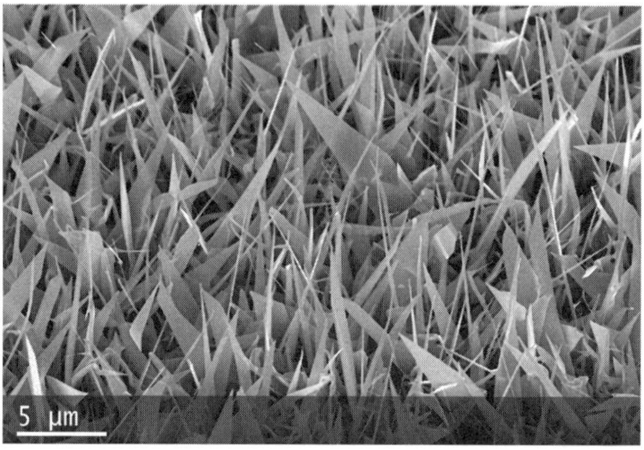

Figure 2. SEM image of nanosized ZnO nanowires

The size and shape of the obtained ZnO nanowires is however sensitive to the condensation condition. Figure 2 shows that ZnO nanowires with different morphology can be produced. The capability to control the lateral dimension of the nanowires will allow the systematic investigation of size reduction effects on the electrical and gas sensing behaviour of ZnO nanowires.

TEM observation confirms the regular crystalline arrangement for the nanowires. No evidences of extended crystal defects governing the growth have been recorded. The high-resolution TEM image and the corresponding digital diffractogram indicate that the lattice symmetry is hexagonal and that the longitudinal direction of growth is parallel to the *c*-axis of the crystal unit cell.

3.2. Electrical characterization

The semiconductor sensing properties are based on reactions between the semiconductor and the gases in the surrounding atmosphere, which cause a change in the semiconductor's resistance due to charge transfer between the adsorbate and the adsorbent.

We have studied the electrical properties of the different nanostructures focusing our attention to nanowire structures with reduced lateral size. In fact it has already been proved that decreasing the grain size of metal oxide thick film gas sensors there is a huge improvement in gas sensing performances, and this remains valid also for nanocrystals [6].

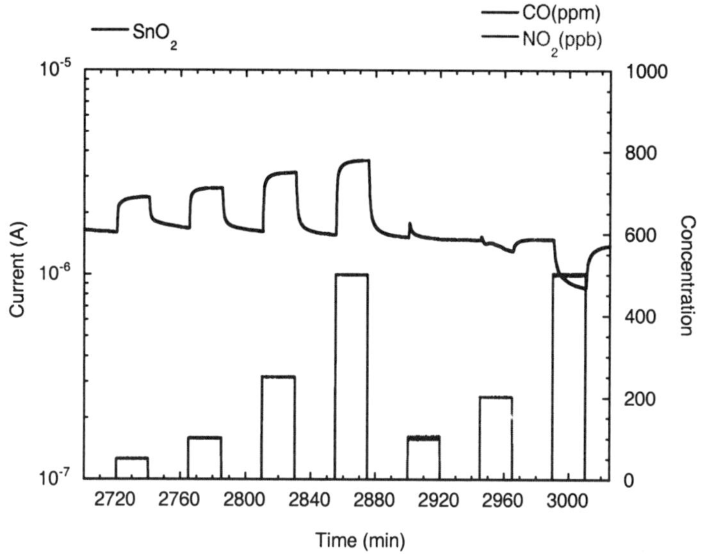

Figure 3. Dynamic response of tin oxide nanowires towards 50, 100, 250 and 500ppm concentration square pulses of carbon monoxide and to 100, 200 and 500-ppb concentration squared pulses of nitrogen dioxide at 300°C working temperature with 40%RH.

Figure 3 illustrates the dynamic response of tin oxide nanowires to 50, 100, 250 and 500ppm concentration square pulses of carbon monoxide and to 100, 200 and 500-ppb concentration squared pulses of nitrogen dioxide at 300°C working temperature with 40% relative humidity (RH). The current flowing through the sample increases for carbon monoxide and decreases for nitrogen dioxide, as was expected for an n-type semiconductor, because of injection and extractions of electrons from conduction band. The response and recovery times are of the order of minutes and the recover of the air conductance value after the gas tests is complete proving that the gas reaction between nitrogen dioxide and tin oxide nanowires surface is reversible at this operating temperature.

Similar behavior is obtained for zinc oxide nanowire based devices.

3.3. Optical characterization

It has been demonstrated before that NO_2 reversibly quenches ZnO PL spectrum, affecting in the same way UV (excitonic) peak and visible peak [15].

In order to test the feasibility of an integrated sensor we have compared results obtained with laser with the ones obtained with UV LED whose power is 10 times smaller than that of laser.

The rationale of this choice is to develop an integrated optical sensor with low cost light source. The use of a commercial LED was possible since there is no need of high power: moreover too high power can results in an unstable signal.

Figure 4 reports the comparison between the signal taken with laser and the one taken with LED as light source. The spectra were not corrected for the response of the optical setup, and the two visible peaks are due to response of the ccd camera. In fact the spectrum is a broad band in the green region. The two spectra have the same shape with a slight difference in intensity of the peak at higher wavelength. An additional peak in the UV region of the spectrum (348 nm) was ascribed to spurious peak coming from LED. So we have demonstrated the feasibility of using a low power low cost source as excitation light for PL measurements.

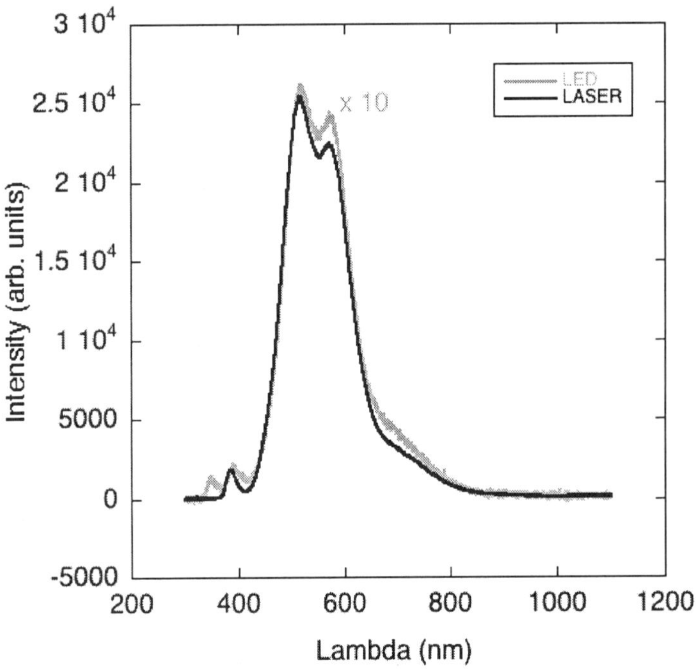

Figure 4. Comparison of the ZnO PL spectrum taken with 330nm LED (gray line, multiplied by 10), and the spectrum taken with He-Ne laser.

In order to perform tests with NO_2 as gas sensor we selected the visible area of the spectrum and acquired a spectrum every 5 seconds. Then we calculated the area under the peak and plotted it against time, during NO_2 introduction. Figure 5 reports the results obtained with 2 ppm of NO_2. If we define the relative variation as $(A_{air}-A_{gas})/A_{gas}$ x100 as is defined for electrical measurements, we calculate a percentage relative variation of 8.6% . Response and recovery times are of the order of minutes (210 s response time, 500s recovery time), this is a good result if we keep in mind that the sensor is working at room temperature and that the recovery to initial value is complete.

The interference of reducing gases like humidity and ethanol has been evaluated. It is not completely absent: strong variation of relative humidity - from dry condition to 70% RH – causes an increase in PL signal equal to 2%, while the introduction of 1000 ppm of ethanol in dry air produces a PL increase of 1.9%.

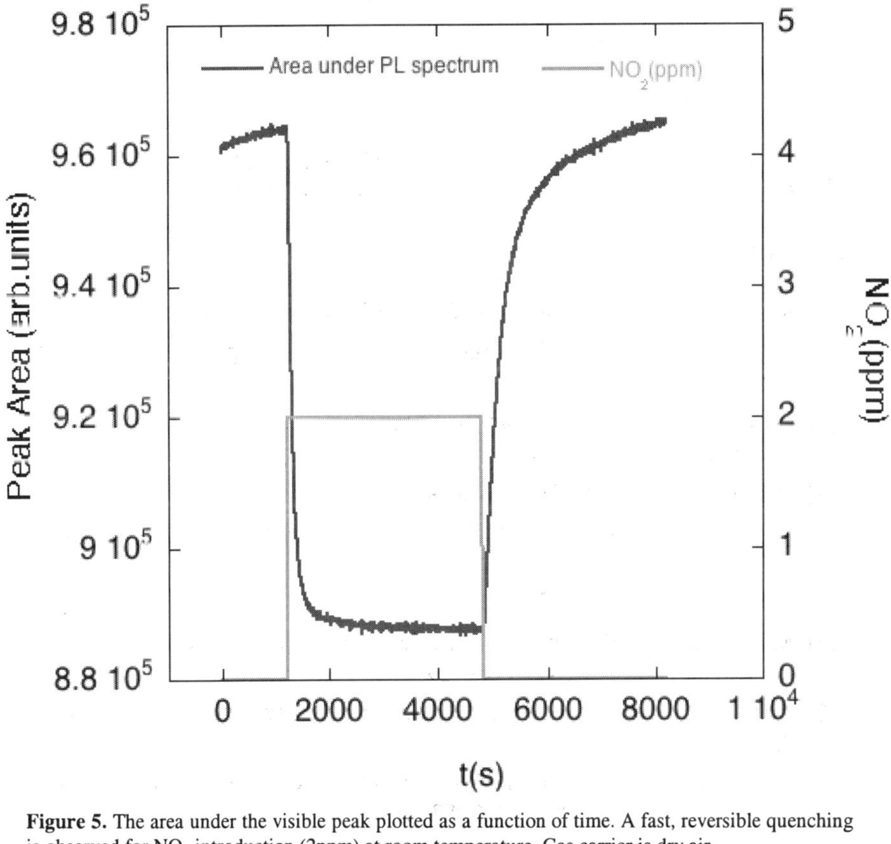

Figure 5. The area under the visible peak plotted as a function of time. A fast, reversible quenching is observed for NO_2 introduction (2ppm) at room temperature. Gas carrier is dry air.

4. Conclusions

A vapor phase deposition procedure has been described. The optimized deposition conditions for the production of uniformly dispersed tin and zinc oxide nanowires, several micrometers and the tents of nanometers wide, were reported. Electron microscopy showed the high degree of crystallinity of these structures and their atomically sharp terminations.

Electrical measurements highlight remarkable response. As far as optical properties are concerned, ZnO nanowires PL spectra are reversibly quenched by the introduction of few ppm of NO_2 at RT. NO_2 adsorbed over the surface introduces non-radiative recombination paths that modify the emission properties of the nanowires. We have also demonstrated that is feasible to use the quenching effect to realize an optical sensor for NO_2, with a LED excitation source. A small interference effect due to humidity and ethanol on optical parameters has been recorded too. Optical and electrical responses are different for kinetics and temperature dependence, but the results obtained make metal oxide nanowires promising for gas sensing applications even at work temperature.

Acknowledgments

This work has been founded by the European Strep project "Nano-structured solid-state gas sensors with superior performance" (NANOS4) no.: 001528.

References

[1] Kong, X.Y Ding, Y. Yang, R.S. Wang, Z.L. Science, 303, 1348 (2004)

[2] Comini, E. Faglia, G. Sberveglieri, G. Pan, Z. Wang, Z. L. App. Phys. Lett., 81, 1869 (2002)

[3] Zhang, D. Li, C. Liu, X. Han, S. Tang, T. Zhou, C. Proceedings of IEEE NANO 2003, San Francisco, 08/03

[4] Hahm J. Lieber, C.M. Nano Lett., 4, 51 (2004)

[5] Li, Z. Chen, Y. Li, X. Kamins, T. I. Nauka, K. Williams, R. S. Nano Lett., 4, 245 (2004)

[6] Yamazoe, N. Sens. Act. B, 3, 147 (1991)

[7] M. J. Madou, S. R. Morrison: Chemical Sensing with Solid State Devices, (Academic Press Inc., NewYork, 1988)

[8] Mina, Y. Tuller, L. Palzer, S. Wollenstein, J. Bottner H. Sens. Act. B 93, 435 (2003)

[9] K. Vanhausden, W. Warren, C. H. Seager, D. R. Tallant, J. A. Voigt, and B. E. Gnade, J.Appl. Phys 79, 7983 (1996)

[10] J. W. Hu and Y. Bando, Appl. Phys. Lett. 82, 1401 (2003).

[11] B. Lin, Z. Fu, and Y. Jia, Appl. Phys. Lett. 79, 943 (2001).

[12] S. A. Studenikin and M. Cocivera, J. Appl. Phys. 91, 5060 (2002).

[13] B. D. Yao, Y. F. Chan, and N. Wang, Appl. Phys. Lett. 81, 757 (2002).

[14] G. Faglia, C. Baratto, and G. Sberveglieri, M. Zha and A. Zappettini Appl. Phys. Lett. 86, 011923 (2005)

[15] C. Baratto, S. Bianchi, E. Comini, G. Faglia, M. Ferroni and G. Sberveglieri, " Metal oxide nanowires for optical gas sensing" Proc. of SPIE Vol. 6474 64741E-1 (2007)].

NANOSTRUCTURED CONJUGATED POLYMERS APPLIED TO SENSORS

IOLE VENDITTI, M. V. RUSSO
Dipartimento di Chimica, Università La Sapienza di Roma,
P.le A. Moro5 , 00189 Roma, Italy

A. BEARZOTTI, A. MACAGNANO
I.M.M.-C.N.R., Research Area Tor Vergata,
Via del Fosso del Cavaliere 100, 00133 Rome, Italy

Preliminary investigations on humidity sensors based on nanostructured PPA (I_2 doped or pristine) and its copolymer P(PA/HEMA) membranes showed fast and reproducible current intensity variations in the range of relative humidity RH 5–90%. The sensor geometry allows its application in miniature devices and the nanostructure enhances the response to humidity variations with respect to previous studies based on amorphous I_2 doped phenylacetylene . Measurements performed on quartz crystal micro balances have demonstrated that these nano structured polymers can be used also devices able to detect mass variations with a good stability and sensitivity.

1. Introduction

Nano-beads of conjugated polymer (PPA = polyphenylacetylene) and co-polymer (P(PA/HEMA) = (poly[phenylacetylene-(co-2-hydroxyethyl methacrylate)])), were prepared by modified emulsion polymerisation technique; the nanostructured morphology enhances the properties of this materials mainly because of their increase of surface/volume ratio. Preliminary investigations on humidity sensors based on nanostructured PPA (I_2 doped or pristine) and P(PA/HEMA) membranes (see Figure 1a,b) showed fast and reproducible current intensity variations in the range of relative humidity RH 5–90%. The sensor geometry (Figure 2) allows its application in miniature devices and the nanostructure enhances the response to humidity variations with respect to previous studies based on amorphous I_2 doped polyphenylacetylene [1] . We observed a sharp increase of the current intensity for a humidity sensor based on nanostructured PPA with respect to data of non nanostructured I_2 doped PPA reported in literature [1]: the best results are obtained for nanostructured PPA that shows an current intensity variation in the range 10-11-10-9A vs RH=5-90%. The reproducibility of the response of

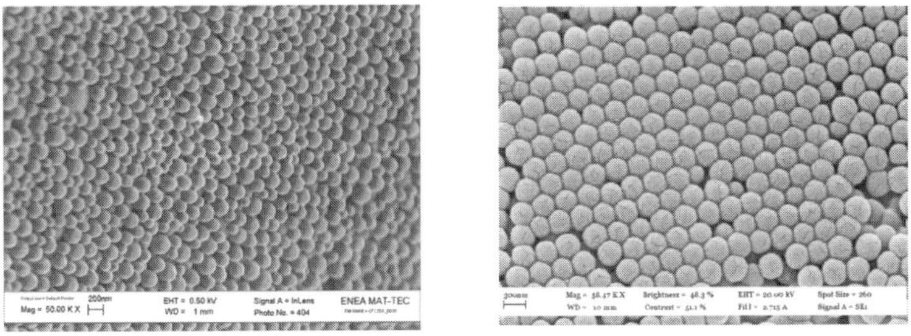

Figure 1. SEM (scanning electron microscopy) image of nanostructured PPA (a) and P(PA/HEMA) (b).

nanostructured PPA and P(PA/HEMA) sensors was tested; current intensity values were reproduced when recorded in different cycles of measurements, in which the same RH percentages (5%) were reached in the test chamber (see Figure 3). The sensors were also tested in subsequent cycles, and in these

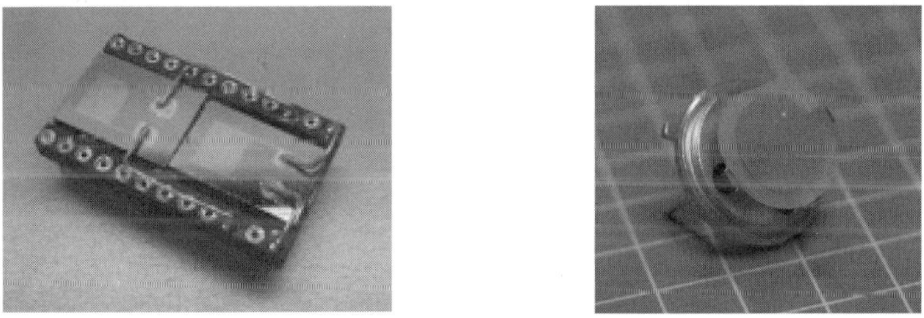

Figure 2. Interdigitated substrate and QCM with nanostructured P(PA/HEMA) film

preliminary studies no significantly aging or degradation of its performance for working times as long as three days were observed.

The mechanism of interaction between PPA, P(PA/HEMA) and H_2O molecules may be interpreted on the basis of previous studies [2]. The nanostructured polymer surface covalently links OH groups deriving from H_2O dissociation and the OH groups are able to promote the hopping of H_3O^+ from a site to another, thus suggesting a ionic conduction. Electrochemical dissociation of water occurs upon applying a voltage, so as suggested by investigations performed on similar materials [3,4]; for these materials which behave as a sponge for humidity, the interpretation of the electrical response is based on diffusion processes of water in the vapour phase inside the nanostructured material.

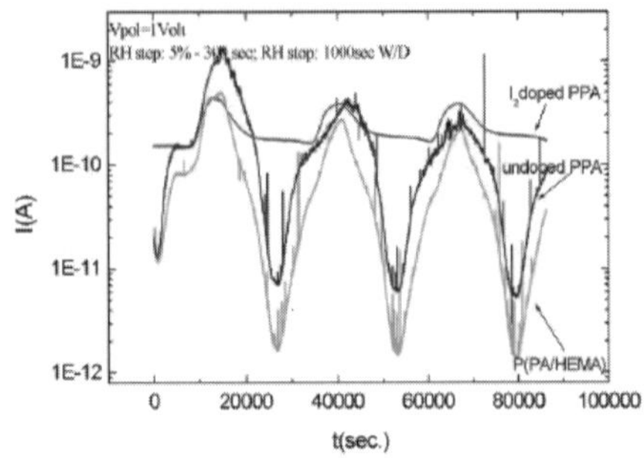

Figure 3. The time response and stability of resistive sensor based on nanostructured PPA (I2 doped or undoped) and PPA/HEMA

Moreover, measures in alternated current point out the important ionic contribute to the conduction. Preliminary results show that the most interesting responses were obtained by sensors based on nanostructered undoped PPA and P(PA/HEMA): the I vs RH curve for these polymers showed a sigmoid curve that confirms the hypothesis of a main ionic contribute to the conductivity of these materials (see Figure 4); it is just important to note that a good response in the whole range low high RH values is observed.

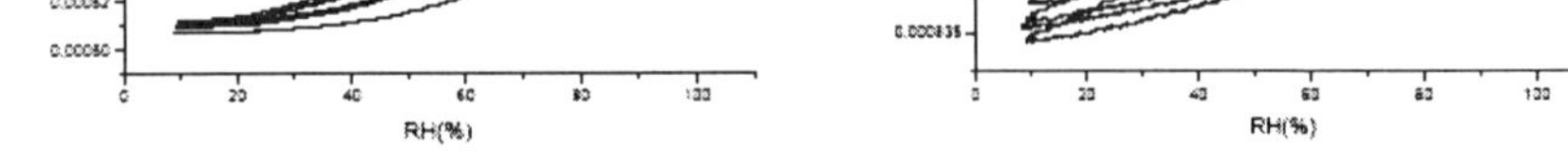

Figure 4. I vs RH curves at 100 KHz for sensor based on nanostructured undoped PPA (ch6) and P(PA/HEA) (ch3).

The work in progress is dedicated to investigate the sensor response in the low RH range and to different gases, whose sensitivity is enhanced by the nanostucured morphology: low levels of water vapours are in fact critical for applications in microelectronics or semiconductor industries, where gases of ultrahigh purity are used; the investigated materials can be sensible also to CO, CO_2 or alcohols, so as reported in the literature for similar non nanostructured polymers [4]. More over, these nanostructured materials show new properties

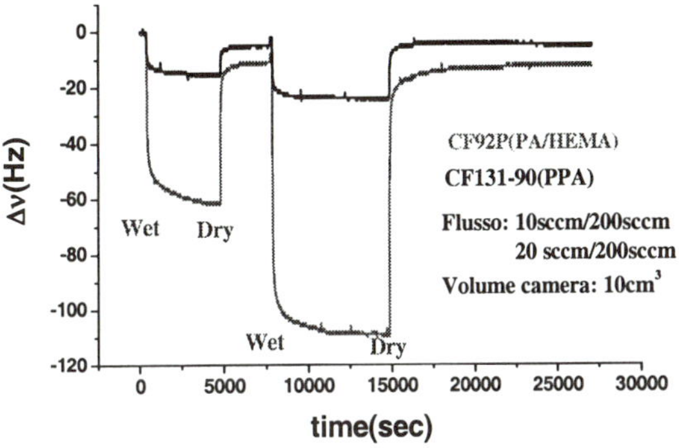

Fig. 5 comparison between the nanostructured sensor responses when two increasing water concentration have been flowed inside the measurement chamber it is depicted.

and functionality that can be modulated by selective control of morphology, dimensions and assembling of particles [5-10]. We have also performed measurements on quartz crystal microbalances (QCM) where no electrical signals or polarization voltages are applied across the material. The Thickness Shear Mode Resonators are collectively referred to as mass-sensitive devices, being able to detect mass changes on their surfaces with high sensitivity. TSMRs are constructed of circular quartz plates (4.9 mm diameter) with thin metal films deposited on each side. These devices generate bulk waves that travel in a direction perpendicular to the plate surfaces. Particle displacements are transverse to the wave propagation direction and parallel to the plate surfaces. The resonant frequency of the fundamental mode is 20 MHz. The responses to added mass in terms of absolute frequency changes (Hz) are ruled by Sauerbrey equation [11]. With this kind of devices is possible to reach a "limit of detection" of 0,5 ng/Hz. In figure 5 are shown the responses for two different RH concentrations of the CF92P(PA/HEMA) and CF131-90(PPA). Both the materials are able to respond to the RH variations but the one with the copolymer is more sensitive and with a fast recover during desorption.

Conclusions

In this work we present preliminary results concerning the preparation and the utilization as humidity sensitive membrane of nanostructured polyphenylacetylene and co-poly(phenylacetylene/2-hydroxyetylmethacrylate). Humidity sensors based on such thin films have shown to be water resistant, able to maintain unchanged their characteristics also under extended exposure to humidity. We have also observed an improvement in terms of conductivity compared to the previously used amorphous PPA and a dependence of the electrical behaviour due to the different metals used as electrodes. We have shown data concerning two different kind of devices (humistor and QCM).

References

1. A.Bearzotti, V. Foglietti, G. Polzonetti, G. Iucci, A. Furlani, M. V. Russo, Mater. Sci. Eng. B 40 (1996), 1
2. G. Polzonetti, MV.Russo, A. Furlani, G. Iucci, Chem. Phys. Lett. 185 (1991), 105
3. L. Palummo, I. Fratoddi, M.V. Russo, A.Bearzotti, Sensor Letters, 2 (2004) 205
4. I. Fratoddi, P. Altamura, A. Bearzotti, A. Furlani, M.V. Russo, Thin Solid Films 458 (2004), 292
5. H.H. Weetall, A. Druzhko, A. Rde Lera, R. Alvarez, B. Robertson, Bioelectrochemistry 51 (2000) 27
6. L. Wang, P. Chow, T. Phan, I. J. Lim, Y. Yang, Adv. Funct. Mater. (2006) 16, 1171
7. M. Garnett, International Journal of Nanoscience (2005) 4, 5&6, 855
8. A. Rae, Advancing Microelectronics (2006) 33, 4,12
9. G. Carbajal, A. Martinez-Villafane, F. G. Gonzalez-Rodriguez, V. M. Castano, Anti-Corrosion
10. Methods and Materials (2001) 48, 4, 241
11. R. D'Amato, L. Medei, I. Venditti, M. V. Russo, M. Falconieri, Mat. Sci. Eng. C (2003) 23, 861
12. G. Saurbrey, Verwendung von schwing quarzen zur Wagung dunner schichten und zur mikrowagung, Zeitschrift fur Physik 155 (1959) 206-222.

METAL FUNCTIONALISED CARBON NANOTUBES THIN FILMS GAS CHEMIRESISTORS

M. PENZA, G. CASSANO, R. ROSSI, M. ALVISI, M. A. SIGNORE, A. RIZZO

ENEA, CR Brindisi, Department of Physical Technologies and New Materials, PO Box 51, Postal Office Br4, 72100 Brindisi, Italy

TH. DIKONIMOS, N. LISI, E. SALERNITANO, E. SERRA, R. GIORGI

ENEA, CR Casaccia, Department of Physical Technologies and New Materials, Via Anguillarese 301, 00060 Rome, Italy

Multi-walled carbon nanotubes (MWCNTs) films have been fabricated by using plasma-enhanced chemical vapour deposition (PECVD) system onto Cr-Au patterned alumina substrates, provided with 3 nm thick Fe growth-catalyst, for NO_2 and NH_3 gas sensing applications, at sensor temperature in the range of 100-250°C. Nanoclusters of noble metals surface-catalysts (Au and Pt) have been sputtered on the surface of MWCNTs to enhance the gas sensitivity with respect to unfunctionalised carbon nanotube films. It was found that the gas sensitivity of Pt- and Au-functionalised MWCNTs gas sensors significantly improved by a factor up to an order of magnitude through a spillover effect for NH_3 and NO_2 gas detection, respectively. The NH_3 and NO_2 gas molecules act as electron-donor and electron-acceptor species respectively doping the unfunctionalised and Pt- and Au-modified MWCNTs, which have *p-type* electrical conductivity in semiconducting character. The gas sensitivity of metal-modified MWCNTs sensors increases with the sensor operating temperature due to better gas adsorption. The metal-functionalised MWCNTs sensors exhibit very high gas sensitivity, fast response, reversibility, good repeatability, sub-ppm range detection limit with the sensing properties of the MWCNTs films tailored by surface-catalyst used to functionalise the MWCNTs sensors.

1. Introduction

Current trend of research in gas sensor nanotechnology is devoted towards the development of sensor nanomaterials which are useful for high-performance gas detection at very low sub-ppm concentrations. Recently, a wide variety of gas sensors based on nanomaterials have been proposed using nanobelts [1], nanowires, nanorods, nanoparticles and nanoclusters, nanohorns, nanotubes [2-6]. Nanostructures [7] based on semiconducting metal oxides of SnO_2, In_2O_3, MoO_3 and WO_3, TiO_2 have been studied as very interesting nanomaterials for gas sensing applications due to their high surface area and porosity which play an key role in the sensing mechanisms.

The carbon-derived nanostructures, in particular carbon nanotubes (CNTs), are widely perceived as one of the most promising nanomaterials to develop high-performance chemical nanosensors. In fact, it has been demonstrated that the carbon nanotubes, at single-walled and multi-walled format, are very attractive for detecting vapour and gas molecules in nanoscale molecular sensors [2-6, 8-9] with high sensitivity and fast response. The CNTs are considered very promising gas sensing nanomaterials because of their outstanding structural, electrical, optical and mechanical properties such as hollow structure, high surface-to-volume ratio, one-dimensional tubular nanostructure, high electron mobility, high chemical reactivity and excellent stability.

Conductometric gas sensors based on CNTs films with remarkable performance have been successfully reported [2-4, 8-9]. The observed behaviour of the conductance changes upon gas exposure can be rationalised based on a charge transfer model between adsorbed gas molecules and carbon nanotubes.

However, a well-defined technique for the tailored fabrication of ppb-level gas sensors composed of carbon nanotubes films functionalised with noble metals nanoparticles (Pt, Pd, Au, Ag) is yet to be reported. In recent years, sensors based on metal oxides semiconducting films doped and modified with noble metals have drawn great attention in order to enhance the gas sensing characteristics.

In this study, carbon nanotubes thin films, deposited by RF-plasma enhanced chemical vapour deposition (PECVD) onto alumina substrates, equipped with a Cr-Au layer strip-pair that serves as electrical contacts, have been functionalised by PVD-deposited nanosized metal clusters of Pt and Au for enhanced NH_3 and NO_2 gas sensitivity, respectively, at operating sensor temperature in the range of 100-250°C. The key role of the metal nanoclusters in RF-PECVD-grown CNTs on gas sensing properties has been experimentally investigated in the chemiresistive transducer demonstrating the enhancement of the gas chemisorption with respect to unfunctionalised CNTs.

2. Experimental details

MWCNTs layers were prepared by a RF-PECVD deposition system for gas detection, at reasonably low growth-temperature of 450°C. The MWCNTs thin films were deposited onto low-cost alumina substrates (10 mm width x 10 mm length x 0.6 mm thickness) provided with vacuum thermally evaporated Cr(20nm)/Au(350nm) metal strips (2 mm width x 10 mm length) for electrical contacts in the two-pole geometry. The cross-sectional scheme of chemiresistor is reported in Figure 1.

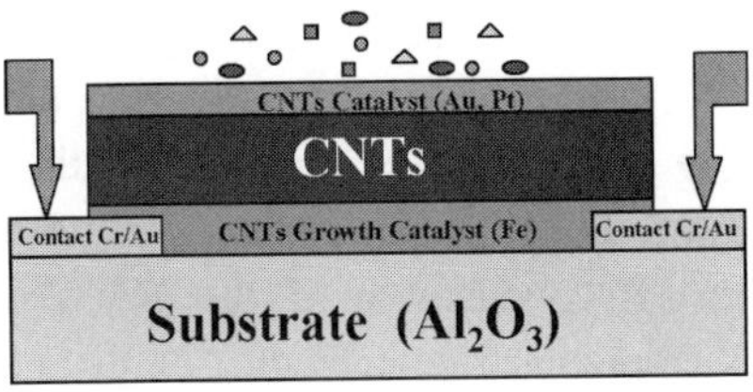

Figure 1. Cross-sectional view of the chemiresistor based on RF-PECVD grown MWCNTs films.

A film of growth-catalyst of Fe nanoclusters with a nominal thickness of 3 nm was vacuum evaporated at 10^{-5} mbar onto substrates for MWCNTs growth. The catalysed-substrates were heated to 450°C with a heating rate of 10°C/min in H_2 atmosphere, with a H_2 flow of 100 sccm at a total pressure of 1.5 Torr.

After the process temperature of 450°C is reached, a H_2 plasma pretreatment is performed at a rf power (13.56 MHz) of 100 Watts for 10 minutes to obtain the metallic nanoclusters necessary for the nanotubes growth. Then, C_2H_2, as carbon precursor, is introduced into the chamber. The gas flow rate ratio between C_2H_2 and H_2 is kept constant at 20/100 sccm, respectively. The MWCNTs PECVD-deposition was performed with constant rf power, pressure and temperature of 100 Watts, 1.5 Torr and 450°C, respectively for 30 minutes by depositing a carbon nanotube networked film that was estimated as about 300 nm thick.

Modification of the MWCNTs involved DC and RF sputtering with a nominal thickness of 5 and 3 nm for Au and Pt nanoclusters respectively over the entire substrate coated by MWCNTs. The electrical resistance, at room temperature, of unfunctionalised and Au- and Pt-modified MWCNTs was measured as 28.8, 13.9 and 16.2 kΩ, respectively; thus, the surface-modification of MWCNTs with metal catalysts is found to decrease the resistance of carbon nanotube films.

Before gas sensing exposure, the unfunctionalised MWCNTs and metal-modified MWCNTs were thermally annealed at 300°C upon dry air flow for 2 hours to purify the MWCNTs removing amorphous carbon and stabilize the sensing properties of catalyst nanoclusters improving their adhesion onto MWCNTs.

The so-fabricated MWCNTs sensors have been located in a test cell (500 ml volume) for gas exposure measurements. The cell case is able to host up to four chemiresitive sensors. Dry air was used as reference gas and diluting gas to air-conditioning the sensors. The gas flow rate was controlled by mass flowmeters. The total flow rate per exposure was kept constant at 1500 ml/min. The gas sensing experiments have been performed by measuring the electrical conductance of MWCNTs thin films in the two-pole format upon controlled ambient of NH_3 reducing gas and NO_2 oxidizing gas in the range of 5-1000 ppm and 100-1000 ppb, respectively, at sensor temperature ranging from 100 to 250°C. The d.c. electrical conductance of the MWCNTs-sensors has been measured by the volt-amperometric technique in the two-pole format by a multimeter (Agilent, 34401A). The sensors were scanned by a switch system (Keithley, 7001) equipped by a low-current scanner card (Keithley, 7158) with a multiplexed read-out.

3. Results and Discussion

Figure 2 displays field emission gun-scanning electron microscopy (FEG-SEM) images of unfunctionalised MWCNTs films and Pt- and Au-modified. The nanostructures show a pronounced tubular structure forming a tangled net of bundled nanotubular chains densely distributed in mats. The average diameter of the MWCNTs seems to be varied in the range of 10-30 nm. The Pt and Au particles partially decorating MWCNTs sidewalls in non-continuous manner form nanoclusters with radius up to 3-10 nm for Pt deposited atoms, while the Au atoms migrate and merge together to form larger nanoclusters (up to 3-30 nm wide) isolated or partially interconnected. The Au nanoclusters are greater than Pt principally due to higher Au surface mobility onto sidewalls. These aspects of catalytic covering of nanotubes strongly affect the properties of gas adsorption of MWCNTs, hence the tailoring of gas sensitivity.

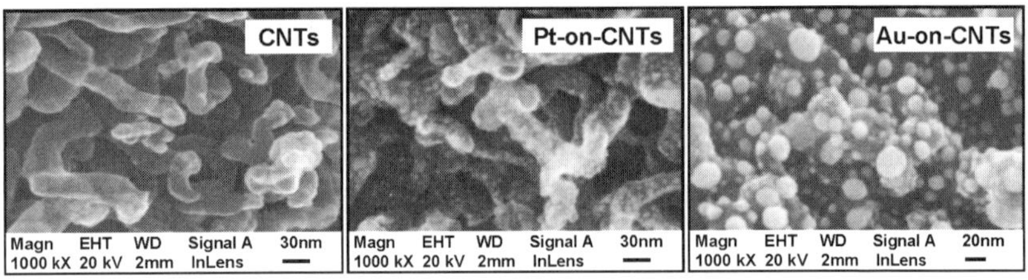

Figure 2. FEG-SEM top-view image at high magnification of (left) unmodified MWCNTs films, (center) Pt-modified MWCNTs films and (right) Au-modified MWCNTs films prepared by RF-PECVD onto alumina substrates. The nominal thickness of MWCNTs is 300 nm. The nominal thickness of Pt- and Au-catalyst is 3 and 5 nm, respectively.

Figure 3 reports the time response in terms of electrical resistance change of the MWCNTs-based chemiresistors unmodified and Pt- and Au-functionalised, exposed to 10 minute-pulses of typical decreasing spot-concentrations of NH_3 reducing gas and NO_2 oxidizing gas in the range 5-1000 ppm and 100-1000 ppb,

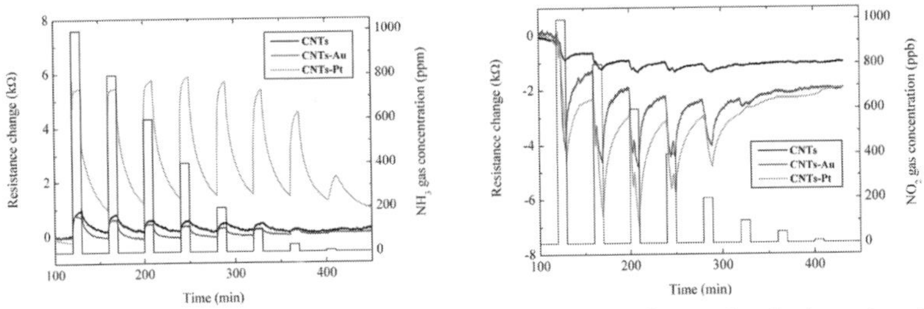

Figure 3. Time response of the electrical resistance change for a chemiresistor based on unfunctionalised MWCNTs films and Pt- and Au-modified MWCNTs films towards (left) NH_3 and (right) NO_2 gas, at working temperature of 150°C. The MWCNTs films thickness is 300 nm and the nominal thickness of Pt- and Au-catalyst is 3 and 5 nm, respectively.

respectively. The working sensor temperature was 150°C with the relative humidity in test cell of about 20%. As observed, the electrical resistance of unfunctionalised and Pt- and Au-modified MWCNTs-based sensors specifically increases upon exposure of NH_3 reducing gas and decreases upon exposure of NO_2 oxidizing gas. In the case of unmodified MWCNTs, their electronic and structural properties are altered by a direct injection of electrical charge with a charge transfer between gas and carbon nanomaterial according to a *p-type* conductivity electrical transport model in semiconductor regime caused by the opposite character of the NH_3 (electron-donor gas) and NO_2 (electron-acceptor gas). As a consequence, electrons are transferred from NH_3 molecule to MWCNTs, decreasing the hole carriers density thus increasing the electrical resistance, shifting the Fermi level away from the valence band. In contrast, electrons are transferred to NO_2 molecule from the valence band of the MWCNTs, increasing the density of holes, thereby decreasing the electrical resistance, shifting the Fermi level towards the valence band.

In the case of Pt- and Au-modified MWCNTs, the test gas is likely dissociated by spillover effect at catalyst surface and the dissociated molecular fragments dissolve into the catalyst bulk. This dissolution of fragmented molecular ionized species should be the main cause for an apparent decrease in the work function of catalyst, in turn a decrease in the barrier height at the interface catalyst/nanotube and an increase in electron transfer. Interface discrete states due to the charge transfer at the catalyst/nanotube contact fill the band gap of the semiconducting MWCNTs. In the case of NH_3-induced catalytic spillover, these energy levels are likely donor gap states; while for NO_2-induced spillover, the energy levels are likely acceptor gap states. The former donor gap states can transfer electrons to conduction band of the *p-type* semiconducting nanotubes with major hole-carrier by decreasing the free net charge density (e.g., increasing electron density in conduction band with unchanged hole density in valence band) and thus increasing the resistance of functionalised MWCNTs, as observed upon NH_3 exposure. The latter acceptor gap states can receive electrons from valence band of the semiconducting nanotubes by increasing the hole density and thus decreasing the resistance of functionalised MWCNTs, as observed upon NO_2 exposure. Hence, the Pt- and Au-modified MWCNTs behave as *p-type* semiconductors as well.

The percentage relative resistance change of the MWCNTs-films upon exposure to two test gases has been registered. Figure 4 shows the calibration curves of the unfunctionalised and Pt- and Au-modified MWCNTs, exposed to a broad range of concentrations of NH_3 and NO_2, at sensor temperature of 150°C. For each gas examined, the sensitivity comparison between unmodified and Pt- and Au-modified MWCNTs sensors demonstrates that the functionalised MWCNTs exhibit higher gas sensitivity than unmodified up to a factor of 6-8 times and 2-3 times for NH_3 and NO_2 using Pt and Au surface-modifications, respectively, at this working temperature. This enhanced gas sensitivity of the metal-modified MWCNTs sensors could be caused by a combination of two additive effects of direct charge injection and catalytically-induced charge into functionalised MWCNTs. The chemical patterns of the mean gas sensitivity (%/ppm), calculated as weighted mean of the ratio between percentage relative resistance change (%) over gas concentration unit (ppm), are shown for NH_3 and NO_2 in the insets of Figure 4. It is found that the gas sensitivity of the MWCNTs-based chemiresistor can be controlled by surface-catalyst used in the MWCNTs functionalization with a tailored selectivity promoted by catalytically metal-induced gap-states with higher NH_3 and NO_2 gas sensitivity for Pt and Au nanosized catalysts, respectively.

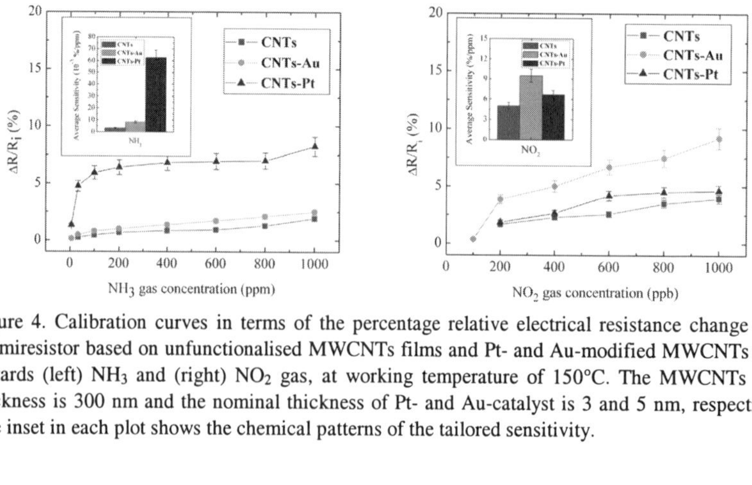

Figure 4. Calibration curves in terms of the percentage relative electrical resistance change for a chemiresistor based on unfunctionalised MWCNTs films and Pt- and Au-modified MWCNTs films towards (left) NH_3 and (right) NO_2 gas, at working temperature of 150°C. The MWCNTs films thickness is 300 nm and the nominal thickness of Pt- and Au-catalyst is 3 and 5 nm, respectively. The inset in each plot shows the chemical patterns of the tailored sensitivity.

Moreover, the calibration curves of the unfunctionalised and Pt-modified MWCNTs, exposed to a broad range of concentrations of NH_3, at sensor temperature ranging from 100 to 250°C, are shown in Figure 5. The results demonstrate that the Pt-functionalised MWCNTs sensors are higher sensitive than unmodified MWCNTs, at any sensor temperature investigated; the unfunctionalised and Pt-modified MWCNTs sensors increase their gas sensitivity with temperature. This means that the gas adsorption capacitance increases with temperature and in particular the Pt-modified MWCNTs sensors exhibit an enhanced gas sensitivity due to temperature-dependent catalytically induced-charge linked to energy levels of gap-states modulated by means of temperature.

Figure 6 reports the continuous NO_2 gas monitoring to 10-minute step-pulses in the range of gas concentrations from 400 to 1000 ppb by using a chemiresistor based on unfunctionalised MWCNTs and Au-modified MWCNTs films, at working temperature of 200°C. The results indicate excellent on-line and real-time gas detection with better sensitivity and resolution for the Au-modified MWCNTs sensor.

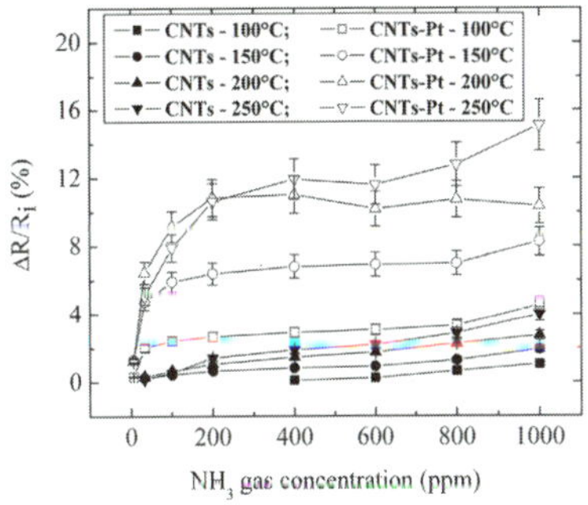

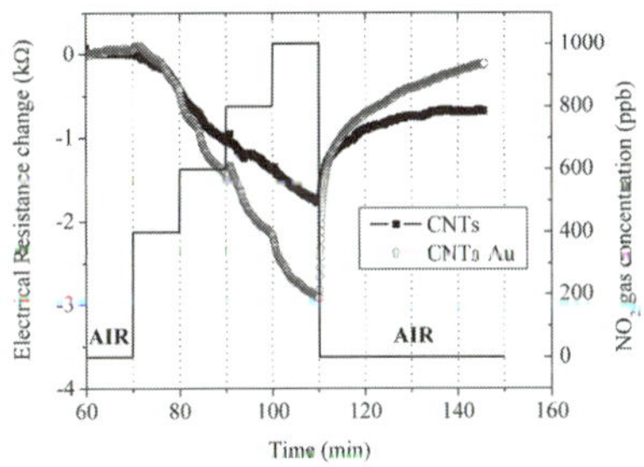

Figure 5. Comparison of sensitivity of the chemiresistor based on unfunctionalised MWCNTs films and Pt-modified MWCNTs films towards NH_3 gas, at different working temperature ranging from 100 to 250°C. The MWCNTs films thickness is 300 nm and the nominal thickness of Pt-catalyst is 3 nm. The exposure time is 10 minutes.

Figure 6. Continuous NO_2 gas monitoring towards step-pulses in the range of gas concentrations from 400 to 1000 ppb using a chemiresistor based on unfunctionalised MWCNTs films and Au-modified MWCNTs films, at working temperature of 200°C. The MWCNTs thickness is 300 nm and the nominal thickness of Au-catalyst is 5 nm.

4. Conclusions

These miniaturized, solid-state sensing elements based on metal-functionalised carbon nanotubes are very performancing devices with improving reliability in low-cost sensor arrays compared to cumbersome, very expensive and time-consuming optical spectroscopy and analytical chemistry techniques with addressed specificity. The surface-catalysts can tailor the gas sensitivity of the high-performance sensors having high sensitivity, fast response, sub-ppm detection level, good reproducibility and reversibility. Experimental evidence of the electrical charge transfer, both direct injection and catalytically-induced charge, between the gas molecules (donor and acceptor) and the unfunctionalised and metal-modified MWCNTs films, including the electrical transport of p-type conductivity are demonstrated. The metallic catalysts surface-modification of carbon nanotubes opens the possibility to make catalytic engineering of sensor nanomaterials for enhanced gas sensitivity and tailored selectivity of nanodevices in nanoarrays for environmental monitoring.

References

1. E. Comini, G. Faglia, G. Sberveglieri, Z. Pan, Z. L. Wang, *Appl. Phys. Lett.* **81**, 1869-1871 (2002).
2. J. Kong, N. R. Franklin, C. Zhou, M. G. Chapline, S. Peng, K. Cho, H. Dai, *Science* **287**, 622–625 (2000).
3. L. Valentini, I. Armentano, J. M. Kenny, C. Cantalini, L. Lozzi, S. Santucci, *Appl. Phys. Lett.* **82**, 961-963 (2003).
4. T. Someya, J. Small, P. Kim, C. Nuckolls, J. T. Yardley, *Nano Lett.* **3(7)**, 877-881 (2003).
5. M. Penza, G. Cassano, P. Aversa, A. Cusano, M. Consales, M. Giordano, L. Nicolais, *IEEE Sens. J.* **6**, 867-875 (2006).
6. M. Penza, G. Cassano, P. Aversa, F. Antolini, A. Cusano, A. Cutolo, M. Giordano, L. Nicolais, *Appl. Phys. Lett.* **85(12)**, 2379-2381 (2004).
7. A. Kolmakov, M. Moskovits, *Annu. Rev. Mat. Res.* **34**, 151-180 (2004).
8. M. Penza, G. Cassano, R. Rossi, A. Rizzo, M. A. Signore, M. Alvisi, N. Lisi, E. Serra, R. Giorgi, *Appl. Phys. Lett.* **90**, 103101 (2007).
9. M. Penza, G. Cassano, R. Rossi, M. Alvisi, A. Rizzo, M. A. Signore, Th. Dikonimos, E. Serra, R. Giorgi, *Appl. Phys. Lett.* **90**, 173123 (2007).

RESISTIVE λ-SENSORS BASED ON Fe-$SrTiO_3$ NANOPOWDERS

G. Neri*, A. Bonavita, G. .Micali, G. Rizzo
Dept. of Industrial Chemistry and Materials Engineering, University of Messina, C/da di Dio, Vill. S. Agata, 98166 Messina, Italy
neri@ingegneria.unime.it

R. Licheri, R. Orrù, G. Cao
Department of Chemical Engineering and Materials, University of Cagliari,Italy

D. Marzorati, E. Merlone Borla
Centro Ricerche Fiat, Torino, Italy.

Summary

Resistive oxygen sensors based on strontium titanate powders are promising alternatives to electrochemical zirconia-based automotive λ-sensors. In this work we investigate a series of STFO ($SrTi_{1-x}Fe_xO_{3-\delta}$) powders, with x ranging from 0 to 0.6, synthesized by Self-propagating High-temperature Synthesis (SHS). A ball-milling (BM) treatment was subsequently carried out for further structure refinement and grain size reduction. The electrical and high-temperature O_2 sensing properties of the synthesized STFO nanopowders are here reported.

1. Introduction

Lambda sensors are high-temperature oxygen sensors widely used in the automotive sector [1]. These sensors convert a chemical signal into an electronic one, which is then relayed to the engine control unit in order to optimize the air-to-fuel ratio (λ) thus reducing deleterious combustion products such as CO, hydrocarbons and NOx [2, 3].

Resistive devices based on titania, TiO_2, or the ternary metal oxide strontium titanate, $SrTiO_3$, show interesting features and have been proposed to replace conventional electrochemical zirconia-based λ sensors. The conductivity of these materials (σ) depends on the oxygen partial pressure following the relation:

$$\sigma\left(P_{O_2},T\right)\propto P_{O_2}^{1/m}\exp\left(-\frac{E_A}{kT}\right) \qquad (1)$$

However, while TiO_2 presents a large dependence of the conductivity with the temperature ($E_A >> 0$), STFO solid solutions ($SrTi_{1-x}Fe_xO_{3-\delta}$) show, at a particular composition, a temperature independent resistive behavior ($E_A = 0$).

In this paper, a series of $SrTi_{1-x}Fe_xO_3$ (with x ranging from 0 to 0.6) powders were prepared by self-propagating high-temperature synthesis (SHS). A ball-milling (BM) treatment has been also subsequently carried out for a further refinement of the structure and grain size reduction. The electrical and sensing properties of thick films of these materials have been investigated.

2. Experimental

$SrTi_{1-x}Fe_xO_{3-\delta}$ powders by SHS were obtained by mixing reactants according to the following reactions:

$$2SrO_2 + Ti + TiO_2 \rightarrow 2SrTiO_3 \quad (2)$$
$$2SrO_2 + (1\text{-}2x)Ti + TiO_2 + 2xFe \rightarrow 2SrTi_{(1-x)}Fe_xO_3 \quad (x = 0.35; 0.5) \quad (3)$$
$$SrO_2 + 0.4TiO_2 + 0.6Fe \rightarrow SrTi_{0.4}Fe_{0.6}O_{2.8} \quad (4)$$

The amount of Fe in the starting mixture was varied in the range 0-18 wt.%, in order to obtain different Fe-doped strontium titanate powders. Samples code and compositions investigated are indicated in Table 1. The SHS products to be used for screen-printing deposition were also further mechanically treated at different milling times (0 – 15 h).

Sample code	Fe/(Fe+Ti) molar ratio (x)	Composition ($SrTi_{1-x}Fe_xO_{3-\delta}$)
STO	0	$SrTiO_3$
STFO35	0.35	$SrTi_{0.65}Fe_{0.35}O_3$
STFO50	0.5	$SrTi_{0.5}Fe_{0.5}O_3$
STFO60	0.6	$SrTi_{0.4}Fe_{0.6}O_{2.8}$

Table 1. Samples ID and compositions investigated

Thick films of STFO powders were screen printed on alumina substrates, supplied with comb-like electrodes and a Pt heater, thus preparing oxygen sensor devices and treated until 1200°C in order to verify their mechanical resistance and moreover to stabilise their composition films. Subsequently, the devices were tested in the thermal range 500-650 °C by using a computer-assisted measurement apparatus, wich measure the resistance values of the sensors maintained under dry flow of nitrogen and pulsing different concentration of oxygen.

3. Nanopowders Preparation and Characterization

$SrTi_{1-x}Fe_xO_{3-\delta}$ powders have been prepared by SHS, according to the chemical reactions (2)-(4). All systems investigated displayed a self-sustaining character due to the high exothermicity of the corresponding synthesis reactions, producing the high temperature necessary for the synthesis [4].

Phase analysis of diffraction patterns showed that the above reactions proceed to completion. Moreover, XRD patterns of Fe-doped samples with different Fe loading show that peaks are shifted toward higher diffraction angles as the iron content increases. Results are in agreement with previous works reporting that $SrTi_{1-x}Fe_xO_{3-\delta}$ crystallizes in a cubic perovskite structure, with lattice parameters decreasing slightly with increasing iron content [5, 6]. This indicates that $SrTi_{1-x}Fe_xO_{3-\delta}$ solid solutions were formed in the corresponding c-$SrTiO_3$-type structure where Ti(IV) is substituted by Fe(III).

After SHS synthesis STFO powders were treated by BM. Diffraction patterns of BM powders are characterized by remarkable peaks broadening, confirming a refinement of the crystal structure, increase of the lattice strain and reduction of grain size. This behaviour is consistent with general models for grain size refinement during ball milling proposed in the literature [7].

4. Electrical and Sensing Test

The effect of the Fe-doping on the electrical behaviour of $SrTi_{1-x}Fe_xO_{3-\delta}$ powders synthesized by SHS and subsequent ball-milling treatment have been investigated in detail, in order to acquire useful information for their possible application in high-temperature oxygen sensors.

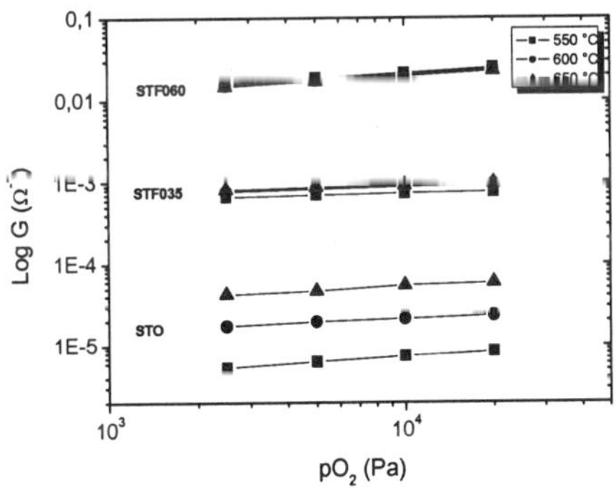

Fig. 1. Conductance as a function of the oxygen partial pressure at different temperatures of 5 h ball-milled STFO samples.

Figure 1 compares the conductance of SHS powders milled for 5 h at different temperatures as a function of oxygen partial pressure. The STO sample shows a low electrical conductance. The addition of iron greatly increases the conductance. It can be noted that, increasing the Fe content, the temperature dependence of the conductance decreases correspondingly. In particular, E_A reaches values around zero at $x = 0.6$, in agreement with literature data [8, 9].

Oxygen sensing tests were carried out to investigate the effect of Fe on STFO behaviour to oxygen concentration variation at high temperature. The relative resistance R_{N2}/R_{O2} is used here to express the sensor response, where R_{N2} is the baseline resistance of the sensor in nitrogen and R_{O2} is the resistance of the sensor at different concentrations of oxygen diluted by nitrogen. As seen in Figure 2, the response increases markedly increasing the Fe loading.

Fe doping contribute then to affect the electrical resistance of STFO materials prepared by SHS. Moreover, sensing tests indicate that this factor also cooperate to increase the sensor response to O_2 at high temperature.

Bench tests carried out under simulated engines conditions have confirmed these good characteristics, showing the negligible interference on the sensor response of high concentrations of CO_2, NO_X and HC, suggesting that resistive oxygen sensors based on nanopowders of STFO are promising candidate as λ-probes for automotive engines.

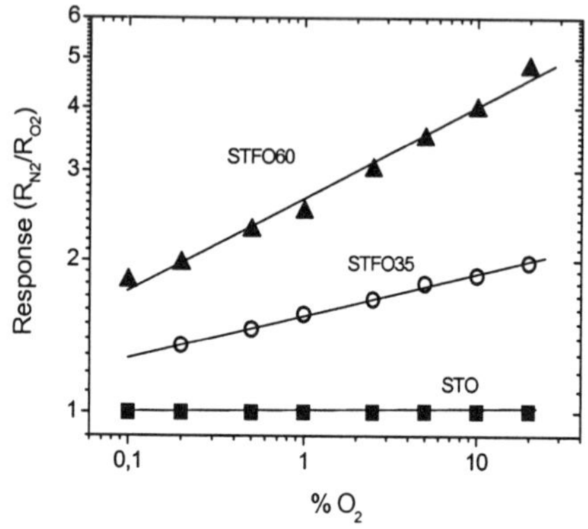

Fig. 2. Sensors response to different O_2 concentrations. T = 650 °C

5. Conclusions

In this work, $SrTi_{1-x}Fe_xO_{3-\delta}$ powders with different Fe content were synthesised by SHS. Results have shown that the SHS technique coupled with high-energy BM is a very effective route to prepare nano-structured solid solutions for oxygen-sensing applications. The effect of the iron content of STFO powders on the electrical characteristics and response to oxygen at high temperatures have been examined in detail. STFO60 thick films have shown the better performances in terms of temperature resistance independence as well as sensor response.

Acknowledgments

The authors gratefully acknowledge the financial support for this work by MIUR under the FIRB-SqUARE project (Contract number RBNE01Y8C3).

References

1. J. Gerblinger, K.H. Hardtl, H. Meixner, R.Aigner, *High-Temperature microsensors*, in: W. Gopel (Ed), Sensors, A Comprensive Survey, 8, Weinheim, 1995, 181-219.
2. H.M. Wiedenmann, G. Ho¨tzel, H. Neumann, J. Riegel, H. Weyl, *ZrO2-Lambda-Sonden fu¨r die Gemischregelung im Kraftfahrzeug*, in: H. Schaumburg (Ed.), Sensoranwendungen, Teubner-Verlag, Stuttgart,Germany, 1995, pp. 371–399.
3. J. Fouletier, *Gas analysis with potentiometric sensors: A review*, Sens. Actuators 3 (1982) 295–314.
4. I. Barin, *Thermochemical data of pure substances*, VHC (1993).
5. T.R. Clevenger, *Effect of* Fe^{+4} *in the system* $SrFeO_3$-$SrTiO_3$, J. Am. Ceram. Soc., 46, (1963) 207-210;
6. L. H. Brixner, *Preparation and properties of the* $SrTi_{1-x}Fe_xO_{3-\delta}$ *system*, Mat. Res. Bull. 3, (1968) 299-308.
7. C. Suryanarayana, C., *Mechanical alloying and milling*, Prog. Mater. Sci., 46, (2001) 1-184.
8. R. Moos, *Donor doped strontium titanate: electrical behavior and modeling*, PhD thesis, Karlsruhe, VDI-Verlag, Dusseldorf, 1994.
9. R. Moos, T. Bischoff, W. Menesklou, K.H. Hardtl, *Solubility of lanthanum in strontium titanate in oxygen-rich atmospheres*, J. Mater. Sci. 32 (1997) 4247–4252.

HYDROGEN SENSOR BASED ON Pd NANOWIRES

BRIGIDA ALFANO

University of Naples, "Federico II", Piazzale Tecchio 80,
80125 Pozzuoli (Na), Italy.

VERA LA FERRARA, MASSERA ETTORE, NASTI IVANA AND
DI FRANCIA GIROLAMO

ENEA Research Center, Località Granatello,
80055 Portici (NA), Italy.

In this work, we will report a brief review about hydrogen sensors based on Pd nanowires and our first approach on fabrication of hydrogen sensor based on array of palladium nanowires. As grown Pd nanowires are broken by an electrical process to modify device resistance. Investigations on treatment's influence are studied using a 4% H_2 mixture in N_2 carrier.

1. Introduction

1.1. *Hydrogen sensor based on Pd nanowires*

Palladium nanowires have remarkable property of adsorbing hydrogen. This absorption modifies the electrical properties of this metal. In particular, hydrogen adsorption decreases the nanowires' electrical resistivity because nanosized Pd particles swell when the gas is introduced in the test chamber, forming new electrical connections.

The resistivity change in palladium by hydrogen absorption has been used to build hydrogen sensors for several decades. Recent advances in the fabrication of nanowires had been achieved through the electrodeposition. In fact, the electrodeposition is a versatile technique that can be used to synthesize nanowires with desired diameter and length. For examples, Walter et al [1] created arrays of mesoscopic palladium wires for hydrogen sensors and hydrogen-actuated switches that exhibit a fast response time for hydrogen detection. Yun et al [2] developed electrochemically grown palladium wires for individually addressable hydrogen sensor arrays. Bangar et al [3] demonstrated a novel electrochemical method for dimensionally controlled growth of a single palladium nanowire between premicrofabricated electrodes. Atashbar et al [4] synthesized palladium nanowires on the 'V' shaped grooves of a highly oriented pyrolytic graphite (HOPG) while Kim et al [5] have electrodeposited Pd nanowires using anodized alumina template (AAO) that is of great interest because of the control over the inter-pore distance and pore diameter by variation of the anodization parameters, such as anodization voltage and electrolyte composition. Cheng et al [6] built up palladium nanowires along the direction of the electric field between the electrodes from an aqueous solution.

In this paper, we report on fabrication and electrical characterization of an hydrogen sensor based on an array of palladium nanowires electrodeposited onto silicon substrate. As grown Pd nanowires do not respond to hydrogen environment, but, after an electrical breakage [7], the array becomes sensitive.

Devices have been tested, at room temperature, in a 4% H_2 mixture in nitrogen using a gas calibration system that is capable to test simultaneously the effect of up to four gases. As a matter of fact, preliminary measurements, towards common interfering gases, show only minor effects.

We have found how the sensor sensitivity and response time change with the breaking parameters.

1.2. *Experimental Setup*

Starting from a silicon substrate coated with 100 nm Si_3N_4, an interdigitated electrode (IDE) pattern, with 8 μm gaps, has been realized by photolithography and Cr/Au e-beam evaporation (Fig. 1).

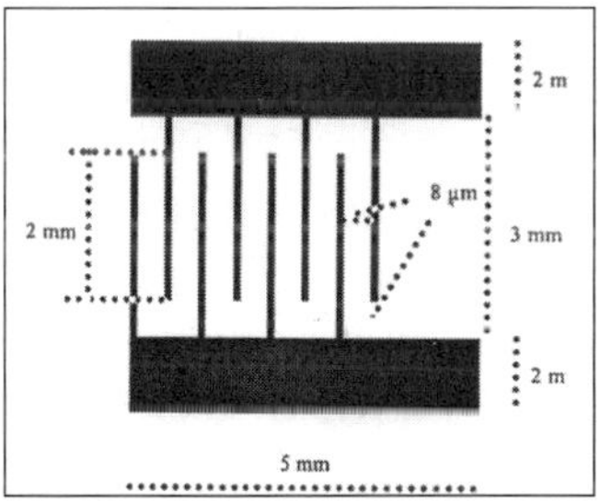

a)

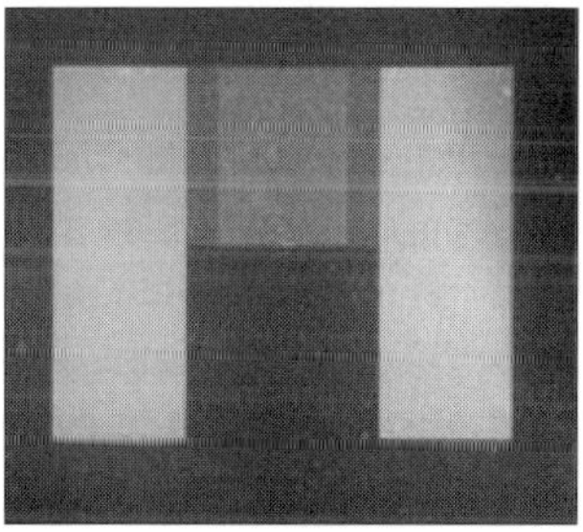

b)

Fig. 1. a) IDE pattern scheme; b) photograph transducer

Here, a saturated aqueous solution has been prepared by dissolution of crystalline Pd(acetate)2 buffered with 10 mM HEPES [6]. Solution has been put in ultrasonic bath for 15 minutes (70W) and then centrifugated for 5 minutes at 13000g. Then, 2 μl of the solution has been deposited by casting onto IDE samples previously washed in isopropyl alcohol, deionised water and dried in nitrogen flux.

The application, for few seconds, of an AC sinusoidal signal at 10 Vpp and 300 kHz, upon the electrodes, results in the formation of Pd nanowires array. Samples have been , then, washed in deionised water and dried under vacuum.

Nanowires morphological characterization has been performed by scanning electron microscopy (SEM). Fig. 2 a) shows how Pd nanowires have been grown perpendicular to the electrodes, as a result of the processing conditions used. In Fig. 2 b) SEM image, at high magnification, of one of the wires is reported. Typically wire width is about 50-100 nm.

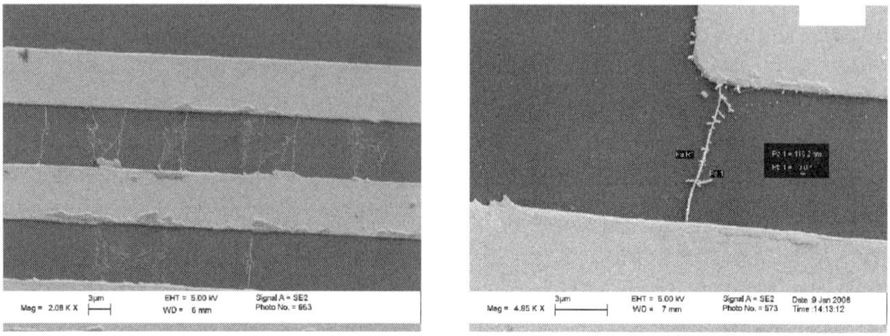

a) b)

Fig. 2. a) SEM photograph of Pd nanowires grow between the Cr/Au electrodes that are spaced 8 µm apart; b) SEM of a single Pd nanowire with a width of about 120 nm.

Devices have been electrically characterized as chemical sensors measuring their response, at room temperature, towards 4% H_2 in nitrogen carrier, produced by an electrolyte high pure hydrogen generator (Claind HG 2400). A volt-amperometric technique, at constant bias, has been employed for sensor dc electrical characterization in a controlled gas-flow environment, pre-mixed with dry carrier in the desired percentage by mass flow meters and continuously controlled by means of an in-line FTIR [8].

Devices have been electrically characterized using a 5mV bias and a total gas flow settled at 500 sccm.

1.3. *Results and Discussion*

Devices have been characterized in DC conditions at room temperature and in ambient air. Current-voltage (I-V) characteristic has been observed to be linear between –0.1÷0.1 Volt, showing an electrical resistance in the range 10-

100 Ω. As-grown Pd nanowires do not respond to hydrogen. In order to activate the response towards H_2, we have set up an electrical treatment [7]. We have applied high voltage to device electrodes in DC condition to obtain the opening of nanoscopic gaps. Each gap acts as a switch that closes in presence of hydrogen since each Pd grain expands its volume when exposed to hydrogen. This is evident in the observed current increasing. Device behaviour is quite reversible but the response time is slower than similar devices [1,3-5]. Further investigations are necessary to improve device perfomance.

We have found that the response time is dependent on the resistance value after breakage treatment. In Fig. 3 e Fig. 4 device response related to two different breakage resistances are reported. In Fig. 3 it has been tested a device after a 20 V breakage resulting in 3 kΩ resistance: it shows a 7 minutes response time with an increase in the current by a factor of up to 4 at 25 °C. In Fig. 4 it has been tested a device after a 30 V breakage resulting in 45 kΩ resistance: it shows a 13 minutes response time and with an increase in the current by a factor of up to 16 at 25 °C

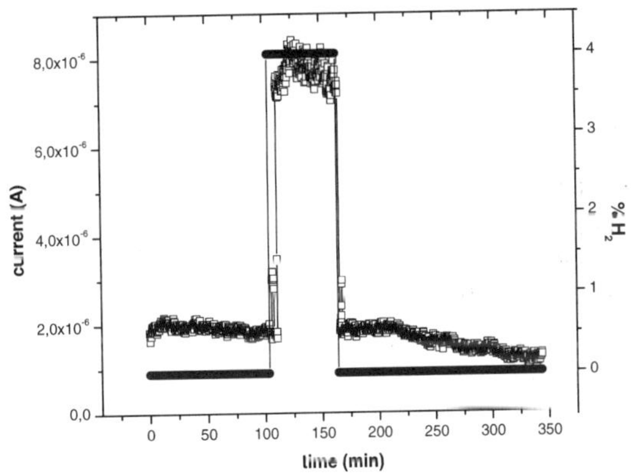

Fig. 3. Electrical response, at room temperature, towards 4% H_2 in nitrogen carrier of a sensor device with a starting breakage resistance of about 3 kΩ

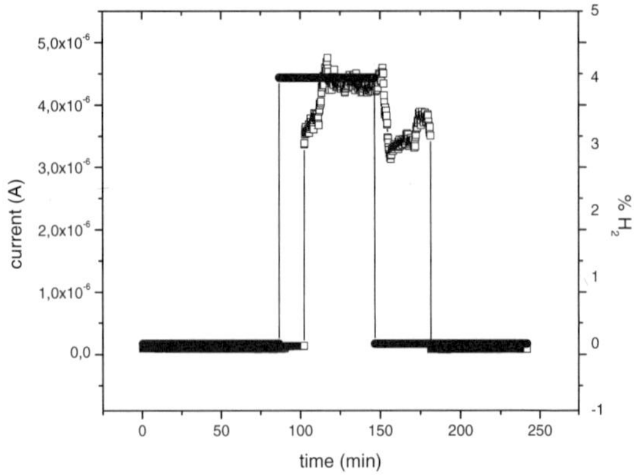

Fig. 4. Electrical response, at room temperature, towards 4% H_2 in nitrogen carrier of a sensor device with a starting breakage resistance of about 45 kΩ.

Then it has been supposed that lower is the breakage voltage, lower is the response time, whereas upper is the breakage voltage, upper is the device sensitivity.

2. Conclusions

We have presented a hydrogen sensor based on a Pd nanowire array grown by single electrochemical process on a pre-patterned silicon substrate. The use of silicon transducer is important for its high compatibility and integrability in microelectronic field.

Response time seems to be at present the major limitation for this type of device. Further investigations are necessary to improve device perfomance.

References

1. E. C. Walter, F. Favier and R. M. Penner, *Anal. Chem.*, **74,** 1546, (2002).
2. M. Yun, N. V. Myung, R. P. Vasquez, C. Lee, E. Menke and R. M Penner, *Nano Lett.*, **4,** 419, (2004).
3. M. A. Bangar, K. Ramanathan, M. Yun, C. Lee, C. Hangarter and N. V. Myung, *Chem. Mater.*, **16,** 4955, (2004).
4. M. Z. Atashbar, D. Banerji, S. Singamaneni, *IEEE Sen. J*, **5**, 792, (2005).
5. K. Kim, S. M. Cho, *Proc. of IEEE Sens.*, **2,** 705, (2004).

6. C. Cheng, R. K. Gonela, Q. Gu, D. T. Haynie, *Nano Lett.*, **5**, 175, (2005).
7. patent pending
8. L. Quercia, F. Cerullo, V. La Ferrara, C. Baratto, G. Faglia, *Phys. Stat. Sol.*, **182**, 473, (2000).

CHEMICAL SENSORS BASED ON CARBON NANOTUBES: COMPARISON BETWEEN SINGLE AND BUNDLES OF ROPES

VERA LA FERRARA†, BRIGIDA ALFANO, IVANA NASTI, ETTORE MASSERA, AND GIROLAMO DI FRANCIA

ENEA Research Center of Portici, 80055 Portici (NA), Italy

A chemical gas sensor based on a single rope of single walled carbon nanotubes (SWCNTs) has been fabricated first isolating the rope on a silicon/Si_3N_4 substrate and then realizing, at its ends, two platinum microelectrodes by means of a Focused Ion Beam (FIB). Its electrical behaviour at room temperature in toxic gas environments has been investigated and compared to sensors based on bundles of SWCNT ropes. For all the devices upon exposure to NO_2 and NH_3 the conductance has been found to increase or decrease respectively. Response time in NO_2 is however faster for device based on the single rope. A mechanism for molecular sensing is proposed.

1. Introduction

Interest in nanostructured materials is growing rapidly even in the field of chemical sensors. Carbon Nanotubes (CNTs) extreme sensitivity to external perturbation has been correlated to their ability to direct the selective uptake of gaseous species. As a result, CNTs based gas sensors show faster response, higher sensitivity, lower operating temperature and wider gases variety that may be detected compared with the other types of gas sensor devices [1].

The effect was first observed by J. Kong and co-workers that reported on a dramatic decrease and increase of the electrical resistance, in single-walled carbon nanotubes (SWCNT) exposed to NO_2 and NH_3 respectively [2-3]. Since then, several different conductometric gas sensor devices designs have been suggested. J. Li and coworkers [4] have, for instance, fabricated simple resistive device by casting a solution of purified SWCNT in dimethylformamide (DMF) on a silicon

†corresponding author: tel.+390817723322, fax.+390817723344,
e-mail: vera.laferrara@portici.enea.it

substrate. The device operates at RT and exhibits a very high sensibility towards NO_2 and nitrotoluene (detection limits in N_2, 44 ppb and 262 ppb respectively). Several other groups have reported on a relevant sensitivity towards methane exhibited by CNT based devices [5,6]. Valentini *et al.* [5] discuss the electrical responses of a device fabricated on silicon with Pt interdigitated (IDE) contacts and a Platinum heater, while Roy *et al.* [6] use electrodeposition to fabricate a CNT films device operating at RT, with methane concentrations around some thousands of ppm in argon. For nanotube bundles, Wongwiriyapan *et al.* [7] have realized, on alumina substrate patterned with Pt
IDE and coated with e-gun deposited Al and Fe as catalysts, NO_2 gas sensors. For nanotube field-effect transistors (NT-FETs) formed by a single tube, Peng *et al.* [8] have conclusively demonstrated that the adsorption of NO_2 (100 ppt) and NH_3 (10 ppm) at the nanotube is responsible for the change in transport properties. Recovery is obtained using UV illumination. For nanotube powder, Nguyen's group [9] has fabricated a gas sensor starting from SWCNT powder deposited by the screen-printing process, followed by annealing pretreatment. The sensor has been tested in 5 ppm NH_3 mixed in 500 sccm N_2 at room temperature. The sensor recovery has been obtained or increasing the carrier flux or heating in desorption time. Physical and chemical properties of carbon nanotube can also be modified by the adsorption of foreign atoms or molecules. This process is usually named functionalization and it is used to increase the nanotube sensitivity and selectivity towards a desired chemical species. Zhang *et al.* [10] has functionalized SWCNT surface with polyaniline (PANI). The device has been tested in NH_3 for which the detection limit is 50 ppb_v. The response time at room temperature is on the order of minutes and the recovery time is a few hours. However the chemical and physical interactions between molecules and sensing nanotubes are only rarely discussed and, up to now, not yet understood. Recently, Strano *et al.* [11] have presented a model on irreversible and reversible binding to a carbon nanotube sensor surface. In fact, the SWCNT based sensors have shown to possess both binding sites. In the case of irreversible adsorption, sensor reaches steady state at saturation when all sites are occupied and the sensor response can't be restored without any regeneration methods.

If the binding to the nanotube surface is reversible, the surface sensor is regenerate by flowing out analyte gas. The ratio of irreversible and reversible sites depends on the typology of the sensor device and the analyte molecules.
In this work, in order to get further insight into the operating mechanism of CNT based chemical sensor mainly as far as device design is concerned, we have fabricated and characterized in controlled environment devices based on a single rope of SWCNTs (SR) and compared their kinetic with devices based on bundles of ropes (BR). SWCNTs used in this work have been prepared by laser ablation arranged in the form of bundles shaping ropes. All of the devices have been realized depositing a SWCNTs suspension in a planar two electrodes configuration using silicon wafer coated with a Si_3N_4 layer as substrate. The series of SR devices has been fabricated isolating a single rope on the substrate and depositing at each end two Pt microelectrodes by means of focused ion beam (FIB). The system (FEI – Quanta 200 3D), used at this aim, integrates an high resolution FIB, a scanning electron microscopy (SEM) and a gas injection system (GIS) composed by a Pt organometallic precursor. The series of BR devices has been fabricated using a photo lithographed and e-beam evaporated substrate. All the devices have been characterized in presence of toxic gasses, such as NO_2 and NH_3, at room temperature. We find that the molecule adsorption on defects sites is the main operating mechanism. In both the reducing (NH_3) and oxidizing (NO_2) environments, the SR series behaviour is different from BR ones. Irreversible adsorption occurs both SR and BR sensors. Diffusion seems to act only for bundles sensors and in oxidizing environments.

2. Experimental

2.1 Preparation of SWCNT ropes/DMF suspension

SWCNTs used in this work have been prepared by laser ablation arranged normally in the form of bundles shaping ropes of about 50-80 nm thickness and 3-4 micron length [12]. To debundle the ropes and to have an homogeneous solution as a feed material for device fabrication, ropes have been dispersed in dimethylformamide (DMF) and sonicated at room temperature, in an ultrasonic bath. Suspension has been drop-deposited (2μL) onto silicon substrate and evaporated under vacuum at 70°C for overnight. Resulting nanotube ropes have an average diameter of about 50 nm and 3-4 μm length.

2.2 Fabrication of chemical sensors

Sensors used in this work have been fabricated starting from crystalline silicon substrate coated with a Si_3N_4 layer (200 nm deposited by PECVD). We have realized several series of sensors based on a single rope (series A). Its electrical behaviour in presence of toxic gasses (NH_3 and NO_2) has been compared with sensors based on several ropes (series B). Two more series (C and D) have been fabricated specifically to investigate the effect of electrode pattern and rope electric contact on the sensor response.

2.2.1 Series A

In order to realize series A, the suspension has been deposited onto silicon substrate coated with a Si_3N_4 layer. After DMF evaporation, we have explored the surface by means of electron microscopy, selecting one of the ropes. Leaving sample in the vacuum chamber, we have started with deposition. By means of the FIB, we have deposited platinum microelectrodes at single rope each end adjusting opportune ion beam work conditions at 30 kV accelerating voltage, 10 pA emission current. Each electrode, 300 nm in diameter and 500 nm in height, has a length of a few microns, ending in 0.25 cm^2 gold pads, previously deposited by e-beam. Emission current has been set at 1 pA near the rope to reduce possible platinum redeposition on its surface.
In Figure 1 a FIB image of a series A contacted by platinum microelectrodes is shown.

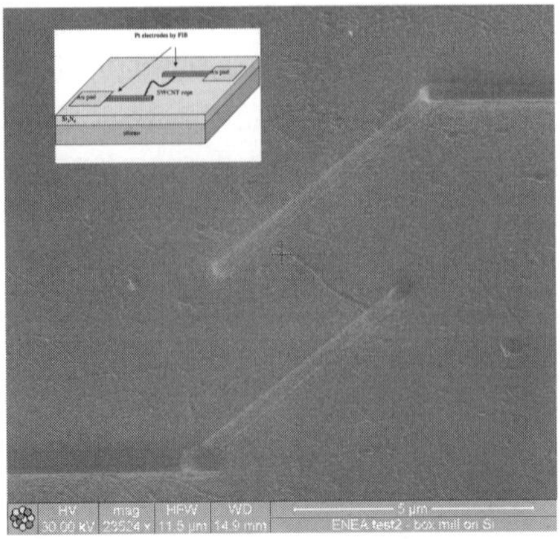

Figure 1. Series A: Ion image of a single rope on the silicon/Si_3N_4 substrate contacted by two Pt electrodes deposited by GIS. In the inset the device scheme is reported.

2.2.2 Series B

Series B has been fabricated using photolithography. An interdigitated electrode (IDE) pattern has been deposited on Si_3N_4/silicon substrate using a lift off process. Suspension has been cast onto the IDE area, obtained by e-beam evaporation of Cr/Au, 20nm/180 nm of thickness respectively. After DMF evaporation, several ropes bridge the electrode fingers as it is shown in Figure 2 [13].

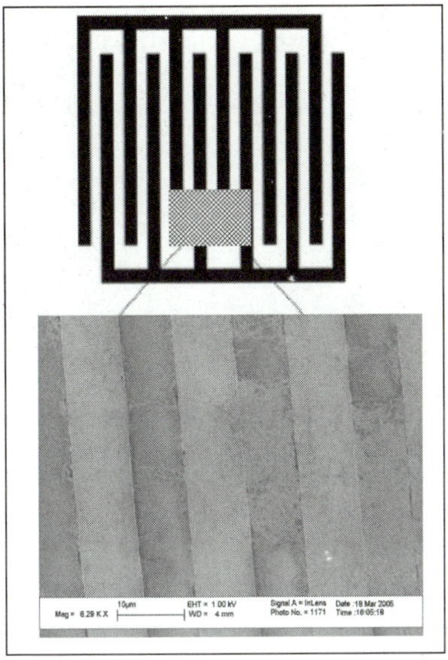

Figure 2. Series B: In the upper part of the figure, a schematic picture of the IDE electrodes pattern is shown. Fingers width is 8 μm, length 2.5 mm and gap size of 8 μm; pattern total dimensions are 7mm x 5mm. In the lower part a SEM image of the deposited ropes is reported.

2.2.3 Series C

A 1 μm width channel has been milled by FIB transversally to a Cr/Au pad, evaporated with the same conditions above reported. FIB operating conditions (accelerating voltage of 30 kV and emission current of 30 pA) have been chosen to obtain a clean channel without damaging nitride layer and avoiding metal redeposition. Few drops (2 μl) of the ropes/DMF suspension have been deposited on the channel. In Figure 3 a FIB image of a series C, after DMF evaporation, is shown.

Figure 3. Sensor C: A channel, 1 µm width, has been milled by FIB transversally to a Cr/Au pad, previously evaporated by e-beam. SWCNT ropes/DMF suspension have been deposited on channel .

2.2.4 Series D

Starting from series C and in order to test the electric contact between rope and pad, we have selected, by SEM, one of the rope depositing at each end, by FIB, two platinum boxes (700 nm x 120 nm x 500 nm). For Pt deposition we have used 30kV as accelerating voltage and 10 pA for current emission have been set. In Figure 4 we report a SEM image of realized device.

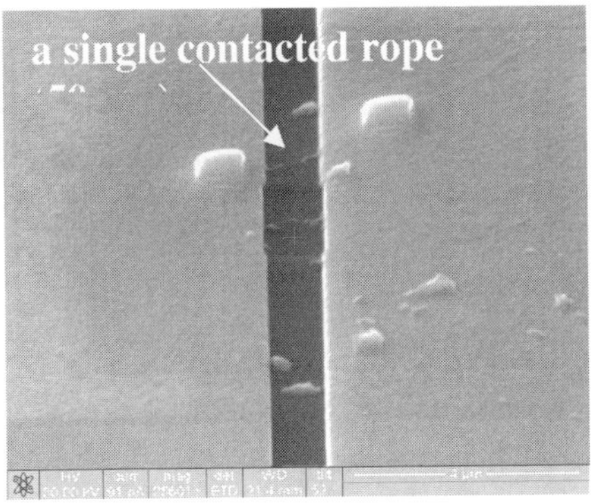

Figure 4. Series D: A channel, 1 μm width, has been milled by FIB transversally to a Cr/Au pad, previously evaporated by e-beam. SWCNT ropes/DMF suspension have been deposited on channel and one rope has been individuated. At each end, Pt boxes have been deposited by FIB.

3. Results and Discussion

3.1 I-V characteristic

All the devices have been electrically characterized at room temperature in ambient air. For all the series, The current-voltage (I-V) characteristic is linear between $-1 \div 1$ volt. The device ohmic resistances are different between series A and the group of series B, C, D as it is reported in Table I. For the latter group, the quite similar values observed suggest that the electronic properties of the deposited film, more than the contacting pattern or the intimate nature of the electric contact itself, play the main role on the device performance.

3.2 Response to toxic gasses

Devices have been electrically characterized as chemical sensors measuring their response, at room temperature, towards NO_2 and NH_3. A volt-amperometric technique, at constant bias, has been employed for sensor dc electrical characterization in a controlled gas-flow environment, pre-mixed with dry carrier in the desired percentage by mass flow meters and continuously controlled by means of an in-line FTIR. All the tested devices have been biased at 0.1 V. Total gas flow has been set to 500 sccm. For the measurements here reported, certified bottles containing mixtures of 30 ppm of nitrogen dioxide in synthetic air and 500 ppm of ammonia in synthetic air have been used [14]. In Figure 5 the normalized conductance is reported towards the observation time when devices are exposed to 500 ppm of NH_3 in synthetic air. The marked decrease in conductance is in agreement with the known electron donor behaviour. The interaction may exist between NH_3 molecules and carbon nanotube through preadsorbed oxygen species on surface, because no binding affinity between NH3 molecules and carbon nanotube has been found. The oxygen molecules could interact strongly with NH_3, this explains the complete irreversibility of the sensor for this gases species [2]. Data have been fitted by a Boltzmann sigmoidal function confirming that NH_3 binding is an irreversible adsorption process.

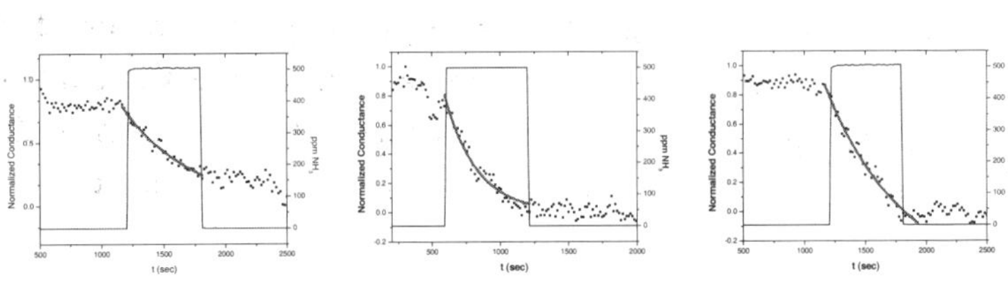

Figure 5: Normalized conductance is reported versus the measurement time under 500 ppm of NH_3 in synthetic air: series A (left), series B and C (centre), whose responses are pratically overlapped. series D (right). Solid line is the fitted by Boltzmann sigmoidal function

When devices are exposed to NO_2 (Figure 6) conductance is shown to increase as expected since the NO_2 molecules behave as electron acceptor when adsorbed to CNT nanotubes. NO_2 is found to bind with nanotube surface with an adsorption energy Ea ~ 0.9 eV [2], resulting in a partial reversibilty.

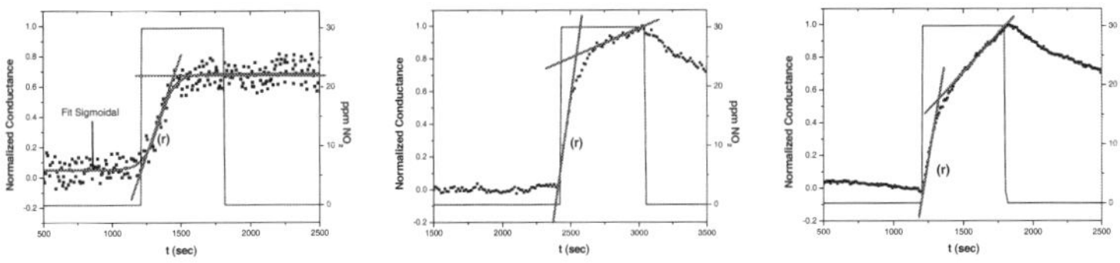

Figure 6. Normalized conductance is reported versus the measurement time under 30 ppm of NO_2 in synthetic air: at the left sensor A with Boltzmann sigmoidal function, at the centre sensor B and C, whose responses are pratically overlapped, at the right sensor D. Graphs show double kinetic for sensor B, C and D. The response time (r) is very similar for all the series in the rinsing part.

In this case the Boltzmann sigmoidal function can be only applied successfully to series A devices, which are characterized by a quite fast kinetic although signal variation is again quite small and in this case noisy too. Devices B, C and D are all characterized by a double kinetic typical of diffusion phenomena. Interestingly enough, the first reative kinetic is quite fast and similar in times to those recorded for series A. In Table I response times and conductance variations in NO_2 and in NH_3, for all the investigated series, are reported. The typology of devices does not seem to play any effect on the device behaviour, the only appreciable difference is in the conductance variation below 1% for the series A devices. The difference in series A conductance variation is readily comprensible recalling that in this case device response is due to gas adsorption on an isolated single rope.

Table I. Resistance, response times and signal variation of series A, B, C and D

SERIES	R (kΩ)	τ_{NH3} (min)	Δ_{NH3} (%)	τ_{NO2} (min)	Δ_{NO2} (%)
A	600	10	1	3	1
B	3	10	15	3 + 5 (double kinetic)	20
C	3	10	15	3 + 5 (double kinetic)	20
D	1.4	10	16	3 + 5 (double kinetic)	20

4. Conclusions

In this paper we have fabricated and characterized in oxidising and reducing environments different types of solid state gas sensors based on a single rope or bundles of ropes of SWCNTs. Single rope devices have been fabricated by means of a FIB to create Pt microelectrodes at the ends of rope on starting substrate. Devices have been tested against traces of NH_3 and NO_2 in dry air. We have found that device geometry and nature of the electric contact do not seem to play any major role in the device behaviour. The most striking difference between SR and BR devices is in the dynamic range of the electrical response, quite small (about 1 %) for devices based on the electrical interaction of a single rope with the surrounding, reducing or oxidizing environment,. In NH_3 the molecules ropes interaction is completely irreversible and follows the model proposed by Strano with quite slow adsorption kinetic. In NO_2 such a behaviour is only observed for the single rope based device. Devices based on BR all show a behaviour typical of diffusion limited process. Interestingly the linear part has characteristic times which are very similar for all the devices in NO_2 environments..

References

1. Sinha n., Ma J., Yeow T.W, *Journal of Nanoscience and Nanotechnology* **6**,. 573, (2006).

2. Kong J., Franklin N.R., Zhou C., Chapline M.G., Peng S., Cho K., Dai H., *Science* **287,** (2000).

3. Collins P.G., Zetti A., Bando H., Thess A. and Smalley R.E, *Science* **278**,. 100, (1997).

4. Li J., Lu Y., Ye Q., Cinke M., Han J., Meyyappan M., *Nanoletters,* **3**,. 929, (2003).

5. Valentini L., Armentano I., Lozzi L., Cantucci S., Kenny J. M., *Materials Science and Engineering*, **C 24**,. 527, (2004).

6. Roy R.K., Chowdhury M.,. Pal A.K., *Vacuum* **77**,. 223, (2005).

7. Wongwiriyapan W., Honda S., Konishi H., Mizuta T., Ohmori T., Ito T., Maekawa T., Suzuki K., Ishikawa H., Murakami T., Kisoda K., Harima H., Oura K. and Katayama M., *Japanese Journal of Applied Physics*, **44**,. 8227, (2005).

8. Peng S., Cho K.J., *Nano Letters*, **3**, 347, (2003).

9. Nguyen H.-Q., Huh J.-S., *Sensors and actuators B* No. **117**, 426, (2006).

10. Zhang T., Nix M. B., Yoo B-Y, Deshusses M. A., Myung N. V., *Gas Sensor, Electroanalysis* **18**, 1153, (2006).

11. Strano S. et al., *Nature Materials* **4** (2005).

12. Dierking I., Scalia G., Morales P, *J. Appl. Phys.* **97**, 44309, (2005).

13. Vacca P., Massera E., Nasti I., Calò O., Polichetti T., ENEA, NOTA TECNICA: 2006.

14. Quercia L., Cerullo F., La Ferrara V., Di Francia G., Baratto C., Faglia G., *Phys. Stat. Sol.* **182**, 473, (2000).

LIQUID PHASE SENSORS

FIBER OPTIC SENSORS BASED ON PARTICLES LAYERS OF TIN DIOXIDE FOR CHEMICAL DETECTION IN WATER AND IN AIR ENVIRONMENTS

M. CONSALES, M. PISCO, P. PILLA, A. CUSANO, A. CUTOLO

Optoelectronic Division – Engineering Department, University of Sannio, Corso Garibaldi 107, 82100, Benevento, Italy

A. BUOSCIOLO

Department of Materials Engineering and Production, University of Napoli, P.le Tecchio, 80125, Napoli, Italy

M. GIORDANO

Institute for Composite and Biomedical Materials, CNR, P.le Enrico Fermi 1, 80055, Portici, Italy

R. VITER, V. SMYNTYNA

Department of Experimental Physics, Odessa National University, 65014, Odessa, Ukraine

In this work, the surprising sensing performances of opto-chemical sensors based on SnO2 particles layers against chemical pollutants either in air and water environment, at room temperature, are reported. The Electrostatic Spray Pyrolysis (ESP) method has been used to deposit the sensing coatings upon the distal end of standard fibers. This technique allows the fabrication of SnO2 layers composed of micron and sub-micron dimensions able to locally modify the profile of the optical near-field collected in the close proximity of the fiber tip. Such layers morphology leads to strong surface interactions between sensing coatings, analyte molecules and the evanescent contribute of the field, resulting in an excellent sensors sensitivity against chemical pollutants, even at room temperature.

1. Introduction

Metal Oxides (MOXs) are interesting materials widely exploited in gas sensing applications where the change in their electrical conductivity is measured to detect the interaction process between the surface complexes such as O^-, O^{2-}, H^+, OH^- reactive chemical species and gas molecules [1]. Thin films of MOXs have been extensively used for the formation of highly sensitive, fast responding, micro-machined and cheap gas sensors [1]. However, the advantages in using such metal oxide-based sensors have some drawbacks which, in some cases, limit their use in practice. In particular, their principle of operation and the high operating temperature at which they typically have to work to obtain the proper sensitivity lead to high power consumptions and to the impossibility to be exploited in aqueous environments for water quality monitoring applications. Here, the experimental results demonstrating the capability of optical fiber sensors based on SnO_2 particles layers to detect very low concentrations of chemical pollutants, both in air and water environments, at room temperature are reported. A simple reflectometric approach has been adopted, using standard single-mode optical fibers coated by structured films of tin dioxide. The layers, composed of micron and sub-micron sized particles are able to strongly perturb the emergent optical near-field profile and induce its local enhancement and focusing [3]. The surprising sensitivities arise from the fact that the interaction between the optical field and the analyte molecules occurs mainly on its surface by means of the evanescent part of the field.

2. Sensors fabrication

The deposition of SnO_2 sensitive layers onto the cleaved end of standard optical fibers, have been arranged in a reflectometric scheme. The main principle of the adopted configuration relies, therefore, on the measurement of the changes in the amount of power reflected at the fiber-film interface occurring as a consequence of the changes in the optical properties of the sensing interface. Such modifications are, in turn, caused by the interaction of the sensing layers with the target analyte molecules. The ESP method has been used for the SnO_2 particles layers deposition, which represents a relatively inexpensive processing method (especially with regard to equipment costs) which offers an extremely easy technique for preparing single or multilayered films of any composition without requiring high-quality substrates or chemicals. To achieve this aim, a specially designed system has been used, a detailed description can be found elsewhere [4, 5]. In particular, here 5 ml of an ethanol solution of $SnCl_4{\cdot}5H_2O$, with concentrations ranging from 0.01 to 0.005 mol/l, have been sprayed upon the sensors substrate, previously heated up to a temperature of 300°C by means of a resistive heater. When droplets of solution reach the heated substrate the chemical reaction of tin chloride with water vapor of solution, stimulated by the temperature, takes place with formation of the tin dioxide film [6]. The deposition process is typically followed by 1 hour of thermal treatment at 500°C in order to transform the eventually present SnO_x to SnO_2 and clean the films surface from other dopants like water or alcohol present in the initial solution.

3. Morphological and optical characterizations

Atomic Force Microscopy (AFM) and Scanning Near-Field Optical Microscopy (SNOM) analyses have been carried out for the morphological and optical characterization of the deposited samples, which provide quantitative information on the surface topography of the sensitive coatings and the knowledge of the relationship between the layer morphology and the optical near field collected in the close proximity of the probes [7]. To this scope, a system capable of simultaneous collection mode SNOM and normal force AFM imaging using the same tip has been used [3]. In Fig. 1.a is reported the 2D image of a SnO_2 layer deposited upon the optical fiber end using 5 ml of precursor solution with a concentration of 0.01 mol/l.

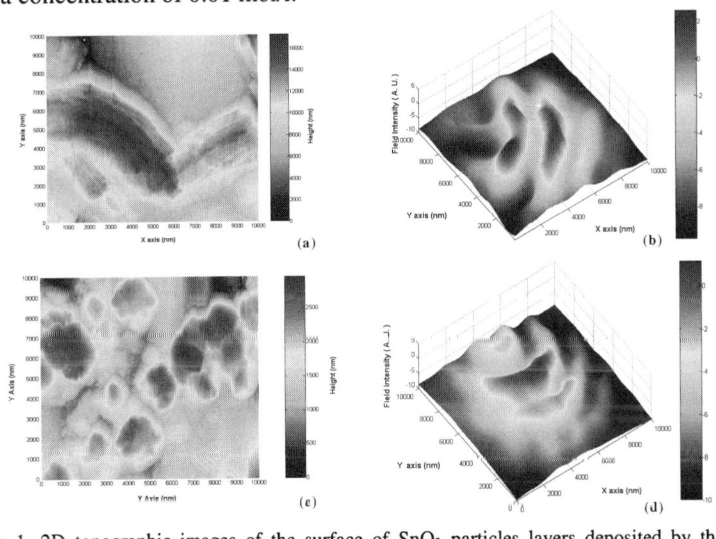

Figure 1. 2D topographic images of the surface of SnO_2 particles layers deposited by the ESP method using 5 ml of ethanol solution of tin chloride with a concentration of (a) 0.01 mol/l and (c) 0.05 mol/l and 3D images of the optical near field simultaneously collected by the SNOM probe in the same region of (b) Fig. 1.a; and (d) Fig. 1.c.

The image refers to a (10x10) μm^2 area centered on the fiber core and reveals an highly rough surface characterized by the presence of a number of SnO_2 peaks. It can be seen from Fig. 1.b that such layer morphology is able to induce strong modifications of the optical profile of the emergent near-field. As matter of fact, the typical Gaussian profile (emerging from uncoated optical fiber) appears strongly perturbed in correspondence of the SnO_2 peaks with lateral dimensions comparable with the light wavelength (λ =1550 nm). This effect, due to the high refractive index of the SnO_2 peaks (approximately 1.967) which, in turn, try to guide the light, is characterized by a localized enhancing of the optical near field combined with a strong increasing of the evanescent wave content [3]. This phenomenon was also observed in the case of isolated particles and confirmed by additional measurements carried out by SNOM technique in the reverse configuration [3]. In addition, by changing the ESP deposition parameters different SnO_2 film topographies and, thus, different optical near field profiles can be obtained (see Fig. 2.c,d). From the results obtained, it is evident that the SnO_2 sensing layers cannot be modeled as a standard Fabry-Perot cavity having a uniform thickness on the core region and, consequently, the reflectivity at the fiber–layer interface simply obtained by the sum of multiple reflections [8]. In fact, due to the local enhancement and focusing of the optical near-field, most of the interaction with the analyte molecules occurs on the sensitive coatings surface by means of the evanescent part of the field and not in its volume. This improves the performances of the proposed metal oxide-based fiber optic sensors since they rely mainly on surfaces interactions. As matter of fact, recently a comparison between the sensing performances of SnO_2-based sensors characterized by almost flat and highly rough surfaces demonstrated that sensitive layers with very rough morphologies exhibited the best sensing characteristic, either in term of sensitivity and responses dynamic [9]. In addition, a sensing mechanism based on surface interactions is also advantageous in terms of sensor response times, since no diffusion of the analyte in the volume of the sensitive layer is needed.

4. Results

The room temperature sensing performances have been tested either in air (against toluene and xylene vapors and gaseous ammonia) and water environments (against aqueous ammonia). To this aim, a simple optoelectronic set up operating at single wavelength (1310nm) was implemented enabling the measurement of the reflected signal at the interface and thus of the overlay reflectance [10, 4].

4.1. *In air chemical detection*

The experimental results obtained by a tin dioxide sensor fabricated with the ESP method using 5 ml of precursor solution with a concentration of 0.01 mol/l are shown in Fig. 2.a and 2.b. The SnO_2 sensor exhibits a surprising capability to detect toluene and xylene vapors at room temperatures, with excellent sensitivities of $4.6 \cdot 10^{-3}$ ppm^{-1} and $1 \cdot 10^{-2}$ ppm^{-1} for toluene and xylene, respectively, leading to very low sensors resolutions (few tens of ppb). The obtained sensitivities are an order of magnitude higher than those obtained with optical fiber sensors coated by other sensitive materials, such as Single-Walled Carbon Nanotubes (SWCNTs) ($2 \cdot 10^{-4}$ ppm^{-1} and $9 \cdot 10^{-4}$ ppm^{-1} respectively) which were simultaneously tested with the SnO_2-based probes.

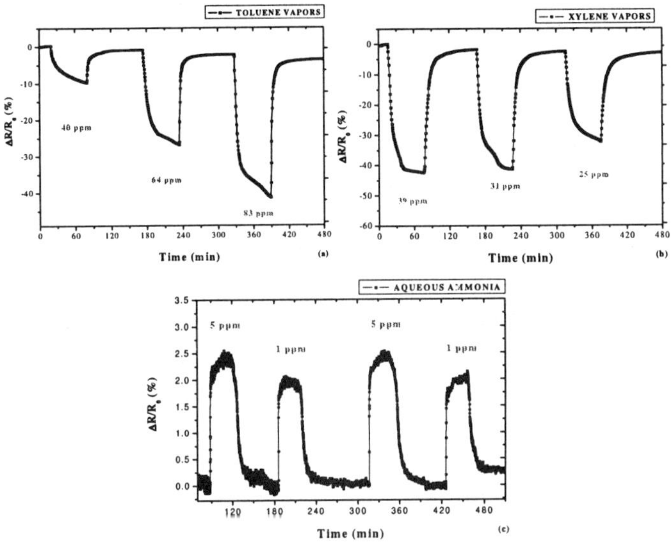

Figure 2. 2D topographic images of the surface of SnO_2 particles layers deposited by the ESP method using 5 ml of ethanol solution of tin chloride with a concentration of (a) 0.01 mol/l and (c) 0.05 mol/l and 3D images of the optical near field simultaneously collected by the SNOM probe in the same region of (b) Fig. 1.a; and (d) Fig. 1.c.

These excellent results can be explained by the fact that, as already mentioned in the previous paragraphs, the interaction between analytes molecules and sensitive material occurs mainly on the SnO_2 surface by means of the optical near-field, with a significant enhancement of the evanescent part of the field. On the contrary, the probe based on the SnO_2 particles layers exhibited slightly high response times (respectively 35 and 25 minutes for toluene and xylene), approximately four times higher than those obtained with the sensor coated by the SWCNTs layer (8 and 9 minutes, respectively). No changes in the sensors responses have been observed upon exposure towards different concentrations of gaseous ammonia (ranging from 10 ppm to 1000 ppm) at room temperature, as well as negligible responses have been obtained as a consequence of changes in the humidity content inside the test chamber. These results reveal some interesting selectivity characteristics of SnO_2-based opto-chemical probes. In addition, further results demonstrate that sensors coated by SnO_2 films with different topographies and, thus, different optical near-field profiles are characterized by very dissimilar sensing performances. This feature is very useful in chemical sensing applications, in fact the correlation between data collected from sensors coated by layers exhibiting highly dissimilar characteristics could enhance the features extraction from a hybrid system by means of pattern recognition methods.

4.2. *In water chemical detection*

Very interesting results have also been obtained in water environments. Fig. 2.c shows the relative reflectance changes of a sensor coated by a SnO_2-particles layer, fabricated using 5 ml of precursor solution with a concentration of 0.01 ml/l, in correspondence with four injections of aqueous ammonia into the test chamber, with repeated concentrations of 1 and 5 ppm. Upon each injection, the SOF sensor exhibits an increase of $\Delta R/R_0$, as high as $2 \cdot 10^{-2}$ and $2.4 \cdot 10^{-2}$ for 1 and 5 ppm, respectively. This is due to the increase of the fiber-film interface reflectance caused by the interaction between analyte molecules and sensitive layer. Also in this case, due to the high surface roughness, a strong superficial interaction between the ammonia molecules an the sensitive layer is expected before they are adsorbed within the film itself. This seems to be confirmed by the response dynamics: the fast variation is attributed to the aforementioned superficial effects, while the slow achievement of the equilibrium condition with the external environment to the successive molecule adsorptions within the SnO_2 overlay. However, further analyses are currently being performed in order to better clarify this aspect. The SOF sensor also exhibits good desorption features, as evidenced by the complete recovery of the steady-state value after each ammonia exposure. Furthermore, the behaviour of the sensor output in correspondence with the same ammonia concentration is highly repeatable, either in terms of response intensity and of response dynamics. An analysis of the mean response and recovery times of the tested SOF sensor revealed their slight dependency on the ammonia concentration, with the mean response time (approximately 6 minutes) lower than the mean recovery time (approximately 14 minutes). In addition, considering the sensor sensitivity in the range of 1–5 ppm ($4 \cdot 10^{-4}$ ppm^{-1}) and the minimum detectable value possible with the employed instrumentation [4], a resolution of approximately 1.5 ppm has been estimated, while considering the sensor sensitivity in the range of 0–1 ppm ($7.7 \cdot 10^{-3}$ ppm^{-1})

a resolution as high as approximately 80 ppb can be obtained. It's worth noting, however, that the SnO_2-based sensors are able to detect ammonia in water but not in air environment. This could be attributed to the first interaction occurring between water molecules and sensing interface when the optoelectronic probes are immersed in water. In fact this could cause superficial modifications allowing the detection of aqueous ammonia. This aspect is also very interesting and should be better investigated.

5. Conclusion

In conclusion, in this contribution, we report on the excellent sensing capabilities of near-field opto-chemical sensors based on SnO_2 particles layers to detect very low concentrations of chemical pollutants in air and water environments, at room temperature. The sensing probes, deposited upon the distal end of standard single-mode optical fiber by means of the ESP deposition technique and characterized by AFM and SNOM analyses, demonstrated their unusual capability of locally enhancing the optical near field collected in the close proximity of the fiber coated end. From this phenomenon arise the excellent sensing performances of the SnO_2-based transducers, as most of the interactions between the sensing coatings and analytes molecules occur on the SnO_2 surface by means of the optical near-field, involving the evanescent part of the field.

References

1. I. Simon, N. Barsan, M. Bauer, U. Weimar, **Sens. Actuators B: Chem. 73** (2001) 1–26.
2. A. Buosciolo, A. Cusano, P. Pilla, M. Consales, M. Pisco, M. Giordano, A. Cutolo, **Optics Express, Vol. 15**, No. 8 (Apr 2007)
3. A. Cusano, M. Consales, M. Pisco, A. Buosciolo, P. Pilla, R. Viter, V. Smyntyna, A. Cutolo, M. Giordano, **Appl. Phys. Lett., 89**, 111103 (2006).
4. M. Pisco, M. Consales, S. Campopiano, R. Viter, V. Smyntyna, M. Giordano and A. Cusano, **IEEE Journal of Lightwave Technology, 24** (12), 2006.
5. Y. Matsui, M. Mitsuhashi , Y. Goto, **Surface and Coatings Technology, 169**, pp. 549-552 (2003).
6. J. Prikulis, H. Xu, L. Gunnarsson, M. Kall and H. Olin, **J. Appl. Phys. 92**, 6211-6214 (2002).
7. H.A. Macleod, **Institute of Publishing**, Bristol and Philadelphia, 2001.
8. M. Consales, M. Pisco, P. Pilla, A. Buosciolo, R. Viter, V. Smyntyna, M. Giordano and A. Cusano, 2005. **Proceedings of IEEE**, 22 Oct. 2006.
9. M Consales, S Campopiano, A Cutolo, M Penza, P Aversa, G.Cassano, M Giordano and A Cusano, **Sens. and Actuators B 118**, 232 (2006).

SYNTHESIS AND CHARACTERIZATION OF A POLYPYRROLE NANOWIRE MODIFIED ELECTRODES FOR AMPEROMETRIC DETECTION OF AMMONIA IN DRINKING WATER *

VANESSA BIAGIOTTI*, FEDERICA VALENTINI, DANILA MOSCONE, GIUSEPPE PALLESCHI

Università degli studi di Roma Tor Vergata, Dipartimento di Scienze e Tecnologie chimiche, via della Ricerca Scientifica 1, 00133 Roma

**corresponding author: vanessa.biagiotti@uniroma2.it*

Higly oriented pyrolitic graphite (HOPG) electrodes were modified with conductive polypyrrole nanowires obtained by chemical oxidation. Morphological characterizations were carried out and the sensor was investigated analytically for ammonia determination in water. Satisfactory results as linear concentration range (10-200 μM), linear regression equation ($y/\mu A = 2.36\ x/\mu M - 1.36$), reproducibility (R.S.D. %= 3 n=3) and limit of detection (LOD=5 μM) were obtained. Finally real drinking water samples were analyzed and the recovery study showed that there was no matrix effect on the sensors performances..

1. Experimental

1.1. *Chemical synthesis of polypyrrole nanowire*

100 ml of a solution containing the monomer (0.18 M pyrrole) and a surfactant as template (0.18 M SDS) were mixed with 100 ml of a solution containing an oxidant (0.26 M $FeCl_3$) and a stabilizer (3% PVA). The final solution was kept at 0 °C for 24 h and then the resulting polypyrrole precipitate was vacuum-filtered and washed copiously with distilled water, methanol and acetone for several times. Finally, it was dried overnight in oven, at 37°C [1].

1.2. *Amperometric study*

According to the literature [2] an applied potential + 0.30 V (vs Ag/AgCl) is sufficiently low to obtain some undoping of the polymer and the undoping is additionally favored by the interaction with ammonia. Measurements were realized, in batch mode, using a 50 mM borate buffer pH 10, containing 0.14 M NaCl. The buffer pH value was fixed at 10 to obtain a large amount of ammonia in solution (pKb= 9.25). In addition, the introduction of chloride in the sample solution resulted in a doping process, which was observed as a current signal. The introduction of chloride into the buffer solution contributed to eliminate this effect.

1.3. *Synthetic water*

Synthetic water was obtained by the addition of several cations and anions normally found in drinking water, to evaluate interference effects on the amperometric NH3 response; it is an aqueous solution containing 1.1 mM of MgSO4, CaCl2, KNO3, Na2CO3

1.4 *Recovery study*

Recovery studies were carried out in amperometric batch mode using 0.5 M borate buffer + 1.4 M NaCl, pH 10, to minimize pH changes in real sample analysis. Buffer (2 ml) was diluted directly in drinking water (8 ml) spiked by ammonium chloride solution to have a final concentration of 20 and 50 μM.

2. Results and Discussion

The properties of the modifying materials depend not only of the chemical composition, but also on their morphologies. On account of their small size, nanoscale materials possess unique properties including electrical conductivity, higher signal/noise ratio and the possibility to be functionalized, so as to make them suitable for a wide range of applications, such as sensor chips, biosensors, nano-arrays, nanomotors [3]. The synthesis of nanoscale materials has attracted great interest in the past ten years; conducting polymers have also been intensively studied for their one-dimensional conjugated structures and adjustable conductivity [4]. Polypyrrole nanofibers, nanotubes and nanowires can be synthesized by an electrochemical polymerization approach using a template method [5]. This method allows to modulate the final polymer properties, such as shape, length, diameter using different electrochemical techniques. Normally polycarbonate or alumina membranes having nanometer pores were used as templates to grow the polymeric nanostructures. The polymers grow into the pores of the template membrane resulting in a cylindrical shape. In this work polypyrrole nanowires were synthesized by chemical oxidation, because it is simpler and cheaper to produce large quantities of nanomaterial. Using this method, FeCl3 was employed as oxidant and PVA as stabilizer while SDS was chosen as surfactant. Polypyrrole nanowires were used to modify HOPG electrodes by casting 6 μl of a 1 mg/ml PPy nanowires dispersion using acetonitrile as solvent. Fig. 1 shows FE-SEM images of the polypyrrole nanowires synthesized by chemical oxidation. It can be noted the characteristic topography of the chemically polymerized conducting polymer: presence of fibrillar structures that confer an interesting coral-like morphology to the deposit. From the SEM pictures it is possible to estimate the nanowire dimensions; in particular the lengths range from 1 to 2 μm whereas the diameters are comprised between 100-150 nm. It is well known that experimental conditions influence the morphology of the polymer: nanowires can be obtained using a 0.07 M and adding an emulsion agent such as PVA 3% [1]. After casting, the electrode can be used to measure ammonium solution by the amperometric method.

FIG. 1

2.1. *Analytical results*

In Table 1 calibration equations and analytical parameters are reported. Measurements were carried out initially in the selected buffer to study ideal conditions and then were repeated diluting a concentrated buffer solution with synthetic water, to evaluate the matrix effect. Good results were obtained in terms of limit of detection, which was found to be comparable to that fixed by Italian law for ammonia (30 μM for drinking water) [6].

Table 1

Measurments performed in	Linear range (μM)	Slope (μM)	[a] R.S.D. % (slope)	Bias (μA)	[a] R.S.D. % (bias)	LOD[b] (μM)	Sensitivity (μA μM^{-1} cm^{-2})
50 mM borate buffer pH 10 + 0.14 M NaCl	10 - 200	2.36	3	-1.36	4	5	33.7
0.5 M borate buffer pH 10 + 1.4 M NaCl (diluite by synthetic water)	10 - 100	16.6	4	99.8	3	10	237

Applied potential: + 300 mV vs Ag/AgCl
[a] n=3
[b] LOD= 3S/N

The results obtained using synthetic water were used to calculate recovery values when real drinking water samples were analyzed. Recovery values are shown in Table 2 and ranged from 92 to 97 %. These results confirm that there was a negligible matrix effect occurred in real samples using our developed sensor.

Table 2

Samples	NH_4^+ added (μM)	NH_4^+ found before spiking (μM)	Expected value (μM)	Measured value (μM)	R.S.D. % (n=3)	Recovery %
Mineral water "Santa Croce"	50	22	72	70	2.8	97
Mineral water "Sorgente"	50	11	61	56	3.2	92

3. Conclusions

A sensor for ammonia detection in drinking water were developed using conducting polypyrrole nanowires. Satisfactory results in terms of sensitivity and detection limit are obtained, because of the higher surface, electrical conductivity and signal/noise ratio. The sensor could be successfully used to detect NH3 in real samples.

4. Acknowledgments

This research was supported by Grants from FIRB N° RBNE01MBTC-002 Italian Project.

References

Chang He, Chunhe Yang, Yongfang Li; Chemical synthesis of coral-like nanowires and nanowire networks of conducting polypyrrole; Synthetic metals 139 (2003) 539-545

M. Trojanowicz, A. Lewenstam, T. Kski, I. Lahdesmaki, W. Szczwpek; Flow injection amperometric detection of ammonia using a polypyrrole-modified electrode and its application in urea and creatinine biosensors; Electroanalysis, 8 (3) (1996) 233-243

Ozin G.A.; Nanochemistry: synthesis in diminishing dimensions; Adv. Mater.,4, (1992), 612

Reynolds et al.; Handbook of Conducting Polymers , 2ed; Marcel Dekker, New York , (1998), 483

Shiratori, Seimei Sha; Mori, Seiji; Ikezaki, Kazuo; Wire bonding over insulating substrates by electropolymerization of polypyrrole using a scanning micro-needle; Sensors and Actuators B, 49, (1-2) (1998) 30-33

D.P.R. 24 maggio 1988, n. 236 Attuazione della direttiva CEE numero 80/778 concernente la qualità delle acque destinate al consumo umano, ai sensi dell'art. 15 della L. 16 aprile 1987, n. 183 Pubblicato nella Gazzetta Ufficiale 30 giugno 1988 n. 152, S.O. "The Italian Official Bullettin"

AZULENE BASED GUEST-HOST POLYMERIC SENSORS

ANNA CASTALDO, LUIGI QUERCIA, GIROLAMO DI FRANCIA

ENEA Enea UTS FIM-MATNANO Centro di ricerche di Portici,
80055 Portici(NA) Italy
anna.castaldo@portici.enea.it

In this study we present a new class of sensors based on polysilsesquioxane-azulene system, that could detect various analytes by changing their conductivity and colour. This system represents one of the first example of guest-host polymer used in the sensor field. Taking advantage of the polysilsesquioxane properties in humid environment and of the brilliant blue azulene, an isomer of naftalene with high dipolar moment, we develop polymeric thin film that could work in high humidity conditions and detect different analytes useful in water analysis in a twofold response in terms of color change and conductivity variation.

1. Introduction

One area of interest in sensors field is the real-time in situ water analysis, that could be performed by electronic noses. Methods for real-time in situ analysis are needed for the monitoring and detection of pollutants and nutrients in water bodies. For the analysis of a given sample with an e-nose, the sample headspace (HS), i.e. the gas volume above the substance to be analysed, or a representative part of it has to be brought into the sensor chamber. The analysis of this headspace using chemical sensors reveals information about the nature or the composition of the sample. An essential property that the constituents of the sensor array need to have is to survive in high humidity environments acting as sensors of certain analytes and this is not simple, especially for polymeric sensors.

A class of polymers extremely useful at this purpose, in our opinion, is that of hybrid organic-inorganic polysilsesquioxanes. Our involvement with these polymers stems from the unique properties of hybrid polymers [4], [5], [6]. In particular we choice the Poly[(propylmethacryl-heptaisobutyl-PSS)-co-styrene], reported in Fig.1.

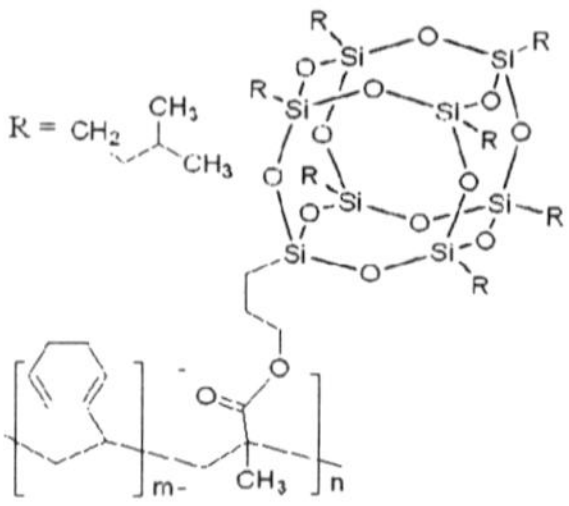

Fig. 1. Poly[(propylmethacryl-heptaisobutyl-PSS)-co-styrene] PSS 25 wt. %

To take advantage of the silsesquioxanic cages we decide to host a small molecule, azulene, with the striking property of a brilliant blue [7] to develop polymeric film, resistant in high humidity environment, that could reveal different analytes in a twofold response in terms of color change and conductivity variation. Until now only few studies have appeared in the literature regarding host-guest polymeric sensors. This is due mainly to the insulating properties of this kind of systems. Clearly, once resolved this problem it is possible to develop films much more homogenous than the poly-composite ones, facing problems as the electrical noise (generally related to the inhomogeneous fillers dispersion [8]) and the selectivity (connected to the proper choice of the guest [9]). Well, polysilsesquioxanes are basically insulating, but in presence of water they can increase conductivity of many magnitude order [10].We use this property to create an innovative guest-host sensor system. At fixed humidity we have a constant conductivity, used as baseline. If the azulene guest interacts with a certain analyte we note an increase or a decrease of current. In literature is well known that also fluorescence of azulene moiety can be used to reveal analytes in wet chemistry, for example a fluoride probe is based on this hydrocarbon [11].

2. Experimental

We have studied poly(propylmethacryl-heptaisobutyl-POSS)-co-styrene with a POSS cage content of 25% in weight. Azulene 99% was purchased by Sigma-Aldrich. Thin films of the guest-host polymer were deposited on glass substrates by means of spin coating technique. The solvent used for dissolving polymer and host, in a 3 % wt ratio purchased by Sigma-Aldrich, was THF. Sensing devices were alumina substrates with gold interdigitated contacts, 3000 Å thick, on which polymeric thin film (500 nm) are deposited. The dc sensing response of polysilsesquioxanes thin film based devices has been monitored in a Gas Sensor Characterization System (GSCS) in which the relative humidity percentage was accurately controlled by a procedure already described [12]. Also ammonia gas has been introduced in the chamber in the desired quantity (1ppm) by a proper dilution starting from 500ppm. Chlorine has been create in situ dissolving sodium hypochlorite in water and then bubbling nitrogen to conduce in gas phase the gas deriving by the equilibria:

$$NaOCl + H_2O \rightarrow Na^+ + HOCl + OH^-$$
$$HOCl + H^+ + Cl^- = Cl_2 + H_2O$$

3. Results and Discussion

POSS sensors based on azulene guest-host system are thin blue films that at fixed relative humidity are conductive. This conductivity and also the colour can be modified in presence of various analytes. In this study we pursue the idea of develop innovative polymeric sensors capable of working in high humidity condition (e.g. head space of a pipe) and to monitor some parameters of interest in water analysis. In particular we have as target ammonia and chlorine.
Ammonia (NH_3) is a colourless gas with a strong pungent odour. When dissolved in water, normal ammonia (NH_3) reacts to form a ionized species called ammonium (NH_4^+):

$$NH_3 + H_2O = NH_4^+ + OH^-$$

Tests for ammonia usually measure TAN total ammonia nitrogen (NH_3 plus NH_4^+) and this value has to be inferior to 0,5mg/L (7,2 ppm) by Italian law (D. Lgs 31/2001). In Fig. 2 is reported the response of our sensor to TAN

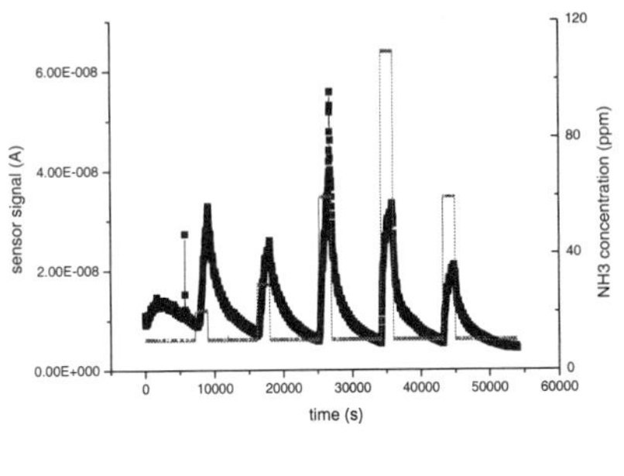

Fig. 2. Response to TAN at 80 RH%

In effect, we observe a current increase in presence of the analyte. There is not a simple correlation with the concentration of the gas NH_3 that we introduce into the system, because equilibrium has to establish in the film. Anyway this effect is due to azulene guest, that interacts with the acid ammonium ion. In Fig. 3 this aspect is cleared.

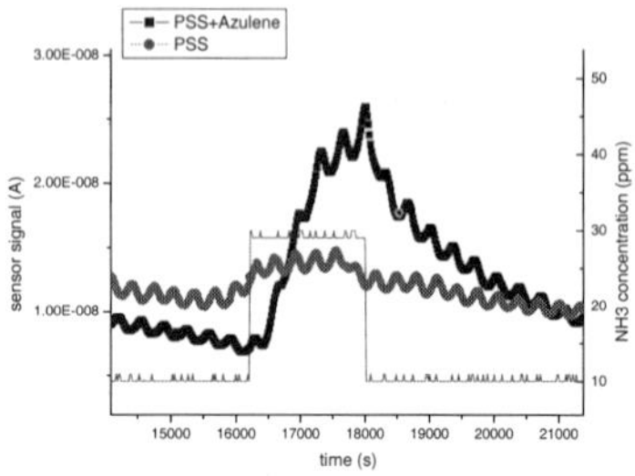

Fig. 3. Azulene role.

The porous PSS in presence of ammonia exhibits only a negligible effect of current increase (red line). The strongest effect (dark line) is mainly due to the azulene ring that interacts with ammonium ion. This effect is subsequent to the formation of this ion inside the film. In other terms, ammonia penetrate in the film and then forms ammonium ion, species interacting with the azulene guest.

Another example of analyte that we have tested is chlorine. Chlorine usually is added to water as the gaseous form or as sodium or calcium hypochlorite. Chlorine gas rapidly hydrolyzes to hypochlorous acid according to the following equation:

$$Cl_2 + H_2O = HOCl + H^+ + Cl^-$$

In general the concentration of hypochlorous acid depends on the pH. The water pH determines if free Cl_2 becomes hypochlorite (OCl-), or hypochlorous acid (HOCl) which kills organisms 40-80 times more effectively.

The two chemical species formed by chlorine in water, hypochlorous acid and hypochlorite ion are commonly referred to as “free available” chlorine. The presence of a free chlorine residual is an indicator of adequate disinfection. Typically, the free chlorine residual is adjusted to maintain a minimum level of 0.2 mg/L Cl_2 throughout the distribution system. Our sensor could detect chlorine in equilibrium with hypochlorite in water increasing its conductivity. The experiment reported in Fig 4 to detect chlorine is a dynamic measure performed comparing responses to RH% of the sensor exposed to vapour obtained bubbling nitrogen in pure water (red line) and in chlorine solution. In this manner we are sure that relative humidity is the same in each point and the difference is attributable only to the chlorine. It is important to note that at high humidity response to chlorine is major and this a good result because in our idea these sensors have to work in high humidity environments.

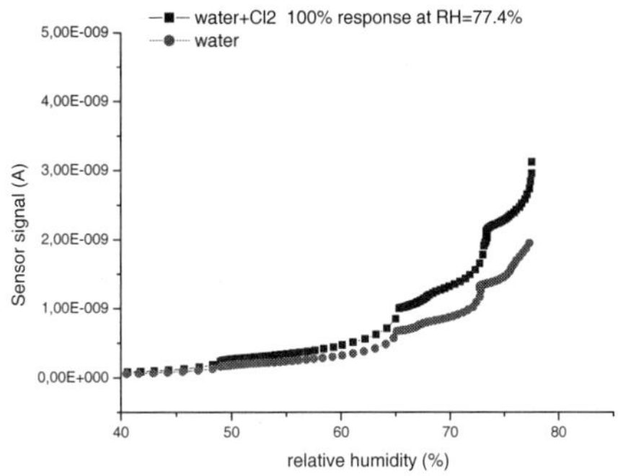

Fig. 4. Chlorine sensing

Conclusions

In this study we have tested sensing properties of polysilsesquioxanic film containing azulene with respect to total ammonia nitrogen, TAN, and with respect to chlorine, two basilar parameters in water analysis. Clearly, it is possible to optimize the system and to work also on the fluorescence variation in presence of other analytes, pursuing the general goal of a multiparametric sensor.

Aknowledgements

We would like to thank Dr Ettore Massera for preliminary experiments on the sensors fluorescence.

References

[1] C. Bastos, N. Magan, Sens. And Act. B 116 1-2 (2006) 151-155.

[2] Y. Sakai, Y. Sadaoka, M. Matsuguchi, Sens. And Act. B 35-36 (1996) 85-90.

[3] C. Zhang, F. Babonneau, C.Bonhomme, R.M.Laine, C. L. Soles, H.A. Hristov, A.F.Yee, J.Am. Chem. Soc. (1998), 120, 8380-8391.

[4] Patel RR, Mohanraj, Pittmann CU, Journal of Polymer Science Part B Polymer Physics 44 (1) (2006) 234-248.

[5] Striolo A., McCabe C, Cummings PT, Journal of Physical Chemistry B
109 (30), (2005), 14300-14307.

[6] Zheng L., Waddon A. J, Farris R. J., Bryan Coughlin E. Macromolecules, 35, (2005) 2375-2379.

[7] T. Zielinski, M. Kedziorek, J. Jurczak Tetrahedron Letters 46 (2005), 6231-6234.

[8] Quercia L.; Loffredo F.; Di Francia G.; Sensors and Actuators B, 109, (2005),153.

[9] David James, Simon M. Scott, Zulfiqur Ali, and William T. O'Hare Microchim. Acta 149, (2005), 1–17.

[10] A. Castaldo, A. Cassinese , P. D'Angelo , L. Quercia, G. Di Francia *submitted to* JAP (2007).

[11] H. Salman, Y. Abraham, S. Tal, S. Meltzman, M. Kapon, N. Tessler, S. Speiser, Y. Eichen, Eur. J. Org. Chem. Vol.2005, 11 Pages: 2207-2212.[12] L. Quercia, F. Loffredo, B. Alfano, V. La Ferrara, G. Di Francia, Sensors and Actuators B, 100 (2004), 22-28.

OPTOELECTRONIC NANOSENSORS BASED ON CARBON NANOTUBES NANOCOMPOSITES FOR THE DETECTION OF ENVIRONMENTAL POLLUTANTS IN AIR AND WATER ENVIRONMENT

M. CONSALES, A. CRESCITELLI, A. CUTOLO, A. CUSANO

Optoelectronic Division – Engineering Department, University of Sannio, Corso Garibaldi 107, 82100, Benevento, Italy

S. CAMPOPIANO

Department for Technologies, University of Naples Parthenope, Via Medina 40, 80131 Napoli, Italy

M. PENZA, P. AVERSA

Materials and New Technologies Unit, ENEA, Strada Statale 7, KM. 706, 72100, Brindisi, Italy

M. GIORDANO

Institute for Composite and Biomedical Materials, CNR, P.le Tecchio, 80, 80125, Napoli, Italy

In this work, the feasibility to exploit optoelectronic chemo-sensors based on cadmium arachidate (CdA)/single-walled carbon nanotubes (SWCNTs) composites for detection of chemical pollutants both in air and water environments has been investigated. The nanocomposite sensing layers have been transferred upon the distal end of standard optical fibers by the Langmuir-Blodgett (LB) technique. Single wavelength reflectance measurements (λ=1310 nm) have been carried out to monitor chemicals concentration through changes in the optical length of the Fabry-Pérot (FP) cavity induced by the interaction of the sensitive layer with the analyte molecules. The preliminary experimental results evidence the good potentiality of these fiber optic nanosensors to detect toluene and xylene at ppm level both in air and water environments at room temperature.

1. Introduction

To date, SWCNTs are building blocks considered as the most promising functional nanomaterial for future miniaturized gas nanosensors due to their hollow nanostructure and high specific surface area which provide attractive characteristics for gas sensing applications[1,2]. In fact, their unique morphology confers to SWCNTs the excellent capability to reversibly adsorb molecules of environmental pollutants undergoing a modulation of their electrical, geometrical and optical properties, such as resistivity, dielectric constant, thickness etc. [3-5]. In particular, the capability of SWCNTs to change their dielectric constant and thickness as consequence of the adsorption of target analyte molecules, has been demonstrated for the first time in 2004 [5]. LB films consisting of tangled bundles of SWCNTs, transferred onto the optical fiber tip by using a linker–buffer material of CdA pre-deposited on the sensor surface to promote their adhesion, were used as sensitive coatings for the development of volatile organic compounds (VOCs) optoelectronic sensors. In 2005, thin films of SWCNTs with different thicknesses were successfully deposited directly upon the optical fiber surface by a modification of the LB process [6], resulting in an improvement of the sensing performances of un-buffered configurations with respect to buffered cases, both in terms of sensitivity and response times. In addition, recently the excellent sensing properties of carbon nanotubes have been also used for the detection of chemical pollutants in aqueous environments at room temperature [7], demonstrating the feasibility of such nanostructured materials to be successfully exploited as sensitive coatings for a wide range of environmental monitoring applications. However, the weak adhesion of the nanotubes to the fiber substrate and the low repeatability of the deposition process, especially concerning the distribution of the tubes upon the fiber tip, represent the two main limitations. The alignment of carbon nanotubes upon the sensors substrate [8] as well as the embedding of controlled quantity of them inside an host-matrix of a foreign material for the synthesis of nanocomposites with tailored amount of nanotubes-filler contents could be two possible ways to overcome these drawbacks. In this contribution, our attention has been focused on the latter solution. As matter of fact, the sensing performances of optical chemo-sensors based on nanocomposite overlays of SWCNTs embedded in a CdA matrix have been investigated against several chemical pollutants both in air and water environments, at room temperature.

2. Sensors Fabrication

The exploited sensor configuration is based on a FP-type sensing cavity realized on the terminal face of a standard optical fiber in order to form a low-finesse interferometer. To this purpose, Langmuir-Blodgett deposition process has been chosen as a way to transfer thin films of SWCNTs-based nanocomposite upon the distal end of properly prepared single-mode optical fibers. The CdA has been chosen as host-matrix material to incorporate the SWCNTs in the nanocomposite due to its peculiar amphiphilic molecular structure suitable for LB deposition process. It has also been chosen due to the know-how already experienced by the authors in the integration of such material and the optical fiber substrates [5,9]. Two separate solutions of arachidic acid in chloroform and SWCNTs in chloroform have been mixed in order to prepare a final solution of chloroform with arachidic acid (0.25 mg/ml) and SWCNTs (0.19 mg/ml). The concentrations and the volumes of the initial solutions were chosen to obtain a weight percentage of the filler-component (SWCNTs) with respect to the matrix-component (CdA) of approximately 75 wt. %. However, different concentrations of arachidic acid and SWCNTs in the final solution could also be exploited for the preparation of composites with different weight percentages. The mixed solution was then accurately dispersed and stirred in an ultrasonic bath for 1 h. Subsequentially, 160 μl of the mixed solution were spread onto a sub-phase constituted by acetate buffer with CdCl2 10-4 M. The pH and the temperature of the sub phase were kept constant at 6.0 and 20° C, respectively. The monolayer of the nanocomposite was compressed with a barrier rate of 15 mm/min up to a surface pressure of 27 mN/m. The single composite monolayer was deposited upon the fiber surface with a dipping rate of 14 mm/min. The optical fibers used for the deposition have been previously accurately polished from the acrylic protection and cleaved with a precision cleaver. Then, they have been washed in chloroform and dried with gaseous nitrogen to be ready for the SWCNTs composite deposition. Repeated dipping of the fiber substrates through the condensed Langmuir layer have been performed, resulting in the deposition of multilayered CdA/SWCNTs films one monolayer at a time. A detailed morphological and structural characterization of the carbon nanotubes-based composites is reported elsewhere [10]. In particular, from X-ray diffraction measurements performed on LB films composed of 20 monolayers of CdA/SWCNTs composites (with SWCNTs-filler weight percentages ranging from 0 to 75 wt.%), deposited on glass substrates, nanocomposite multilayer periods in the range 5.51-5.56 nm have been estimated, which is in agreement with the total length of the CdA molecule reported in literature [11]. The obtained values indicate that the embedded SWCNTs do not influence the periodicity of the CdA molecule and that the structural order of the CdA host-matrix is maintained also in the nanocomposite, even at high content of SWCNTs-filler.

3. Results

The investigation of the room temperature sensing capabilities of the fabricated nanosensors has been carried out both in air and water environments. Here, the attention is focused on toluene and xylene detection by using a sample fabricated by 20 monolayers of nanotubes-based composite with a SWCNTs-filler weight percentage of 75 wt.%, whose thickness is estimated to be of approximately 56 nm. To the aim, the reflectometric probe has been optically interrogated exploiting a proper optoelectronic set-up operating at single wavelength (1310nm) enabling the continuous monitoring of the chemicals concentrations within the test ambient through the changes in the fiber-film reflectance [5,12].

3.1. *Chemical Detection in Air Environment*

For the detection of toluene and xylene vapors in air environment at room temperature, the fabricated sensor has been settled in a properly designed and realized cylindrical test chamber, having a volume of approximately 1500 ml. The bubbling method has been exploited for the generation of the analytes vapors and dry-air has been used as carrier gas to transport the individual VOC. The total flow rate per exposure has been held constant at 2000 ml/min, and the gas flow rate has been controlled by a mass flow-meter driven by a controller-unit. Preliminary results are shown in Fig.1.a where the relative reflectance changes ($\Delta R/R_0$) occurred as a consequence of the exposure to xylene vapors have been reported. It can be observed that the analytes adsorption within the SWCNTs-based nanocomposite overlay leads to changes in the optical length of the FP sensing cavity, resulting in a decrease of the fiber-film reflectance. The results also reveal the capability of the optoelectronic transducer to detect very low concentrations of the tested pollutants at ppm levels, as well as its good attitude to recover the initial baseline signal upon the complete analyte molecules desorption. Furthermore, a good linearity in the sensor response was observed (reported in Fig. 1.b) towards the two organic vapors in the investigated range (0-83 ppm and 0-44 ppm for toluene and xylene, respectively). Higher sensitivity against xylene vapors ($1.2 \cdot 10^{-3}$ ppm^{-1}) was obtained in respect to toluene ($5 \cdot 10^{-4}$ ppm^{-1}). Also, an analysis of the response (t_{10-90}) and recovery (t_{90-10}) times revealed that the opto-chemical probe coated by 20 monolayers of CdA/SWCNTs nanocomposite is characterized by a faster dynamics in the case of toluene exposure (32 and 39 minutes, respectively) than in the case of xylene (36 and 41 minutes, respectively). It is worth noting that the relatively slow response times can be attributed to the high number of composite monolayers (and thus to the overlay thickness used for sensor fabrication). Also, the content of SWCNTs can influence the diffusion times as well as the sensor sensitivity. These aspects are actually under investigation. In addition, the sensing performances of the fabricated probe can be compared with the results obtained with 4 monolayers of SWCNTs directly deposited upon the fiber end, demonstrating toluene and xylene sensitivities of approximately $4 \cdot 10^{-4}$ ppm^{-1} and $8 \cdot 10^{-4}$ ppm^{-1} and response times of 9 and 11 minutes, respectively [12]. The results obtained reveal that slightly higher sensitivities have been obtained by using CdA/SWCNTs composites, combined with longer response times, probably due to the different thickness in the two cases (approximately 4 nm and 56 nm for un-buffered and composite-based configuration, respectively).

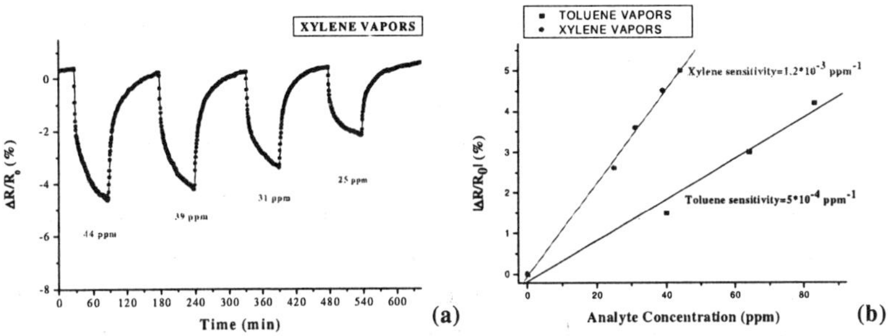

Figure 1. (a) Time response of the opto-chemical nanosensor coated by 20 monolayers of CdA/SWCNTs nanocomposite (75 wt. %), exposed to different concentrations of xylene vapors, at room temperature, and (b) comparison between the sensor characteristic curves obtained for toluene and xylene.

3.2. *Room Temperature Detection in Aqueous Environment*

Once verified their excellent VOCs adsorption capabilities in air at room temperature, the sensing characteristics of the LB CdA/SWCNTs nanocomposite have been investigated also for toluene and xylene traces detection at room temperature in water. To this aim, the same sensor used in air environment was inserted in a Pyrex beaker containing pure water. The presence of toluene and xylene within the test ambient has been promoted by their injection inside the beaker. The injected volumes have been selected, each time, in order to obtain the desired analyte concentrations. The polluted water has been continuously stirred to ensure the maximum dispersion of the analyte. In addition, after each analyte exposure, the capabilities of the reflectometric sensor to recover the initial steady state level have been investigated by restoring the initial condition of uncontaminated water. Pure water, in fact, was continuously injected in the test chamber, while the contaminated water, previously present in it, contemporarily stilled out. Fig. 2.a reports the transient responses of the fiber optic probe as a consequence of several toluene injections with concentrations ranging from 20 to 100 ppm (µl/l). From the figure, as the air case, a reflectance decrease occurs as a consequence of the analyte adsorption within the sensing layer. In addition, a good repeatability can be observed in the sensor response when exposed to two successive 100ppm xylene injection. Furthermore, Fig. 2.b demonstrates that also in the case of detection in aqueous ambient a linear dependence exists between the fiber-film reflectance change and the concentrations of the two organic analytes, with an higher affinity towards xylene (the sensor sensitivity is $1.0 \cdot 10^{-3}$ ppm^{-1}) than toluene ($4 \cdot 10^{-4}$ ppm^{-1}). These sensitivities are slightly lower than those obtained for the chemical detection in air; however strong differences have been observed between the sensor dynamics in the two environments, with the response and recovery times against toluene (11 and 7 minutes, respectively) and xylene (14 and 7 minutes) in water considerably lower than the ones evidenced by the CdA/SWCNTs–based sensor in air (this effect could be due to the difference of the two test chambers and of the diffusion coefficients in the two environments). Finally, in aqueous ambient the proposed sensor exhibited enhanced performances with respect to un-buffered SWCNTs films (both in term of sensitivity and response time) [7], revealing its potentiality, especially for practical water quality monitoring applications.

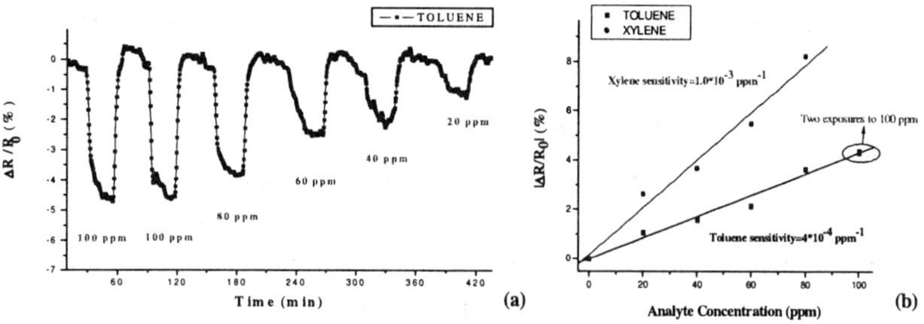

Figure 2. (a) Time responses of the opto-chemical nanosensor coated by 20 monolayers of CdA/SWCNTs nanocomposite (75 wt. %), exposed to different concentrations of toluene in water at room temperature, and (b) comparison between the sensor characteristic curves obtained in correspondence of toluene and xylene exposures.

4. Conclusion

In conclusion, in this contribution we demonstrated the possibility to exploit CdA/SWCNTs nanocomposites as highly sensitive materials to be integrated with the optical fiber technology. This combination would enable the fabrication of optical chemo-sensors for chemical detection in air and water environments at room temperature. Further works will be focused to the characterization of the fabrication process in terms of repeatability and to the investigation of the sensing performances dependence on the nanotubes content within the composite.

References

1. J. Kong, N.R. Franklin, C. Zhou, M.G. Chapline, S. Peng, K. Cho and H. Dai, **Science 287**, 622 (2000).
2. S. Chopra, K. McGuire, N. Gothard, A. M. Rao and A. Pham, **Appl. Phys. Lett. 83**, 2280 (2003).
3. C. Cantalini, L. Valentini, L. Lozzi, I. Armentano, J. M. Kenny and S. Santucci, **Sens. Actuators B 93**, 333 (2003).
4. O.K. Varghese, P.D. Kichambre, D. Gong, K.G. Ong, E.C. Dickey and C.A. Grimes, **Sens. Actuators B 81**, 32 (2001).
5. M. Penza, G. Cassano, P. Aversa, F. Antolini, A. Cusano, A. Cutolo, M. Giordano and L. Nicolais, **Appl. Phys. Lett. 85**, 2378 (2004).
6. M. Consales, S. Campopiano, A. Cutolo, M. Penza, P. Aversa, G. Cassano, M. Giordano and A. Cusano, **Measurement Science and Technologies 17** 1220, (2006).
7. M. Consales, A. Crescitelli, S. Campopiano, A. Cutolo, M. Penza, P. Aversa, M. Giordano and A. Cusano, **IEEE Sensors Letters (To be published)**.
8. L. Valentini, I. Armentano, J.M. Kenny, L. Lozzi and S. Santucci, **Mater. Lett. 58**, 470 (2004).
9. M. Penza, G. Cassano, P. Aversa, A. Cusano, A. Cutolo, M. Giordano and L. Nicolais, **Nanotechnology 16**, 2536 (2005).
10. M. Penza, M.A. Tagliente, P. Aversa, G. Cassano and L. Capodieci, **Materials Science and Engineering C 26**, 1165 (2006).
11. J.B. Peng, G.T. Barnes and I.R. Gentle, **Adv. Colloid Interface Sci. 91**, 163 (2001).
12. M Consales, S Campopiano, A Cutolo, M Penza, P Aversa, G.Cassano, M Giordano and A Cusano, **Sens. and Actuators B 118**, 232 (2006).

CHEMICAL SENSOR ARRAYS AND NETWORKS

A MULTICHANNEL QUARTZ CRYSTAL MICROBALANCE FOR VOLATILE ORGANIC COMPOUND ANALYSIS

S. PANTALEI, E. ZAMPETTI, A. MACAGNANO, E. PROIETTI
Microelectronics and Microsystems Institute of the National Research Council
Rome, Italy

C. DI NATALE, A. D'AMICO
Electronic Engineering department, Tor Vergata University
Rome, Italy

Sensor arrays based on Quartz Crystal Microbalances (QCM) are widely used in volatile organic compound analysis. These systems typically employ a number of quartzes coated with different chemically interactive materials (CIM). In this work we have studied the possibility of including four different QCM in the same quartz plate and performed preliminary measurements.

1. Introduction

Quartz crystal microbalance based Electronic Noses employ a number of quartzes coated with different chemically interactive materials. Usually, single QCM are packaged in a rather large holder, then the total volume of the array is rather large. The reduction of measurement cell volume may be desirable in terms of better homogeneity of sample concentration and response time. In this work the performances of four QCM fabricated on a single quartz substrate have been studied in some details. In particular we have experimentally examined, from the point of view of the cross talk, a single quartz plate coated by four CIM which where four different Metalloporphyrin. The multichannel quartz crystal microbalance (MQCM) consists of a single quartz plate where four chromium-gold electrodes couples have been deposited, each couple is a resonator with a suitable oscillation frequency. The quartz plate thickness was 160 μm corresponding to 10 MHz of fundamental oscillator frequency. The multisensor has been accommodated in a suitable package, comprising the test cell, whose overall dimensions were about 35 mm x 35 mm x 10 mm **(Fig. 1)**.

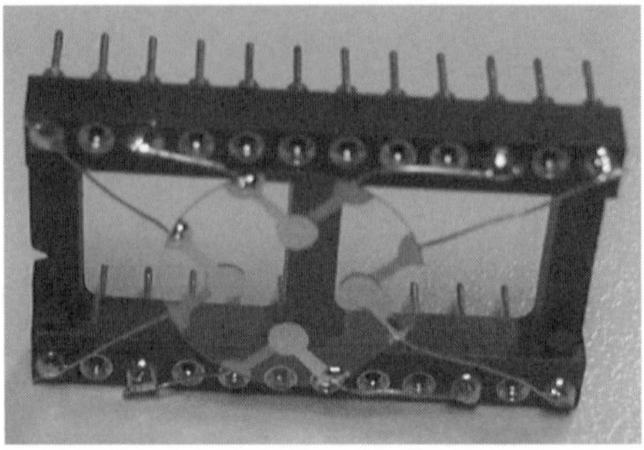

Figure 1:Multichannel Quartz Crystal Microbalance bonded in a standard DIP package for prototype purposes.

2. Characterization

A network analyzer has been used for the complete electric characterization of the quartz multisensor. The scattering matrix model has been used to characterize the multichannel quartz; in this model each port represents a single resonator. Using the Butterworth Van Dyke model for each resonator and considering no channel to channel cross talk, the electrical elements R,L,C has been calculated and shown in **Tab.1**. The results obtained for the scattering coefficients |Sii| and |Sij| are shown in **Fig. 2,3**. Resonators fundamental frequencies were 10164188 Hz, 10140562 Hz, 10115906 Hz, 10088438 Hz. They resulted almost independent one each other in the sense that the worst attenuation among the four frequencies was less than 40 dB.

Pad	f_0	$f-f_0$	R	L	C	Q
1	10164188		58.8 Ω	57.2 mH	4.8 fF	64330
2	10140562	23626	52.2 Ω	50.6 mH	4.9 fF	61833
3	10115906	48282	91.8 Ω	53.1 mH	4.7 fF	36785
4	10088438	75750	58.4 Ω	51.0 mH	4.9 fF	55431

Table 1: Electrical parameters of the Butterworth Van Dyke model for the four resonators.The term f-f0, is the shift frequency of 2th, 3th, 4th resonator frequencies with respect to 1th resonator one.

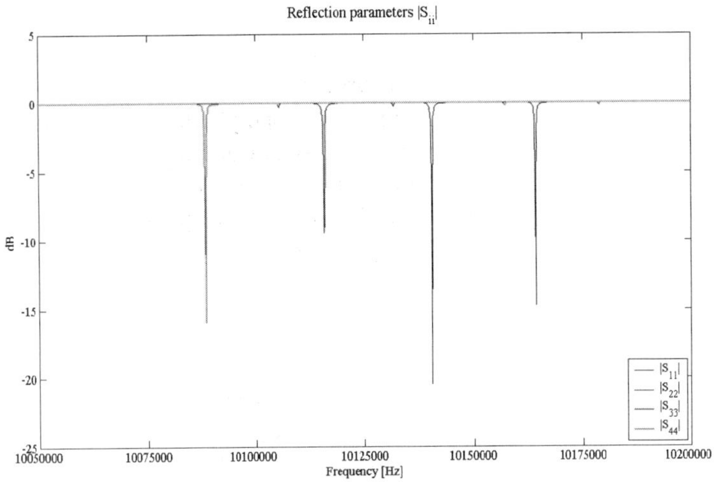

Figure 2: Scattering coefficients |Sii|, , where $1 \leq i \leq 4$.

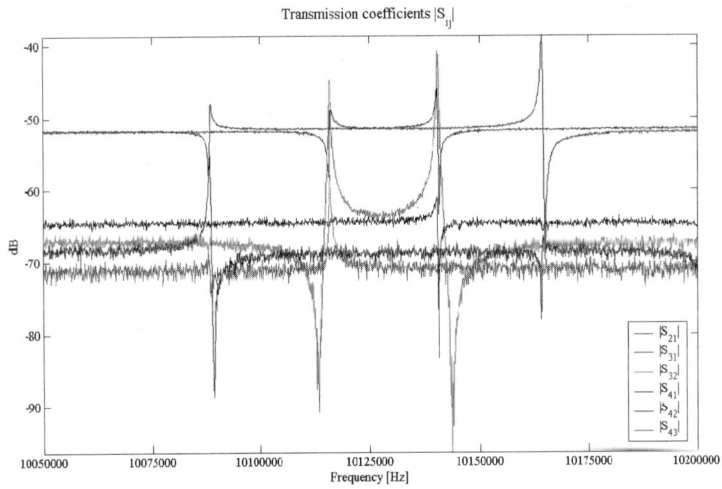

Figure 3: Scattering coefficient |Sij|, where $1 \leq i, j \leq 4$ with $i \neq j$.

3. Chemical Interactive Material Deposition

Four different metalloporphyrin layers where deposited by spray casting technique on the gold electrodes (**Fig. 4**). The MQCM sensor array was placed in a suitable test chamber endowed with sample inlet and outlet (**Fig. 5**). To have an idea of the effective cross-talk existing between all the four channels, the

frequency shift induced on each resonator by the mass load due to the polymer deposition on all the others has been calculated.
For each resonator the frequency shift during the deposition has been monitored. At the end of the deposition of the four metalloporphyrins, the total frequency shift from the initial resonant frequency has been evaluated. The difference from the total shift and the shift caused by the deposition is due to the mass load on the other channels. **Tab. 2** shows the results obtained in this way. The cross frequency response results to be less than 2.4 10^{-3}.

Figure 4: Four QCM sensors with different four metalloporphyrins: Mn-TPP, Co-TPP, Ru-TPP, Zn-TPP .

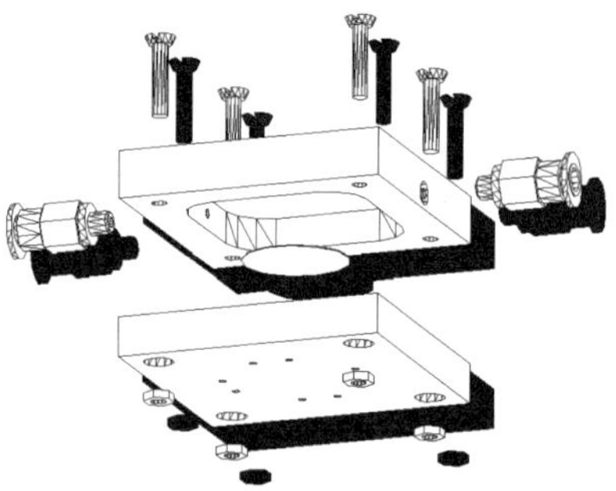

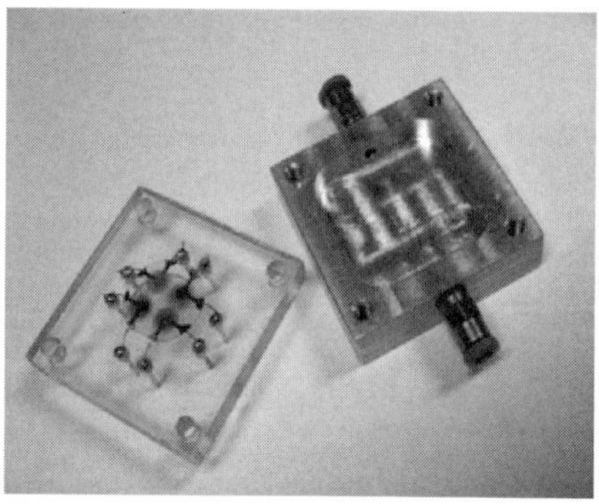

Figure 5: The MQCM has been accommodated into a suitable test chamber. In the left, the mechanical layout of the test chamber is depicted with its inlet and outlet to flow the analytes. In the right, the realized sensor array in the flow cell.

	A	B	C	D
S1	10200 [Hz]	10170 [Hz]	30638 [Hz]	0.98‰
S2	10216 [Hz]	10194 [Hz]	30614 [Hz]	0.72‰
S3	10492 [Hz]	10420 [Hz]	30388 [Hz]	2.37‰
S4	10057 [Hz]	10024 [Hz]	30784 [Hz]	1.07‰

Table 2: Frequency shifts obtained from the deposition of the four metalloporphyrins on each resonator. For each port of the device has been calculated: A) the frequency shift from the fundamental one due to overall mass deposition on all the four ports, B) the frequency shift due to mass deposition on the port, C) the sum of the frequency shifts on the other three ports and D) the ratio of the frequency shift due to cross-talk and the frequency shift that caused it ((A-B)/C).

4. VOC Analysis

The array was tested with vapours of toluene (aromatic), ethanol and methanol (alcohols), hexane (hydrocarbon). For each volatile compound three concentrations in nitrogen carrier were delivered into the measurement cell. Each exposure condition was repeated several times to test the array reproducibility. Steady-state frequency shifts were analyzed by Principal Component Analysis (PCA). The different sensitivities of the four channels are shown in **Fig. 6**.

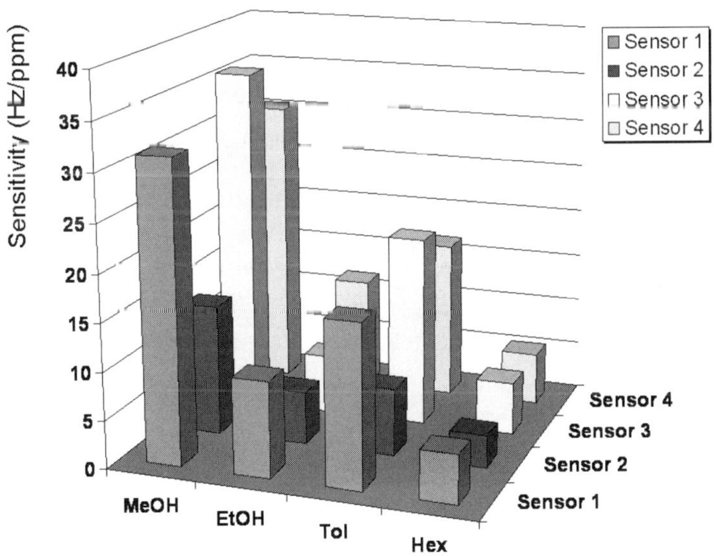

Figure 6: Sensitivities of the sensor array for three gases.

Due to the limited cross talk each resonator exhibits different sensitivities only dependent from the Metalloporphyrin coated on it. As shown in **Fig. 7** the scores plot shows the expected separation between vapours and concentrations.

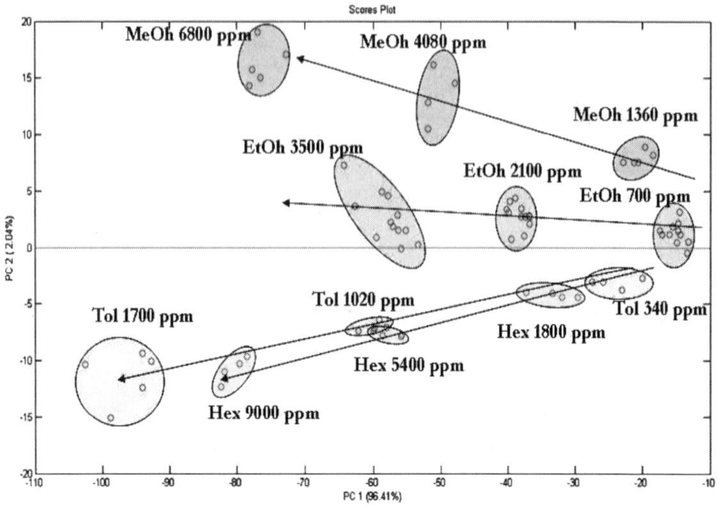

Figure 7: Principal Component Analysis: score plot of the first two principal components.

References

1. T. Abe, M. Esashi, *Sensors and Actuators B*, 82 (2000) 139-143
2. P. Boeker, G. Horner, S. Rösler, *Sensors and Actuators B*, 70 (2000) 37-42
3. D. W. Dye, *Proc. Phys. Soc.* ,38:399-457 (1926)
4. S. Butterworth, *Proc. Phys. Soc.*,27:410-424 (1915)
5. K. S. Van Dyke, *Proc. I.R.E.*,16:742-764 (1928)
6. G. Sauerbrey, Verwendung *von Schwingquarzen zur Wagung¨ dunner¨ Schichten und Microwagung,¨* Z. Phys. 155 (1959) 206–222

DEVELOPMENT OF A NEW PORTABLE MICROSYSTEM FOR WINE QUALITY ANALYSIS*

D. S. PRESICCE, L. FRANCIOSO, P. SICILIANO

Istituto per la Microelettronica ed i Microsistemi (C.N.R.- I.M.M.)

Via per Monteroni, 73100 Lecce, Italy

E-mail: dominique.presicce@le.imm.cnr.it

ADAMI. A., LORENZELLI L., MALFATTI M., GUARNIERI V., ZEN M.

ITC – IRST – Microsystems Division,

via Sommarive 18, Povo, 38050 Trento, Italy

Chromatography is an analytical technique whereby a mixture of chemicals may be separated by virtue of their differential affinities for two immiscible phases and this involves traditional analytical instruments of significant size and cost, such as gas chromatograph mass spectrometer (GC-MS), and therefore on-line, real-time analyses are difficult to realise. In this paper we report on the development of a portable microsystem and the evaluation of innovative micromachined gas sensor array performance for wine quality analysis, monitoring different blend of Apulian wine, looking towards new applications into fast and cheap miniaturized multisensor systems in a more general food quality scenario.

1. Introduction

The monitoring of the quality and safety control in the food industry is an up-to-date topic, because of the important issues of quality of life and health care as well as its possible industrial applications. Recent years have been characterized by a growing interest focused on food analysis. The organoleptic

---

* * This work has been funded by the "Programma Operativo Nazionale Ricerca, Sviluppo ed Alta Formazione"— MINICONTAL Project.

analysis, based on both analytical methods and trained inspectors who use odour evaluation, is the wider used method to define quality and safety in foods.

In many cases, monitoring and determining the constituents of a sample gas typically involves collecting samples and analyzing them in traditional analytical instruments such as gas chromatograph–mass spectrometer (GC–MS) of significant size and cost. In addition, in some cases the sample preparation is time consuming and thus on-line, real-time analysis cannot be easily performed. In fact many applications, such as the detection of volatile organic compounds (VOCs) generated from food in agro-food industry, need smaller, more portable, cheaper, and even disposable sensor-based systems designed to analyse such complex mixtures [1].

The aim of the work is the evaluation of innovative micromachined gas sensor array performance into field of wine quality monitoring. These systems can constitute a pre-screening step in the quality control system, where in depth analyses with traditional laboratory instrumentation may be performed on a reduced number of samples, only where they are really needed, or for a periodical benchmark of portable systems, leading to a more detailed quality control during production and supply chain, and to a decrease of costs.

A classic gas chromatograph has been modified for this aim through the use of a splitter after thc column, so the detection of volatile compounds in the headspace of wine sample has been simultaneously detected by MS detector and sensors array. This is a preliminary approach for the development of a microsystem, composed by a microcolumn and a sensor array, and for the realization of a lab on chip.

In the last decade, really, the micro-machined gas sensors based on semiconducting metal oxides (MOX) have benefited from a parallel evolution in which micromachining technologies for MEMS, aimed to improve the thermal response and power consumption of the sensors, have been added to on -chip sensors and electronics. In this work a monolithic multi – sensor chip, consisting of an array of sensors elements in a very small area implemented on membrane – based microhotplates, has been developed and tested.

2. Experimental

Tungsten oxide (WO3) has been deposited on the sensor active area by RF sputtering starting from pure 99.99% 4 in. target at different oxygen partial pressure and supplied power density.

Deposition process has been carried out with the substrate at room temperature, followed by an annealing step at 500 °C in air for an hour. In order to pattern the sensitive layer on the active area, the deposition procedure of the sensitive metal oxide layer has been performed at wafer level by a lithographic step using a positive resist. Operative deposition parameters for WO3 films are listed in Table 1.

Table 1. Deposition process parameters for WO_3 sensitive film.

Materials	WO_3/20% O_2	WO_3/30% O_2	WO_3/40% O_2
Deposition pressure (mbar)	5,5 x 10-3	5,5 x 10-3	5,5 x 10-3
Substrate Temperature (°C)	17-26	15-31	29-48
Air flow (sccm)	24	21	18
O2 (sccm)/ %	6/20	9/30	12/40
Bias (V)	-309	-301	-296
Rate (Å/s)	2,5-2,9	1,2-1,3	0,8-0,9

So a MEMS-based Chromium doped WO3 sensors arrays, realized by the authors [2], has been mounted in a dedicated chamber with reduced dead volumes (50μl, equivalent to 3% of the column volume), connected to the splitter output and properly heated. Signal read-out has been provided by a dedicated electronic board, allowing the sensor bias and read-out and microheater temperature control [3] and data acquisition to PC by a standard USB connection with a dedicated software. Array working conditions has been set, on the basis of preliminary tests, to the optimal temperature range (300÷400°C).

Testing of system performances has been performed by comparing results of SPME/GC/MS (Solid Phase MicroExtraction) and SPME/GC/MOX analysis, simultaneously performed with a traditional chromatograph equipped for this purpose with a purged splitter. Scheme of system is reported in Figure 1.

In order to validate the proposed system in wine quality monitoring, three different typical wines of Salento region (Lecce, Italy) have been sampled: “Canonico” (100% Negramaro), “Tenuta Albrizzi” (Cabernet Sauvignon and Primitivo) and “Selva Rossa” (Negramaro and Malvasia) producted by Cantine Due Palme placed in Cellino S. Marco (Brindisi). Preliminary tests for a portable system for wine quality application, based on MOX (Metal Oxide Semiconductors) sensor arrays and dedicated control board, have been carried out.

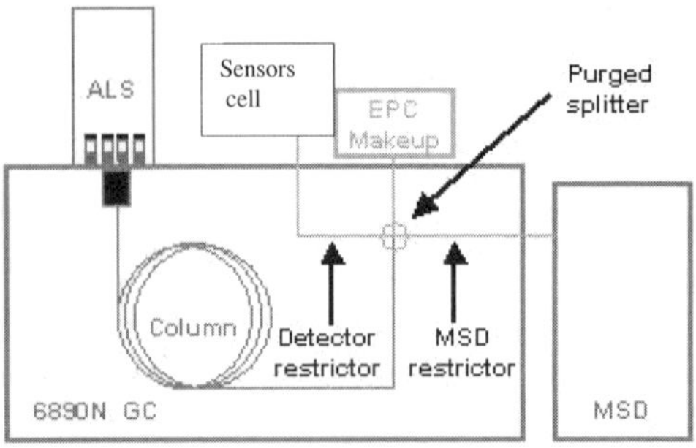

Figure 1. Scheme of the testing set – up.

3. Results and Discussion

An analytical method has been developed in the aim to analyse flavour of typical wines of the South of Italy. Headspace solid phase micro extraction (HS-SPME) has been studied by gas chromatographic analysis of major compounds in red wines. Volatile compounds belonging to different chemical families such as alcohols, phenols, terpenols, ethers, ketones, aldehydes, esters, lactones have been identified. SPME/GC/MS and SPME/GC/MOX measurements have been simultaneuosly carried out and the main compounds have been individuated and compared by retention times.

The main peaks and compounds have been listed in table 2.

They are the same for different kind of wine, so it shows how the consequent aroma is the result of different ratio of concentration of the same compounds.

In figure 2 we have the comparison of both chromatograms obtained by MS and MOX detector, where we can highlight the tailing of various peaks in the last chromatogram due to the size of the measurement sensor array cell and to the gas carrier flow. In fact, the cell refilling time is 3 s and the dilution factor is 10 times in 27 s, so the recovery time of the sensors have been conditioned. This tailing is also ascribable to the slow desorption rate of the volatile compounds on the surface of sensitive material, in particular, the recovery time for ethanol has been evaluated 30÷40 s in the presence of minor peaks that can be highlighted from MS – chromatogram.

Table 2. Main compounds in the headspace of wine.

Peak number	Main Peak
1	ethyl acetate
2	ethanol
3	2-methylpropanol
4	3-methylbutanol
5	ethyl 2-Hydroxipropanoate
6	ethyl octanoate
7	ethyldecanoate
8	diethyl butandioate
9	2-phenylethylacetate
10	bis -2-ethylesil hesandioate
11	2-ethylphenol

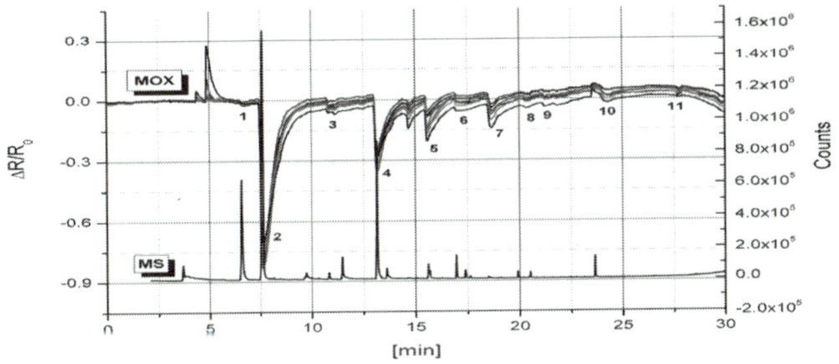

Figure 2. Comparison of MS and MOX system results for a sample of "Canonico" wine.

3. Conclusions

The presented microsystem based on MOX sensor array demonstrated a good sensitivity to specific analytes for wine quality applications (as shown in figure 2), allowing the development of low-cost modules for portable chromatographic systems. Systems based on MOX sensors, working with a "fingerprint recognition" approach, that can also take advantage from array sensor data analysis techniques, such as PCA, allow obtaining results comparable with traditional systems in term of sensitivity, with lower costs, higher portability and ease of use. However, work is in progress in order to optimize the deposition process of the sensing layer and to improve the microfluidics of the device for overcoming the threshold of recovery time of sensors.

References

1. S. Capone, A. Forleo, L.Francioso, R. Rella, P.Siciliano, J. Spadavecchia, D.S. Presicce, A.M. Taurino. *J. Optoelectron. Adv. Mater.*, 5 (2003), 1335
2. A. Adami, L. Lorenzelli, V. Guarnieri, L. Francioso, A. Forleo, G. Agnusdei, A. M. Taurino, M. Zen, P. Siciliano, *Sensors and Actuators B*, 117 (2006), 311
3. M. Malfatti, M. Perenzoni, D. Stoppa, A. Simoni, A. Adami, - IMTC 2006, Sorrento, Italy

POLY-PYRROLE DERIVATIVES USED AS COLORIMETRIC SENSORS FOR VOLATILES DETECTION

F. OLIMPICO*, A. SCARPA, O. CATAPANO, L. FACHECHI, S. GRECO

Biological Division Technobiochip Scarl, Via Provinciale per Pianura, 5 (Loc. San Martino), 80078, Pozzuoli (Na), Italy

Technobiochip has recently patented a series of poly-pyrrole derivatives. Here we report, for the first time, the development of a low-cost and sensitive colorimetric sensor array for the analysis of different volatile compounds in spirits and liquors using that derivatives. The identification is based on color intensity changes of poly-pyrrole derivatives sensor elements upon ligand binding. The data obtained revealed the sensor's ability for sensitive and specific detection of different type of VOCs, so providing a useful test for foods analysis and environmental monitoring.

1. Introduction

A colorimetric sensor device represents a powerful approach toward the detection of chemically diverse analytes, in a wide range of applications including foods analysis and environmental monitoring. The technology we here have used, reported previously by Janzen et al [1,2], is based on different dyes that change color, in either reflected or absorbed light, upon changes in their chemical environment or analyte interaction. Dye has to contain a center to interact strongly with analyte (not simply by physical adsorption) and to be strongly coupled to an intense chromophore. The numerous and diverse molecular interactions between dye and analyte cause both different degrees of color or intensity color changes. A distinct pattern of responses produced by the array provides a characteristic fingerprint for each analyte. Moreover, by using properly dyes and substrates, the sensor could result less sensitive to humidity, one of the most important trouble encountered with electronic nose. Here, we attempted to evaluate the capacity of patented poly-pyrrole derivatives, synthesized at Technobiochip, to change color subsequently to the interaction with different volatile compounds, and so the ability of our colorimetric sensor arrays to distinguish among a large family of volatile organic compounds (VOCs) in commercial spirits.

* Address Correspondence to: Dr. Francesco Olimpico, BSc, Researcher Biological Division Technobiochip Scarl, Via Provinciale per Pianura, 5 (Loc. San Martino), 80078, Pozzuoli (Na), Italy. E-mail: f.olimpico@technobiochip.com - Phone: +39 081 5264315 – Fax: +39 081 5265116

2. Materials and Methods

2.1. *Reagents*

All chemicals were purchased by SIGMA (Milan, Italy). Thin-layer chromatography silica gel 60 F_{254} plates were purchased by MERCK (Darmstadt, Germany).

2.2. *Poly-Pyrrole derivatives preparation*

Poly-pyrrole polymers were obtained from polymerization of equimolar pyrrole and aldehyde solution reacted in a saturated BF_3 environment acting as catalyst [3,4,5].

2.3. *Dye spotting*

Poly-pyrrole polymers were spotted onto thin-layer chromatography silica gel plates using a microlitre syringe (Figure 1). After spotting, sensor array was dried under vacuum at 50 °C for one hour before using. The array was incubated in a Couplin's jar, saturated with analyte vapour for 60 minutes at room temperature.

2.4. *Image Analysis*

Array images were acquired by scanning (400 dpi resolution) before and after incubation with liquid compounds. Finally, images were analyzed by a software developed on MatLab 7.1 platform (Figure 2). Spot in the array is described uniquely by RGB color values. Analyzing the difference of RGB values between the "before" and "after" image, a merged map was obtained. This color profile is useful for a rapid visualization of color changes and provides a signature for each analyte. In a more detailed analysis, difference of averages (Δ Average = $\Sigma_i n_i c_i / \Sigma_i n_i$ where c_i = color values (0, 255) and n_i = number of color c_i pixel) was calculated.

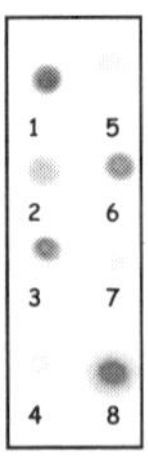

Figure 1. Scheme of colorimetric sensor array.

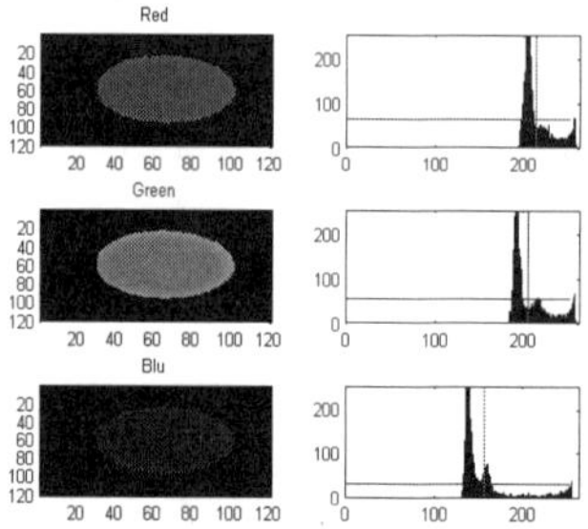

Figure 2. Image analysis.

3. Results and Discussion

Different liquors and spirits were tested by the colorimetric sensor array to demonstrate its ability to discriminate among different type of compounds. A colorimetric array constituted by poly-pyrrole derivatives (shown in Table 1) was incubated with *San Marzano* and *Strega* liquors, *Crown*, *Ballantaine*, *Long John*, *Chivas*, and *Langs Supreme* whiskeys.

Table 1. Poly-pyrrole derivatives.

Sensor	Polymer
1	Poly [2-4 (methoxybenzyl)]-1H-pyrrole
2	Poly [2-(-9 phenanthry-ylmethyl)]-1H-pyrrole
3	Poly [(-2-ylmethyl)-2-ethoxyphenol] 1H-pyrrole
4	Poly{2-[2-(2E)-3-phenylprop-2-enyl] }-1H-pyrrole
5	Poly [1-acetyl-1*H*-indole]-1H-pyrrole
6	Poly[2-(thien-2-ylmethyl)]-1H-pyrrole
7	Poly [ferrocene]-1H-pyrrole
8	Poly[2-(benzyl)]-1H-pyrrole

Histograms showed in Figure 3 represent data expressed as Δ average for RGB values for some of liquors analyzed. On Y axis change of color was indicated as increase or decrease of Δ average for each spot, indicated on X-axis. As shown in figure each sensor had a different reactivity and responses for the different type of liquors or spirits.

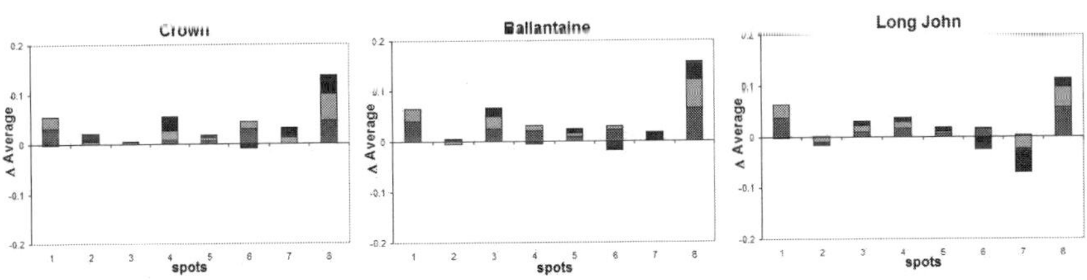

Figure 3. Histograms represent data expressed as difference of average for RGB color, where red, green and blue bar represent the three component of RGB respectively.

Obtained data were also analyzed using a principal component analysis (PCA), created by linear combinations of the RGB responses of the 8 dyes used in these arrays. As shown in Figure 4, PCA analysis clearly divided data, where data from negative control (acetic acid) and those from liquors or spirits are well separated into two groups. Nevertheless, a good separation between liquors with different percentages of alcohol was obtained.

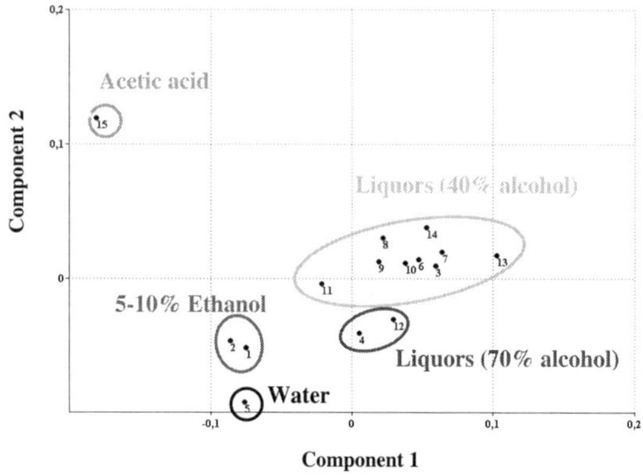

Figure 4. PCA plot of image analysis data from spirits and liquors.

These preliminary data indicate that colorimetric array sensor, based on a Technobiochip's patented poly-pyrrole derivatives might be a useful tool for the discrimination of different liquors and spirits.

4. Conclusions

A colorimetric sensor device is suitable for numeroús and several applications. The present study regards the development of a disposable colorimetric sensor array based on an array of Technobiochip's patented poly-pyrrole derivatives. The experiments performed with the colorimetric sensor array have shown its ability to discriminate among the different liquors and spirits with different percentage of alcohol. These data demonstrated the possibility of the sensor to be a useful and rapid test for identification of a VOCs complex mixture.

References

1. M.C. Janzen, J.B. Ponder, D.P. Bailey, C.K. Ingison, and K.S. Suslick, *Anal Chem.* **78**, 3591 (2006).
2. N.A. Rakow & K.S. Suslick, *Nat.* **406**, 710 (2000).
3. A. D Alder, F. R. Longo, F. Kampus, J. J. Kim, *Inorg. Nucl. Chem.* **32**, 2443 (1970).
4. A. D Alder, F. R. Longo, V. Vardi, *Inorg. Synth.* **16**, 213 (1976).
5. S. Shambayati, S.L. Schreiber. B.M. Trost, I. Fleming, L and A. Paquette (Eds.) in *Comprehensive Organic Synthesis*, Vol. 1, Chapter 1.10, p. 283, Pergamon Press, New York (1991).

ANALYSIS OF NH_3/DMA/TMA MIXTURES BY A MULTISENSOR MINIATURISED GAS CHROMATOGRAPHIC SYSTEM

S. CAPONE*, M. ZUPPA*, L. FRANCIOSO*, I. ELMI°, S. ZAMPOLLI°, G.C. CARDINALI°,*P. SICILIANO**

**Institute of Microelectronics and Microsystems I.M.M.-C.N.R., Lecce, Via Monteroni, 73100 Lecce, Italy*

°Institute of Microelectronics and Microsystems I.M.M.-C.N.R., Bologna, Via Gobetti, 101, 40129 Bologna, Italy

Summary

In this work we report the functional characterization of a multisensor miniaturised gas chromatographic system prototype for a specific application in food analysis, i.e. for the evaluation of fish freshness. A μ-machined GC column and an array of four micromachined gas sensors based on SnO_2 and In_2O_3 were the basic elements of the prototype. The system was tested to different mixtures of ammonia (NH_3), trimethylammine (TMA) and dimethylammine (DMA), which are volatile species typically used as markers of fish deterioration, showing appreciable properties of gas separation and detection. A specific data analysis method for this application, suitably developed to process the gas sensor chromatograms acquired by the system, allowed the identification and quantification of the components in NH_3/TMA/DMA mixtures.

1. Introduction

Fish quality is a complex concept involving a whole range of factors, freshness being one of the most important. There is an increasing demand for quality assurance protocols for the European fish factory, due to export reasons and European fish control regulations. Hence, it's fundamental to develop methods and devices able to rapidly evaluate the products history and their storage

conditions from "harvest-to-home". When fish spoils it releases a variety of basic volatile amines such as trimethylamine (TMA), ammonia (NH_3) and dimethylamine (DMA), collectively known as Total Volatile Basic Nitrogen (TVB-N).
Some devices proposed rely on the use of polymer matrix, a pH sensitive dye that responds, through visible colour changes, to the spoilage volatile compounds [1]. Other recent works proposed an optical system based on a chemical sensor array where the optical features of layers of metallophorphyrins, sensitive to volatile compounds typical of spoilage processes in fish, are interrogated by a very simple platform based on a computer screen and a web cam [2]. Other works proposed the use of electronic noses based on semiconducting metal oxides array [3,4] or multi-sensor-devices for defining the quality of fish [5].
In this work we report about the performance of a multisensor miniaturised gas chromatographic system prototype, developed in IMM-CNR laboratories, to assess fish freshness. In particular, the work, is part of an ongoing European Integrated Project, Project n° 508774-IP GoodFood, Food Safety and Quality Monitoring with Microsystems, that aims at the development of Microsystem based solutions for the agrofood sector. The use of simplified gas chromatographic (GC) separation techniques together with an array of sensors constitutes an innovative system which can reliably perform quantitative analyses. GC column provides selectivity enhancement through temporal separation of the different compounds, whereas partially selective solid state gas sensors produce different patterns from the same sample. Here, we show the first test results of the prototype to mixtures of ammonia (NH_3), trimethylammine (TMA) and dimethylammine (DMA), which are volatile species markers of fish deterioration. The GC system showed appreciable properties of gas separation and detection. In particular, the work was addresses to develop a specific suitable data analysis method for this application to process the gas sensor chromatograms acquired by the system, allowing the identification and quantification of the components in NH_3/TMA/DMA mixtures.

2. The multisensor miniaturised gas chromatographic system prototype

The key system components of the multisensor miniaturised gas chromatographic system prototype (fig. 1a) are the μ-machined GC column and the gas sensor arrays connected to a compact, specifically developed pneumatic circuit involving a minipump and a microvalve. These key components allow to combine the temporal separation typical of chromatography with the spatiochemical separation typical of e-noses, allowing an improvement in system analysis capability.

Fig. 1a: System prototype picture

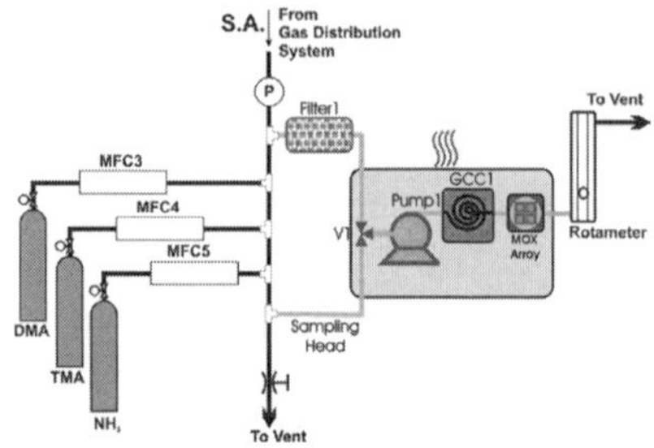

***Fig. 1b**: Measurement setup scheme in controlled environment*

The system is designed to work with different GC columns (filled with different stationary separation phases) and different gas sensors, allowing a wide variety of possible application fields. For the specific application of fish freshness evaluation through TMA, DMA and NH_3 identification and quantification, a suitable stationary separation phase and sensors with satisfying gas-sensing properties were selected. The best results obtained on separation among the target gases were provided by a packed C+0.2% Carbowax 20M operated on programmed temperature step mode. With respect to gas sensors, SnO_2- and In_2O_3- thin film based sensors deposited by sol-gel technique were selected as detectors of the system. The gas-sensing layers have been deposited on micromachined hotplate arrays integrated in a single 5x5 mm^2 Si chip [1] The two chips bounded on a TO-8 socket were located in the system chamber, which can host up to two chips arranged frontally each to one another within a small volume ($V=0.5cm^3$).

3. Gas-sensing tests

In order to test the functionality of the system, a systematic measurement campaign was carried out. Different NH_3/DMA/TMA gas mixtures variable in composition and concentration have been provided to the system to evaluate its separation and quantification capabilities (table 1). The measurement setup scheme used to evaluate the system is reported in Fig.1b. The 4-sensor array consisted of SnO_2 and In_2O_3–based sensors at different working temperatures (SnO_2 @250°C and @400°C, In_2O_3 @300°C and @400°C).

Gas sensing protocol	Nr Mixture	$[NH_3]$ (ppm)	[DMA] (ppm)	[TMA] (ppm)
	1	4	0	0
NH_3	2	1	0	0
	3	0	0	0
	4	0	1	0
DMA	5	0	0.25	0
	6	0	0	0
	7	0	0	1
TMA	8	0	0	0.25
	9	0	0	0
	10	4	0.24	0
NH_3/DMA	11	4	1	0
	12	0	0	0
	13	4	0	0.25
NH_3/TMA	14	4	0	1
	15	0	0	0
	16	4	0.25	0.25
	17	4	1	1
NH_3/TMA/DMA	18	0	0	0
	19	4	0.25	0.25
	20	0	0	0

Table 1. *Gas sequence of the NH_3/DMA/TMA mixtures*

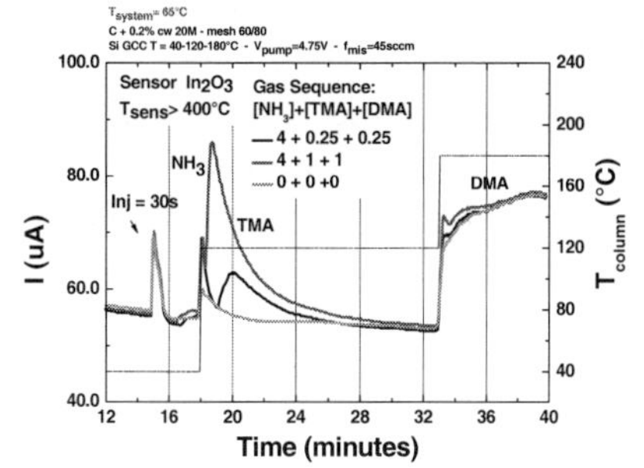

Fig. 2: *Dynamic response of a In_2O_3-based sensor, working at a temperature of about 400°C, towards the NH_3/DMA/TMA mixtures highlighted in the table.*

The system works on measurement cycles following a step sequence which has been set up after long test. Step duration and column temperature has been chosen to optimize separation between TMA, DMA, NH_3 and other interferents (as water for instance). The total sequence duration is 1 hour; this means that, during this prototypal laboratory characterization phase, the system is able to perform 1 measure each hour, during which a gas mixture of NH_3, DMA and TMA was analized. As example, fig.2 shows the chromatograms of a In_2O_3-based sensor working at T=400°C towards the NH_3/DMA/TMA mixtures. It can be observed that a first rough separation of the two gases is possible.

4. Data analysis

The aim of data analysis is to extract information from the sensor array chromatogram patterns useful to identify and quantify the components of the different NH_3/DMA/TMA gas mixtures analyzed by the prototype. The 4-gas sensor array chromatograms were analyzed according to the following steps:

1) Signal processing
Signal processing counteracts the baseline instability effects and removes noise or spikes. The technique is based on Discrete Wavelet Analysis; in this case the signal is decomposed at different resolutions. Those signal components showing drift or noise contamination are discarded and then the signal is reconstructed.

2) Fourier Analysis
In this work the Fourier analysis was applied only to gas sensor chromatograms related to single gaseous substances in study (NH_3, DMA, TMA). Information

on the signals' frequency content are useful to perform more appropriate the Continuous Wavelet Transform, as described in the next section.

3) Continuous Wavelet Transform analysis (CWT)

The wavelet transform provides a multi-resolution signal decomposition, i.e. it analyses the signal at different frequency bands with different resolutions. It is a windowing technique with variable-sized regions. Wavelet analysis allows the use of long time intervals where we want more precise low-frequency information, and shorter regions where we want high-frequency information. CWT is defined as the sum over all time of the signal multiplied by scaled, shifted versions of the wavelet function ψ:

$$C(scale, position) = \int_{-\infty}^{+\infty} f(t)\Psi(scale, position, t)dt \ .$$

The results of the CWT are wavelet coefficients C, which are a function of scale and position. The scale is related to the frequency of the signal: high (low) scale corresponding to low (high) frequency. CWT is used to analyse all 4-gas sensor array chromatograms at appropriate scales, selected by means of the previous Fourier Analysis. The computed wavelet coefficient vectors are the feature sets which are analysed in the next step of data analysis.

4) Identification and quantification of the chemical components in a gas mixture

In this step a method to identify and quantify the components of the analyzed gas mixtures was developed. The wavelet coefficient vectors extracted from the measurements of single gases (NH_3, DMA, TMA) at different concentrations were chosen as the reference vectors, Z_j (j=1,.., c), of the respective substance c classes. The identification of these three gases present in a binary or ternary gas mixture relies on the Euclidean distance between the input wavelet coefficient vectors and the reference vectors. The used algorithm is a fuzzy version of the nearest prototype classifier, where the degree of membership u of each input vector x in each of the c classes is assigned according to following relation: $u_i(x) = \dfrac{1/\|x - Z_i\|^2}{\sum_{j=1}^{c}\left(1/\|x - Z_j\|^2\right)}$. The results are shown in Table 3.4. The CWT decomposition of the gas chromatograms at high scales allows to identify TMA in a mixture, while the CWT decomposition at low scales allows to identify NH_3 in a mixture. The presence of DMA in a mixture is not well recognized.

	CWT DECOMPOSITION AT LOW SCALE							CWT DECOMPOSITION AT HIGH SCALE						
Mixtures	NH_3 4 ppm	NH_3 1 ppm	TMA 0.25ppm	TMA 1ppm	DMA 0.25ppm	DMA 1ppm	NO GAS	NH_3 4 ppm	NH_3 1ppm	TMA 0.25ppm	TMA 1ppm	DMA 0.25ppm	DMA 1ppm	NO GAS
NH_3 4 ppm/ TMA 0.25 ppm	19.6%	7.1%	16.6%	13.5%	14.4%	14.2%	14.5%	16.1%	13.2%	30.4%	9.6%	10.3%	9.2%	11.2%
NH_3 4 ppm/ TMA 1 ppm	65.8%	3.1%	7.2%	5.0%	6.5%	6.2%	6.2%	9.6%	8.4%	9.5%	50.7%	7.1%	7.2%	7.6%
NH_3 4 ppm/ DMA 0.25 ppm	54.6%	4.4%	8.9%	6.9%	8.3%	8.5%	8.4%	45.5%	16.4%	5.0%	1.4%	11%	9.3%	11.3%
NH_3 4 ppm/ DMA 1 ppm	76.9%	1.8%	4.5%	2.8%	4.7%	4.6%	4.6%	45.6%	16.3%	5%	1.4%	11%	9.3%	11.4%
NH_3 4 ppm/ DMA 0.25 ppm/ TMA 0.25 ppm	17.3%	15.4%	13.8%	8.4%	15.5%	14.7%	14.9%	14.6%	12.2%	24%	20.4%	9.4%	8.9%	10.4%
NH_3 4 ppm /DMA 1 ppm /TMA 1 ppm	16.4%	8.3%	17.9%	11.2%	15.6%	16.0%	14.6%	11%	9.5%	10.3%	44.4%	8.2%	8.1%	8.7%
NH_3 4 ppm/ DMA 0.25 ppm/ TMA 0.25 ppm	12.7%	20.4%	13.8%	8.2%	1%	14.7%	14.8%	15.1%	13%	25.7%	16.3%	9.7%	9.1%	11.1%
no gas	2.2%	2.6%	16.1%	2.2%	24.0%	25.2%	27.6%	5%	17%	17.5%	0.3%	7.9%	11.8%	56.4%
no gas	1	1.9%	1%	0.6%	24.4%	22.1%	44.2%	3%	14.7%	11.4%	0.2%	6.8%	14.6%	69%

***Table 3.4**: The membership of each wavelet coefficient vector in each class assigned based on Euclidean distance from the reference vector of the class.*

Sammon's Mapping algorithm was used to visualize the inherent structure of the feature sets During the mapping on a lower dimensional space. The algorithm preserves all inter-point distances, in this way it is possible to find clusters, correlations or underlying distributions. Fig.3 and fig.4 show the results of the Sammon's mapping based on Euclidean metric. In fig.3 the feature sets are obtained by CWT decomposition of the gas sensor chromatogram at low scale (a=9). In this case the wavelet coefficient vectors corresponding to binary or ternary mixture of three gases (NH_3\TMA\DMA) are closer to the NH_3 reference vector. Otherwise, in fig.4 the feature sets are obtained by CWT decomposition of the gas sensor chromatogram at high scale (a=105). The wavelet coefficient vectors corresponding to binary or ternary mixture are closer to the TMA reference vector. The identification of DMA fails; it is clear that the corresponding chromatograms are similar to ones obtained from measurements of the reference gas.

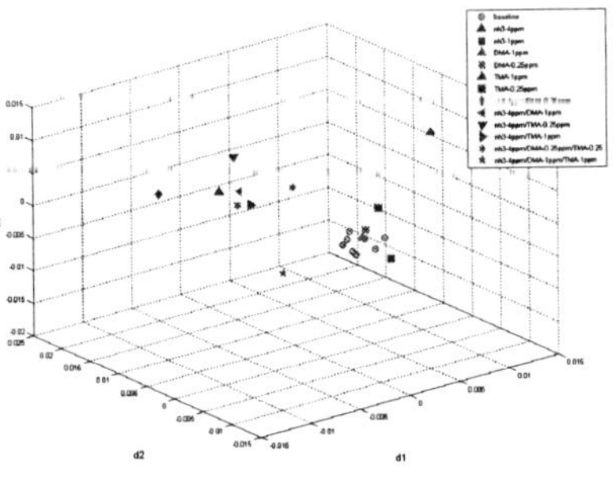

***Fig. 3:** Sammon mapping of feature set obtained by CWT decomposition of the signals at low scale (a=9).*

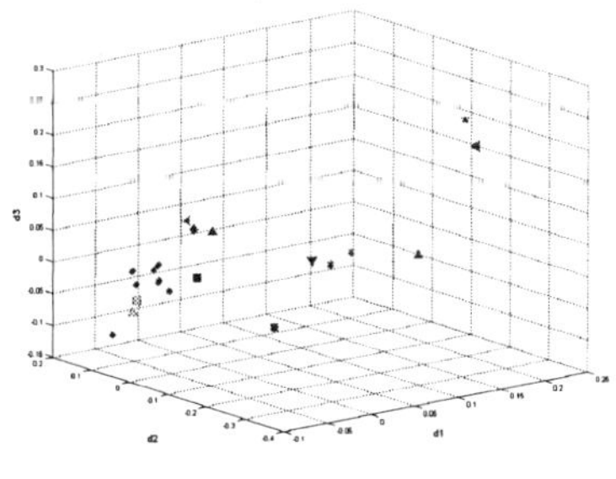

***Fig. 4:** Sammon mapping of feature set obtained by CWT decomposition of the signals at high scale (a=105).*

5. Conclusions

The results obtained show that the system is able to identify and quantify the gases in their mixtures in dry air, except for DMA. The recognition of TMA didn't have any problem and it was enough to use one sensor (based on In_2O_3), while the sensor based on SnO_2 is more suitable to identify NH_3. The next step of the data analysis is to check the validity of the developed method to identify the presence of NH_3, DMA and TMA in a mixture. For this aim different gas-sensing tests will be carried out by optimizing the working temperatures of the 4-sensor array. The measurement protocol will involve a larger number of gas mixtures NH_3/DMA/TMA in different concentrations.

Acknowledgments

This work was funded by European Integrated Project, Project n° 508774-IP GoodFood, Food Safety and Quality Monitoring with Microsystems.

References

1. A. Pacquit, J. Frisby, D. Diamond, K. T. Lau, A. Farrell, B. Quilty, D. Diamond, Food Chemistry Volume 102, Issue 2, 2007, Pages 466-470
2. A. Alimelli, G. Pennazza, M. Santonico, R. Paolesse, D. Filippini, A. D'Amico, I. Lundström, C. Di Natale, Analytica Chimica Acta,Volume 582, Issue 2, 23 January 2007, Pages 320-328
3. J. Hammond, B. Marquis, R. Michaels, B. Oickle, B. Segee, J. Vetelino, A. Bushway, M. E. Camire, K. Davis-Dentici, Sensors and Actuators B: Chemical Volume 84, Issues 2-3, 15 May 2002, Pages 113-122
4. M. O'Connell, G. Valdora, G. Peltzer, R. Martín Negri, Sensors and Actuators B: Chemical Volume 80, Issue 2, 20 November 2001, Pages 149-154
5. G. Olafsdottir, P. Nesvadba, C. Di Natale, M. Careche, J. Oehlenschläger, S. V. Tryggvadóttir, R. Schubring, M. Kroeger, K. Heia, M. Esaiassen, A. Macagnano, Bo M. Jørgensen, Trends in Food Science & Technology Volume 15, Issue 2, February 2004, Pages 86-93
6. http://www.goodfood-project.org/

A GAS MICROSENSOR ARRAY AS NEW METHOD TO ANALYSE THE PRESENCE OF UNBURNED FUEL IN ENGINE OIL

S. CAPONE, M. ZUPPA, D. S. PRESICCE, F. CASINO, L. FRANCIOSO, P. SICILIANO

Institute of Microelectronics and Microsystems I.M.M.-C.N.R., Lecce, Via Monteroni, 73100 Lecce, Italy

Summary

We developed a novel method to detect the presence of unburned diesel fuel in lubricating oil for internal combustion engine. The method is based on the use of an array of different gas microsensors based on metal oxide thin films. The sensor array, exposed to the volatiles of different engine oil samples contaminated in different percentages by diesel, resulted to be appreciable sensitive to them. Principal Component Analysis (PCA) applied to the sensor response data-set gave a first proof of the sensor array ability to discriminate among the different contaminated oils. Moreover, in order to get information about the headspace composition of the fuel-contaminated engine oils samples, we analyzed the engine oil samples by Static Headspace Solid Phase Micro Extraction/Gas Chromatograph/Mass Spectrometer (SHS-SPME/GC/MS).

1. Introduction

Fuel engine oil is indispensable in automotive engines for lubrication, prevention of corrosion and transport and dispersion of heat, but inevitably, it gradually deteriorates with use in an engine. Causes of degradation of engine oil are: a) oxidation and sedimentation, b) thermal degradation, c) corrosion, d) shearing (oil breakdown due to shear forces), d) contamination by suspended insoluble matter, dissolved resinous material, water and fuel. Engine oil contamination by unburned fuel is a widely diffused and largely underestimated phenomenon as compared to other degradation causes. Engine oil may result diluted by fuel and its lubricant ability reduced much earlier other degradation causes occur.

At present, there is no device able to measure the presence of fuel in lubricant oil. Engine oil condition can be checked by a number of laboratory tests on used oil samples by measurements related to: chemical and/or physical properties of oil. Other on-board approaches are based on advanced algorithms taking into account driving parameters such as, e.g., the elapsed mileage since the last oil change, the number of cold starts, the oil temperature, and the engine speed [1]. Some of these algorithms are supported by sensors measuring the oil's permittivity [2], conductivity [3], viscosity [4], but all of them don't provide information about the possible presence of fuel.

In this work we focused the attention on the external contamination of motor oil by unburned diesel. We propose the use of an array of metal oxide (MOX) gas sensors as innovative approach to the analysis of possible contaminations of engine oil by diluting diesel fuel. Since MOX-based gas sensors are generally sensitive to hydrocarbons, that are typical components of fuel, this could constitute a potential for a gas sensor array to distinguish between engine oil samples contaminated by fuel and not-contaminated samples. Different engine oil samples contaminated in different percentages by diesel were prepared and analysed by the system. Data analysis performed on the sensor responses gave a first proof of the system ability to detect the presence of fuel contamination in lubricating oils. The sensor testing was also supported by chemical analytical analysis carried out by Static Headspace Solid Phase Micro Extraction/Gas Chromatograph/Mass Spectrometer (SHS-SPME/GC/MS).

2. Experimental

2.1 *Measurement procedure*

Two generic different types of commercially available engine oils (labeled in the following as A and B) were tested. By adding suitable liquid amounts of diesel, different not-used engine oil samples contaminated with 0.1%, 5% and 10% v/v of diesel were obtained. In particular, 20 ml vials were filled with 10 ml of fuel-contaminated oil. Uncontaminated engine oil samples were also considered for comparison. All the prepared engine oil samples were analysed both by SHS-SPME/GC/MS and gas sensor array.

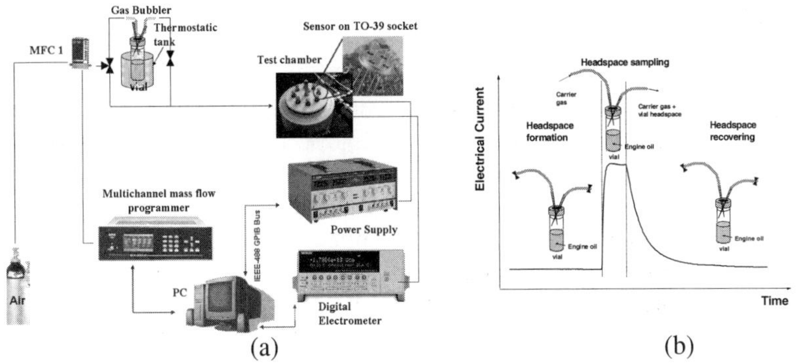

Fig. 1 (a) Experimental apparatus; **(b)** typical sensor electrical signal acquired by using a dynamic stripping of headspace as measurement mode. A constant air flow is drawn by the use of two electrovalves, first into the vial, where the liquid phase of the fuel is in equilibrium with the vapor phase, and next into the test chamber for sensor recovering in dry air.

2.2 *Gas sensor array*

An array of eight different metal oxides-based (undoped, Pt-, Pd, Rh-doped SnO_2, In_2O_3 and mixed In_2O_3-SnO_2) gas sensors have been used for the analysis of the different engine oils samples [5]. These sensing elements have been synthesized by the sol-gel method and deposited by spin-coating onto suitable functional 2mm x 2mm silicon substrates. The sensors, working at a constant temperature of 350-400 °C, were put into a suitable airtight test chamber. Their electrical current signals (under constant polarization) were monitored under exposure to the headspace of the engine oils samples contaminated in controlled way by diesel fuel and under next recovery in dry air (Fig.1).

2.3 *HS-SPME/GC/MS method*

A new SHS-SPME/GC/MS method has been optimized for the qualitative determination of volatile diesel components into the engine oil samples. A 75 μm CAR (Carboxen)-PDMS (polydimethylsiloxane) (black code, Supelco) fiber was experimentally found to be the best fiber that is apt to determine different volatile compounds in diesel fuel contaminated engine oil. The fiber was desorbed in a split/splitless injector, equipped with deactivated SPME glass inserts, and analyses were carried out on a 30m x 250 μm ID x 0.25 μm HP INNOVAX polyetilenglycol column. A GC system HP 6890 Series, Agilent Technologies, was coupled with HP 5973 mass selective detector, Agilent Technologies.

3. Results and Discussion

3.1 *Gas sensor analysis*

In this work we applied the following typical steps for data analysis on sensor responses: 1) signal pre-processing to remove noise signal and frequency components where is present drift contamination; 2) feature extraction to extract useful information less redundant than the original data set and to speed up processing and pattern classification; 3) normalisation to remove as much as possible any concentration effects in the sample; 4) Application of Principal Component Analysis (PCA) to analyze the data structure.

Signal pre-processing was performed by means of discrete wavelet transform (DWT). DWT provides a multi-scale processing analysis where the signal is split into low- and high-frequency components at different frequency bands with different resolutions. Once the signal components, where drift contamination was present, were selected and discarded, the drifting signal was recovered to subsequently data analysis. As feature extraction, the DWT approximation coefficients, related to the lowest frequency components of the signal, were considered. Next, PCA were carried out on processed data.

(a)

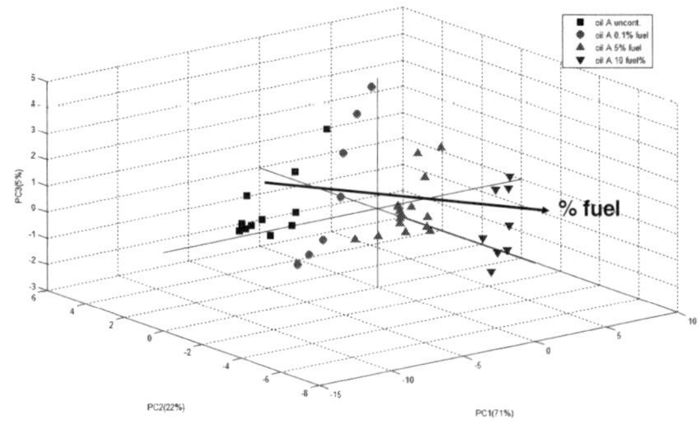

(b)

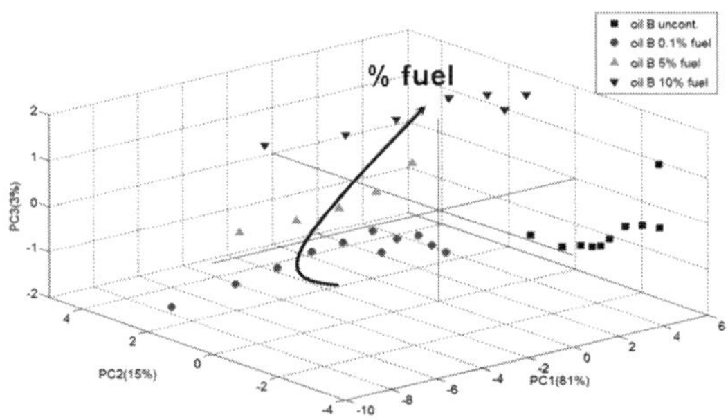

Fig. 2 PCA score plot for (a)oil A and (b)oil B, based as extracted feature on the DWT approximations coefficients.

The discrimination among the different diesel contaminated engine oil samples is evident in Fig. 2a, where the PCA results for the engine oil samples A are reported. In each PCA score plots the data clusters related to the engine oils samples contaminated with 0.1%, 5% and 10% v/v of diesel fuel are separated from the uncontaminated engine oil. Moreover, a direction of increasing of the percentage of fuel can be found. Similar results were obtained for engine oil samples B (Fig. 2b). In fact, also the samples of oil B with different levels of diesel contamination can be distinguished. They result placed in progression along a preferential direction of fuel contamination increasing and separated from the uncontaminated oil samples. However, for this oil B, the data cluster related to the uncontaminated oil sample doesn't seem to follow this progression.

PCA performed on the whole dataset of oils A and B showed an interesting data structure (Fig. 3). We can observe that there is a only direction along which all the engine oil samples contaminated with an increasing percentage of diesel fuel are progressively distributed. Moreover, areas, in which oil samples of both types A and B with the same level of fuel contamination are placed, can be marked off. In general, data clusters seem to arrange according to a geometric structure reflecting the different fuel contamination level and the different headspace composition of engine oil of different types. Taking into account that it is desirable that the system recognizes the presence of fuel in the lubricating oil independently from the type of oil, such geometry in data structure could be useful for the definition of a classification algorithm. However, it can be noticed that only the cluster related to uncontaminated oil B places itself not closed to the uncontaminated oil A, but in an area of fuel contaminated oils. Even if this may be imputable to a wrong contamination or some experimental problems, it could cause a false detection of fuel contamination. The complexity of the data in the specific application may make difficult the data clustering process.

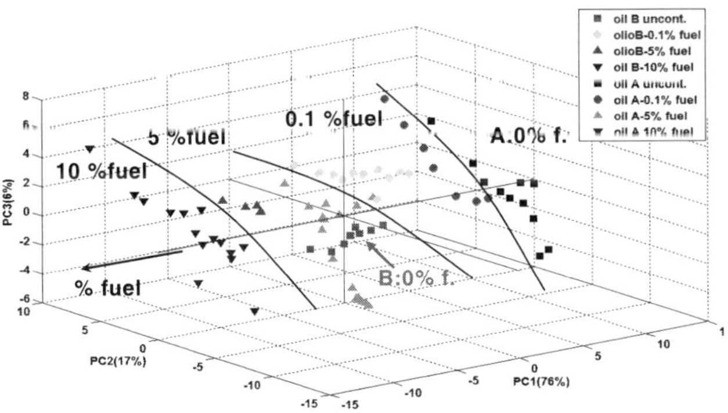

Fig. 3 PCA score plot for oil A and oil B based as extracted feature on the DWT approximations coefficients.

3.2 *SHS-SPME/GC/MS analysis*

In Fig. 4a, the different composition of engine oils A and B are reported, in fact there is not p-xylene in oil B. After the identification of the typical volatile compounds of the uncontaminated A and B engine oils samples, the composition of the engine oil samples contaminated with diesel has been carried out. Some volatile compounds as decane, toluene, ethylbenzene, p-xylene and o-xylene were chosen as markers of contamination and, as expected, it has been found that, in particular for the decane, the ethylbenzene and the o-xylene, the concentration increases when increasing the percentage of contamination by diesel fuel in the engine oil B. The trend not increasing, observed for toluene, can be due probably to the saturation of the absorption of this compound by the fiber (Fig. 4b). The same results have been observed with A engine oil.
A different composition found by GC/MS analysis in the headspace of the two kinds of engine oil confirms the different response of the sensors array to these oils, shown in Fig. 3. In fact, in PCs' space the data clusters related to the uncontaminated oils A and B are separated between them. Morever, a composition richer in species typical of fuel, found by GC/MS analysis, in engine oils samples contaminated by increasing percentage of fuel reflects into the separation of the related data clusters in the PCA score plot based on the sensor responses.

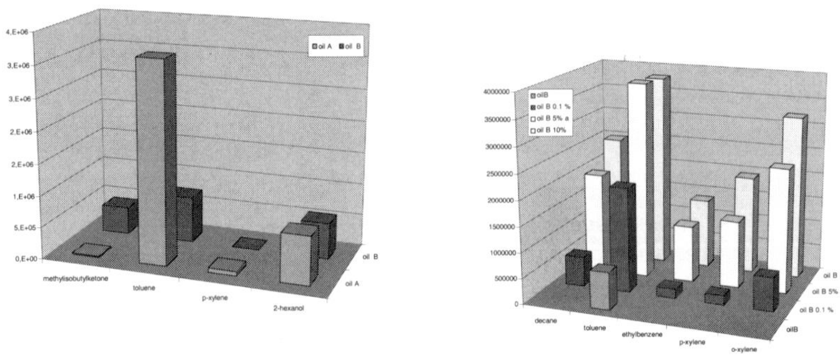

(a) (b)

Fig. 4 (a) Comparison of peak average area of volatile compounds in A and B labeled engine oils and (b) in oil B at different percentages of diesel fuel contamination.

4. Conclusions

A novel method for analysing the contamination by unburned fuel in engine oils based on an array of resistive gas sensors was here evaluated. The results showed that different fuel contaminated engine oils can be discriminated by the sensor array. However, a lot of work has to be done in order to optimize the array configuration. The sensor array seems to have an intrinsic capability to detect the presence of fuel in engine oil, but a wider set of data on different types of oils are necessary in order to test the effective performance of the sensor array. Next work will be addressed to study the effect of engine oil matrix on the detection of fuel contamination by the system. The engine oil ageing with mileage will be also taken into account in order to study how the other causes of oil degradation may influence the detection of fuel in oil matrix.
Moreover, a new analytical method has been developed for the qualitative determination of diesel contaminated engine oil with SHS-SPME-GC/MS technique. Some volatile compounds have been individuated to discriminate different contaminated samples of engine oil. Future work will also consider data fusion between sensor responses and gas-chromatographic data.

Acknowledgments

This work was fully funded by Italian Government MIUR FAR Project.

References

1. European Patent Application, EP 1 363 123 A2, Bulletin 2003/47, Application number: 03076292.6
2. E. Irion, K. Land, T. Gürtler, and M. Klein, SAE, Tech. Paper 970 847.
3. S. S. Wang, Sensors and Actuators B, vol. 73, no. 2-3, pp. 106–111, 2001.
4. A. Agoston, C. Ötsch, B. Jakoby, Sens. and Act.A, 121 (2005) 327–332
5. M. Epifani, R. Díaz, J. Arbiol, P. Siciliano, J.R. Morante, Chem. Mater. 2006, 18, 840-846

ENABLING DISTRIBUTED VOC SENSING APPLICATIONS: TOWARD TINYNOSE, A POLYMERIC WIRELESS E-NOSE

S. DE VITO, E. MASSERA, G. BURRASCA, A. DI GIROLAMO DEL MAURO, D. DELLA SALA, G. DI FRANCIA

ENEA FIM-MATNANO, C.R. PORTICI, Loc. Granatello, 80055 Portici, Napoli, Italy

In this work, we present the development of a novel wireless e-nose platform designed for indoor distributed VOC detection and quantification. The proposed w-nose, called TinyNose, rely on a small polymeric sensor array that is connected to a commercial wireless mote by means of custom developed electronics. A custom developed software architecture allow for signal acquisition, processing and transmission to a data sink where data are stored and/or presented to the remote user. In this work a preliminary assessment of TinyNose capabilities to operate in open air configuration is conducted by using different source of indoor VOC pollution to be detected and classified by the developed architecture.

1. Introduction

Distributed sensing applications (DSAs) represent the basis of the development of the so called smart environments, probably the most important evolutionary step for the today building, industry, military and shipping automation [1].

DSAs such as distributes air quality monitoring, microclimate monitoring, NBC attack detection, etc., are built around wireless sensor networks that represent their infrastructure. The inner core for this kind of applications are motes which are basically dedicated sensing platforms equipped with data processing and transmission units and capable of hosting limited computing activities, such as those needed to run sensor fusion and networking components (formation and routing protocols). By running these software components, several motes, deployed into the measurement area, became capable to interact with each other enabling the DSA to build a comprehensive image of the perceived environment.

E-nose unique capabilities to classify and quantify complex mixtures can be brought in these scenarios by designing a wireless e-nose (w-nose) capable to deal with their demanding requirements: low dimension profile, rapid deployment, easy relocability or even mobility, low power operations and ability to integrate in a mesh shaped, re-configurable network.

Actually, the w-nose should be able to digitalize sensors responses, support local processing of sensor fusion algorithms, transmit the processed data to the other w-nose constituting the network, finally routing the perceived environment data to the data sink where further processing and presentation will take place.

Today, most commercial and research electronic noses are designed for fixed applications. Some of them are portable and allow for limited wireless capabilities so they can be used for mobile application (e.g. [2]), but very few designs are actually suitable for self powered distributed sensing applications (see for example [3]). The low energy requirements represent a limit for the class of sensors platforms to be used, denying, for example, the usage of high temperature operating sensors. Room temperature operating sensors, either nanostructured or polymeric, instead, seems to be a promising choice for the building of a w-nose platform. Polymer-carbon black sensors are, already used in electronic noses (see [4][5]) and their selectivity can be tailored at hand using the wide range of polymeric materials available. In particular this devices could enable the distributed indoor monitoring of VOCs, a primary and very dangerous pollution source for laboratories, industries (solvents) and even homes, where furniture glues, drying paints, cosmetics and special (dry) cleaning products could represent a serious danger. Their effects includes eye, nose, and throat irritation; headaches, loss of coordination, nausea; damage to liver, kidney, and central nervous system; furthermore some organics can cause cancer in animals and suspected or known to cause cancer in humans. However, as with other pollutants, the extent and nature of the health effect will depend on many factors including exposure length and concentration levels at present, not much is known about what health effects occur from the levels of organics usually found in homes.

Pursuing the objective of empowering a distributed VOC sensing application, we are developing a polymeric wireless nose prototype, relying on a five stages architecture. Actually, we propose a wireless e-nose prototype based on a polymeric array coupled with a commercial mote platform featuring zigBee communications and running TinyOS operating system.

2. Experimental and Results

The proposed w-nose platform is based on the five stages architecture depicted in Fig. 1. The sensor array is coupled with a signal conditioning stage that allow for the connection to a commercial mote platform, the Crossbow

TelosB / Moteiv TMote Sky [6], that implements analog to digital conversion, local processing and data transmission by hosting ENEA developed components.

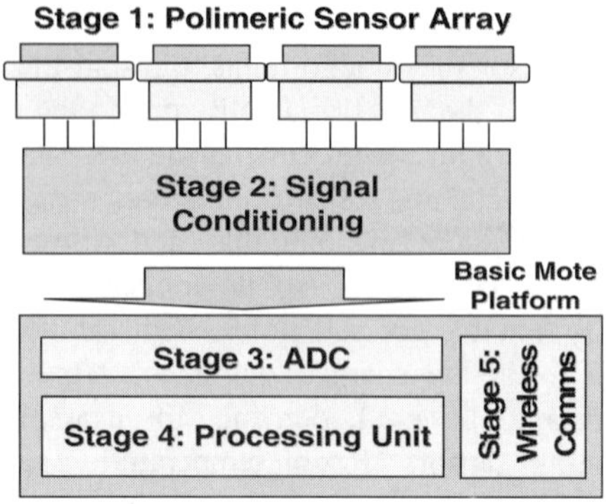

Figure 1: The five stage design of the prototype w-nose.

In particular, the sensor array subsystem has been equipped with four different polymeric-nanocomposite based devices developed at ENEA using a carbon black conducting phase (Black Pearls 2000) dispersed into different organic polymers and dissolved in different solvents. Nanocomposites physical and chemical properties have been suitably adjusted in order to optimise device response and stability. Devices have been fabricated by dropping the suspensions onto a pre-patterned Al2O3 substrate (0.5cm x 0.5cm).

POLYMERS	CHEMICAL STRUCTURE
Poly-(methyl-methacrylate) (PMMA)	$[CH_2C(CH_3)(CO_2CH_3)]n$
Poly-(2hydroxy-ethyl-methacrylate) (PHEMA)	$H(NHCH_2CH_2)_nNH_2$
Poly-(styrene) (PS)	$(C_6H_{10}O_3)n$
Poly-(ethylenimine) linear (PEI)	$[CH_2CH(C_6H_5)]n$

Figure 2: Sensor Array composition and chemical structure of the polymeric matrix.

The array has been coupled to the signal conditioning subsystem, which is based on commercial integrated operational amplifiers, in order to convert resistance changes in the [0, 3] V range required by the ADC subsystem (see Fig. 2) of the selected mote platform.

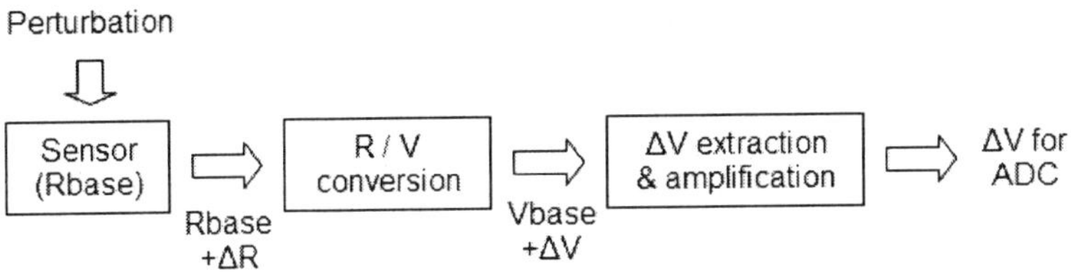

Figure 3: Signal conditioning stage simplified schematics.

The data processing subsystem of this platform features a Texas Instruments inc. TIMSP430 microcontroller and a communications subsystem complying with 802.15.4 zigBee recommendations. The hosted microcontroller features eight 12bits-ADC inputs, four of which has been directly connected to the output of the signal conditioning subsystem. It is also capable of a standby mode with power consumption in the μW range with fast recovery, this characteristic is fundamental for our self powered applications.

The software architecture enabling the DSA, pictured in figures 4 and 5, works on a stack of stand alone components running on the deployed motes (i.e. sensor drivers and routing components), the data sink (data collecting and processing components) and user terminals (presentation components). Custom drivers for sensor array data retrieving, have been developed at ENEA using NesC, a C-like language designed for embedded applications relying on TinyOS operating system [7]. The prototype has been loaded with the developed software and immersed in a simple network (mesh shaped) where repeaters based on the same mote platforms provided a route towards the data sink, hosted on a PC. At the data sink, a server component, written in Java language and acting as a re-broadcaster, provide a means for client located everywhere on the network to connect via a TCP socket to the data stream coming from the deployed motes. Finally, a custom developed component has been given the responsibility to either visualize and/or record the data stream.

Preliminary observation have shown the feasibility of the proposed approach by measuring and recording response patterns modifications in the selected array when exposed to VOC mixtures. In order to assess to capability to operate in open air configurations, a measurement campaign has been set up exposing, in a cyclic way, the proposed system to four typical sources of indoor VOC pollution: a male perfume, a female perfume, a special cleaning product and a commercial anti-dust containing dangerous VOC. Finally a cleaning product containing NaOH has been used as a control.

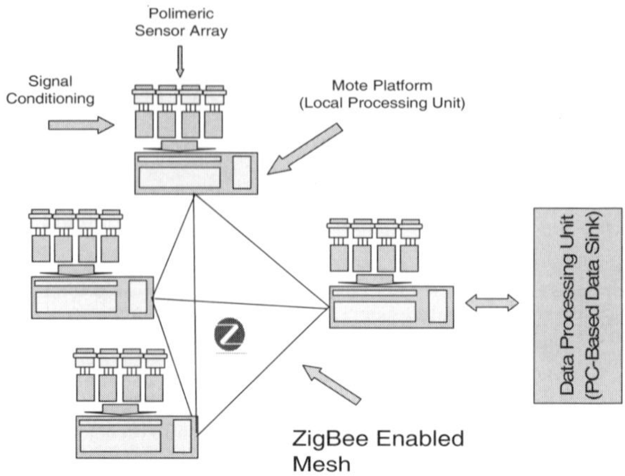

Figure 4: W-nose network architecture with overall sampled data received by a data sink on which data fusion take place to build to cooperative olfactive image of the sampled environment.

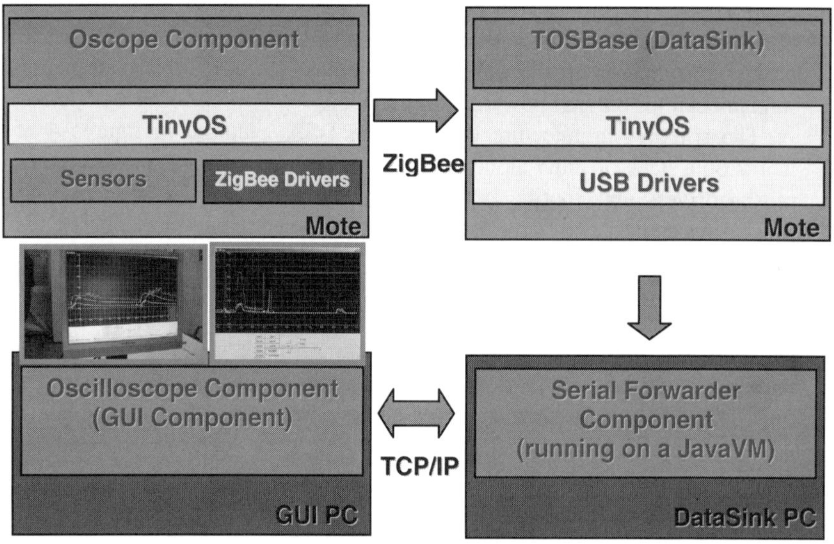

Figure 5: W-nose prototype software architecture : the data sink hosts a server component that receive data from the multiple motes and act as a rebroadcaster (via TCP/IP) towards any PC hosting a GUI component. This provides user interface and data processing, storing and retrieving features.

Recorded data has been then processed by means of a sensor fusion algorithm to be operated on the data sink to assess detection and classification capabilities. The observed response profile is characterized, as expected, by a slow raising segment on the absorption phase, a plateau and a slow descent in the desorption phase. The data set has been populated by averaging the sensor response in each of the three response phases during the 21 exposure cycles obtaining a total of 84 samples.

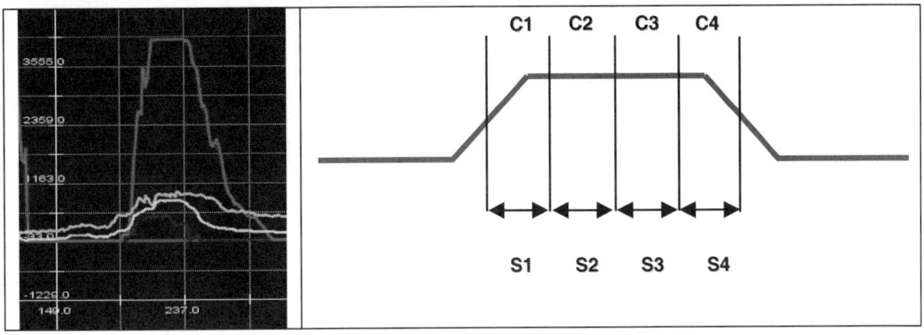

Figure 6: Data preprocessing scheme. The mean response is computed during the raising edge (C1 - absorption phase), the falling edge (C4 - desorption phase) and the plateau phase (divided in two intervals, C2 and C3) building up four preprocessed samples for each exposure cycle.

A PCA analysis, conducted on all the samples, allowed for sufficient cluster separation when using PC1 and PC4 (see fig. 7) suggesting the possibility to use a neural network based approach to sample classification. A classic back-propagation network with 5 inputs, 10 logsig hidden neurons and 5 logsig outputs has been trained choosing 70% of the data set for training purposes, 20% of the data set as validation set and the remaining samples as test set. The chosen learning algorithm was the Conjugate Gradient learning. Results have gone through a 15-fold cross validation process to get a reliable confusion matrix.

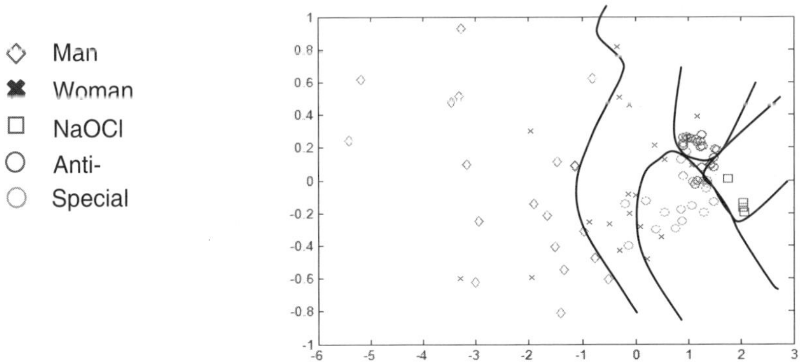

Figure 7: PCA plot of first and fourth principal component, with superimposed semantic region edges showing an interesting mixture discriminating capability.

The overall correct classification score was found to be 83.9% that is a rather interesting value. Analysing the confusion matrix, we can observe that female perfume samples were frequently misclassified as male perfumes and vice versa while misclassification among perfumes and cleaning products were rare. Furthermore, Anti-dust samples were always correctly classified while 4 male perfume samples and 1 special cleaning sample were classified as anti-dust. Control samples were almost always correctly classified because of their rather characteristic pattern.

	DF	DM	AD	SC	IS
DF	26	7			
DM	4	20	4		
AD			24		
SC		2	1	30	
IS				1	3

Figure 8: Confusion Matrix of the featured discrimination experiment. The samples belonging to each of the five mixtures (y-axis) are classified by the sensor fusion algorithm to one of the possible classes (x-axis). Misclassification occurs primarily between female and male perfumes while rarely occurs between perfumes and cleaning products.

3. Conclusions

We have proposed a prototype wireless e-nose architecture capable to act as a single networked sensing element in a distributed machine olfaction framework. When acting together these w-nose could be able to cooperatively build an olfactive image of the sensed environment. The proposed applicative scenario was the VOC distributed detection for indoor environments.

In order to evaluate the capabilities of the prototype for this kind of scenario we have exposed the prototype w-nose to four different sources of VOC pollution in houses. Results shown that the single w-nose, when equipped with the selected polymeric sensor array, was able to detect and discriminate the different sources with interesting accuracy, despite the simplicity of the design. The sensor fusion algorithm, chose for this application, a classic back-propagation network will be in the near future hosted on board as an integer weights neural network.

Further works will include the integration of signal conditioning subsystem and the development of a suitable measurement chamber for simulating indoor environments and analysing the cooperation strategies to be implemented by the networked w-noses.

References

1. D.J. Cook, S.K. Das, Smart Environments: Technologies, Protocols, and Applications, John Wiley (New York, 2004).
2. S. De Vito, E. Massera, L. Quercia, G. Di Francia, "Analysis of Volcanic Gases by means of E-nose", *Proceedings of XX Eurosensors*, Goteborg, SE, Sep. 2006.
3. A. Zhao, L. Wang, C.H. Yao, "Research on Electronic Nose based on Wireless Sensor Networks", *Journal of Physics: Conference Series* 48 (2006), pp. 250–254
4. E. J. Severin, B. J. Doleman, N. S. Lewis, "An investigation of the concentration dependence and response to analyte mixtures of carbon black/insulating organic polymer composite vapor detectors", *Anal. Chem.*, 72 (2000), pp. 658-668.
5. S. C. Ha, Y. Yang, Y.S. Kim, S.H. Kim, Y. J. Kim, S. M. Cho, "Environmental temperature-independent gas sensor array based on polymer composite", *Sensors and Actuators B.*, 108 (2005), pp. 258-264.
6. J. Polastre, R. Szewczyk, D. Culler, Telos: Enabling Ultra-Low Power Wireless Research *Proceedings of IPSN/SPOTS,* Los Angeles, CA, April 25-27, 2005
7. P. Levis, S. Madden, J. Polastre, R. Szewczyk, K. Whitehouse, A. Woo, D. Gay, J. Hill, M. Welsh, E. Brewer, and D. Culler, TinyOS: An operating system for wireless sensor networks, *Ambient Intelligence,* Springer-Verlag, (New York 2005).

POLYPYRROLE-DERIVATIVES SENSOR FOR TRADITIONAL ITALIAN CHEESES DISCRIMINATION BY LIBRA NOSE

ANTONIO SCARPA*, LUCA TORTORA, SIMONA GRECO

Technobiochip S.C.aR.L., Via Provinciale Pianura, 5 (Loc. S. Martino)Pozzuoli, Napoli 80078 Italy

We report the use of a new patented polypyrrole derivatives synthesized at Technobiochip for the development of novel series of nano-gravimetric sensors showing a high affinity for many classes of volatile organic and inorganic compounds (VOCs). 20 MHz AT-cut resonant quartzes were coated with the polypyrrole derivatives, applied to the Technobiochip Libra Nose 2.1 and used to distinguish different Italian cheeses. The different types of cheese aromas were evaluated by inspecting the headspace. Finally, we used Principal Component Analysis (PCA) to discriminate samples on the basis of the VOCs released.

1. Introduction

The evaluation of ripening process, shelf life, or simply the inspection of counterfeited foods and agriculture typical products is a key area in the modern production and distribution.

The study of these phenomena using traditional and innovative methods is a useful tool to characterize the kinetics of the quality decay and to define the acceptability or stability time for their marketing and storage.

In this case, sample classification through the aroma fingerprint allows recognizing of different types of cheese or an original cheese from an imitation.

An electronic-nose (e-nose) system requires poor skilled users, lower investments, automated artificial intelligence and can be easily used as a screening method. It has been applied far several years in agriculture and food industry to characterize the odors of several products [1-3].

The electronic nose is a device equipped with an array of weakly specific and broad-spectrum chemical sensors that should mimic the human olfactory perception and provide a digital fingerprint of the odorant, which can be analyzed with appropriate statistical software.

* Address Correspondence to: Dr. Antonio Scarpa, BSc, Senior Researcher Biological Division Technobiochip Scarl, Via Provinciale per Pianura, 5 (Loc. San Martino), 80078, Pozzuoli (Na), Italy. E-mail: a.scarpa@technobiochip.com - Phone: +39 081 5264315 – Fax: +39 081 5265116

Such characteristics greatly facilitate the application of the electronic nose to rapid monitoring of the volatile components of foods, providing real-time information.

2. Materials and Methods

2.1. *Sample*

Three different Italian cheeses (Parmigiano Reggiano®, Grana Padano®, Pecorino Romano®) and a commercial Mix constituted by 10% Parmigiano Reggiano® and 30% Grana Padano®, were purchased in a market and stored at 4°C in their original packaging until using.

2.2. *Thin film deposition*

20 MHz AT-cut resonant quartzes (*Gambetti Kenologica, Italy*) were coated with different Technobiochip's patented polypyrrole-derivatives (Table 1) deposited *via* Langmuir–Shaeffer method (LS) using the *KSV LB-5000* (KSV Instrument Helsinki, Finland). The Langmuir-Shaeffer method for the deposition of thin films is an easy and efficient way to deposit ordered layers of molecules on solid substrates. In this method, the substrate is aligned almost parallel to the air-water interface and is lowered to touch the compressed monolayer, until the latter adheres to the surface. The functionality of the sensor is based on the mass variation (Δm) onto the quartz surface, due to a direct interaction between sensor and the analyte. As a consequence, it is possible to observe a frequency variation (Δf) of the quartz fundamental oscillation frequency (f_0), as explained by the Sauerbrey's law [Eq. 1].

$$\Delta f = -\frac{C_f f_0^2}{A}\Delta m \qquad \text{[Eq. 1]}$$

Where A is the area of the sensitive layer and C_f the mass sensitivity constant.

Table 1: deposited polymers.

Sensor	Polymer
1	Poly [1-acetyl-1*H*-indole]-1H-pyrrole
2	Poly[2-(thien-2-ylmethyl)]-1H-pyrrole
3	Poly[2-(benzyl)]-1H-pyrrole
4	Poly [(-2-ylmethyl)-2-ethoxyphenol] 1H-pyrrole
5	Poly[2-(-9 phenantrhryl-ylmethyl)]-1H-pyrrole
6	Poly{2-[2-(2E)-3-phenylprop-2-enyl] }-1H-pyrrole
7	Poly [ferrocene]-1H-pyrrole
8	Poly[2-4(methoxybenzyl)]-1H-pyrrole

2.3. *Electronic Nose Apparatus*

The sensors obtained from LS deposition are applied in a Libra Nose 2.1, an electronic nose developed at Technobiochip, for the determination of gas compound analytes. LibraNOSE 2.1 is a compact, easy-to-use instrument that has been thought to perform reversible measurements and to distinguish different odors classified on a qualitative basis. The instrument consists of a thermostatic measuring chamber, where the eight sensors are positioned, a pump with adjustable flow rate and an electrovalve, that can be controlled by software.

All samples were kept at room temperature (25±1°C) for 30 min before analysis. Five grams of each sample were placed in a 250 ml-capped Pyrex bottle. Measurements started after the gases into the bottle being equilibrated.
Active carbon-filtered room air (reference air) was conveyed over the sensors at constant rate (0.15 l/min) for 5 min to stabilize the baseline. An automatic pump then aspirated the cheeses' headspace and conveyed it over the sensors surface for 3 min using a recycling system for sample enrichment. The sensors were exposed again to the reference air to eventually recover the baseline and calculate Δf. The total measuring cycle was 8 min. Each sample was evaluated in triplicate and Δf average was used for statistical analysis.

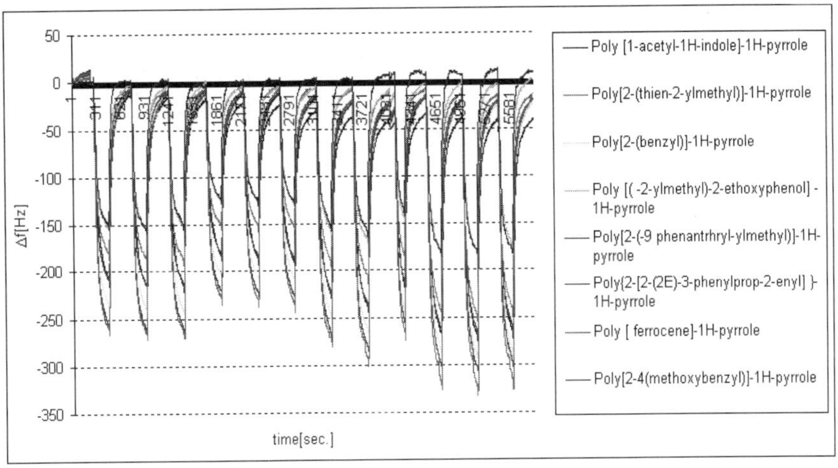

Figure 1: typical sensors signals of the microbalance quartzes coated with polypyrrole derivatives.

2.4. *Data Analysis*

The data obtained from the sensor array of the electronic nose were analyzed by the Principal Component Analysis (PCA) using the "*NasoStat Analisi Statistica Dati per Libra Nose*" by SIGEDA (Milan, Italy).

Principal Component Analysis is used for the explorative data analysis as it identifies orthogonal directions of maximum variance in the original data, in decreasing order. Data are translated into a lower-dimensionality space formed of subset of the highest-variance components. The orthogonal directions are linear combination (principal components) of the original variables and each component explains, in turn, a part of the total variance of the data; in particular the first significant component explains the largest percentage of the total variance, the second one, the second largest percentage, and so on.

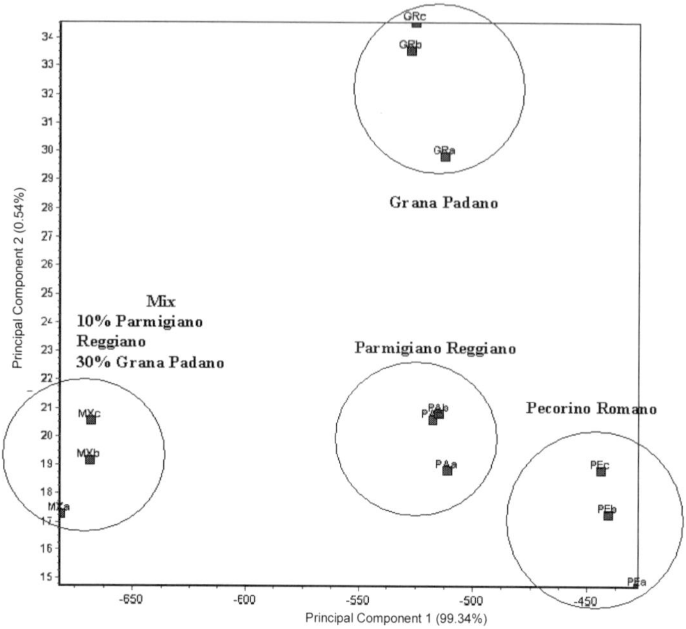

Figure 2: Principal Component Analysis of traditional Italian cheeses.

3. Results and Discussion

As shown in Figure 2, PC1 and PC2 explain 99.34 and 0.54% of the variance in the data, indicating that Polypyrrole-polymers-modified sensors are able to discriminate among different cheeses' aromas.

In conclusion, Technobiochip's Libra Nose 2.1 may be a useful tool for the analysis of features of different cheeses.

References

1. C.Di Natale, F. A. M. Davide, A. D'Amico, P. Nelli, S. Groppelli, and G. Sberveglieri, *Sens. Actuators B* **33**, 83 (1996).
2. A. Guadarrama, M. L. Rodríguez-Méndez, J. A. de Saja, J. L. Ríos, and J. M. Olías.. *Sens. Actuators B* **69,** 276 (2000).
3. T. Pearce, C. J., W. Gardner, and S. Friel, *Analyst*, **118** 371 (1993).
4. G.Sauerbrey, *Z.Phys.*, **155**, 206 (1959).

NEURAL CALIBRATION OF PORTABLE MULTISENSOR DEVICE FOR URBAN ATMOSPHERIC POLLUTION MEASUREMENT

S. DE VITO, G. DI FRANCIA

ENEA FIM-MATNANO, C.R. PORTICI, Loc. Granatello, 80055 Portici, Napoli, Italy

L. MARTINOTTO

Pirelli Labs, Viale Sarca 222, 20126 Milano, Italy

In this work we analyze the feasibility of using on-field data to train a sensor fusion subsystem coupled to a gas multisensor device for urban atmospheric pollution measurements. A gas multisensor device has been co-located with a conventional fixed monitoring station in order to collect a suitable training data set for a neural network operating in regression mode. Benzene concentration estimation performance are evaluated by comparing them with conventional station output. Performance relationship with training set length has been also explored showing that 10 days training length is sufficient to obtain a less than 2% error with respect to conventional station measurements for benzene concentration estimation.

1. Introduction

Urban air pollution is a major concern for public health in most countries, its impact on public health is growing higher in terms of direct and indirect costs [1]. Continuous monitoring of pollution gases and particulate density is decisive for both short term and strategic decision making regarding cars traffic management.

Nowadays, urban air pollution monitoring is carried out by means of networks of fixed stations hosting laboratory scale equipment. These equipments are mostly based on industrial analyzers that can selectively estimate the concentration of many atmospheric pollutants minimizing interferents driven errors. Unfortunately, their cost (including maintenance), and dimensions heavily limits the density of the network. This may cause to misinterpret the real distribution of pollutant concentrations in a turbulent environment such as a city.

The micrometeorology of an urban environment, which heavily influence pollutant diffusion, is in fact very complex, depending on building density and height, street width and orientation, and featuring difficult to model effects such as the "canyon" one. This pose significance threats to the possibility for local sampled data to be representative of the surrounding area situation [2].
Portable e-noses and solid state multi-sensor devices, could have a positive impact on this scenarios thanks to their low cost and portability, effectively helping to rise the density of the measurement mesh. Anyway, a limited number of studies have been conducted regarding solid state devices capabilities for gas concentration estimation in such complex gas mixtures [3] and very few attempts to verify the possibility to obtain a calibration with on-field campaign [4-6]. However, the attempt to extends calibrations obtained via measure campaigns conducted with synthetic mixtures to the field revealed impracticable, due to the low selectivity of solid state transducers that heavily affected their performances in the harsh traffic environment. In this study, we have pursued a multivariate calibration of a portable solid state multisensor device, using a sensor fusion subsystem based on the artificial neural network paradigm for on field estimation of atmospheric pollutants. The calibration was obtained using the concentration estimations of a traditional fixed station as ground truth for the supervised training of the sensor fusion subsystem.

2. Experimental and Results

The setup was built up by a portable multisensor device developed at Pirelli Labs hosting up to seven solid state gas sensors and a conventional monitoring station. The device is characterized by a 31x26x12cm metallic case inside which are stored a power management unit, the signal conditioning and acquisition electronics, a microcontroller board hosting a microprocessor eventually capable to run simple sensor fusion algorithms, GSM data transmission unit and of course, the sensors array subsystem. Total weight is 2.5 Kg. In particular, the microcontroller board take care of first stage data processing operations, storing up to 72 hours measurements at 8 seconds sample rate and controls the communication unit in order to transfer processed data (8 , 15 or 60 minute mean sampled values) to data sinks. In this work, the proposed multisensor device has been equipped with 5 independent sensor slots for easy management and replacement, hosting solid state sensors depicted in table 1. Further two slots has been devoted to host commercial temperature and humidity sensors.
In order to validate its response, the device has been co-located with a fixed conventional monitoring station equipped with spectrometer analysers owned by

Lombardy Regional Agency for Environment Prevention (ARPA). Measurement campaign took place, in facts, at a main street located in the centre of an Italian city characterized by heavy car traffic. The fixed station output was the hourly average concentration of each pollutant while the multi-sensor device output data was averaged to reach the same rate.

Table 1: Selected sensor array principal parameters as obtained in single specie laboratory based characterizations.

Sensor Type /Parameters	**CO**	**NO_X**	**O_3**	**Non-Metanic Hydro-Carbons**	**NO_2**
Concentration Nominal Range	0.1 ÷ 100 mg/m^3	20 ÷ 1000 μg/m^3	20 ÷ 600 μg/m^3	-	-
Detection Limit	0.1 mg/m^3	20 μg/m^3	20 μg/m^3	-	-
Expected Operating Life	12 months	12 months	12 months	12 months	12 months
Sensor Response Time (T_{90})	60s	90s	90s	90s	90s
Operating Temperature Range	-20 ÷ 50 °C	-20 ÷ 50 °C	-20 ÷ 50 °C	-20 ÷ 50 °C	-20 ÷ 50 °C
Operative Relative Humidity Range	10% ÷ 90%	10% ÷ 90%	10% ÷ 90%	10% ÷ 90%	10% ÷ 90%
Output Drift	<2% (6 months)	<2% (6 months)	<2% (6 months)	<2% (6 months)	-

The fixed station provided the ground truth for the concentration of five different atmospheric pollutants, i.e. CO (mg/m3), NMHC* (μg/m3), C6H6 (μg/m3), NOx(ppb), NO2(μg/m3).

After the measurement took place, the output of the multisensor device, namely the hosted sensor resistivities, was coupled with back propagation neural architectures in order to exploit the information content about the perceived environment and estimate pollutant concentrations. Each network was used to estimate a single pollutant by using all the sensor responses. For each network, multiple hidden layer complexities have been considered, and the network obtaining the best estimation error performances was selected.

In this study we present, in particular, the results obtained for the estimation of Benzene concentrations. The overall measurement campaign covered 12 months (March 2004 – April 2005) and in a preliminary run the training set was built up using the first 8 months observations period while the test set covered the remaining 4 months. The results obtained by the proposed architecture was evaluated by using multiple statistics reported in table 2.

---

* Non Metanic Hydrocarbons

Table 2: Best Performances of preliminary run neural network architecture.

MRE =6.90e-03	MAE/Range=6.74e-04	MSE=9.10e-04	SCC=0.9998
σ_{RE}=9.91e-02			

For the Benzene concentration estimation, the square correlation coefficient (SCC) was found to be 0.9998 while the mean relative error was 6.90e-03. Fig. 1 show the empirical cumulative distribution function (CDF) of the relative and absolute error, it can be seen that in the 99% of test samples the absolute error was less than 0.3 $\mu g/m^3$.

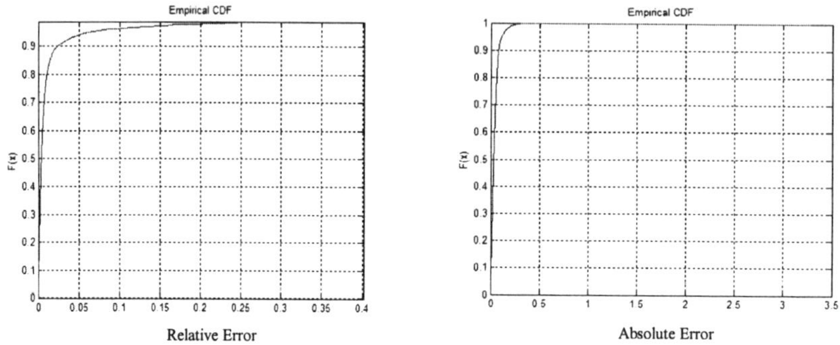

Figure 1: Relative and absolute error CDF. Y axis shows the fraction of test samples which showed an error figure less than what reported in X axis.

In Fig 2, where we show the system performances over a test subset covering a week, it is possible to note that the ground truth estimation performed by the traditional analyzer is perfectly overlapping the estimation carried out by the multisensor device when coupled with the sensor fusion subsystem.

Of course, the amount of time requested to build such a huge training set is too much for considering the practical use. As such we tried to explore the ratio between the length of the campaign segment to be used for training purposes and the performance expressed by the trained fusion subsystem in order to estimate a useful training length. We found that, for Benzene concentration estimation, in order to keep the relative error of the overall system under a 2% threshold with respect to the traditional fixed station output, a training length of 10 days proven sufficient (see table 3). These results show that is possible to obtain, using a 10 day training period, a neural calibration capable to let the featured multisensor device to operate in field successfully carrying out Benzene concentration estimation. It is worth to note that these results were achieved using no specific sensor for benzene.

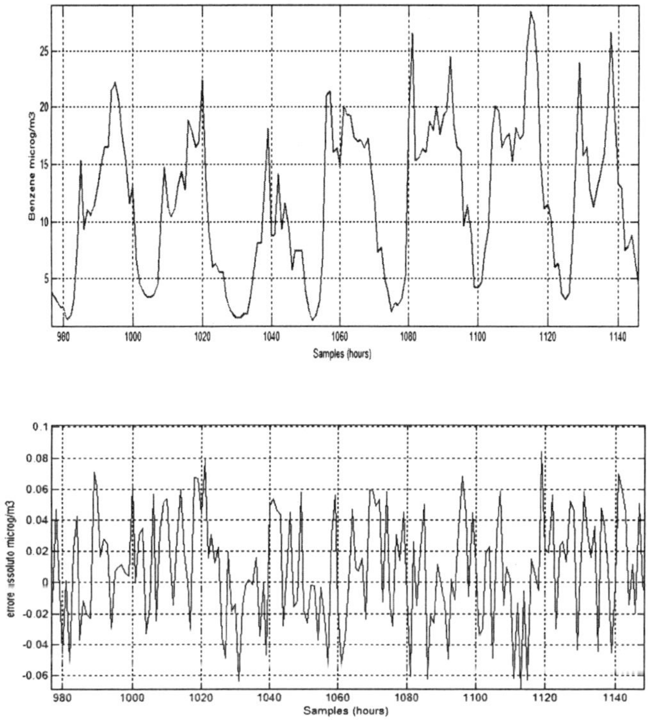

Figure 2: Benzene concentration estimation in the preliminary run: [a] Fixed station output (red) totally overlaps network estimations (blue) over a one week test subset length. [b] Absolute error ($\mu g/m^3$) computed over the same period.

Table 3: Network performances against training period length. A 10 days length training set produce a less than 2% MRE.

Training Length	MRE	STD_RE	MAE $\mu g/m^3$	STD_AE $\mu g/m^3$	SCC
24 hours	0.50	1.07	5.09	5.63	0.4383
96 hours	0.16	0.24	1.31	1.65	0.8849
10 days	0.020	0.11	0.13	0.25	0.9938
25 days	0.017	0.071	0.081	0.19	0.9986
50 days	0.012	0.057	0.050	0.19	0.9984
100 days	0.009	0.076	0.050	0.16	0.9995

Actually, the proposed sensor fusion architecture exploited the combined functional correlation with benzene concentration expressed by all sensor array responses (especially the sensor) and the intrinsic concentration ratios found in city traffic pollution.

Analyzing the weekly scores of the MRE for the above mentioned architecture along the test set duration, we found also that the performance indexes values are constantly good for more than six months. After the 30th week a definite worsening trends take place (see Fig. 3). One of the reason for this is likely the change in the composition of the city pollution that occurs in the winter season because of house heating systems emissions. These changes probably hamper the capability of the network to correctly evaluate Benzene concentrations disrupting the relationships the network has been trained to build between the array response patterns and the polluting specie in the spring time. In this case, a further calibration to occur shortly after the beginning of the winter season or at the end of autumn should help to recover the optimal performance.

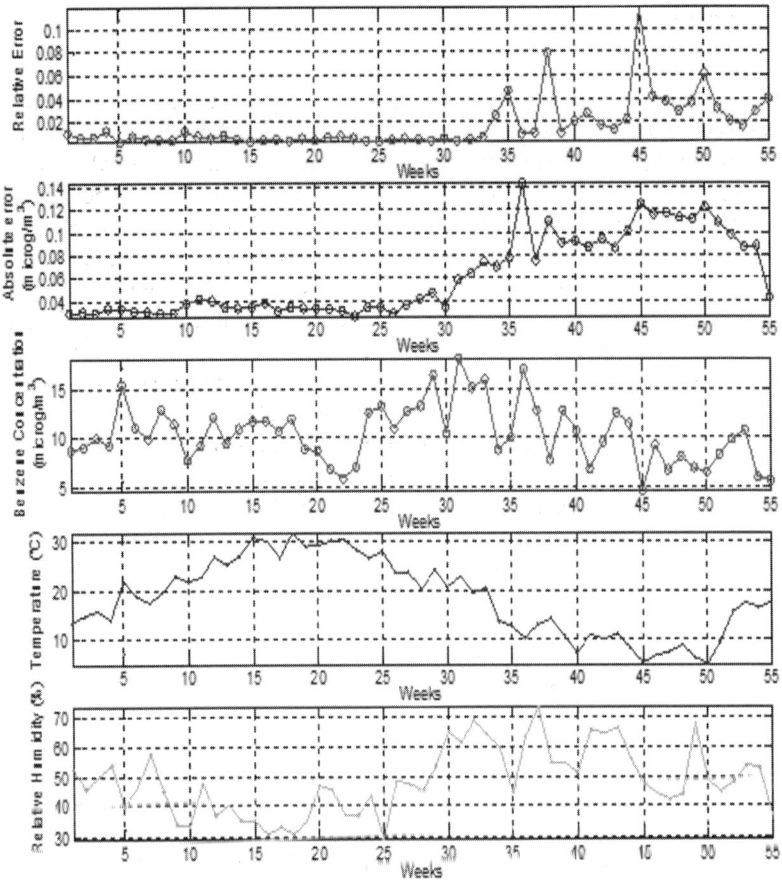

Figure 3: Benzene concentration estimation results in the 10 days training length run (weekly averages): [a] Mean relative error; [b] Mean absolute error; [c] Benzene concentrations as measured by conventional station; [d] Temperature measured by multisensor device; [e] Relative humidity measured by multisensor device. Results are affected by what we expect to be seasonal metereological effects, evident after the 30th week (starting of November) superimposed to a slow degradation due to sensor aging effects (drifts). Peaks of mean relative error seem to occur at low temperatures after a rapid fall of relative humidity. After the 50th week the absolute error show a definite recovery trend.

3. Conclusions

We show that the proposed sensor fusion scheme can effectively make the developed multisensor device able to correctly evaluate Benzene concentrations. Optimal results have been obtained by using only ten days of on-field measurements for training purposes. We have explored the performance evaluation during time, highlighting seasonal issues and proposing a recalibration strategy for the recovering of optimal performance indexes.
The overall architecture could hence be used for densening the sparse atmospheric pollution measurement matrix for the evaluation of Benzene concentrations. Furthermore, this architecture could also be used as a redundant system able to enforce graceful performance degradation in case of conventional analyzers faults (Instrument Fault Accommodation scheme).
However, further works are needed to explore the performance degradation induced by the use of a calibration obtained for a particular city location in others ones i.e. for understanding the portability of the obtained calibration.

Acknowledgments

This work is partially funded by Pirelli Labs.

References

1. D. Dockery et al.. New England *J.* Med. (1993) 329 1753-1759
2. N.A. Mazzeo, L. E. Venegas, Int. J. of Environment and Pollution 25 (2005) 164-176
3. C. Pijolat, C. Pupier,M. Sauvan, G. Tournier, R. Lalauze, Sens. Actuators B 59 (1999) 195-202 .
4. M. Kamionka et al., Sens. Actuators B 118 (2006) 323-327.
5. M.C. Carotta et al., Sens. Actuators B 68 (2000) 1-8.
6. W. Tsujita, A Yoshino, H. Ishida, T. Moriizumi, Sens. Actuators B 110 (2005) 304-311.

OPTIMIZATION OF SUPPORT VECTOR MACHINES FOR QUANTITATIVE E-NOSES

D. ESPOSITO°, S. DE VITO*, E. MASSERA*, G. DI FRANCIA*, F. TORTORELLA°

**ENEA FIM-MATNANO, C.R. PORTICI, Loc. Granatello, 80055 Portici, Napoli, Italy*

° Università di Cassino, via Marconi 10, 03043 Cassino (FR), Italy

In this work we analyze the optimization of tapped delay support vector machines (TD-SVMs) for analyzing quantitative e-nose data. Here, an array of nanostructured and polymer based sensors is exposed to several NO2-NH3-RH mixtures in order to built a suitable data set for testing its real time concentration estimation capabilities. TD-SVM performance depends on both SVM and TD lines parameters. The partial knowledge about their mutual relationships and availability of a GRID infrastructure made a brute force approach on performance optimization feasible. Results indicate that while it is not advisable to optimize SVM and TD lines parameters separately, for this problem a region of quasi optimality is detectable for SVM parameters.

1. Introduction

In the last decades, sensor fusion architectures have been used for analysing sensor array signals in order to carry out classification and, less frequently, quantification tasks on gas mixtures [1][2]. Real time quantification involves the fast estimation of single analyte concentrations by means of sensor array responses, that is, to solve a point by point regression problem. This class of problem can be very tricky to solve, in most e-nose scenarios, due to complex responses of sensors caused by their lack of selectivity. However, particularly for classification tasks, this non-specificity can be turn in an advantage by the use of sensor fusion techniques.

Room Temperature (RT) operating sensors have the advantage of low power consumption and are suitable for operating in flammable atmosphere. They are particularly suited for cheap wireless sensor networks implementations or for portable devices due to the very limited power they need to operate. Unfortunately, it is well known that they usually show severe lack of selectivity performances especially when operating in their maximum sensitivity range;

often, they can also suffer from instability and eventually from complex dynamic response.
To boost both overall precision and responsiveness indexes, the number of sensor response samples with which to feed a sensor fusion subsystem can be extended with past samples so to provide to the system significant information about sensor responses dynamics [3][4]. In this case, tapped delay architectures can represent a valid solution because of their simple and easy to integrate structure. Tapped delay lines allow regression models to take into account feature vector dynamics in the multivariate space. However, tapped delay sensor fusion subsystem performances depends on several parameters values, e.g. the length of tapped delay or the number of hidden neurons in a tapped delay neural network scenario.
This multiple optimization problem can be faced with different heuristic approaches but they often requires a preliminary knowledge about relationships among parameters, if existing. In the e-nose community, hyper parameter selection is mainly conducted with empirical methods that are strongly dependant on a priori knowledge and on the data analyst expertise. Support vector machines are emerging as a powerful sensor fusion tool also for e-nose applications but their usage expertise is still in a development phase in this field with a limited number of papers addressing the parameter values selection problem [3]. For example, in [4] feature selection problem and SVM parameter values selection is approached separately. The risk is to select the best feature vector composition matching the previously selected SVM parameter values. A brute force approach, instead, although very time consuming, could help to shed a light on some unsolved issues, e.g. the stability of SVM optimal kernel parameters with respect to the tapped delay line related ones and the relationship between length of observation and regression performance.
In this work, a tapped delay (TD) line is coupled with an array of three ε-SVM based on RBF kernel, each one designed to estimate in real time, by using a sensor array response, the concentration value of the different species in a NH3-NO2 gas mixture at several RH levels. We face a brute force approach on the overall optimization of a TD-SVM subsystem using ENEA GRID facilities and obtained results are discussed.

2. Experimental and Results

The operative scenario is represented by the concentration estimation of the analytes in a ternary gas mixture. We have selected an array of RT operating sensors for the purpose of analysing a NH3-NO2 in humid air gas mixture, in order to quantify absolute concentrations of each analyte. Sensors and sensing materials have been completely devised and developed at the ENEA Research Centre in Portici (Naples). The sensor array has been equipped with a Polimeric sensor, a Carbon Nanotubes based sensor, two slightly different Porous Silicon membrane sensing units and a Au-Porous Silicon-cSi-ITO Heterojunction [5].

Table 1: Concentration ranges used to mix species in the 64 mixtures.

NH3	NO2	RH
[0,10,15,20] ppm	[0,300,600,900] ppb	[30,40,50,60] %

Measurement campaign has been conducted using 64 different exposure cycle to different mixtures of the above mentioned species, with 60 min. exposure time and 90 min. as desorption time, in order to build a suitable dataset for SVM training. The array has been exposed to the mixtures and conductivity has been sampled at the sampling frequency of 2 samples/min.

NO2 concentration values (see Table 1) have been chosen to be close to alarm levels imposed in many European country. The association with NH3 in the selected values, is also of environmental interest because of the strongly increased production of this gas in urbanized area due the introduction of catalytic exhaust treatment systems on cars. Further details on sensor array fabrication and measurement campaign can be found in [5].

Table 2: Sensor array composition

Sensing Unit	*Sensitive material*
VOC	**Polymeric nanocomposite (PMMA/Carbon Black)**
MEM1	**Porous Silicon Membrane**
CNA	**MWCNT/DMF dispersion**
MEM2	**Porous Silicon Membrane**
POR	**Au-Porous silicon-cSi heterojunction**

The proposed tapped delay fusion architecture rely on the use of a RBF kernel based SVM with a fixed epsilon parameter. Cost parameter C and RBF γ hyper-parameters have been left free to be optimised. Tapped delay length N and distance D between first and last sample with which to feed the SVM are the two additional free parameters with each point in this 4- dimensional parameter space characterizing a different regression model (see Fig.1).

Initially, we have partitioned the 64 exposure cycle dataset reserving 51 cycles for training purposes and leaving the remaining for testing purposes, i.e. for computing performance indexes. A 5-fold cross validation scheme has been applied for the sake of boosting results reliability. The Squared Correlation Coefficient (SCC) has been selected as a performance estimator of the different regression models.

It can be foreseen a relationship between SVM parameter and tapped delay parameter values, so in this work we tried both to find the optimal design and to assess the stability of optimal C and γ with respect to tapped delay parameter values variation.

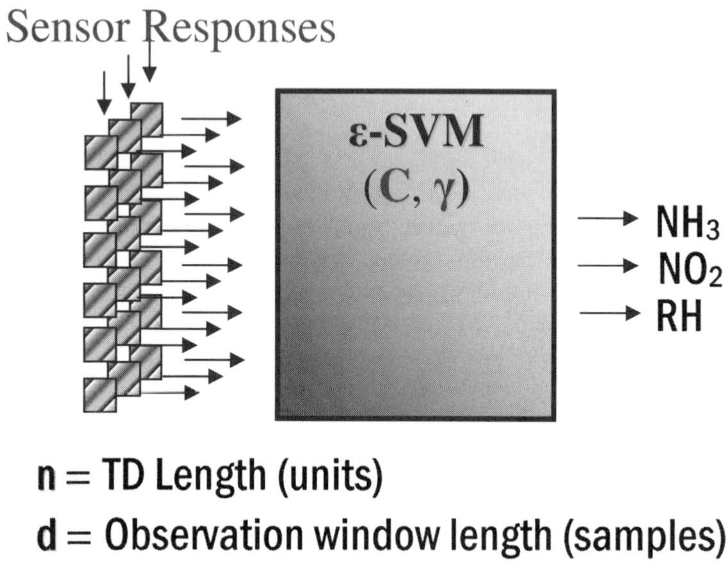

Fig. 1: The proposed tapped delay based sensor fusion architecture.

In the preliminary run, we tried to optimise the (C,γ) couple (SVM parameters) and the (N,D) couple (Tapped Delay parameters) separately. We have choosen to ground to an arbitrary, although likely, value the (N,D) couple choosing N=10 and D=10, that is letting the SVM evaluate the concentration of the three compounds by analysing the current value of each sensor response together with the nine consecutive past values (5 minutes observation window length). We have then optimised the value of C and γ by means of the Multilevel Coordinate Search [6] algorithm using square correlation coefficient (SCC) between expected and guessed value as performance estimation index. By using the obtained C and γ values we have then proceeded to evaluate the best (N,D) couple by exploring the [2,10] X[4,200] values matrix.

Results for NH3 concentration estimation problem shown a SCC, between estimated and ground truth values, greater then 0.94 at N=10, D=180, when choosing optimal values for (C, γ, N, D) parameters with a significant advantage with respect to our previous work reported in [5] (SCC = 0.87). As expected, the SCC value grows while N and D approaches their maximum.

However, if the optimal C and γ values vary significantly with the chosen N,D couple, the optimisation procedure reported above may led to sub-optimal overall performance. In order to reach maximum SCC performance we need to explore relationships between all parameters, for this reasons a brute force approach exploring the parameter space has been conducted by using ENEA Grid computing power. Performance estimation has been conducted using leave-one-cycle out methodologies thus using 63 complete exposure cycles for training, leaving the remaining cycle for testing purposes with a 64-fold cross-validation scheme. The brute force approach was definitely computationally expensive. In order to explore a 5x5x6x4 [N,D,C,γ] sampled parameter space, we have endured 115200 (3 lanes x 64 cycles x 25 N,D matrix point x 24 C,γ matrix point) SVM training runs on a 18900 sample data set using feature vector lengths that varied form 10 to 50. Computation has been carried out using ENEA computing grid, based primarily on AFS (Andrew File system) and the LSF Scheduler, in particular, a computing cluster, based on IBM p6xx AIX machines powered by 16 1.1 GHz CPUs (4 GFlops per CPU) was in use for more than one week.

Again, we report the results for the NH3 estimation problem. In particular, in Fig.2 and Fig. 3 we report the best SCC and MAE values obtained among all C,γ parameters values for each different TD-SVM architecture characterized by a point in the N,D sampled space.

A first glance, the SCC values reach their best while N,D both approach their maximum values in the selected grid. However, it is very interesting to note that it is possible to obtain significantly good results limiting the observation limit to 60 minutes while using only 6 samples for each sensor lane. By using this value we can reasonably avoid the problem of the SVM learning the sequence pattern with which the different mixtures have been presented to the e-nose during the measurement campaign. The boost performance of this grid search with respect to SCC values presented in [5] is 11%. MAE/Range parameter reach 3% while in [4] it was 6.6%.

A closer look reveals also that the relationship among N and D is quite interesting, the descriptive power of the feature vector using N samples initially increase with the observation length until a maximum after which it seems to slightly fade. This agree with a simple observations, i.e. a small observation window retains few information on sensors dynamics but a large observation window require an adequate number of sampling points in order to correctly capture the optimal description power.

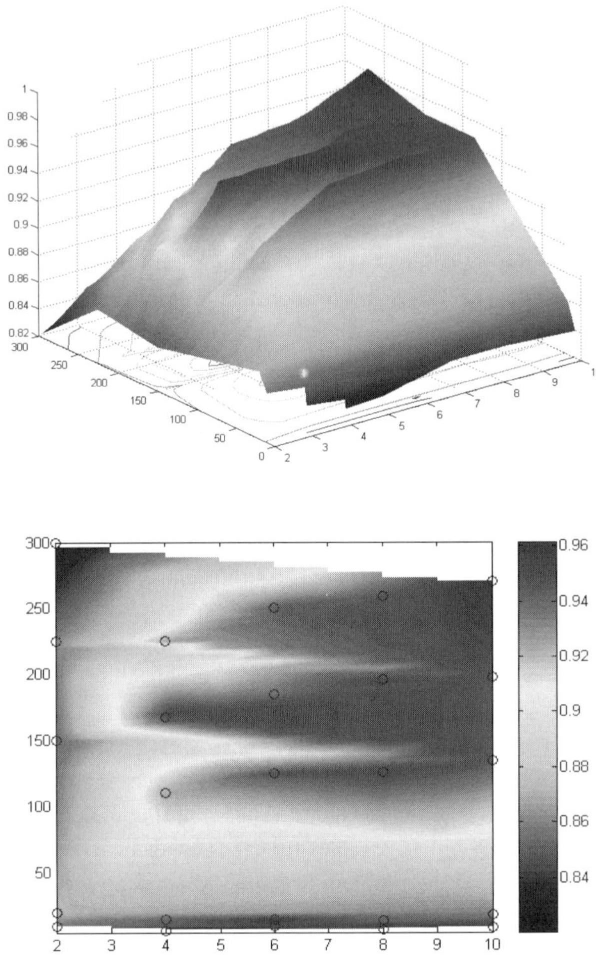

Fig. 2: Best SCC values surface and pseudo-color plot obtained for each N,D couple searching through the C,γ space. Circles depict actual performance evaluation point corresponding to different TD architectures. Highest N,D values scores best performance but interesting performance but can be obtained with 60 minutes wide observation window if adequately sampled.

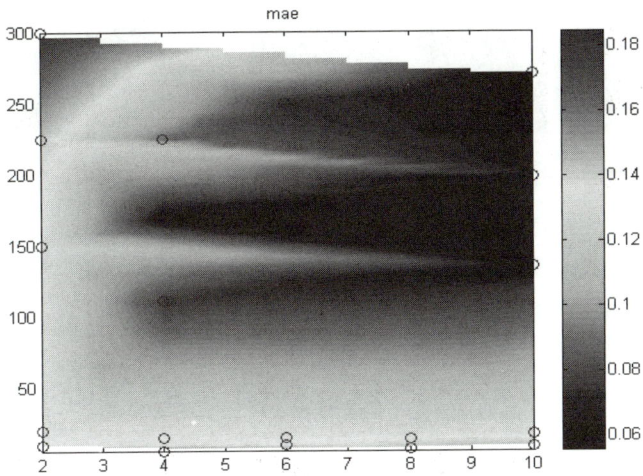

Fig. 3: Best MAE values pseudo-color plot, computed on [-1,1] normalized concentrations, obtained for each N,D couple searching through the C,γ space. Circles depict actual performance evaluation point corresponding to different TD architectures.

In Fig. 4 we depict the SCC values against C and γ values for each of the analyzed N,D couples. Optimal C and γ values varies among the different N,D couples although it is possible to recognize a region where quasi-optimal performance can be obtained whichever couple is chosen. This performance stability in the C, γ space with respect to N,D is the cause of the similarity in the results obtained by the two optimization procedures. It is worth to note that C value became more critical at high T and D values.

Fig. 4 : SCC performances of C (y-axis) , γ (x-axis) dependent SVM architectures computed on the samples matrix (blue circles) for different N,D couples. Low C, γ values guarantee the optimal performance in most of the cases but C value became more critical at high N,D values.

However, regarding the preliminary experiment, if we initially relayed on a different [N,D] couple, e.g. N=4 D=6, we would have selected a different C, γ couple that would have probably led us to a different best N,D couple, i.e. the one that reach the best score with C=10, γ=10 choice.

Finally, we report SCC best values computed on the N,D space for RH and NO2 concentration estimation problem in Fig. 5 and 6. RH estimation problem express a very similar behavior with respect to NH3 estimation one with best overall scores obtained at high N,D values even if best SCC scores are slightly worse. NO2 estimation problems gave significantly worse SCC figures and above hall express a different behavior in which best SCC scores are obtained even at low N,D values. This agree with the possibility that the highest information amount is embedded in the short term sensor dynamics with long term dynamics giving no significant contribute to NO2 concentration estimation problem. Furthermore, it is likely that the lack of an adequate number of samples describing short term dynamics for high D values causes the poor SCC scores the proposed TD-SVM express in top right regions of Fig. 6.

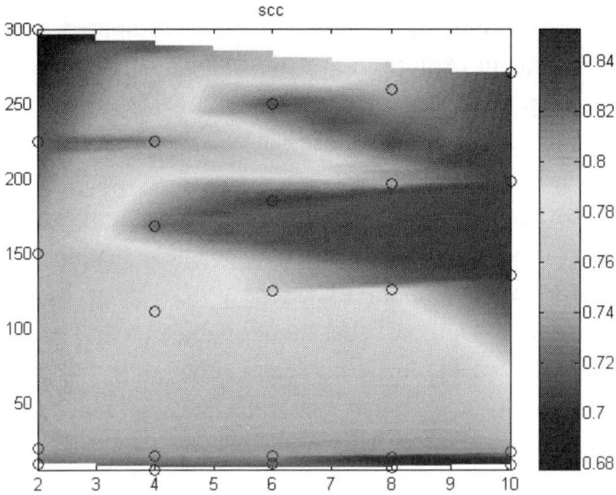

Fig. 5: Best RH estimation SCC values pseudo-color plot obtained for each N,D couple searching through the C,γ space. Circles depict actual performance evaluation point corresponding to different TD architectures.

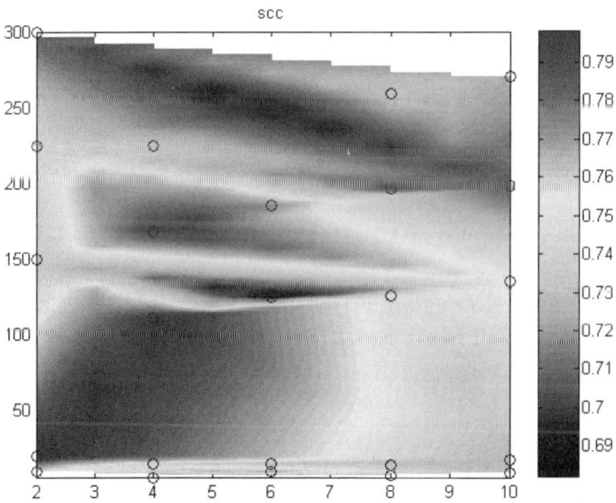

Fig. 6: Best RH estimation SCC values pseudo-color plot obtained for each N,D couple searching through the C,γ space. Circles depict actual performance evaluation point corresponding to different TD architectures.

3. Conclusions

We have reported the solution of parameter values selection problem for a sensor fusion architecture by means of brute force approach. This brute force approach allowed us to explore the relationships among model parameters for a tapped delay support vector machine. As regards as tapped delay related parameters, for NH3 concentration estimation, we found that a 60 minutes wide observation window sampled each 12 minutes is sufficient to obtain very interesting performances boost with respect to a previous work. We found also that wide observation windows should be adequately sampled in order to extract the needed information amount to assure good sensor dynamics descriptive power to the TD line. Optimal inner model parameters (C and γ) revealed to be slightly dependant from TD line parameters although it was possible to design a limited region in the C,γ space that guarantee quasi optimal performances almost for all TD parameters values. This region is localized at low C and γ values revealing a regression problem needing high generalization capabilities and simple kernel structure. However, this preliminary results confirm that separate optimization of parameters values could lead to sub-optimal performance scores.

References

1. C.Di Natale et al., Sensors and Actuators B 24-25 (1995) 808-812.
2. M. Penza et al.,, Meas. Sci. Tech. 13 (2002) 846-858
3. M. Pardo et al., *Sens. Act. B,* 65 (2000), 267-269.
4. S. De Vito et al., doi:10.1016/j.snb.2006.12.039 , to be published in *Sens. Act. B*.
5. O. Gualdron et al., doi :10.1016/j.snb.2006.05.029, to be published in *Sens. Act. B*.
6. Huyer, W. and Neumaier, A., Journ. Global Optimization, 14, 331--355 (1999)

MICROFABRICATION AND MICROSYSTEMS

EXPERIMENTAL STUDY OF WETTING PHENOMENA IN POROUS SILICON BY RAMAN SCATTERING

M. A. FERRARA[a), b)], L. SIRLETO[a)], G. MESSINA[c)], M. G. DONATO[c)], S. SANTANGELO[c)], AND I. RENDINA[a)]

[a)] *Istituto per la Microelettronica e Microsistemi – CNR*
Via P. Castellino 111 - 80131 Napoli
[b)] *DIMET - Università "Mediterranea" Reggio Calabria*
[c)] *MECMAT - Università "Mediterranea" Reggio Calabria*
antonella.ferrara@na.imm.cnr.it

ABSTRACT

The stress in porous silicon during exposition to a liquid is investigated by an approach based on Raman scattering. When the porous silicon structure is exposed to isopropanol or ethanol, a reversible blue shift of the Raman spectra is observed. The blue shift of Raman scattering is ascribed to the contraction induced by the liquids that fill the pores.

1. Introduction

Porous silicon (PS) exhibits a very special structure, characterised by the presence of interconnected pores in a single crystal. The large internal surface (600 m^2/cm^3 for low level doping, *p*-type samples[1]) and the great reactivity of PS layer allow its use as a base for detection of vapours and liquids[2,3].

However, PS is a material of great interest for the study of fluid-solid interfacial phenomena. Adsorption and wetting phenomena are due to molecular interactions between a fluid and an adsorbent, which is usually considered to be rigid. Nevertheless, the adsorbent also experiences the action of the molecular forces, causing substrate deformations, as indeed revealed by observations of strains induced by adsorption and wetting[4].

In PS sample strain depends on the porosity of the PS layer, the type and the level of doping of the silicon substrate[5,6]. The PS lattice parameter is slightly expanded in direction perpendicular to the surface[6], while in direction parallel to the surface, the porous layer has the same lattice parameter as the substrate.

Due to the large internal surface, the strain is also very sensitive to oxidation, to the presence of a fluid or to impregnation with various substances. Since the pores are in the nanometer range, the presence of any fluid inside the pore network leads to a lattice variation detectable, for example, by X-ray diffraction[4-7,8]. When a p^+ type porous silicon layer is filled with a liquid, a small increase of strain is observed[8], while for vapour adsorption a small decrease of strain has been observed when capillary condensation occurs in the pores[4]. After evaporation, these effects are generally reversible.

On the other hand, strains and stress produce a modification of the vibrational properties of the materials, which can be observed by Raman scattering. Compressive stress results in the shift of the position of the Raman peak towards higher wavenumber (commonly called "blue shift"), while tensile stress results in a shift towards lower wavenumber ("red shift")[9].

On this line of argument, in this paper the modifications induced in porous silicon by wetting phenomena are investigated by Raman spectroscopy. Raman spectra measured on unperturbed PS layers and in PS layers wetted by two different liquids. We observe a reversible blue shift of the PS Raman peak, due to compressive stress induced by wetting in porous silicon.

2. Spontaneous Raman Scattering

In order to obtain spontaneous Raman scattering, a beam of light has to illuminate a material (solid, liquid or gas). Measuring the spectra of the scattered light, two weak sidebands are observed beyond to the strong signal corresponding to the frequency of the incident light. The Raman bands arise from changes in the polarization density in a crystal or molecule during vibrations. The difference of frequency among such components and the input source depends on the characteristics of diffusing means. These two sidebands are called Stokes component, shifted down by increments equal to vibrational frequencies of the material irradiated, and anti-Stokes component, shifted to frequencies equal to the sum of the incident wave frequency and the vibrational frequencies. The Stokes component is typically orders of magnitude more intense than the anti-Stokes component and usually dominate in measurements[10]. In Raman scattering both energy and momentum are conserved. Since the induced polarization for Stokes and anti-Stokes scattering differ only in their frequencies and wavevectors, we will restrict ourselves to Stokes scattering.

The real structure of PS is spongelike, *i.e.*, it is composed of wires and/or dots of non-uniform sizes interconnected in a single crystal. In PS, the location of the Raman peaks and the shape of crystallites are obtained through a fitting procedure of the experimental results using the model proposed by Campbell and Fauchet[9,11]. When the particle size reduces to the order of the nanometers, the wavefunction of optical phonons will be no more a plane wave. If PS is modelled as an assembly of quantum wires, the phonon confinement is assumed to be two dimensional, while if PS is modelled as an assembly of quantum dots, the confinement is three dimensional. The localization of the wavefunction leads to the relaxation of the selection rule of wavevector conservation in the Raman scattering process in a crystal. Not only the phonons with zero wavevector **q**=0, but also those with **q**≠0 take part in the Raman scattering process, resulting in red shift and broadening of the Raman line[9,11].

3. Experimental results

Porous silicon can be produced by electrochemical etching of a crystalline silicon wafer in concentrated hydrofluoric acid solution under a controlled current. The pore size and porosity can vary over a wide range depending on the electrochemical parameters and the doping conditions of the initial Si wafer. The pores preferentially propagate in directions specified by the crystal axes or the direction of current flow. High porosity silicon may have a pillar-like structure (1-D) while a low porosity must be regarded as an interconnected nanocrystals structure (0-D). Because of the great number of pores in the porous silicon, the surface area can become very large.

PS is basically a mixture of air and Si, therefore the refractive index of PS is expected to be lower than that of bulk Si. Of course, as the porosity is increased, the refractive index of the PS layer tends to that of air. The refractive index is also sensitive to the aging or treatments the PS samples can suffer. For example, it decreases during oxidization, because of the lower refractive index of Si-oxide compared to that of Si. Since the size of the etched pores is much smaller than the wavelength of visible and infrared light, PS behaves as a homogeneous dielectric layer with an effective refractive index. Regarding the transmission spectra, they are shifted towards higher energy compared to that of bulk Si and this shift increases with increasing porosity.

Moreover, PS is CMOS compatible, it has a very large internal surface which is highly chemically reactive[12] and, being the formation of PS a wet etch process, the apparatus to form PS is simple and cheap.

In this experiment a porous silicon monolayer obtained by electro-chemical etching on p^+ type (ρ=8-12mΩ cm) standard silicon wafer is used. The total porous silicon sample thickness is 20 micron and the porosity is 70%.

The Raman scattering measurements were carried out, at room temperature, by using a Jobin Yvon Ramanor U-1000 double monochromator, equipped with a microscope Olympus BX40 for micro-Raman sampling and an electrically cooled Hamamatsu R943-02 photomultiplier for photon-counting detection. A Coherent Innova 70 Ar^+ laser, operating at 514.5 nm wavelength was utilized as excitation source. In order to prevent laser-annealing effects, the laser power was about 2 mW at the sample surface. Using a 50X objective having long focal distance, the laser beam was focused to a diameter of about 1 μm. Its position on the sample surface could be monitored with a video camera. All components of the micro-Raman spectrometer were fixed on a vibration damped optical table.
The peak position of Stokes component and the shape of crystallites are obtained fitting the experimental results by a curve obtained with the model proposed by Campbell and Fauchet[9,11]. We start our investigation measuring the normalised Raman spectra in unperturbed porous silicon monolayer. The results are shown in Fig. 1; the measured Raman peak is at about 519.2 cm^{-1}, corresponding to a red-shift of 15.7 THz with respect to the pump wavelength. We note that the observations are in excellent agreement with the value of thc optical phonon frequency in porous silicon[13,14]. The estimated crystal size is of 6.0E-7 cm.

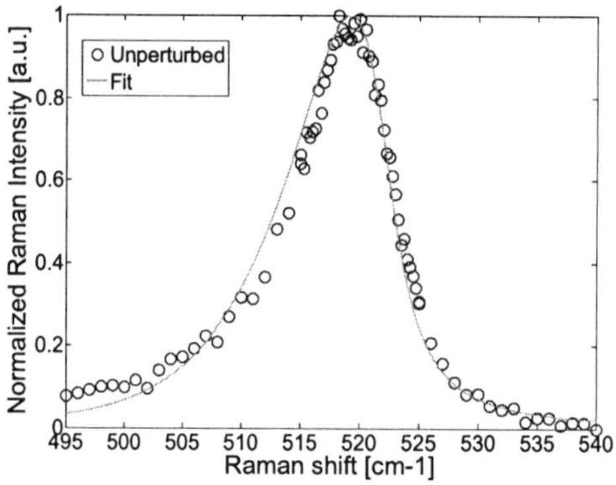

Figure 1. Unperturbed Raman Spectra.

Afterwards, with the aim of study the influence of the chemical species infiltration in PS, a small amount of volatile liquids were added on the PS sample. The liquids fill rapidly pores, after that the acquisition of spectra were carried out.
The experimental Raman spectra for isopropanol is shown in Fig.2, the Raman peak is obtained at about 519.7 cm^{-1}. A shift of the Raman

spectrum, with respect to the unperturbed case, of 0.5 cm^{-1} was observed. Isopropanol partially fills the pores, causing a compressive strain in the porous structure. The induced strain is pointed out by the blue shift of the Raman spectra. The observed blue shift is reversible when the sample is set again in an unperturbed condition.

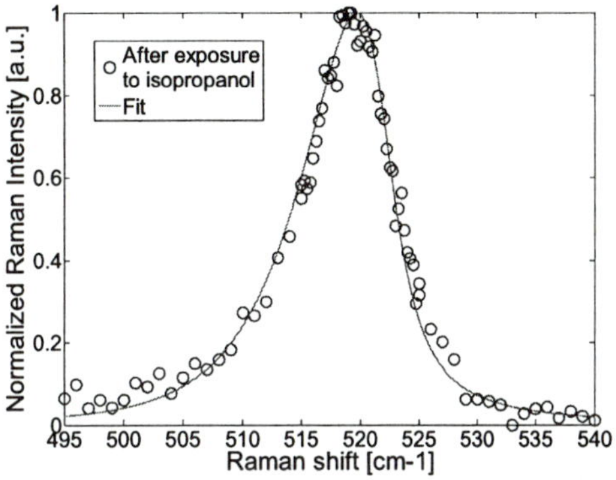

Figure 2. Spontaneous Raman spectra in porous silicon monolayer after exposure to isopropanol.

Results similar to that one obtained for isopropanol were carried out by exposing the same PS sample to ethanol (see Fig. 3). In this case the Raman peak is obtained at about 519.8 cm^{-1}, with a blue shift, with respect to the unperturbed case, of about 0.6 cm^{-1}. The Raman shift is reversible also in this case.

In both cases, the strain is compressive because an increase of the Raman frequency shift is observed.

In unperturbed conditions, the position of the Raman peak is the result of a competition[15] between confinement effects[9] and built-in strain[16], which cause a red shift of the peak position with respect to bulk Si, and compressive stress[7], induced by lattice mismatch with Si substrate, which causes a blue shift of the peak. An increase of the lattice mismatch is observed[4] when PS layers are wetted by liquids. Thus, the compressive stress which the PS layer is subjected to must increase, causing the blue shift of the peak with respect to unperturbed conditions. The use of two liquids having quite similar density and surface tension resulted, as expected, in quite comparable blue shift of the peak. This effect may be conveniently used in sensing applications of liquids on PS layers.

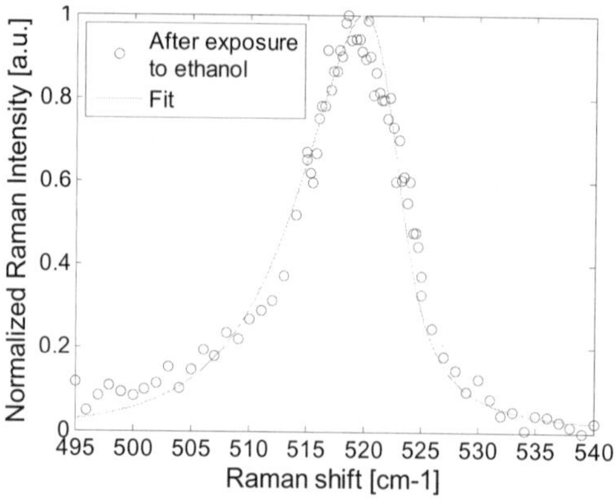

Figure 3. Spontaneous Raman spectra in porous silicon monolayer after exposure to ethanol.

4. Conclusions

In this paper, experimental results proving adsorption strains in porous silicon by Raman measurements are reported. We prove that when the porous silicon structure is exposed to isopropanol or ethanol, a reversible increase of the Raman shift is observed.

We believe that these results are very interesting, in fact they could open the way to a new family of optical sensors for chemical species. Moreover, this effect could be also used to get a tunability of Raman spectra in porous silicon amplifier based on Raman scattering[17].

References

1. R. Herino, G. Bomchil, K. Barla, C. Bertrand, and J. L. Ginoux, J. Electrochem. Soc. **134**, 1994 (1987).
2. A. G. Cullins, L. T. Canham and P. D. J. Calcott, J. Appl. Phys. **82**, 909-965 (1997).
3. W. Thei B., Surf. Sci. Report **29**, 91-192 (1997).
4. G. Dolino, D. Bellet, and C. Faivre, Phys. Rev. B **54**, 17919-17930 (1996).
5. D. Bellet, G. Dolino, Thin Solid Films **276**, 1-6 (1996).
6. I. M. Young, M. I. J. Beale, J. D. Benjamin, Appl Phys. Lett **46**, 1133-1135 (1985).
7. U. Gruning, A. Yelon, Thin Solid Films **255**, 135-138 (1995).
8. D. Bellet, G. Dolino, Phys. Rev. B. vol .**50**, 17162-5 (1994).
9. I. H. Campbell and P. M. Fauchet, Solid State Communications **58**, 739-741, (1986).

10. P.Y. Yu, M. Cardona, "Fundamentals of semiconductors – Physics and materials properties",Springer-Verlang Berlin Heidelberg New York (2003).
11. H. Ritcher, Z. P. Wang and L. Ley, Solid State Communications **39**, 625-629, (1981).
12. T. Canham. "Nanostructured silicon as an active optoelectronic material. In Frontier of Nano-Optoelectronic Systems", Edited by L. Pavesi and E. Buzaneva. Kluwer Academic Publishers (2000), .85-98.
13. F. Kozlowski, W. Lang, Appl. Phys. Lett, 5401-5408, 1992.
14. R. Tsu, H. Shen and M. Dutta, Appl. Physics Lett. **60** (1), 112-114, 1992.
15. D. Papadimitriou, J. Bitsakis, J.M. Lopez-Villegas, J. Samitier, J.R. Morante, Thin Solid Films **349**, 293-297 (1999).
16. M. Yang, D. Huang, P. Hao. F. Zhang, X. Hou, X. Wang, J. Appl. Phys. **75**, 651-653 (1994).
17. L. Sirleto, V. Raghunatan, A. Rossi and B. Jalali, Electron. Lett. **40**, (2004).

HIGH FLOW RATE PERMEATION MEMBRANE ON POROUS SILICON FOR HYDROGEN FILTERING DEVICES

R. AINA, U. MASTROMATTEO*
STMicroelectronics, Via Tolomeo 1 , 20010 Cornaredo, Italy

F. BELLONI, V. NASSISI
Laboratory of Applied Electronics, Department of Physics, I.N.F.N., University of Lecce, 73100 Lecce-Italy.

M. RENNA, A. ROMANO
STMicroelectronics, Statale primosole 50, 95121 Catania, Italy

A porous silicon based hydrogen filter has been designed in STMicroelectronics R&D centers with the purpose to increase hydrogen permeation in systems where high purity gaseous hydrogen is required.
The use of Pd alloy membranes is quite well established when hydrogen separation from other gases is needed, as, for example, in hydrogen generation systems like electrolysers. However, such kind of membrane efficiency is very low at room temperature; so a working temperature ranging between 200 and 500 °C is requested to obtain proper flow rates. Moreover, the cost of Pd alloy membranes as purifiers is quite high, due to the thickness of the Pd alloy foil able to withstand at high permeation pressure gradients.
In order to improve the performance of a device like that described above, in the STMicroelectronics it has been proposed a new design in which a very thin Pd layer (500 nm) on a porous silicon substrate (200 μm thick) allows to obtain high flow rates using a very small amount of the precious metal (Pd), with no mechanical filter robustness degradation. The thickness of the porous substrate, as well as the design of the polymer sustain, will be evaluated.

Introduction

Hydrogen filtering basics

Nowadays the so called "Hydrogen Technologies" are under the attention of many researchers in several application fields. The main reason of that is the possibility to use this energy carrier to produce clean energy. It is very well known that the combustion of hydrogen (in appropriate conditions) produce only water and so no pollution at all.

A common characteristic of these technologies is the high purity level required for the hydrogen as for the systems able to generate low cost hydrogen gas, like electrolysers, deliver a gas mixture that needs purification.

A typical example application that needs very pure hydrogen gas is the fuel cell, because of the possibility to poison the catalyst with a degradation of the yield in the electrochemical activity of the cell, when the fuel is not pure.

The active sites of the catalyst may be compromised by the presence of carbon oxides (CO, CO_2) coming from reforming processes of hydrocarbon compounds used to supply the fuel cell. As a consequence, very frequently, it is necessary to perform purification processes by selective gas filtering.

For hydrogen the need to have selective filtering is particularly important.

The main properties of a filter, usually manufactured in a membrane shape, are the permeability and the selectivity. The permeability assures the flux amount and the selectivity the separation and blocking for undesired gas species. The practical objective is to have a membrane with both of these two parameters maximized (high permeability and high selectivity).

By the way, while for the flux amount may be enough the have large filtering areas, for the selectivity it is very important the choice of a proper material or alloy.

In order to obtain a filtering device it may be enough a porous membrane, but in this way it is not possible to have very high selectivity when the molecular weight of the mixed gases are close. This is true even decreasing the pores size to arrive in conditions where the Knudsen's regime applies. To have the best selectivity it is necessary to eliminate as much as possible the pores and activate selective absorption and diffusion of suitable materials.

The material mostly used to filter hydrogen is palladium. Palladium absorbs spontaneously many gases and it is extremely permeable to hydrogen: the ratio of the absorbed gas volume and palladium volume is about 900 at ambient temperature. However

the physical mechanisms that allow such huge and easy absorption are not completely understood.
One important characteristic of palladium is the practical absence of oxidation and this property put it in a privileged position versus other materials with a similar affinity in hydrogen absorption (zirconium, niobium, vanadium, tantalum). Since the hydrogen diffusion in the metallic lattice changes the reticular distances between atoms, a certain fragility of a pure palladium membrane may occur during the device life. To avoid that, a palladium-silver alloy is often preferred to the pure palladium membrane.
A typical metallic membrane for hydrogen filtering is usually made with two parts connected together to form a single unit: a porous substrate on witch is deposited the filtering material. As an example the substrate can be steel and the filtering material palladium-silver alloy: the filter guarantees the selectivity and the substrate, usually porous versus the hydrogen, the high permeability and so high fluxes [1].
The deposition of the material on the substrate may be done with several techniques like PVD (Physical Vapor Deposition), laser ablation, electron beam evaporation and others. Particular care has to be placed in taking into account the effect of the porosity of the substrate on the integrity of the thin filtering layer.
The membranes made by palladium deposition on porous ceramic are able to guarantee 100% of hydrogen selectivity, giving a pure hydrogen permeate even though not all the hydrogen goes through the membrane. With the improvement of the technology the thickness of the deposited palladium has been reduced from 5-10 micrometers to fraction of a micrometer as in the devices described also here.

Thin film filtering devices

Under typical operating conditions (i.e., temperatures of 200 °C and above), the "rate limiting" step is the diffusion of hydrogen atoms through the bulk metal. In such conditions, other physical processes can be assumed to exist in a state of dynamic equilibrium and, under this assumption, the following expression for hydrogen permeation rate through a Pd alloy membrane of thickness d is derived from diffusion Fick's Law [2].

$$J = D/d \left(P_{in}^{n} - P_{out}^{n}\right)$$

Where ***J*** is the flux, ***d*** the thickness of the membrane, and ***P*** the inlet/outlet pressure with the exponent ***n*** depending from the hydrogen permeation mechanism (***n*** is close to one when surface phenomena are prevailing). On purpose some inventions have been generated in relation to gas separation membranes including a metal-based layer having sub-micron scale thicknesses. In one case the metal-based layer is a palladium alloy supported by ceramic layers such as a silicon oxide layer and a silicon nitride layer. By using MEMS, a series of perforations (holes) can be patterned to allow chemical components to access both sides of the metal-based layer. Heaters and temperature sensing devices can also be patterned on the membrane. A possible application of such a device may be in portable power generation system or in chemical microreactors including the gas separation membrane. In the figure 1. it is illustrated an integrated MEMS device showing the above mentioned application [2], [3].

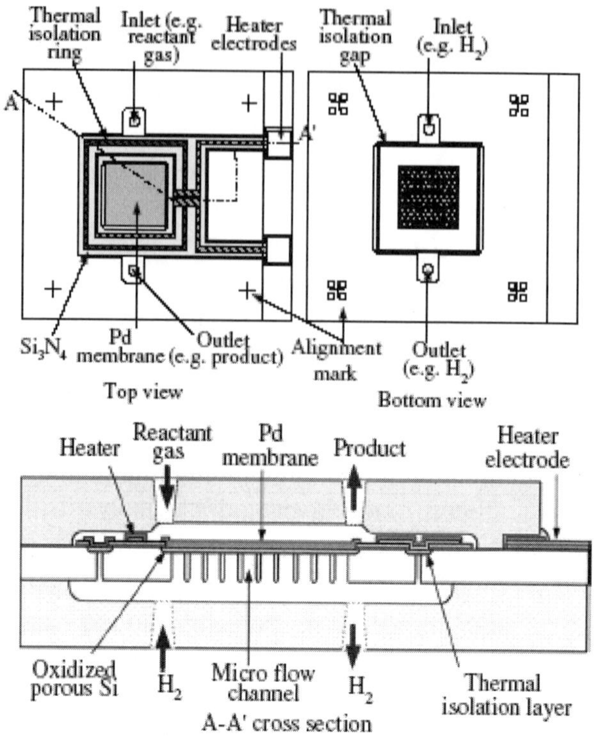

Figure 1. Integrated microreactor using purified hydrogen by filtering

The method for fabricating this kind of gas separation membrane is very common in the microelectronics industry. Due to the ability to make chemical microreactors of very small sizes, a series of reactors can be used in combination on a silicon surface to produce an integrated gas membrane device.

Simplified filtering membrane

In STMicroelectronics R&D labs, it has been designed a specific process (compatible with the manufacturing techniques used in the semiconductor industry), able to deposit on a porous substrate a very thin palladium layer with the purpose to make an high

efficiency hydrogen filter. The porous substrate being ceramic or silicon. In spite of more complex structure can be made using microstructured silicon wafers with the same technology used to make MEMS and wafer to wafer bonding for the filter encapsulation, in this work we are proposing a very simplified structure of the membrane in order to use the filter also with the deuterium isotope and be able to observe the LENR (Low Energy Nuclear Reactions) induced surface changes claimed by Japanese researchers [4]. In order to do that the embodiment shown in figure 2. will allow an easy membrane removal from the chamber to make proper surface analysis; moreover the very simplified manufacturing process has been conceived to avoid any possible contamination that could affect the analysis of appearing transmuted elements.
The idea, (see figure 3.), consists in the use of a small piece of porous silicon substrate on witch to deposit the thin palladium layer. The area of the chip where the silicon should be porous may be determined by photolithographic processes or by mean of the tool configuration in the electrochemical bath. This second option is preferred to achieve very low contamination levels.

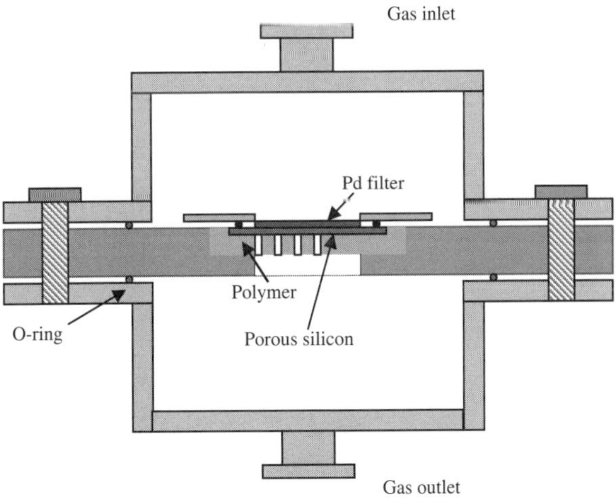

Figure 2. Embodiment for an easy membrane testing and removal for surface analysis.

As mentioned above the process to manufacture the porous silicon membrane has been designed to reduce as much as possible the source of contamination, so the porous area is fixed by the a proper design of the handling tool during the electrochemical process: by mean of a simple o-ring only a well defined portion of the silicon chip is exposed to HF chemical solution and to the anodic current that generates the porosity.

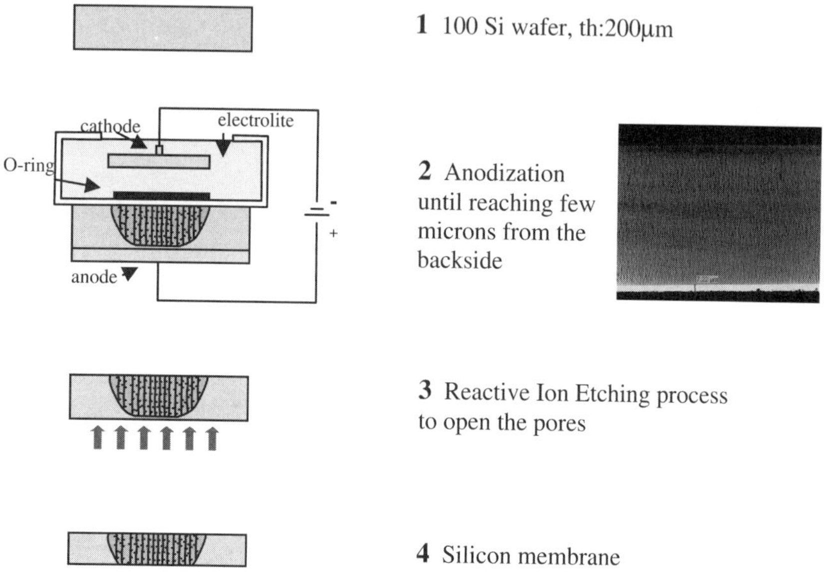

Figure 3. Schematic view of the porous silicon membrane process

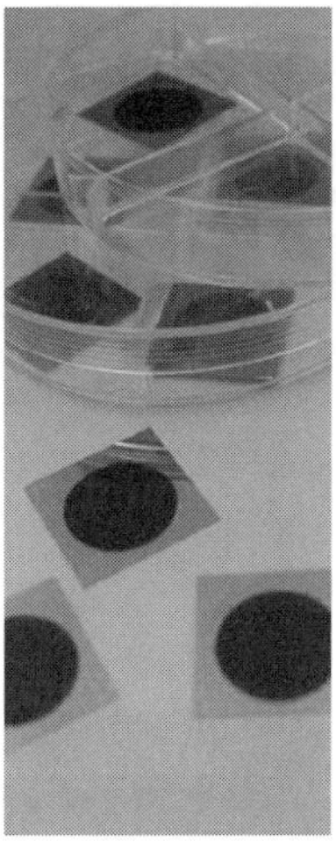

Figure 4. Picture of the porous silicon (black area) chips

Multilayer filter with quantum effects

As mentioned before, Japanese researchers using stacks of alternate palladium and metal oxides thin layers claimed a quantum behavior versus hydrogen isotopes (deuterium) flowing through such kind of filters, resulting in unforeseen nuclear transmutation of elements on purpose implanted into the Pd layer [4].
Also in this case the high flow rate may be an optimum requirement to enhance this very intriguing phenomenology.

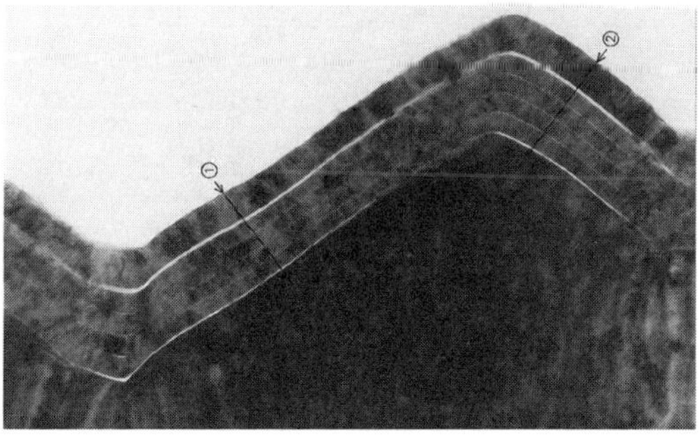

Figure 5. SEM picture showing the layer stack used in LENR experiments

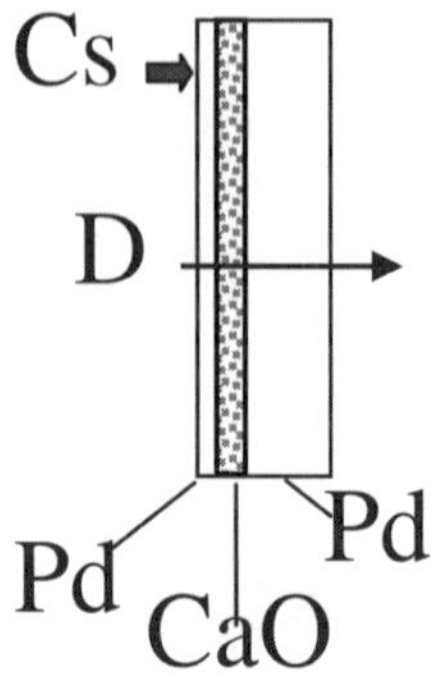

Figure 6. Sketch of the membrane structure submitted to deuterium permeation

The figure 6 shows a sketch of a palladium filter permeated by deuterium gas where on the surface from where the gas enters a certain quantity of cesium (Cs) has been introduced by ion implantation. Due to the peculiar structure of the stacked layers as shown in thc figure 5. picture, transmutation of cesium into praseodymium occurs with a rate proportional to the deuterium flux amount.

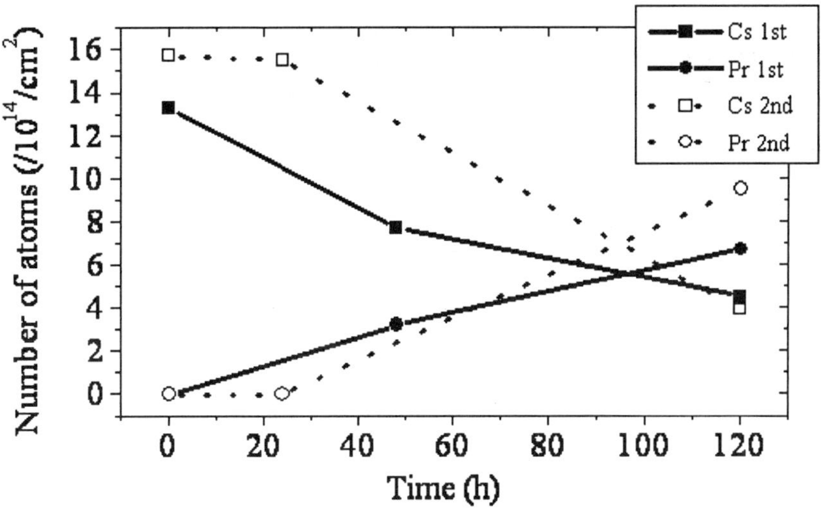

Figure 7. Transmutation rate for the filter sketched in figure 6.

Conclusions

With the purpose to make a filter for hydrogen and its isotopes able to maximize the permeation gas flux even at temperatures below 100 °C, a specific approach has been studied in STMicroelectronics R&D sites. The difficulties of matching the potentiality of the MEMS technology, (allowing the integration of heaters an temperature sensor on the filtering membrane), and the need to have a very low level of contaminants introduced during the manufacturing process, forced the working team to choose the simplest approach, sacrificing the possibility to have an higher integration level. As shown in the figures (3,4), the process to make the membrane has been fixed with satisfying results. Also the chamber were the filters performances will be evaluated is now ready. If in the evaluation phase the flux will exceed of one or two orders of magnitude what has been claimed by Iwamura's team, it would be possible to perform transmutation experiments on a very short time base, with a formidable enhancement in the LENR phenomena comprehension.

References

1. www.hy9.com/hypurifier.html
2. Shu – Ying Ye et al.; *Thin palladium membrane microreactors with oxidized porous silicon support and their application*, J. Micromech. Microeng. 15 (2005) 2011-2018.
3. AJ Franz, KF Jensen, MA Schmidt, S Firebaugh - US Patent 6,810,899, 2004
4. Y. Ywamura, T. Itoh, M. Sakano, N. Yamazaki, S. Kuribayashi,, *Proceedings of the* 11th *International Conference on Cold Fusions;* Condensed Matter Nuclear Science, J. P. Biberian (Editor), World Scientific 2006, 339

RF- MEMS COPLANAR SHUNT SWITCHES BASED ON SU-8 TECHNOLOGY

ANDREA LUCIBELLO, EMANUELA PROIETTI, SIMONE CATONI
ROMOLO MARCELLI, LUCIANO FRENGUELLI
CNR-IMM, Via del Fosso del Cavaliere 100 , 00133 Roma, Italy

GIANCARLO BARTOLUCCI†
Dip. Iing.elttronica, Università di Roma "Tor Vergata", Via della ricerca scientifica, 00133 Roma, Italy

Abstract - RF MEMS switches are currently considered for isolation and transmission of RF signals as the ideal next generation devices with respect to PIN diodes, because of the very low losses, high reliability, no signal distortion and no power consumption. In this work, RF MEMS shunt switches in coplanar waveguide (CPW) configuration have been designed, realized and tested for wideband isolation purposes. SU-8 negative resist technology has been introduced for improving the bridge mechanics and the RF performances of the device. The polymeric material is used to elevate the ground planes of the CPW structure, with minor consequences on the electrical matching and an improvement in the bridge ends definition. The EM design has been followed by a six-step photo-lithographic process on a 4" oxidized high resistivity silicon wafer, up to the release of the bridge by using a plasma etching technique.

1. Introduction

1.1. *RF MEMS*

During the paste decade, several new fabrication techniques have been proposed in the field of the micro-electromechanical systems (MEMS), and numerous novel devices have been reported in diverse areas of engineering and science. One such area is that for microwave applications, for solving many intriguing problems of high-frequency technology for wireless communications. The recent developments of personal communication devices forced the market to introduce miniaturized efficient devices, and this is currently possible only by the development of radio frequency (RF) MEMS. The term RF MEMS refers to the design and fabrication of MEMS for RF integrated circuits. MEMS devices in RF MEMS are used for actuation or adjustment of a separate RF device or component, such as variable capacitors, switches and filters. Silicon micromachining has been a key factor for the vast progress of MEMS and RF MEMS. Silicon micromachining refers to the fashioning microscopic mechanical parts out of a silicon substrate or on a silicon substrate. Silicon micromachining comprises of two technologies[3]: bulk micromachining, when micro-structures are etched into the silicon substrate, and surface micromachining, when the micromechanical layers are formed from layers and films deposited on the surface. Bulk micromachining and surface micromachining are the two major micromachining processes of silicon.

1.2. *RF MEMS switches*

The development of digital switches (transistors) in logic devices has proceeded at an incredible speed over the last decades in terms of components per chip, cost per function, clock rates, power consumption, compactness and functionality. However, the limits of digitally controllable analog signal switches have not the same speed advancing that fast, and electronic switches based on PIN diodes and field effect transistors (FET) can hardly meet the performance requirements of today's communication systems, especially concerning the isolation, the insertion loss (on resistance) and the signal linearity. Even though also semiconductor-based switches have been improving over the last decade, their RF signal performance still decreases drastically with frequency above 1 GHz where they are limited either in power handling or they show a very large insertion loss, poor isolation and high signal distortion. However, MEMS switches perform very well over an extremely large bandwidth with very uniform characteristics, even above 100 GHz. That is the advantage of having a purely mechanical element opening and closing or short-circuiting an almost unimpaired transmission line. In fact, RF MEMS Switches are devices processing RF signals via a Transmission (TX) Line by means of an electrostatic actuation (but it can also be magnetic, piezoelectric, thermal and so on) due to a voltage applied between membrane and electrode, and they pass from an UP state to a DOWN state generating an open-circuit or short-circuit on the TX line. The switches can be categorized by the following three characteristics (Figure 1):

1) RF circuit configuration (series or shunt);
2) mechanical structure (cantilever or air bridge);
3) form of contact (metal-metal or capacitive).

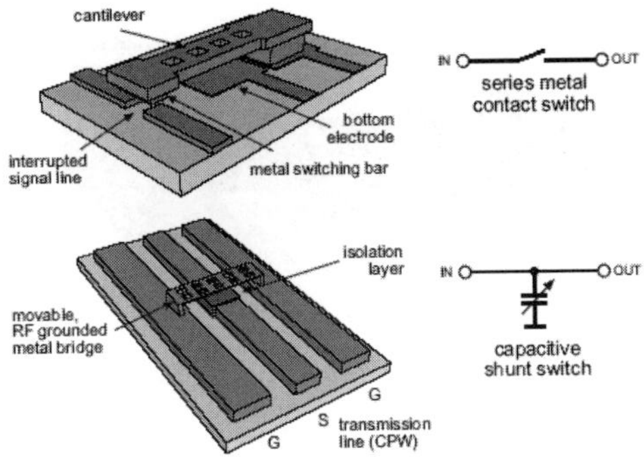

Figure 1. The two most common electrostatically actuated MEMS switch types.

2. Objectives

Recent studies performed at CNR-IMM of Rome lead to encouraging results by using photolithographic processes and SU-8 thick negative photo-resist on low resistivity silicon [1]. CPW lines elevated with respect to the substrate have been obtained, with advantages in the signal transmission due to the propagation almost on-the-air. The above technology has been extended to the realization of RF MEMS Switches in Shunt configuration by means of surface micromachining. The advantage of a structure build in this way (Figure 2) is mechanical rigidity greater than the normal MEMS with lateral supports galvanically grown, and the low cost for the realization. The SU-8 material is a negative, epoxy-type, near-UV photo-resist based on EPON SU-8 epoxy resin. It is a multi-use kind of polymer, suitable of applications in electronics, coating as well as in integrated optics. The idea of a dielectric or polymeric layer deposited on the top of a semiconductor in order to reduce the losses has been initially stated in several works. In this paper the SU-8 material was chosen for its promising characteristics in micro-electromechanical systems (MEMS). Actually from an optical standpoint SU-8 is also characterized by a very low absorption coefficient.

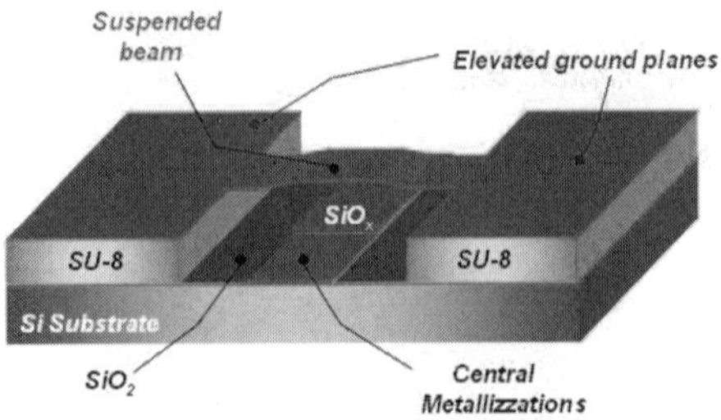

Figure 2. RF-MEMS coplanar shunt switch based on SU-8 technology.

3. Design and simulation of MEMS switch with SU-8

3.1. *Simulation*

The project foresee the realization of:

1. 9 capacitive shunt switches technologically actuated to analyze the switch behaviour in the contact metal/dielectric/metal
2. 9 capacitive shunt switches

All the structures have the same geometry (Figure3). Only the lateral dimensions of the bridge are changed, the area of the oxide, and the number of the teeth (from 2 to 6). The teeth should prevent problems of adhesion in the moment of the release of the structure. On the beams there are holes necessary for the removal of the sacrificial layer.

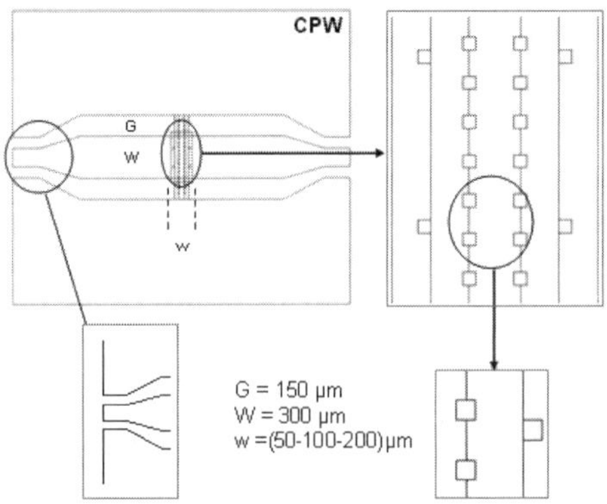

Figure 3. Devices Design

3.2. *Simulation*

Ansoft HFSS 3D was used to predict the electrical performances of the devices and in the following Figure 4 is shown the comparison between the isoltation of the MEMS switches technologically actuated by using the bridge width "w" as a parameter: 50 μm, 100μm, and 200 μm. As it is possible to see, the resonating response of the RLC equivalent circuit is drifted downwards in frequency when the width of the bridge is wider, because of the decreased value of the capacitance.

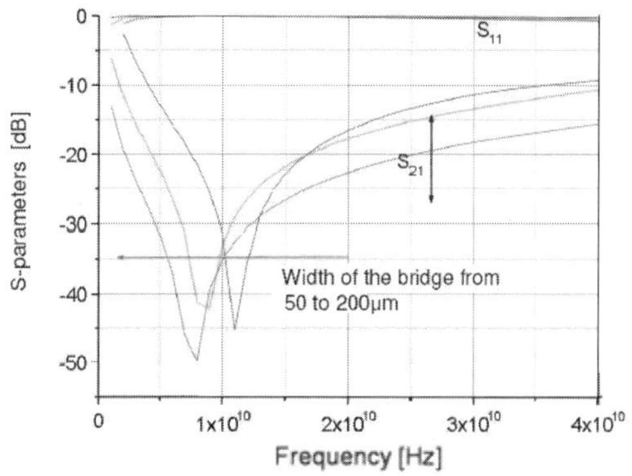

Figure 4. EM simulation for the reflection parameter (S_{11}) and for the transmission parameter (S_{21}) of the RF MEMS switch when actuated.

4. Fabrication process

The realization of the devices has been performed by using high-resistivity Silicon wafer having: a diameter of 4 inches and a thickness of 400μm. For the realization of the device, 5 masks have been necessary. The flow chart of the fabrication process is divided in five steps. In the first step (Figure 5a) the realization of the central conductor of the CPW is considered:

1. Wafer-cleaning in de-ionized water + Thermal Oxidation of the wafer
2. Cr/Au deposition by thermal evaporation
3. Spinning resist + Exposure to UV-ray (1st Mask) + Development
4. Wet etching of the Cr/Au

The second step (Figure 5b) foresees the realization of the S_iO_x necessary in the capacitive configuration to get a high ratio in the ON/OFF states:

1. Spinning of Resist AZ5214 + Exposure to UV-ray (2nd Mask)
2. Development + Deposition of Cr and S_iO_x by thermal evap. + Lift off

The third step (Figure 5c) foresees the creation of the lateral supports in SU-8. The ground planes of the CPW line are used to support the bridge for the switch:

1. SU-8 spin-coating + soft bake
2. Exposure to UV-ray (3rd Mask) + post bake
3. Development of the resist + hard bake

In the fourth step the realization of the switches technologically actuated (Figure 5d) is performed:

1. Spinning of Resist AZ5214 + Exposure to UV-ray (4th Mask)
2. Development + Deposition of Cr and Au +Lift off

The fifth step comprises the realization of the switches(Figure 5e and 5f):

1. SU-8 spin-coating + soft bake + Exposure + post bake
2. Deposition Cr/Au + Spinning resist + Exposure to UV-ray (5th Mask)
3. Development + Etching Cr/Au + Release of the structure (Plasma etching)

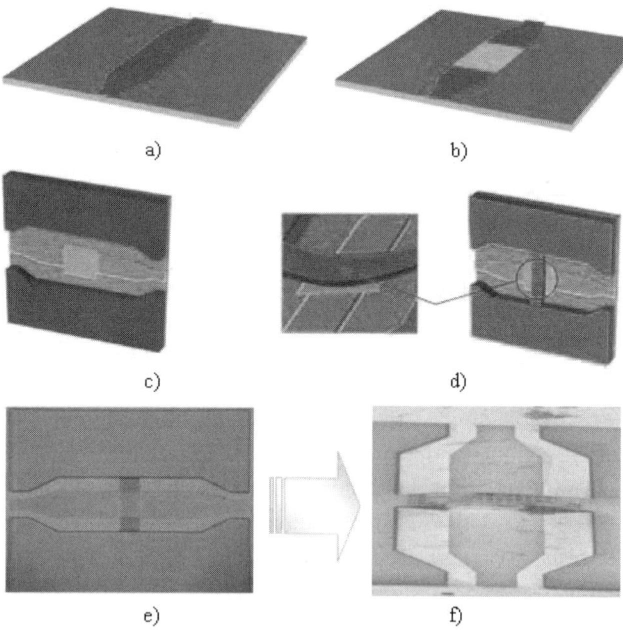

Figure 5. a) Central conductor of the CPW, b) S_iO_x area for the capacitive contact, c) lateral supports in SU-8, d) switch technologically actuated, e) coplanar shunt switch based on SU-8 technology, f) SEM photos of the device.

5. Experimental measurements and comparison with simulations

The trend of the scattering parameters S_{11} (reflections) and S_{21} (isolation) obtained with the simulator, and the measured data obtained on the technologically actuated bridge are in good agreement between them (Figure 6). The amplitude of S_{21} allows an isolation of 30 dB or better as expected from simulations and the Reflection turns out to be very high, i.e. the signal is properly grounded.

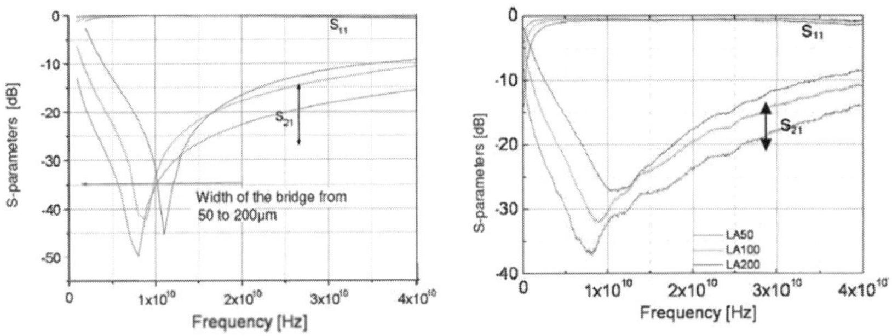

Figure 6. Comparison between Experimental measured (6b) and simulations of the fully actuated switch (6a).

6. Conclusion

The technology for the realization of RF MEMS switches performed at CNR-IMM Roma has been described. In the first run, technologically actuated switches as a reference for the expected electrical performances have been obtained. Few prototypes of suspended structures have been also realized by plasma etching.

It exists a good agreement between simulated and experimental data, and it has been demonstrated that the SU-8 improves the electrical and mechanical characteristics of the switch. Wet and Dry etching will be both considered in the second run, to get results on a cost-effective technological solution with the maximum yield.

Acknowledgments

We kindly acknowledge for their technical assistance: M. Maiani, and C. Risi for thin film depositions, C. Biagiolini, F. Fontana for mask design and realization, from the Rome Section of IMM. We are also grateful to R. Buttiglione from SELEX S.I for her helpful suggestions about SU-8 technological processes and all SELEX S.I.

References

1. Romolo Marcelli, Simone Catoni, Luciano Fringuelli "Low loss coplanar lines on low resistivity silicon by SU-8 thick negative photoresist", CAS 2005, *Proceedings of the INTERNATIONAL SEMICONDUCTOR CONFERENCE, 28th Edition*, October 3-5, 2005, Sinaia, Romania, pp.107-110 (2005).
2. Dimitrios Peroulis, Sergio P. Pacheco, and Linda P. B. Katehi, "RF MEMS Switches With Enhanced Power-Handling Capabilities", *IEEE Trans. Microwave Theory Tech.,* Vol. 52, No. 1, January 2004, pp.50-68
3. R.Marcelli, G. Bartolucci, G. Minucci, B. Margesin, F. Giacomozzi, F. Vitulli: "Lumped element modelling of coplanar series RF MEMS switches", *Electronics Letters*, Vol. 40, No. 20, September 2004, pp. 1272-1273.
4. E. Pettenpaul, H. Kapusta, A. Weisgerber, H. Mampe, J. Luginsland, and I. Wolff: "CAD Models of Lumped Elements on GaAs up to 18 GHz", *IEEE Trans. On MTT*, Vol. 36, No. 2, pp.294-304 (1988).
5. G. M. Rebeiz, *RF MEMS Theory, Design, and Technology*, Hoboken, John Wiley and Sons, 2003
6. I. Wolff and N. Knoppik: "Rectangular and Circular Microstrip Disk Capacitors and Resonators", *IEEE Trans. Microwave Theory Tech.,* Vol. MTT-22, No. 10, pp.857-864, October 1974.
7. T. Edwards, *Foundations for Microstrip Circuit Design*, Second Edition, John Wiley and Sons, Chichester, 1992.
8. Rottenberg, R P Mertens, B Nauwelaers, W De Raedt and H A C Tilmans, "A distributed RF-MEMS capacitive series switch" *J. Micromech. Microeng.*, Vol. **15** S97-S102 (2005).

PHASE SHIFTERS BASED ON RF-MEMS COPLANAR SHUNT SWITCHES

DANIELE POCHESCI, SIMONE CATONI AND ROMOLO MARCELLI
CNR-IMM, Via del Fosso del Cavaliere 100, 00133 Roma, Italy

GIANCARLO BARTOLUCCI
Università degli Studi di Roma "Tor Vergata", Dip. Ing. Elettronica, Via della Ricerca Scientifica, 00133 Roma, Italy

FLAVIO GIACOMOZZI AND BENNO MARGESIN
ITC-irst, Via Sommarive 18, 38050 Povo (TN), Italy

Abstract - This paper presents the experimental results obtained on a digital phase shifter based on RF MEMS coplanar shunt switches for radar, beam forming applications. A new design approach is proposed for the design of digital distributed MEMS phase shifters, and the image parameter representation of two port networks is used to develop an analytic model for this component. Vector Network Analyzer measurements have been performed by recording the scattering (S) parameters of the reflected and transmitted signals, and they have been elaborated to get the signal phase shift around the frequency F = 13.7 GHz used for the design. Actually, good performances of the phase shifter have been obtained with respect to the expected ones.

1. Introduction

Among many applications the MEMS switches are widely used for the realization of distributed digital microwave phase shifters. In particular, binary devices with a phase shift of 180 degrees have been considered in literature, and the lowest number of switches utilized in these components is typically 10 [1]. The aim of this paper is to present the realization and the experimental characterization of a 180 degrees phase shifter composed by only 6 MEMS switches, a very low number of MEMS devices in comparison with that one employed for phase shifters synthesized by means of the conventional procedure. Actually, a phase shifter is a device critical from the design point of view, because of the necessary precision in defining its performances, for the frequency at which the shift has to be imposed as well as for the value of the phase shift. The design of this structure has been performed by means of an approach previously proposed [2], and based on the image parameter representation of two port networks. This method is here improved, obtaining a numerical tool for the synthesis of MEMS distributed digital phase shifters. Experimental results for a component with a differential phase shift of 180 degrees are reported, and confirm the validity of the proposed approach. The devices have been fabricated in coplanar technology, which is currently considered the future trend in high frequency systems, due to the improved EM boundary conditions for the electromagnetic field confinement and for packaging purposes. Measured results for this phase shifter are presented, which confirm the validity of the proposed approach.

2. Modeling

The distributed phase shifter is obtained by the cascade connection of identical elementary cells. Each cell (figure 1) is composed by two transmission line sections separated by a shunt connected MEMS switch, represented in figure with a capacitor. For such a structure the image parameters are the natural approach for the development of a simple and rigorous model. A design method, based on this kind of modelling, has been presented in [2] for distributed phase shifters using semiconductor devices as switch elements. In this paper this method is modified in order to satisfy the matching condition in both the states of the binary phase shifter. To this end, because of the particular circuital topology of this structure, the following two conditions must be imposed on the image impedance Z_{ic} and on the image phase φ_{ic} of the basic cell:

$$Z_{ic} = Z_0 \quad , \quad N\varphi_{ic} = m\pi \tag{1}$$

According to [2] for Z_{ic} and φ_{ic} we have:

$$Z_{ic} = Z_c \sqrt{\frac{2 - Z_c \omega_0 C \tan\theta}{2 + Z_c \omega_0 C / \tan\theta}} \tag{2}$$

$$\varphi_{ic} = 2 \arctan\left(\tan\theta \sqrt{\frac{2 + Z_c \omega_0 C / \tan\theta}{2 - Z_c \omega_0 C \tan\theta}} \right) \tag{3}$$

being ω_0 the operating frequency and C the value of capacitance.

Equations (1) (2) and (3) have been implemented in the MATLAB commercial software package to have a numerical tool for the design of the structure. For fixed N, m, Z_c and θ, two values C_{UP} (UP state of the bridge) and C_{DOWN} (DOWN state) for the unknown C must be found in such a way that the difference of the output signals in the two states is the required differential phase shift.

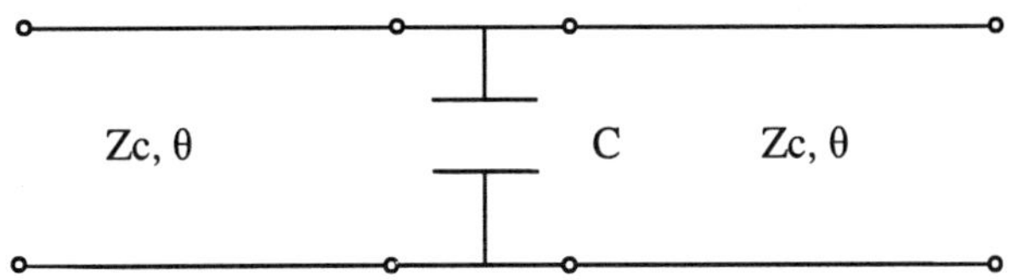

Figure 1. Elementary cell

3. Simulations

The simulation has been carried out with Agilent ADS circuital CAD supported by a 3-Dimensional CAD: Ansoft HFSS. First of all it has been taken into account the single switch (MEMS): By starting from measurements, in order to obtain the equivalent circuit between measured and simulated data, a fitting process on the recorded S-Parameters has been performed at a selected frequency.

This equivalent circuit is composed by a two transmission line section and a resonant LC circuit, at a shunt position with respect to the lines, in such a way that a T network is formed. The losses are taken into account in the values of the inductance to get the real and the imaginary part of an impedance.

The equivalent circuits of the UP and DOWN states are different between them mainly for the value of the capacitance, that is greater in the DOWN state than in UP one, taking into account the operating frequency f_0

By analyzing the equivalent circuit, it is possible to obtain the values of the susceptance in the UP and DOWN states used to design the single cell of the phase shifter related to values of the inductance and of the capacitance. The values of the susceptance are obtained considering the admittance at the input of the 2-port network and closing the output port with its image impedance.

Such switch is formed by a micro-mechanical system and I/O lines with a characteristic impedance Zcs.

Once the values of the susceptances are known and the operation frequency is imposed, it is possible to create the elementary cell placing at both sides of the switch two transmission line sections, in coplanar configuration, with electrical length θ and characteristic impedance Zc. The difference between Zcs e Zc causes an abrupt step of impedance which involves electromagnetic effects. For the above reason, two black boxes including the electromagnetic effects of the steps between the switch and the transmission lines have been used. Then, it is possible to obtain the values of the elements in such a way that the structure is considered a cascade of N identical elementary cells, where N is chosen to verify the values of the required shifting.

4. Fabrication

In order to fabricate micromechanical switches together with integrated resistors and DC blocking capacitors an eight mask process has been developed. Two electroplated gold layer of different thickness are provided for the realization of highly complex movable bridges and the co-planar waveguides. The process sequence is summarised in figure 2.

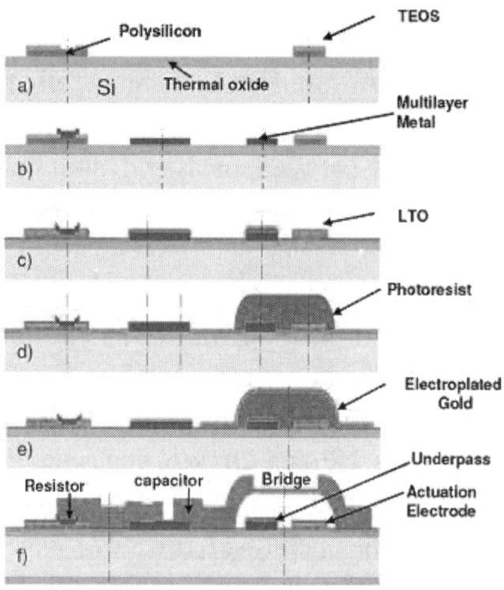

Figure 2. Process sequence of the MEMS fabrication.

The substrates are p-type, <100>, 525μm thick, 5 kΩ·cm high resistivity silicon wafers.

(a) As a first step 1000 nm thick thermal oxide is grown as an isolation layer. In order to reduce the fixed oxide charge the oxide is then annealed at 975°C in nitrogen for one hour. Next a 630-nm thick un-doped poly-silicon layer is deposited by LPCVD at 620°C and slightly doped with Boron by ion implantation. At this point the first photolithography is performed in order to define the resistors and actuation electrodes and the poly-silicon layer is selectively dry etched. The previously implanted Boron is diffused at 925°C for 1 hour in nitrogen and a 300 nm thick silicon oxide layer is deposited from TEOS at 718°C by a LPCVD process. This oxide provides the high isolation needed for the actuation electrodes. Contact holes are then defined and etched by a plasma process.

(b) After ashing the photoresist mask, a multilayer underpass metal is deposited by sputtering: 30nm of Ti and 50 nm of TiN (reactively sputtered), both deposited at 400°C, act as a diffusion barrier. Next 450 nm thick Al 1%Si and 30 nm thick Ti are sequentially deposited at room temperature. Finally an 80 nm capping layer of TiN is deposited at 300°C. The total thickness of the multilayer has to be the same of the polysilicon, in order that metal underpass and actuation electrodes are at the same level. Underpass lines and diffusion barriers on the poly-silicon contacts are defined and etched.

(c) The wafer front side is then covered with 100 nm of low temperature oxide (LTO) to provide an insulating layer for capacitive shunt switches. The vias in the LTO are defined by masking and dry etching. The over-etch time of the dry etching is determined in the order to allow for the removal of the TiN capping of the multilayer metal and to expose the underlying aluminium layer. Next 10 nm thick Chromium adhesion layer followed by 150 nm Gold layer are deposited by PVD. This Chrome/Gold layer is defined by lithography and etched wet. Main purpose of this layer is to cover with a noble metal the exposed electrical contacts of the series ohmic switches in order to provide low resistive electrical contacts. This layer can also be used to build a floating electrode for the shunt switches.

(d) The sacrificial layer required for the construction of the air gap is formed by 3 μm thick photoresist, hard baked at 200°C for 30 minutes to obtain well-rounded edges. As a seed-layer for electrochemical Au deposition a 10/150nm thick Cr/Au layer is deposited by PVD.

(e) The movable air bridges are defined using a 4 μm thick positive resist. After an exposure to oxygen plasma at 80 °C a 1.8 μm thick gold layer is selectively grown in a gold sulphite bath.

(f) The first plating mask is removed with an appropriate solvent and the CPW lines and anchor posts for the movable air bridges are defined with 5 μm thick positive resist and then a 4 μm thick gold layer is selectively grown. The last plating mask and the seed layer are then removed wet. The seed layer removal is completed by a dip in diluted (1:3) *acqua regia* in order to remove Cr/Au whiskers. At this point a sintering in nitrogen at 190 °C for 30 minutes is performed in order to provide the gold layers with the appropriate tensile stress. If required the wafers are coated with a 2.5 μm photoresist layer and pre-diced, i.e. with the dicing saw the wafers are diced about half through. This allows handling the wafer still as a whole during the next step while the dies can be easily separated by breaking along the pre-diced scribelines. Finally the air bridges are released with a modified plasma ashing process (20 minutes oxygen plasma at 200°C) in order to avoid sticking problems.

5. Measurements

The measurements on a single switch have been performed to obtain the equivalent circuit (see the chapter of the simulations). The actuation voltage of the switch is about 74 V changing the state from UP to DOWN. The inverse process (de-actuation from DOWN to UP) is obtained around 60 V.
Then the whole structure (Phase shifter formed by 6 elementary cells) shown in Figure 3 has been measured getting the S-Parameters. The actuation voltage is the same for all switches composing the structure. This structure changes the state (from UP to DOWN) at 50 V and its de-actuation voltage is around 39 V. The measurements of both states UP and DOWN are presented in Figure 4 and in Figure 5.
By analyzing the above figures, the reflection S-Parameter is about -25 dB and the losses are 1.8 dB, thus demonstrating the good quality of the lines.
In the DOWN state the values of the S_{11} is around -19.4 dB and losses 2.5 dB, which fulfil the good design for the required operating frequency as well as for an accurate shifting.
In the last Figure 6 is shown the most important data: the phase difference from DOWN state to UP state is exactly 180° at the desired operating frequency.

6. Conclusions

The design and test of the phase shifters for microwave signal processing based on RF-MEMS switches has been presented. The analysis of the device has been performed by means of the Image Parameter Method to get the best electrical matching for the exploited structure. A Reverse Engineering Approach has been used for modelling the Phase Shifter by means of the equivalent circuit of the actual switch.

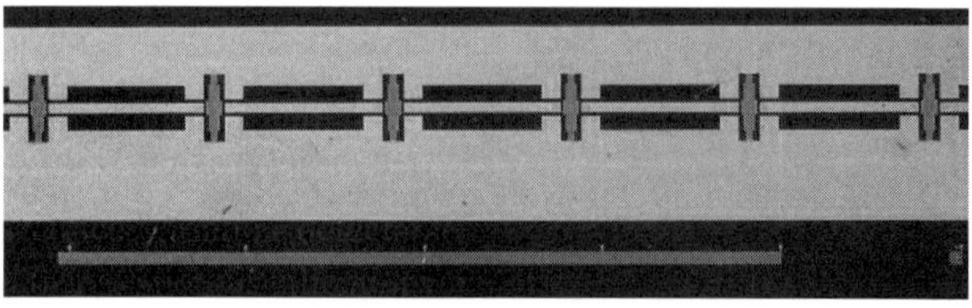

Figure 3. Whole structure of the phase shifter

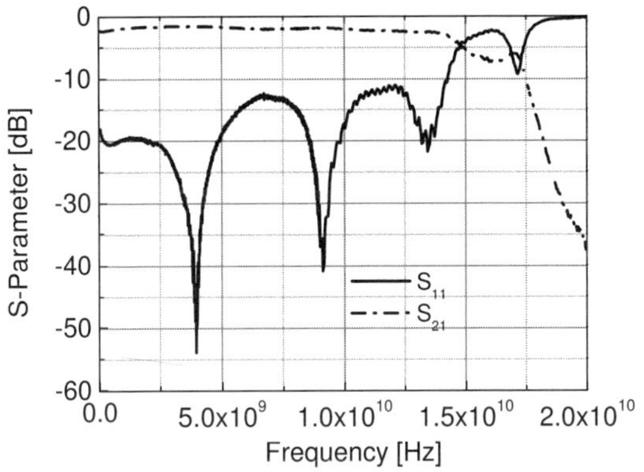

Figure 4. S-Parameters of the DOWN state

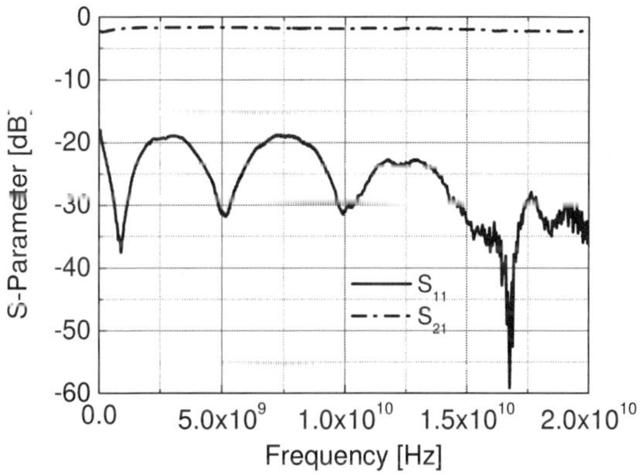

Figure 5. S-Parameters of the UP state

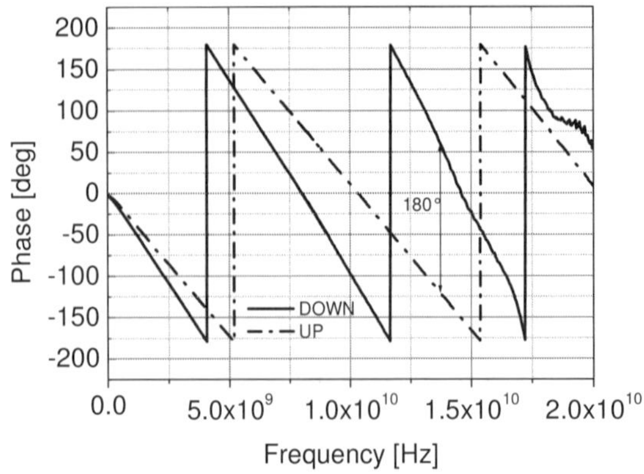

Figure 6. Phase difference at 180°

References

1. Joseph S. Hayden, Gabriel M. Rebeiz,: Very low-loss distributed X-band and Ka-band MEMS phase shifter using metal-air-metal capacitors", IEEE Transactions on. Microwave Theory and Techniques, 2003, 51, (1), pp. 309-314.
2. G. Bartolucci,: "Image parameter modelling of analog travelling-wavc phase shifters", IEEE Transactions on Circuits and Systems, -I: Fundamental Theory and Applications, 2002, 49, (10), pp. 1505-1509, October 2002.
3. Joseph S. Hayden, Gabriel M. Rebeiz, "2-bit MEMS distributed X-band phase shifters" IEEE Microwave Guided Wave Lett., December 2000, Vol. 10, pp. 540-542.
4. A. Ocera, R. Sorrentino, P. Mezzanotte, "Design of tunable phase shifters by the image-parameters method", IEEE Transactions on Microwave Theory and Techniques, vol. 54, no. 6, pp. 2383-2390, June 2006.

MEMS ACCELEROMETER CALIBRATION AT LOW FREQUENCIES

F. LO CASTRO [1,*], G. BRAMBILLA [1], P. VERARDI [1], A. D'AMICO [1,2]

[1] *CNR – Institute of Acoustics, Via del Fosso del Cavaliere 100, 00133 Rome, Italy.*
[2] *University of "Tor Vergata" - Dep. of Electronic Engineering, Via del Politecnico 1, 00133 Rome, Italy.*
* *fabio.locastro@idac.rm.cnr.it*

The accelerometer calibration at low frequencies and low acceleration values requires a specific instrument not easy to find on the market.
To solve this problem a system based on a pendulum has been realized in order to reach frequencies less than 1 Hz and low acceleration values down to $0.01 ms^{-2}$.
The system does not require a reference accelerometer, as traditional calibration systems need. The fundamental equation of a physical pendulum is used to obtain the reference acceleration values, which are function only of the angle ϑ (between the gravity direction and the rod).

1. Introduction

Today MEMS accelerometers are more and more used in automotive and in electric household appliances. The precision required depends on the applications, for example in vibration measurements the accelerometer must be calibrated. Usually two ways are used to calibrate the accelerometer: by an interferometer (absolute method) or by comparison with a calibrated accelerometer.
Performing an absolute calibration by interferometer can be expensive and a lengthy process. This is mainly due to the fact that an interferometer measures the acceleration (after integration from displacement) at the precise target on the transducer's surface where the laser beam impinges.
To save time and cost, a reference accelerometer, periodically calibrated by absolute techniques, is usually used to calibrate any accelerometers.
During the calibration process shakers are used, but those available on the market show problems to reach low frequency (less than 5 Hz) and small acceleration (less than 1 ms^{-2}); so it is necessary to use a special designed shaker to obtain better performance.
In order to perform calibration at low frequency and with small acceleration values a physical pendulum have been designed and realized. The advantage of this system is that a reference accelerometer on board is not necessary, being the motion of the pendulum based on the Earth's gravity.

2. Experimental

The calibration system developed and realized consists of a physical pendulum (0.58 m length), an A/D acquisition board (12 bit), a software to analyze the acquired data and an STM LIS3L02AS4A MEMS capacitive accelerometer used as test devices to be calibrated.
The block diagram of the whole system is shown in figure 1.
The output voltages of the accelerometer $V(a_n)$ and $V(a_\tau)$, corresponding to the centrifugal and tangential acceleration respectively, and the output $V(\vartheta)$ of the potentiometer, corresponding to the angular position of the pendulum, are acquired by the proper developed acquisition board.

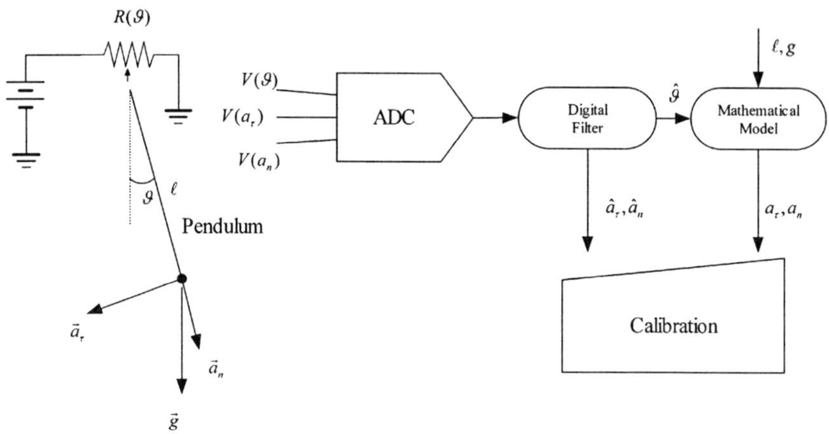

Figure 1 – The pendulum and the system developed.

The digital data, before mathematically processed [1], need to be filtered, particularly the ϑ values because they are differentiated one or two times.
The filter [2,3] acts on the FFT [4,5,6] spectrum smoothing the high frequency (figure 2 and figure 3). The start frequency is selectable according to the period of the pendulum and the level of the noise.
The mathematical model block evaluates the theoretical acceleration values of a_n and a_τ[7], given by

$$a_n = l\dot{\hat{\vartheta}}^2 \qquad 1)$$

$$a_\tau = l\ddot{\hat{\vartheta}} \qquad 2)$$

using the filtered angular position $\hat{\vartheta}$ and the distance l of the accelerometer from the pivot.

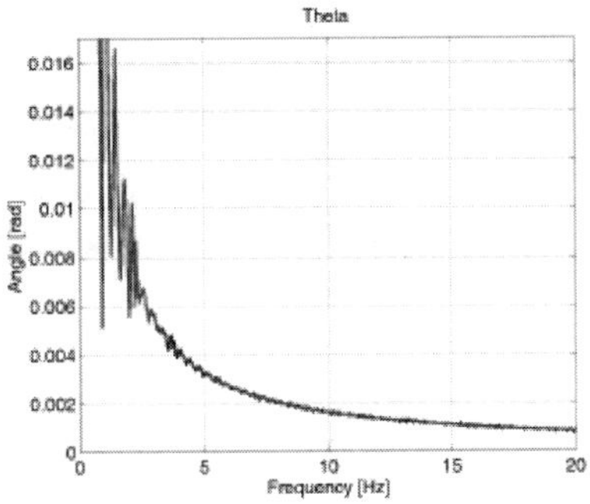

Figure 2 – The black line corresponds to the original angular signal spectrum while the red line is the filtered one.

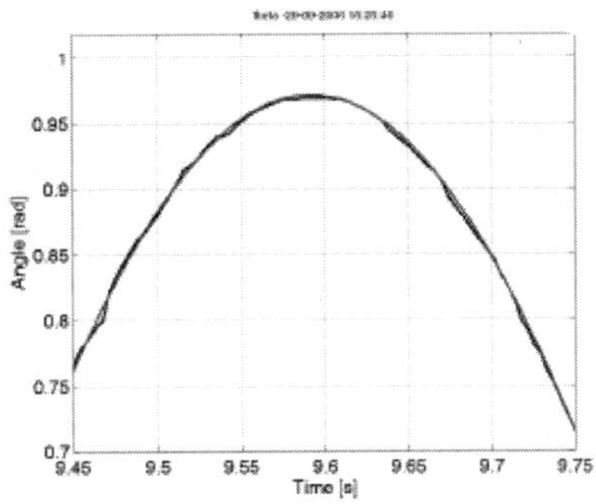

Figure 3 – The black line corresponds to the original angular signal while the red line is the filtered one.

The output calibration block compares the theoretical acceleration given by the eq. 1 and eq. 2 with the sampled values (denoted by caret symbol) of the acceleration (equation 3 and 4).

$$\hat{a}_n = l\hat{\dot{\vartheta}}^2 + g\cos\hat{\vartheta} \qquad 3)$$

$$\hat{a}_\tau = l\hat{\ddot{\vartheta}} + g\sin\hat{\vartheta} \qquad 4)$$

Equations 3 and 4 also consider the sensitivity to the gravity of the used MEMS accelerometer.

3. Results and discussion

The realized calibration system (figure 4) allows to achieve very low frequencies and small acceleration values according to the physical structure of the pendulum.

In the experiment the built pendulum oscillates between 0.5 Hz and 1 Hz changing its centre of mass and reaches small acceleration values of about 0.01 ms^{-2} before the end of its oscillation.

The measurement of the angle sweep is useful to eliminate all the parameters related to the pendulum, as dimension, friction, etc.

Figure 4 – Realized pendulum.

As the sampled values of the acceleration and the angular position are affected by noise it is necessary to filter it, before applying any further processing.
The setup needs to calibrate the angle ϑ at its minimum and maximum aperture ($\Delta\vartheta$ = 180°) and, if the accelerometer is sensible to the DC acceleration, it is possible to measure the acceleration at ϑ=+90° and ϑ=-90° (corresponding to test sensor at +9.81 ms^{-2} and -9.81 ms^{-2}) in order to calibrate the offset and the sensitivity of the sensor.
For acceleration different from 9.81 ms^{-2} the centrifugal, a_n, and tangential acceleration, a_τ, at the distance l from the pivot of the pendulum can be measured.
In the figures 5 and 6 it is shown respectively the sampled acceleration a_n and a_τ in red and the theoretically acceleration in blue. At the beginning and at the end of the curve it is visible the offset error of the accelerometer, while during the motion can be observed the difference between the calibrated curve (blue) and the sampled one (figure 7).

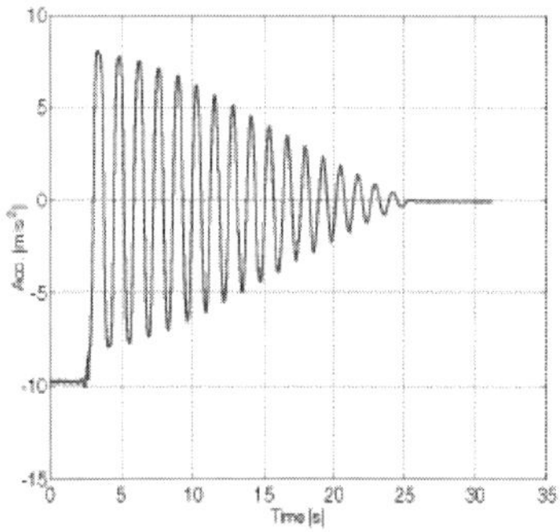

Figure 5 – Tangential acceleration; theoretical values are plotted in blue, while the sampled values in red.

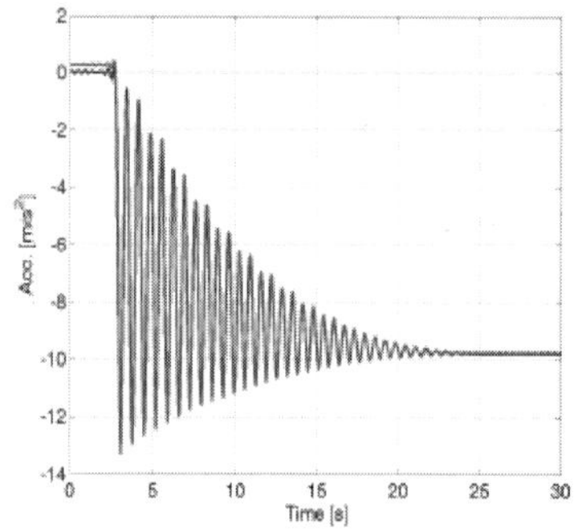

Figure 6 – Centrifugal acceleration; theoretical values are plotted in blue, while the sampled values in red.

The comparison between the calibrated curve and the sampled one shows that the resulting accuracy for the tangential acceleration values is less than 3% full range (19.62 ms^{-2}), as shown in figure 8, and less than 5% for the normal acceleration values (figure 9).

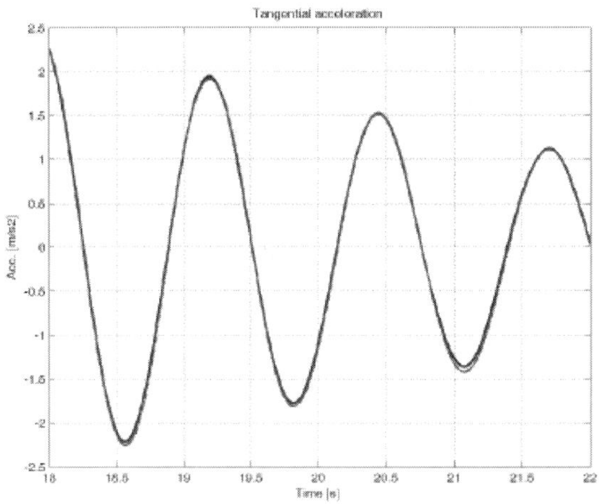

Figure 7 – Particular of the calibrated (solid blue) and the sampled (dotted red) curve.

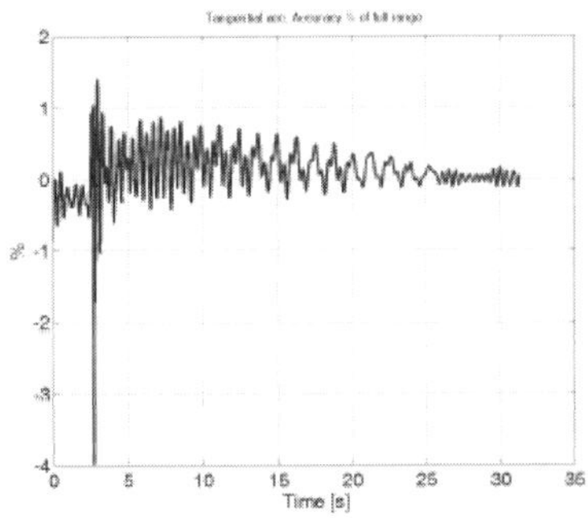

Figure 8 – Tangential acceleration accuracy.

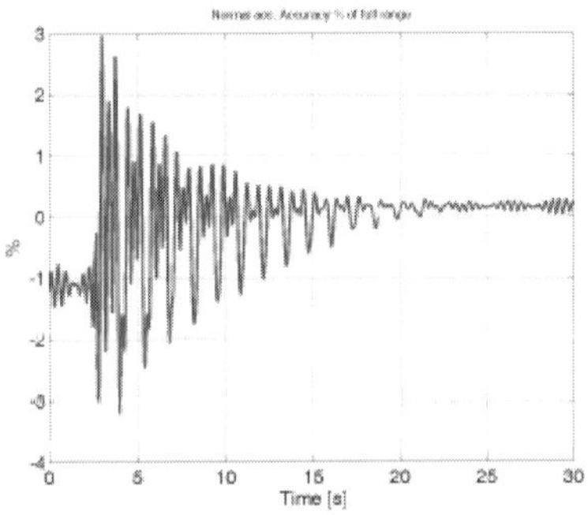

Figure 9 – Centrifugal acceleration accuracy.

4. Conclusions

This work describes a method to calibrate accelerometers alternative to those using a shaker. It is based on a physical pendulum having the advantage to reach low frequencies (less than 1 Hz), low acceleration values (down to 0.01 ms^{-2}) and does not need a reference accelerometer. The full range accuracy is less than 3% for the tangential acceleration and less than 5% for the normal acceleration.
Different settings of the pendulum (the accelerometer position along the rod and the length of the rod) allow to calibrate on a wide range of frequencies and accelerations.

References

1. William T. Vetterling , “ Numerical Recipes” - William M. Press 1986.
2. Hewlett-Packard, Application Note 243 - 1985, “The Fundamentals of Signal Analysis”.
3. Michael Cerna, Audrey F. Harvey, “The Fundamentals of FFT-Based Signal Analysis and Measurement” – National Instruments AN041.
4. Harris, Fredric J., Proceedings of the IEEE Vol. 66, No. 1, January 1978.
5. Bruël and Kjær, Technical Review No. 3, 1987 - “Windows to FFT Analysis part I”.
6. Bruël and Kjær, Technical Review No. 4, 1987 “Windows to FFT Analysis part II”.
7. D. Sette, “Lezioni di Fisica Vol. 1” – Editoriale Veschi Milano 1988.

POROUS SILICON MEMBRANES FOR DRUG DELIVERY

ILARIA DE SANTO, FILIPPO CAUSA, PAOLO NETTI
Department of Materials and Production Engineering
University Federico II of Naples,
Piazzale Tecchio 80, 80125, Naples, Italy

VERA LA FERRARA, IVANA NASTI; GIROLAMO DI FRANCIA
Centro Ricerche ENEA,
Via Vecchio Macello,
Portici, NA, 80055, Italy

Nano porous silicon (PS) membranes, obtained by electrochemical etch in fluoridric acid, were used to study the release characteristics of biomolecules through nanochannels. The release of dextrans biomolecules from the nanoporous membranes was deeply affected by material structure parameters, as well as molecular chemistry and concentration.

1. Introduction

Drug delivery systems have already had an enormous impact on medical technology, improving the performance of many existing drugs and enabling the use of new therapics. Efforts to miniaturize drug delivery devices from the macroscale (>1mm) to the micro-nanoscale (100μm – 1nm) ultimately promise integrated systems that combine device technology with therapeutic molecules (small molecules, nuclei acids, peptides, proteins) to allow the creation of implantable devices that can monitor health status and provide therapeutic treatment *in situ*. In this frame, nanostructured porous silicon is becoming an interesting biocompatible drug vector because of large surface area (>500 m^2/cm^3) controllable pore sizes (2 to 2000 nm) and suitable surface chemistry which can be modified with a wide range of biological or organic molecules. Possible applications include micromachined silicon membranes to create implantable biocapsules for the immunoisolation of pancreatic islet cells – as a possible treatment for diabetes[1] as well as sustained release of injectable drugs needed over long time periods [2] or silicon based microparticles with nanoporosities for the delivery of cytotoxic substances [3].

The interest for manufacture micro and nanofabricated membranes are of scientific and technological importance because of the presence of voids of controllable dimensions at the atomic, molecular, and nanometer scales, enabling them to discriminate and interact with selected molecules allowing greater control of dose profile and, in the case of nanoporous membranes, permit the suppression aspects of the immune response. However little is know about the transport mechanism behind the release of therapeutic molecules and a compelling need is to optimize release conditions for several drug delivery applications from such devices.

2. Experimental

Porous silicon is obtained by an electrochemical etch in fluoridric acid based solution starting from silicon substrate. For manufacturing porous silicon membranes, initially an etching area has been delimited through a photolithographic process on silicon substrates, then 8 mm diameter porous silicon membranes have been obtained by electrochemical etch and lift off technique. Membranes have been morphologically characterized to control pore dimensions. SEM images were performed (see fig. 1). Nitrogen adsorption/ desorption curves were acquired by a surface area and pore analyser (Quantachrome Autosorb-1).

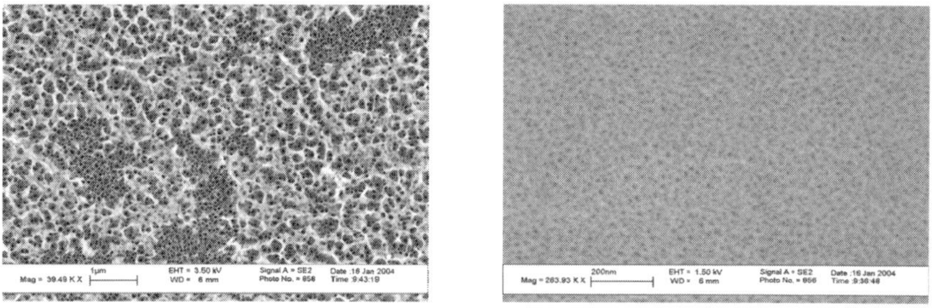

Figure 1. Left) SEM image of top side of a nanoporous silicon membrane obtained from electrochemical etch and lift off technique- average radii between 20-100nm. Right) SEM image of bottom side of a nanoporous silicon membrane obtained from electrochemical etch and lift off technique-average radii of 10nm .

Adsorption-desorption isotherms showed silicon membranes being formed by slit shaped mesopores. Using BET (Brunauer–Emmett–Teller) equation a surface area of 442 m^2/g was found. An average pore diameter was deduced from the desorption branches of isotherms applying BJH (Barrett–Joyner Halenda) model (see Figure 2).

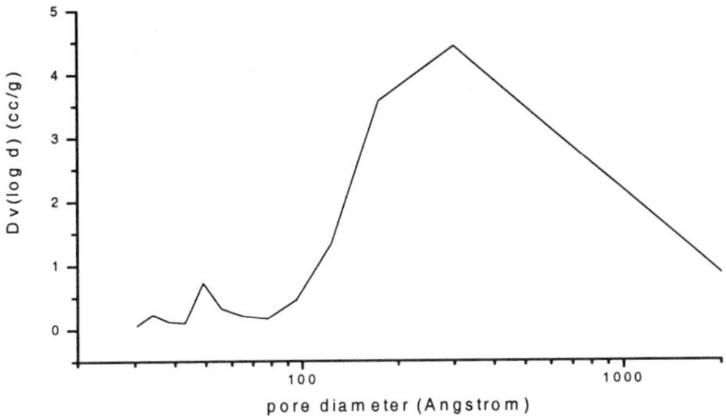

Figure 2. Pore diameters evaluated from the desorption branches of isotherms based on BJH (Barrett–Joyner Halenda) model. Maximum value in the range of 30-40 nm was found.

An average membrane thickness of 25 μm was measured by means of a Profilometer. In order to characterize biomolecules release and to study their diffusion, porous silicon membranes have been glued by a polymeric solution onto Transwell and between two reservoirs. The proposed experimental system envisages the uniaxial flow of biomolecules through the membrane nanochannels. The membrane separates an initially filled biomolecules-containing reservoir from a water solution environment where biomolecules are depleted. The volume above the membranes was filled with a fluorescent dextrans-FITC macromolecules solution, whereas the lower volume contained bi-distillated water. Dextrans concentration in the lower reservoir was measured by means of a Spectrofluorimeter (485-535 nm). Two different molecular weights have been used (10kDa and 500kDa) at two different concentrations (0.1 and 0.2 mg/ml for 10kDa MW and 0.4 and 0.8 mg/ml for 500kDa MW). Measurements were carried out every two days on average. The two MWs taken into account have been chosen as they both have hydrodynamic radius of about the same order of dimension as the membranes pores.

In order to overcome evaporation and prevent diffusional layer formation a diffusion cell was specifically designed to properly perform permeation measurements. In this scenario membranes were glued onto plastic support discs and placed in between two O-rings. The holding system was then mounted between two chambers. Each volume was stirred by means of a tangential flux. Experiments were carried out taking advantage of low concentration, 0.1mg/ml,

10kDa Dextrans-FITC. Permeated concentration was measured trough a Spectruofluorimeter.

3. Results

Release measurements showed that the 10kDa dextran at 0.1mg/ml concentration permeated the silicon membrane and that in 700 hours a value of C/C_{eq} up to 80% was reached (see Figure 3).

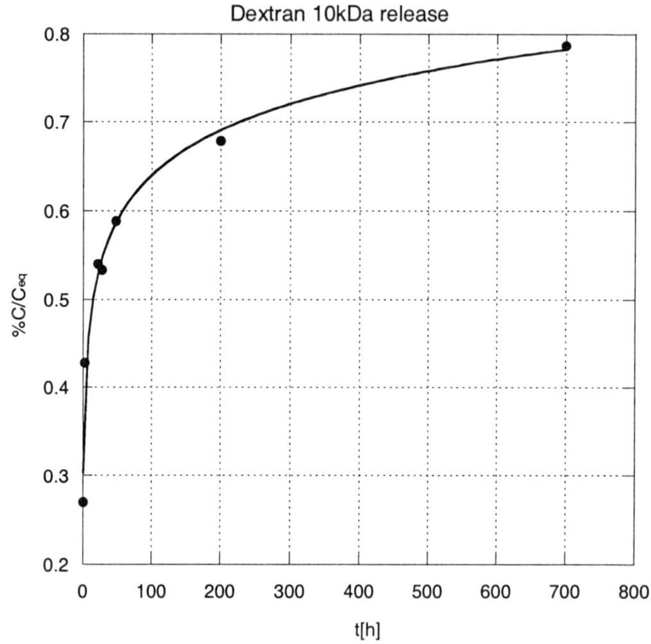

Figure 3. Dextran-FITC 10kDa release from Transwell over time. Dots are experimental values, line works as an eye-guide.

At 0.2 mg/ml the permeation through membranes was impaired. The results clearly showed that the diffusion mechanism deeply depend on concentration. The same results could be found for the 500kDa dextrans which permeated the silicon membrane at low concentration, whereas at higher concentrations did not (see Table 1).

Table 1. Summary of the release results of 10kDa and 500kDa dextrans from Transwell devices at two different initial concentrations.

Probe	Dextran 10kDa		Dextran 500kDa	
Concentration (mg/ml)	0.1	0.2	0.4	0.8
Released Amount (% C/Ceq)	0.6 (7 days)	0 1 month	0.2 (7 days)	0 (7 days)

A mutual diffusion coefficient can be derived from release data shown in Fig. 3. Under the hypothesis of constant concentration in the reservoir, moreover neglecting both evaporation effects and effects due to the presence of a diffusional boundary layer, a mass balance can be applied to the uniaxial system described above. Pore geometry was assumed straight and cylindrical, and the number of pores was derived from BET data. The mass balance which applies under the assumption of quasi-stationary process [4] is reported below (Eq. 1):

$$\ln[1 - C(t)/C(\infty)] = -(DnA_p/L) \times (1/V_1 + 1/V_2)t \qquad (1)$$

where C(t) is the dextran concentration in the lower reservoir at a time t, C(∞) is the equilibrium concentration of the system, D is the dextran diffusion coefficient in the pores, n the number of pores, A_p and L are the membrane's section and thickness respectively, V_1 and V_2 are the reservoir's volumes.
The diffusion coefficient derived is equal to 5E-12 cm^2/s.
In order to better describe the permeation process a diffusion cell equipment was designed. The results found for this experimental set-up are shown in Figure 4. Lag time of the permeation process was about 150 hours. The diffusion coefficient in an experiment of permeation can be derived from (Eq. 2):

$$t_D = \frac{L^2}{6D} \qquad (2)$$

where t_D represents the characteristics time lag of the permeation process. From Equation 2, a diffusion coefficient of about 2E-12 cm^2/s was derived.

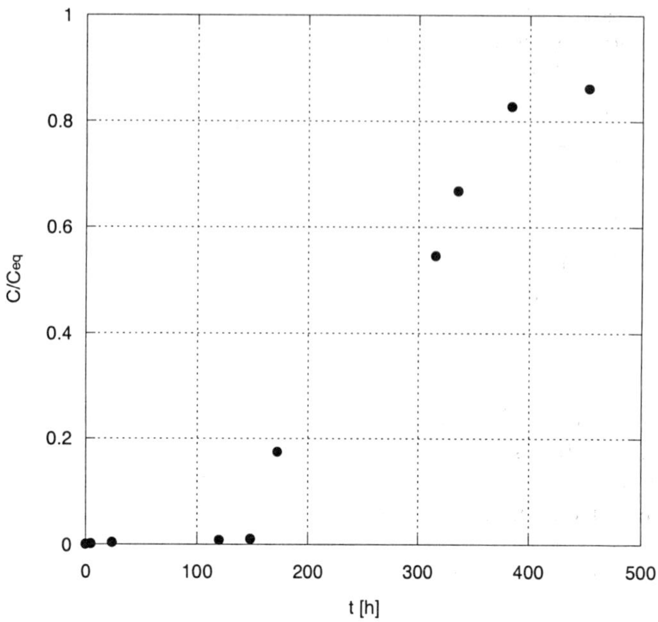

Figure 4. Dextran-FITC 10kDa permeation in a diffusion cell.

4. Conclusions

Nano porous silicon (PS) membranes have been used to study the release characteristics of biomolecules through nanochannels. The release of biomolecules from nanoporous membranes results to be deeply affected by material microstructure parameters, as well as molecular chemistry and concentration.

However, these findings claim a dipper insight into transport phenomena through nanochannels in order to better design a drug delivery systems.

References

1. L. Leoni and T.A. Desai, *Ad. Drug De.l Rev.* **C56**, 211 (2004)
2. F. J. Martin and C. Grove, *Biomed. Microdev.*, **3:2**, 97 (2001)
3. M.H. Cohen, K. Melnik, A. Boiarski, M. Ferrari, F.J. Martin, *Biomed. Microde.* **5:3,** 253 (2003)
4. D.S .Cannell and F. Rondelez,,*Macromolecules* **C 13**, 1599 (1980)

SILICON BASED TRANSDERMAL DRUG DELIVERY SYSTEM

IVANA NASTI*, VERA LA FERRARA, GABRIELLA RAMETTA AND GIROLAMO DI FRANCIA.

ENEA Research Center of Portici, 80055 Portici (NA), Italy

First results on fabrication of a Transdermal Drug Delivery device are presented. Device consists of two regions with different functions: one used for scrabbling the stratum corneum and injecting the drug and the other one used as drug reservoir. Device is manufactured starting from a silicon substrate where a porous silicon layer is obtained. Silicon substrate permits the integration of our device with a future electronic controller while biodegradable porous silicon is a good choice as drug reservoir. In order to test the device as drug delivery system, L-Ascorbic Acid impregnation has been evaluated by means of FTIR and BET analysis.

1. Introduction

Transdermal drug delivery (TDD) is an alternative and innovative, rapidly increasing, research area because it offers a lot of benefits in terms of drug-controlled release into the patient since it is user-friendly, convenient, painless and feasible for multi-day dosing. Moreover it is possible to avoid the problem of degradation of oral drugs by enzymes in the gastrointestinal tract.

The first problem in transdermal drug delivery is represented by the stratum corneum (SC) of the skin (a 50 μm lipophilic layer with bricks of corneocytes in a lipid rich matrix) that has the barrier function against foreign materials like chemicals and microbes and prevents water loss. This skin barrier limits the number of drugs that can be delivered by passive diffusion from an adhesive patch (only drugs with molecular weight <500Da and lipophilic molecules are in fact suitable).

* Corresponding author. Tel.:+390817723326; fax:+390817723344

E-mail address:nasti@portici.enea.it

Recently various passive and active methods to deliver drugs are under investigation. Passive methods include liposomes, other vesicles, supersaturated systems, eutectic systems, penetration enhancers and SC hydration that slightly increase the permeability of the skin and offer an improvement in dose control.
Electrical and mechanical methods (as for example iontophoresis, electroporation, abrasion, perforation) together with ultrasound, magnetophoresis, laser radiation, thermophoresis and needless injection are classified as active methods. They greatly improve delivery of molecules with high weight (> 600 Da like proteins and peptides) and also hydrophilic molecules. Some of the active TDD electro-systems are currently commercialized specially for monitoring purpose and delivery of local anesthetics.
The microneedles methods use microfabrication techniques, including semiconductor fabrication techniques and several works are reported on the topic and present them as a very promising technology. Yuzhakov et al. have for example realized an apparatus with two array of microneedles, one for transfer the drug from a reservoir into the skin and one for extract body fluids for analysis and monitoring drug using iontophoresis technique [1-4].
In this work we present a TDD device realized starting from a silicon substrate where, microneedles have been realized to scrabble the SC and a porous silicon (PS) layer is obtained for using as drug reservoir. L-Ascorbic Acid (LAA) impregnation has been investigated by means of FTIR analysis, showing that a better absorption is possible when PS is oxidized. BET analysis have showed a clear drug loading.

2. Experimental

Linear microneedles, for scrabbling the SC, have been realized by scribing a silicon substrate (p-type <100> oriented Si with 0,01 ohm*cm resistivity) with a Karl Suss scriber with a diamond tip. We make some batches separated by 50 millimeter, each group is constituted of 20 linear incisions with a separation distance of 50 micron. Then the substrate was subject to an isotropic etch with a HNA solution (5 mL HF 50%, 10 mL HNO3 70%, 16 mL CH3COOH) for few minutes, microneedles height is 5-10 μm. This step improves the shape and the depth of incisions.
To preserve these areas we cover them with a positive photoresist Shipley 1818.
Drug reservoir consists of a porous silicon structure realized between the batches. PS has been prepared by anodic etch in HF: ethyl alcohol (1:2 by volume) solution with a current density of 68 mA/cm^2 for 25 minutes (PS

thickness 60 μm). To realized a porous silicon layer on a lower height level as regards the level of micro-incisions, we make an anodic etch followed by an electropolishing step before realizing the porous silicon reservoir. The photoresist was after removed by acetone.
L-Ascorbic Acid reagent powder (Aldrich) was used to study drug loading into porous silicon substrate. LAA is hydrophilic and was dissolved in distilled water, the solution (0,6M, pH=2,2) was deposited on porous silicon substrate at room temperature for 90 min in dim light. The substrate was dried with nitrogen. A thermal treatment to improve hydrophilicity of porous silicon was performed at 300°C for 1 h in air [7].

3. Results and Discussion

3.1. *SEM analysis*

Characterization of device cross section has been acquired by SEM (model number LEO 1530), shown in fig.1a. Two different layers are evident: the upper layer made of incisions and the lower layer made of porous silicon. In fig. 1b and 1c particulars of incisions and PS are reported respectively.

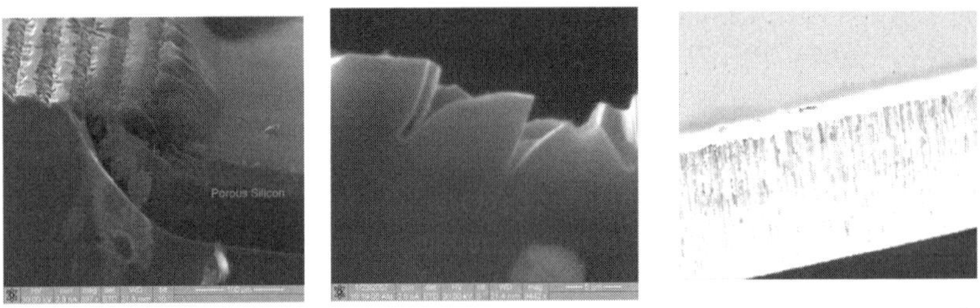

(a) (b) (c)

Fig. 1: (a) Cross section of our device with a porous silicon reservoir and with an upper layer with incisions; (b) Incisions; (c) Porous silicon.

3.2. *FTIR analysis*

The presence of LAA in porous silicon was investigated by FTIR analysis. When LAA is deposited onto as-prepared porous silicon, spectra present all of its characteristic peaks confirming the hypothesis that Ascorbic acid doesn't penetrate into porous silicon (Fig. 2a). In order to enhance LAA adsorption, PS was thermally treated. FTIR analysis on LAA loaded oxidized PS (LOPS) shows only two characteristic peaks of the C=O group (1756 cm^{-1}) and C=C group (1685 cm^{-1})(Fig.2b) demonstrating that LAA is included into pores. Fig.2b moreover shows that when LOPS is exposed to pure water, LAA can be replaced with water presenting a reversible adsorption. [5,6]

Two images recorded from a digital camera show as treatments change surface morphology. It has shown a crystallization on PS when LAA is deposited on as prepared PS (Fig.3a) while PS is lucid when LAA is deposited on oxidized porous silicon (OPS) (Fig.3b).

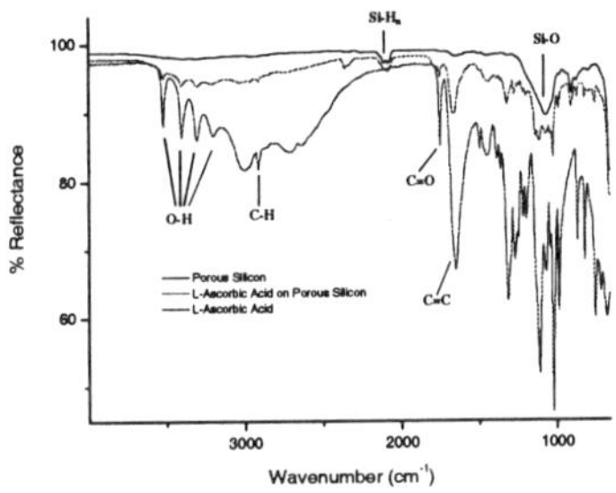

(a)

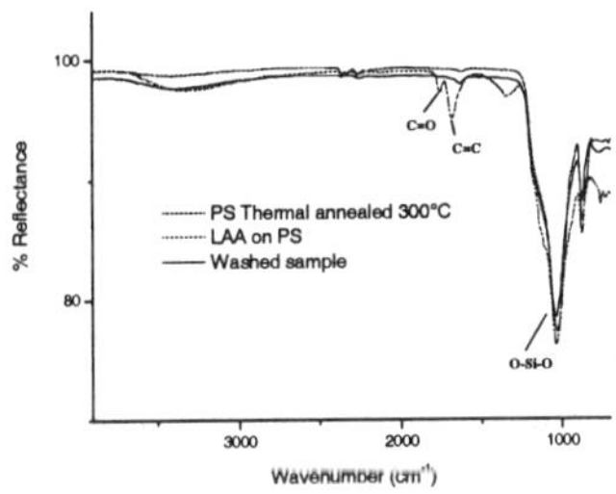

(b)

Fig. 2: (a) ATR-FTIR spectra of: as anodized porous silicon, LAA in KBr and porous silicon loaded with LAA; (b) ATR-FTIR spectra of: thermally oxidized PS, thermally oxidized PS with LAA and washed loaded sample.

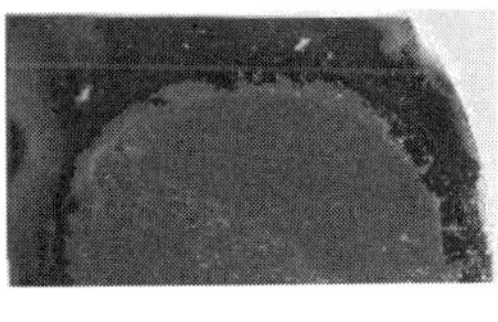

(a)

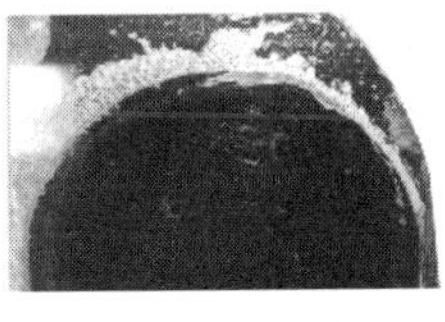

(b)

Fig. 3: (a) A freshly porous silicon loaded with LAA; (b) Oxidized porous silicon loaded with LAA.

3.3 *Surface Area and Pore dimensions of PS Reservoir*

Both oxidized porous silicon and loaded oxidized PS were characterized by a gas porosimeter, Autosorb-1 by Quantachrome. The surface area of OPS is 186 m2/g, the average pore diameter is 156 angstrom and the pore shape is cylindrical while surface area of loaded oxidized porous silicon is 140 m2/g, the average pore diameter is 36 angstrom and the shape is wedge as evident from the Adsorption/desorption isotherm (fig.4). The surface area and the pore volume both decrease as the result of the drug loading (fig.5).

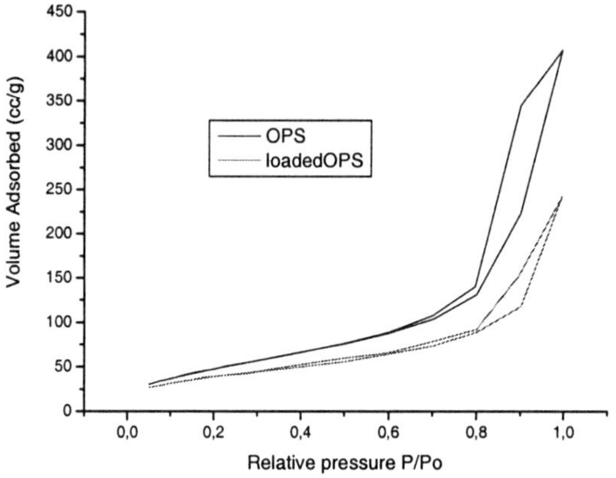

Fig. 4: Nitrogen adsorption/desorption isotherms for oxidized porous silicon before and after L-ascorbic acid impregnation.

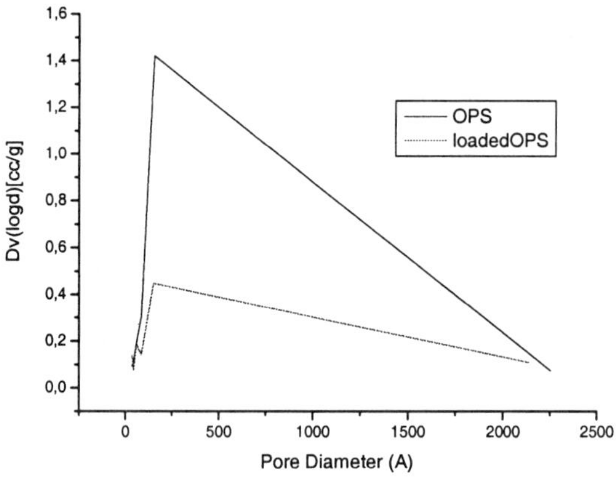

Fig. 5: Pore distribution before and after loading.

Conclusions

In this paper a simple transdermal drug delivery device has been presented. The first results underline the feasibility to realize a low cost TDD technology . Microincisions are realized on silicon without techniques like anisotropic silicon etch that require a temperature control or RIE that involve the use of very expensive PECVD systems. The realization of porous silicon, moreover is a simple and well known procedure where is possible modified pore dimensions by changing anodization parameters (like solution concentration and current density). First results on LAA impregnation have been evaluated by means of FTIR and BET Analysis.

Acknowledgment

This work has been supported by TRIPODE project.

References

1. B.W.Barry *European Journal of Pharmaceutical Sciences* **14**, 101-114 (2001).
2. R.D.Gordon, T.A.Peterson, *Drug Delivery Technology* **Vol. 3 No. 4** (2003).
3. V.V.Yuzhakov, F.F.Sherman, G.D.Owens, V.Gartstein, *Patent serial number* **U.S. 6,256,533** (2001).
4. M.B. Brown, G.P. Martin, S.A. Jones, F.K Akomeah *Drug Delivery*, **13**:175-187(2006).
5. T.I.Gorbanyuk, A.A.Evtukh, V.G.litovchenko, V.S.Solnsev, E.M.Pakhlov, *Thin Solid Films* **495**,134-138(2006).
6. W. Lohmann, D.Pagel, V. Penka, *Eur. J. Biochem.* **138**, 479-480 (1984).
7. J.Salonen et al. *Journal of Controlled Release* **108**,362-374 (2005).
8. Nicholson et al. *Rep.Prog.Phys.* **64,** 815 (2001).
9. Limbach et al., *AIChE J.* ***36***, 242 (1990).
10. K.Khanafer, K.Vafai *Heat Mass Transfer* **42**:939-953(2006).

SEMICONDUCTING NANOPARTICLES IN POLYMER FILMS: SYNTHESIS, CHARACTERIZATIONS, APPLICATIONS

T. DI LUCCIO

ENEA, Centro Ricerche Portici, Via Vecchio Macello
Portici (NA), I-80055, Italy

D. CARBONE

European Synchrotron Radiation Facility (ESRF), 6 Rue Jules Horowitz
Grenoble, I-38000, Italy

M. PENTIMALLI

ENEA,Centro Ricerche Casaccia, Via Anguillarese 301
S. Maria di Galeria (Roma), I-00060, Italy

E. PISCOPIELLO

ENEA, Centro Ricerche Brindisi, SS 7 Appia
Brindisi, I-72100, Italy

Films of CdS and ZnS nanoparticles embedded in a COC polymer matrix were prepared by spin coating cadmium and zinc thiolate precursors dispersed in the polymer solutions. Thermal annealing of the spin coated films at temperatures between 200 and 300°C induces the growth of the nanoparticles in the polymer. The nanoparticle structural and optical properties show a simple dependence on the annealing temperature. This effect can be of interest for sensing and electronic devices based on nanostructured materials.

1. Introduction

In the recent years a wide part of the material science research has been focused on the development of polymer-based nanocomposites for different technological applications, such as catalysers, sensors, organic electronic devices [1]. Nanocomposite films of polymers combined with conducting nanoparticles have shown interesting properties as gas sensors [2]. New challenges in the field of sensors are represented by polymers filled with semiconducting nanoparticles. Some of them have shown temperature dependence of electronic properties that makes them very promising temperature sensors [3]. Very recently, "touch sensors" thin films devices have been realised by alternating Au and CdS nanoparticles separated by a polymer dielectric layer [4]. Other examples of application of nanoparticle/polymer nanocomposites are biosensors [5] or optical sensors [6]. In this scenario the synthesis of semiconducting nanoparticles within polymer matrices together with a careful structural characterization can be very useful to tailor the structural and chemical properties of the nanocomposites towards specific applications.

We present our work on thin films of CdS and ZnS nanoparticles of different size and crystal structure in a thermoplastic cyclic-olephin copolymer (COC). The nanoparticles are synthesised by thermal decomposition of a (Cd,Zn) thiolate precursor previously dispersed in the COC solution [7]. The polymeric films are obtained by spin coating the precursor/polymer solution on a Si substrate. The final precursor/polymer composite film is then annealed to obtain the nanoparticles embedded in the polymer. To grow nanoparticles of diameter between 1 and 8 nm we use annealing in vacuum (pressure ~ 10^{-2} mbar) at temperatures between 200 and 300°C for 10 minutes. The structure of the samples has been investigated by x-ray diffraction using synchrotron radiation, and transmission electron microscopy (TEM).

2. Experimental Description

The CdS and ZnS nanoparticles within thermoplastic polymers have been prepared by thermal decomposition of metal thiolates precursors [7]. $(Zn,Cd)(SR)_2$ bis-thiolates, with R = $-(CH_2)_{11}CH_3$, were used as precursors while a COC, topas (grade 50.13, purchased by Ticona) was used as polymer matrix. The samples were prepared both in bulk and thin film form. The bulk samples were obtained by adding 100mg of precursor to 250mg of COC dissolved in 10mL toluene. The solution was spin cast and the resulting precursor/polymer foil was annealed at temperatures between 200 and 300 °C. The COC films with embedded nanoparticles were obtained by spin coating the precursor/polymer solution on a silicon substrate and successively annealed. The solutions were prepared at concentrations of 28 % w/w for the bulk and between 10% and 24% for the films. The spin coating was performed by dropping 100 μlt of the precursor/polymer solution on the substrate and spinned at 4000rpm for 10 s.

High resolution HRTEM images were acquired by a FEI TECNAI G^2 F30 transmission electron microscope operating at 300 KV with point-to-point resolution of 0.205 nm. For the HRTEM analyses the films were spin coated on NaCl substrates (IR Select, 13x2 mm^2). The NaCl substrates were dissolved in distilled water and the TEM specimens were prepared by collecting the floating film on standard carbon-coated grids. Two films deposited at the same spinning conditions on silicon and NaCl substrates were annealed together in the same annealing run for each temperature.

X-ray diffraction experiments were performed at the beamline ID01 of the European Synchrotron Radiation Facility (ESRF), Grenoble (FR). The energy of the x-ray beam (selected by a Si(111) crystal monochromator) was 9.592 keV (λ = 1.29Å) just below the measured Zn K-absorption edge at 9.673 keV, to reduce the background generated by the Zn fluorescence. The structural properties of the films were investigated using x-ray diffraction techniques. Powder diffraction spectra were measured using a Grazing Incidence Diffraction (GID) geometry to enhance the signal coming from the polymeric film with respect to the Si substrate. Measurements were performed at an incident angle $\alpha_i = 0.18^o$ (corresponding to the critical angle measured on the films). The scattered signal was collected by a position sensitive detector with area detection of $(50x1)mm^2$, mounted perpendicular to the sample surface to get the maximum scattering signal. The incidence slits were set at $(0.35x0.35)\mu m^2$ and the detector slit was set at 1mm vertical opening (corresponding to an angular resolution of 0.06^o). Using these conditions, detector scans were performed in the direction parallel to the sample surfacc.

TEM and diffraction data were supported by optical measurements of photoluminescence under UV irradiation.

3. Results and Discussion

The size of the nanoparticles synthesized by thermal decomposition of metal thiolates depends on the annealing conditions, mainly the temperature and to less extent on the duration of the process. The dependence of the size upon the annealing temperature is reported in Fig. 1 (right axis). The data refer to CdS nanoparticles in COC bulk samples annealed by Joule effect under vacuum (10^{-2} mbar) for 10 min. The mean size was computed by simulation of the x-ray diffraction curves considering that the CdS nanoparticles have both wurtzite and zinc-blende crystal phases [8]. In the temperature range between 230 and 300°C the nanoparticles size increases from 1.8 to 8 nm. Correspondingly, a red shift of the photoluminescence peak is observed as reported in Fig. 1 (left axis). In these samples the photoluminescence mechanism is due to charge recombination in trap states [9]. The inset of Fig. 1 is a picture of four CdS in COC bulk samples annealed at different temperatures when they are irradiated by a UV lamp. The temperatures in °C are indicated for each sample. They emit light of different colour from blue to dark orange as the temperature used for the annealing is increased (small to larger size of the nanoparticles). This is a clear evidence of the so called *quantum size effect* [10]. Further details about the temperature dependence of the structural and optical properties are reported in ref. [8]

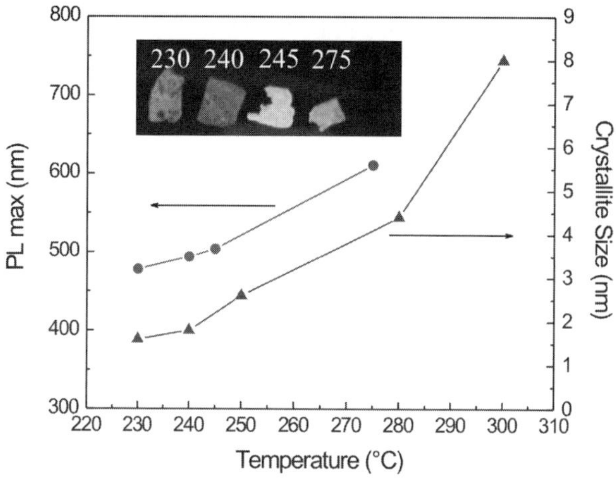

Figure 1. Dependence of the crystal size and the photoluminescence peak position on the annealing temperature of CdS nanoparticles dispersed in a polymer (COC). The nanoparticles are grown by thermal decomposition of cadmium thiolate. The inset shows a picture of the samples illuminated by a UV lamp. Above each sample the corresponding annealing temperature (in °C) is reported. Reprinted with permission from (Di Luccio et al. 2006). Copyright © 2006 American Chemical Society.

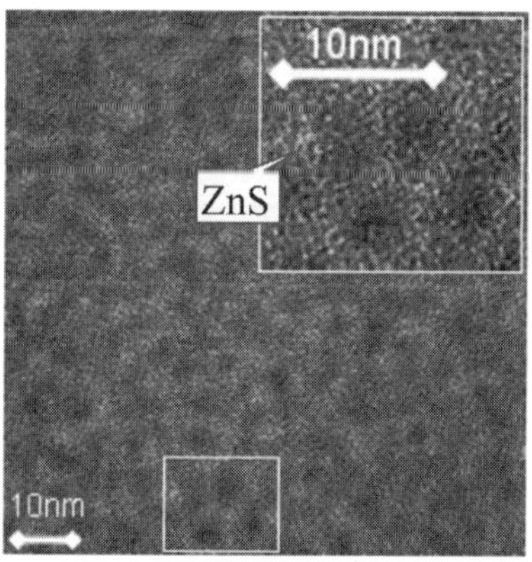

Figure 2. HRTEM image of a polymer (COC, topas) film containing zinc sulphide (ZnS) nanoparticles. The dark spots evidenced by the red circles represent the nanoparticles, while the clear zones are the polymer matrix. The average size of the nanoparticles is 4 nm.

In Fig.2 we report a HRTEM image of a COC film containing ZnS nanoparticles. In this sample the precursor/polymer concentration was 25% w/w and the film was annealed at 250°C for 10 minutes. From the image we can observe some dark circular areas (the nanoparticles) dispersed inside clear regions (the COC matrix). The nanoparticles appear homogenously dispersed in the matrix. A zoom of the region marked by the white rectangle is shown in the inset where the nanoparticles are highlighted by the red circles. The average size was evaluated about 4nm.

The GID data as a function of the in-plane diffraction angle 2θ, shown in Fig. 3, refer to a film of CdS in COC (concentration about 10% w/w) annealed at 232°C for 10 min. The Bragg peaks that are indicated in the figure between 20° and 50° are due to nanometric CdS. The average size of the nanoparticles is about 1-2 nm. Due to such a small size, the nanoparticles may have both wurtzite and zinc-blende crystalline phases. The COC matrix has an amorphous reflection at 2θ = 14° while residuals of the Cd thiolate precursor are mainly positioned below 10°.

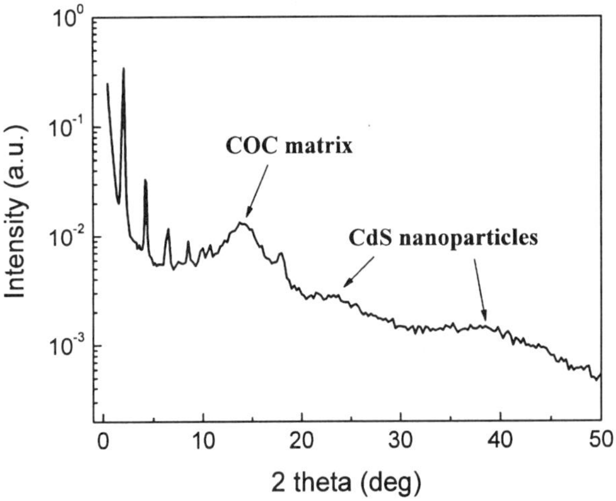

Figure 3. Grazing incidence diffraction (GID) data measured at the ESRF (wavelength = 1.29Å) on a CdS nanoparticle in COC film. The signal from the nanoparticles is comprised in the angular range between 20 and 50°. The COC matrix has an amorphous reflection at 14° while residual precursor peaks are located mainly below 10°.

4. Conclusions and Perspectives

CdS and ZnS nanoparticles were grown inside polymer films by spin coating the precursor/polymer solutions on silicon substrates that were successively annealed. Different concentrations and annealing temperatures were used. At temperatures of 250°C the ZnS particles present a homogeneous dispersion in the matrix and a size of about 4 nm as measured by TEM analyses. The precursor decomposition and the nanoparticle growth were studied by synchrotron x-ray GID experiments. The diffraction spectra evidenced the formation of the nanoparticles from the presence of the Bragg reflections due to CdS and ZnS nanocrystals after the annealing process. The nanoparticles possess interesting optical properties since they show photoluminescence upon irradiation by UV. The main interest of our research consists of the possibility to control locally the size of the particles. We have shown that the particle nucleation and growth proceed from the thermal decomposition of the precursor. The precursor molecules must be heated above their decomposition temperature that must be sufficient to favour the aggregation of sulphur and cadmium (or zinc) atoms to form CdS and ZnS nanocrystals. Possible thermal treatments might be induced by an electric current passing through the precursor/polymer film deposited between two electrodes as shown in Fig. 4a and 4b. In this way nanoparticles of the desired colour can be located where needed for sensing, catalyst, electronic applications (Fig. 4c).

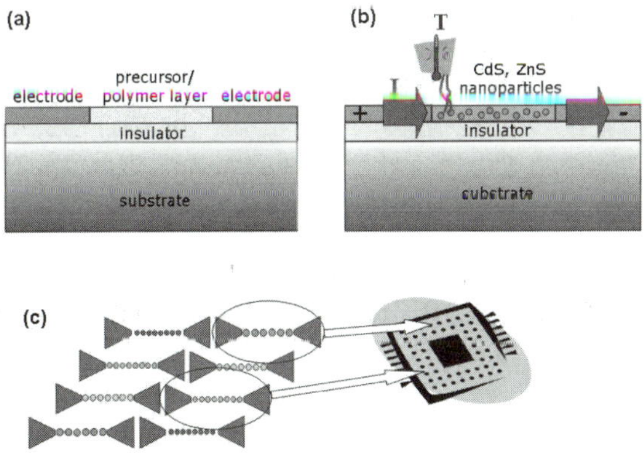

Figure 4. (a,b) Scheme of a possible way to grow the nanoparticles within a polymer by electric current induced decomposition of the precursor. (c) By this method a sequence of nanoparticles of the desired size and colour emission can be created for applications.

Acknowledgments

Dr. A. M. Laera is acknowledged for the synthesis of the thiolate precursors. TDL thanks dr. D. Della Sala for suggestions and discussions about applications of nanoparticle nanocomposites.

References

1. A. N. Shipway, E. Katz, and I. Willner, *Chemphyschem* **1**, 18 (2000), Wiley-VCH-Verlag, Weinheim.
2. X. Chen, Y. Jiang, Z. Wu, D. Li, J. Yang, *Sensors and Actuators B* **66**, 37 (2000).
3. A. E. Varfolomeev, A. V. Volkov, D. F. Zaretskii, M. A. Moskvina and V. Z. Mordkovich, *Technical Physics Letters* **30**, 663 (2004).
4. V. Maheshwari and R. F. Saraf, *Science* **312**, 1501 (2006).
5. J. Travas-Sejdic, H. Peng, R. P. Cooney, G. A. Bowmaker, M. B. Cannell, C. Soller, *Current Applied Physics* **6**, 562 (2006).
6. H. Du, S. H. Ng, K. T. Neo, M. Ng, I. S. Altman, S. Chiruvolu, N. Kambe, R. Mosso, K. Drain, *Proceedings of MN2006 Multifunctional Nanocomposites 2006*, p. 1-6.
7. F. Antolini, M.Pentimalli, T. Di Luccio, R. Terzi, M. Schioppa, M. Re, M. Marenghi, L. Tapfer, *Mater. Lett.* **59**, 3181 (2005); M. Pentimalli, F. Antolini, E. M. Bauer, D. Capitani, T. Di Luccio, S. Viel, *Mater. Lett.* **60**, 2657 (2006).
8. T. Di Luccio, A. M. Laera, L. Tapfer, S. Kempter, R. Kraus, B. Nickel, *J. Phys. Chem. B* **110**, 12603 (2006).
9. C. Petit, P. Lixon, and M. P. Pileni, *J. Phys. Chem.* **94**, 1598 (1990); L. Spanhel, M. Haase, H. Weller, and A. Henglein, *J. Am. Chem. Soc.* **109**, 5649 (1987); H. Weller, *Angew. Chem. Int. Ed Engl.* **32**, 41, (1993); N. Pinna, K. Weiss, J. Urban, and M. P. Pileni, *Adv. Mat.* **13**, 261 (2001); J. W. M. Chon, M. Gu, C. Bullen, P. Mulvaney, *Appl. Phys. Lett.* **84**, 4472 (2004).
10. R. Rossetti, L. Ellison, J. M. Gibson, and L. E. Brus, *J. Chem. Phys.*, **80**, 4464 (1984); R. Rossetti, R. Huli, J. M. Gibson, and L. E. Brus, *J. Chem. Phys.* **82**, 552 (1985).

PACKAGING METHODS FOR INTEGRATED THERMAL GAS FLOW SENSORS

P. BRUSCHI, M. DEI, M. SCHIPANI

Dipartimento di Ingegneria dell'Informazione, Università di Pisa, Via G. Caruso 56122, Pisa – Italy. E-mail: p.bruschi@iet.unipi.it

M. PIOTTO

IEIIT –Pisa, CNR, Via G. Caruso, 56122, Pisa – Italy

Two different packaging solutions for integrated thermal gas flow sensors are proposed. The sensors are based on the differential temperature configuration and are made up of two heaters placed between an upstream and a downstream temperature probe. The chip was designed using a commercial CMOS process and thermal insulation of heaters and temperature probes from the substrate was obtained by means of a post-processing technique. The two packaging solutions, corresponding to placing the whole chip inside the flow channel or adapting the pipe to a part of the chip itself, have been compared in terms of device sensitivity increase.

1. Introduction

Packaging is one of the most important research area of the MEMS field. In the integrated circuit technology, package must provide reliable electrical interconnections and heat dissipation, as well as protect the chip from the external environment. By contrast, MEMS packaging has to take account of a more complex scenario. It must protect the fragile micromachined structures and, at the same time, allow the access to or the interaction with the physical measurand. In consequence, standards for MEMS packaging do not exist and the design of the package becomes important as the design of the MEMS itself because it affects the functionality and the performances of the device [1-2].

As far as gas flow sensors are concerned, several strategies have been proposed in literature. The simplest solution is that of placing the whole chip, including bonding wires, inside a channel where the fluid is made to flow [3]. In this case, the channel cross-section can not be reduced below several mm^2. It should be noted that thermal flowmeters are actually velocity sensors and it is often requested that the gas flow is accelerated in proximity of the sensing structure in order to detect very low flow rates. In addition, flow disturbance caused by the bonding wires and the chip borders must be taken into account and, eventually, minimized. Flow disturbance is avoided by separating the chip and the flow by means of a thermally conductive membrane [4]. Unfortunately, since the whole chip or large portions of the membrane should be heated for correct operation of this kind of sensors, an important degradation of the sensor speed and power consumption should be expected. Furthermore, this solution is clearly not optimized for device sensitivity so that it is usually employed in anemometric applications. Better sensitivity is achieved with sensors based on micro-channels etched either on the same silicon substrate as the sensing structures or on a silicon/glass cover, bonded to the main chip [5]. Nevertheless, this approach is applicable only to real microfluidic applications, since only very reduced channel cross-sections can be obtained.

In this work, a packaging method aimed to obtain a sensitivity increase is proposed. The sensing chip is based on the classic calorimetric configuration with the exception of the heater which has been split into two thermally insulated sections, in order to allow constant temperature difference operation. A metal pipe was shaped so that it could be positioned directly over the chip with its sidewalls between the bonding pads. The reduced pipe section over the sensor was exploited to locally increase flow velocity at constant mass flow rate. A sensitivity increase with respect to that obtained with a solution based on the whole chip inside the channel has been demonstrated.

2. Device Design and Fabrication

The sensor is made up of two heaters positioned between an upstream and a downstream temperature probe. Each heater consists in a 5.3 kΩ polysilicon resistor positioned over a dielectric rectangular membrane suspended by means of four 45 degrees arms. The temperature probes are two thermopiles made up of 20 poly n+/Al thermocouples with the hot contacts at the edge of a dielectric cantilever beam and the cold contacts on bulk silicon.

The chip was designed with the BCD3s process of STMicroelectronics and a detailed description of the structure and fabrication process has been reported in Ref. [6]. The heaters and the hot contacts of the thermopiles were thermally insulated from the substrate by means of a cavity in the bulk silicon obtained using a post-processing technique, consisting in an anisotropic silicon etching applied to the front side of the chip.

To this purpose, openings in all the oxide layers had to be included in the layout, stacking passivation opening, vias, contact and active area layers. The resulting deep holes could produce problems of photoresist accumulation during the chip fabrication. In order to avoid this drawback, patches of the two more upper metallization layers (i.e. metal2 and metal3) of the BCD3s process were placed over the openings, partially filling the holes. The patches were easily removed in the post-processing phase using 4 μm resolution photolithography to protect the pads and a standard H_3PO_4, HNO_3 and CH_3COOH water mixture as the aluminium etchant. Alignment of the mask to the chip was not critical due to the large distance between the pads and the active structures.

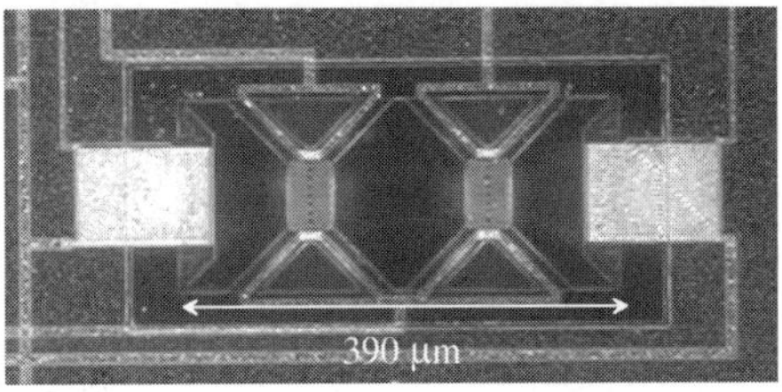

Figure 1. Photograph of the sensing structure after the silicon removal.

After the patch removal, the silicon substrate, directly accessible through the designed dielectric holes, was etched by means of an EDP solution type “S” [7] at 115 °C for 105 minutes with an etch rate of about 45 μm/hr. This solution has a good selectivity towards dielectric layers and aluminium allowing silicon etching without any additional mask. A photograph of the sensing structure after the silicon etching is shown in Fig. 1. The die was finally glued to a DIL case by means of an epoxy resin while the pads were wedge-bonded to the copper leads.

3. Device Packaging

Two different packaging solutions, schematically shown in Fig. 2, have been experimented: (a) whole the chip inside the channel (CIC) and (b) a metal pipe adapted (MPA) to the chip itself.

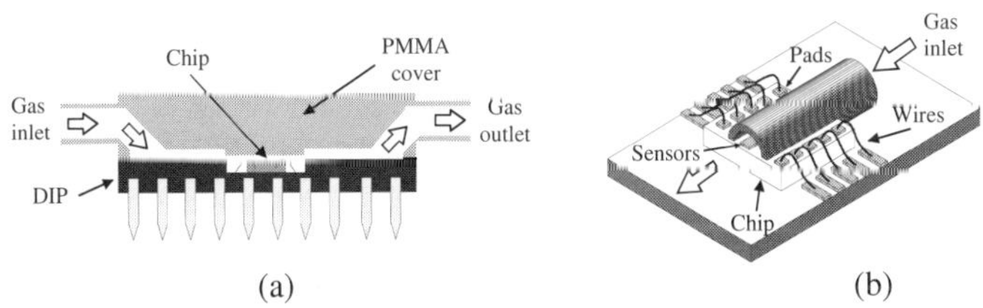

Figure 2. Two different packaging solutions: (a) the whole chip inside the channel (CIC); (b) a metal pipe adapted (MPA) to the chip.

In CIC solution, a PMMA cover has been glued to the top of the DIP case by means of epoxy resin. The cover includes two channels, used as inlet and outlet for the gas stream, which convey the gas into the small chamber that hosts the chip. Upstream and downstream horizontal sections, each 1 cm long, have been

included to make the gas flow parallel to the chip surface. A channel with a cross section of about 8 mm^2 was obtained around the sensing structure.
The MPA solution is based on a metal pipe, with an external diameter of 2.4 mm, which was milled in order to obtain a groove where the chip was positioned. In this configuration, the chip is connected directly to the pipe and a very small channel, with a section of about 0.8 mm^2, was obtained over the sensing structure.

Figure 3. Photograph of the device in the MPA configuration.

The pipe was carefully aligned to the chip using a x-y-z micrometer stage assuring that the sidewalls of the pipe fell between the pads. It should be pointed out that some pads perpendicular to the pipe axis become unusable because they are enclosed by the pipe itself. For this reason, in the chip layout all the pads of the sensing structures of interest must be placed along the chip borders parallel to the pipe axis. After the alignment, the pipe was glued and sealed to the chip and DIL case with an epoxy resin. In Fig. 3 a photograph of the final device in this configuration is shown.

4. Device Characterization

Measurements were performed at room temperature connecting the device to an electropolished stainless steel gas line equipped with a mass flow regulator (MKS 1179B). The read out electronics is similar to that described in Ref. [6] and is made up of only one operational amplifier powered with a 10 V single supply voltage (V_{DD}), as schematically shown in Fig. 4.

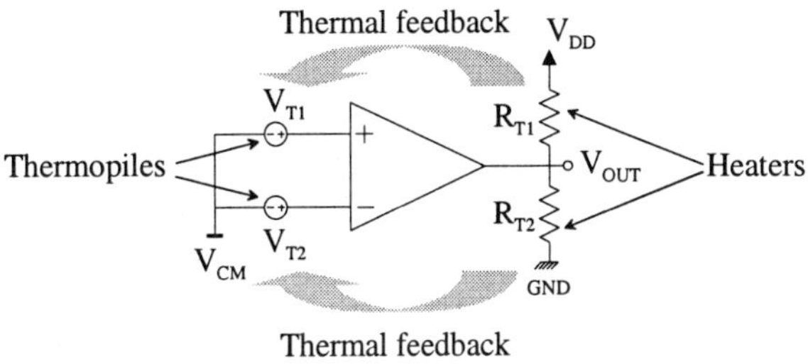

Figure 4. Schematic view of the read-out circuit.

The signals V_{T1} and V_{T2} are the thermopile outputs, resistors R_{T1} and R_{T2} indicate the thermopile resistances, and V_{CM} the common mode voltage required by the op-amp. The feedback is the heat transfer from the heaters to the thermopiles and, if the thermopiles are correctly connected, a stable circuit is obtained.

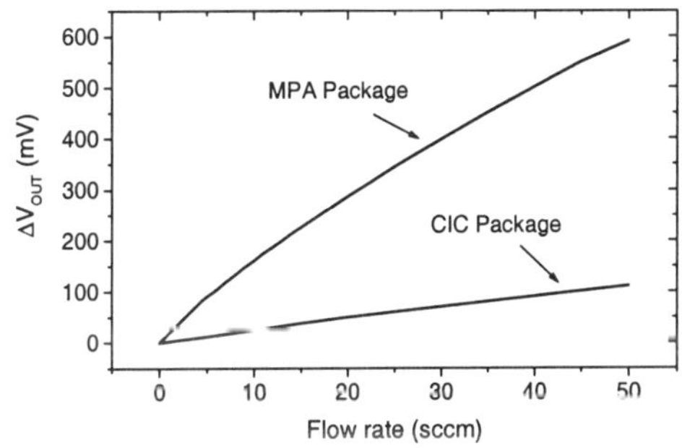

Figure 5. Response of the sensor in CIC and MPA configuration to nitrogen.

With the gas at rest, symmetry arguments suggest that equal powers are dissipated in the two heaters and, consequently, the two thermopiles are at the same temperature. In this condition, the op-amp input is null and $V_{OUT}=V_{DD}/2$, neglecting op-amp offset. When a flow is applied, the downstream thermopile tends to reach a temperature higher than the upstream one. The resulting differential signal at the op-amp input produces a power unbalance to the heaters which tends to re-equilibrate the temperatures. The difference ΔV_{OUT} between the actual output voltage and the value it assumes in rest conditions represents the output signal.

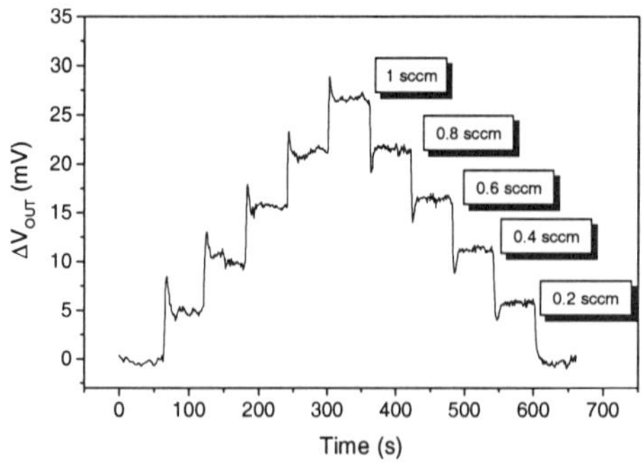

Figure 6. Sensor response in MPA configuration to a stair-step gas flow in nitrogen at room temperature.

In Fig. 5 the signal ΔV_{OUT} as a function of the nitrogen flow rate in the range 0-50 sccm is shown for both packaging solution. An almost linear behaviour has been obtained in the investigated flow range. An evident sensitivity increase of about six times has been achieved with the MPA package solution. The gain can be ascribed to the decrease in the channel cross-section over the sensor that, for a fixed mass flow rate, causes the gas velocity to increase.
The device resolution in the MPA configuration has been tested by measuring the sensor response to a stair-step gas flow and a typical result for nitrogen at room temperature is shown in Fig. 6. The stair-step starts at 0 sccm and increases up to 1 sccm in 0.2 sccm increments. Each increments is held for 60 seconds. The stair-step flow was obtained by driving the MKS flow regulator by means of a personal computer through the serial port. From the data of Fig. 6, we estimated a resolution of about 0.05 sccm.

Acknowledgements

This work was financed by the Fondazione Cassa di Risparmio di Pisa (Italy) and by the company Laben, Florence. The authors thank the STMicroelectronics of Cornaredo (Italy) for fabricating the chips.

References

1. N. Maluf, K. Williams, *An introduction to microelectromechanical systems engineering*. Artech House 2nd edn. (2004) 217
2. S. Beeby, G. Ensell, M. Kraft, N. White, *MEMS Mechanical Sensors*, Artech House, 1st edn., (2004), 57
3. F. Mayer, O. Paul, H. Baltes, *Procs. Transducers 95*, Stockholm, (1995), 528-531.

4. B. W. van Oudheusden, *Meas. Sci. Technol.*, **1**, (1990), 565-575.
5. C. H. Mastrangelo, R. S. Muller, *Tech. Dig. IEEE Solid-State Sensor and Actuator Workshop*, Hilton Head Island, (1988), 43-46.
6. P. Bruschi, A. Diligenti, D. Navarrini, M. Piotto, *Sensors and Actuators A*, **123-124**, (2005), 210-215.
7. A. Reisman, M. Berkenblit, S. A. Chan, F. B. Kaufman and D. C. Green, *J. Electrochem. Soc.*, **126**, (1979), 1406-1415.

ELECTRICAL DETECTION OF CELL ADHESION IN A SINGLE-CELL ELECTROPORATION BIOCHIP

A. DE TONI, G. CELLERE, M. BORGO, E. ZANONI

Department of Information Engineering, Padova University, Italy

L. SANTONI, L. BANDIERA

Biosilab srl, via Zeni 8, 38068 Rovereto (TN) Italy

L.LORENZELLI

ITC-IRST, Microsystems Division, Trento, Italy

Working a Single cells represents an important type of analysis for the study of the living organism, so new technologies and new methodologies have been developed in these years with this aim. In this work, electrochemical impedance spectroscopy (EIS) measurements are used to detect the presence of a cell above a biochip microelectrode used for single-cell electroporation, thus allowing fully automatic, multisite single cell, electroporation.

1. Introduction

The progresses in the field of biological sciences have opened the doors to in depth studies of the fundamental structural unit of the living organism: the cell. So, new technologies and new methodologies have been developed to allowing the analysis of new specific cellular phenomena. In particular, transfection represents a fundamental methodology with which it is possible to study the cellular reactions and transformations, consequent to the introduction through its membrane of different molecules. Different methodologies can be used in order to obtain the transfection of a cell with a molecule, including mechanics (ballistics [1], microinjection [2]), chemical (calcium phosphate [3], cationic liposome [4]), electrical (electroporation), or viral [5].

Electroporation is a particular method of transfection obtained by a transitory and significant increase in the electrical conductivity and permeability of the cell plasmatic membrane caused by an externally applied electrical field. Among the advantages of electroporation in comparison to the other methods are: 1) it is possible to transfect cells which cannot be easily transfected with different methods, such as primary cells; 2) high efficiency; 3) the biological system is not perturbed if not with the molecule of interest. Electroporation is commonly performed by immersion of two macroelectrodes in a test-tube containing 15-20 millilitres of solution composed by a buffer salt, the cells and the molecule to transfer [6]. The main drawbacks of this technique are: 1) the high voltages used lead to a high mortality; 2) cells must be detached from the culture substrate; 3) the method operates randomly on the while population, that is, the electroporation degree and molecule uptake strongly vary among the same population.

On the other side, by using a microelectronic device it is possible to realize an array of small electrodes, each individually addressable, which can be used as electroporators small enough to operate with high selectivity on one or few cells, thus enabling the study of cell-cell interactions. This single cell approach allows for example: 1) operating on a cell subgroup; 2) transfecting different polar molecules into different cells of the same population; 3) transfecting different cells of the same population with the same molecule, possibly even with different electrical protocols.

To operate such a device, the biologist must determine the presence of a cell above a given device electrode and evaluate the quality of cell adhesion by using optical methodologies. This study shows the possibility to detect a cell above the surface of a gold micro-electrode and to estimate the electroporation efficiency by monitoring the adhesion through purely electrical measurements.

2. The Device

The device used in this work is a multi electrode array (MEA) with 61 round gold microelectrodes, spatially distinct, and with a surface suitable to growth of adherent cells. The microelectrodes have a diameter of 20 μm to match the typical cell dimension. Electrodes are connected to the external circuitry by way of metal wires passivated with a silicon nitride layer [7].

A dedicated instrumental bench coupled with the biochip, is able to both stimulate the cell over the electrode by electrical pulses (to obtain the electropermeabilization of the cell membrane), and to measure the electrochemical impedance between the selected electrode and an internal reference (to predict the electroporation efficiency based on cell adhesion characteristics).

3. Cells and Solution

Chinese Hamster Ovary (CHO) are an immortalized cell line obtained from the ovary of the homonymous mammalian. CHO cells are cultured directly on multi electrode array surface, maintained in an incubator at 37°C and 5% CO_2. The cell diameter is typically comprised between 12 and 30 μm, thus matching the gold electrode diameter. Cultured medium is an HAM: F12, supplemented with 10% inactivated Fetal Bovine Serum (FBS), 100 ug/ml penicillin, and 100 mg/ml streptomycin. The buffer used during the electrical measurement is a physiological solution (Phosphate Buffered Saline, PBS), with pH=7.4, containing NCl (137 mM), KCl (2,7 mM), Na_2HPO_4 (10 mM), and KH_2PO_4 (2 mM).

4. Results and Discussions

In order to investigate the electrical property of the cell-electrode adhesion we measured the electrochemical impedance between the electrode (with and without adherent cell) and an internal gold reference electrode embedded in the biochip (reference electrode area is 1,5 x 2,5 mm) [7].The electrochemical impedance spectroscopy (EIS) measurements were carried out by using a impedance analyzer (Schlumberger 1260) with frequency ranging between 100 Hz to 500KHz, voltage amplitude of 10 mV and null (0V) voltage bias. The presence of a cell above the gold microelectrode is estimated by visual inspection at an optical microscope.

At first, we have characterized the biochip electrodes without adherent cells. The boxes in figure 1 represent the EIS averaged over the 61 electrodes of the same biochip. Bars give the standard variation (electrode-to-electrode). Notice that data follow a capacitance–like curve, in fact the electrolyte/electrode interface can be modelled by an RC parallel circuit, with $R=10^7\Omega$ and C=50pF [7]. After measurement, biochips were plated with cells and kept in incubator for at least 48 hours. At this point, we measured them once again.

The second curve in the figure 1, shows what happen when a cell adheres over an electrode. We observe an increase in the impedance over the entire frequency range. The difference between the two curves reaches its maximum around 66 kHz, where the standard deviations of the two curves are not overlapped.

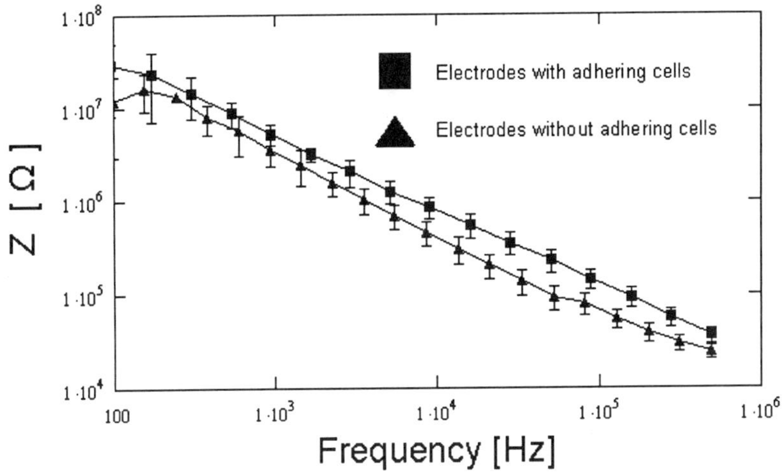

Figure 1 : Average behaviour of electrodes with and without adhering cells on a wide frequency spectrum (100 Hz – 500 KHz).

Therefore, this same 66kHz frequency appears like a good candidate frequency to characterize the cell/electrode impedance while optimizing the experimental procedure (by reducing the measurement time, one also reduces the time spent by cells at room temperature outside the incubator).
For each electrode, we defined the normalized impedance variation:

$$\Delta Z\% = \left(\mathrm{Im}_{cell} - \mathrm{Im}_{free}\right) \Big/ \mathrm{Im}_{free} \tag{1}$$

where Im_{cell} and Im_{free} are the measured impedance at 66kHz with and without the cell. Figure 2 shows ΔZ% as a function of the electrode number, for all electrodes of the same biochip As we can observe from the figure, ΔZ variation ranges between 0 (this happens for a free electrode, as checked by optical microscopy) to almost 100 % (which happens for totally covered electrode, as checked, once again, with optical microscopy).
These data confirm that it is possible to detect the presence of the cell over the electrode through purely electrical measurements. In particular, by measuring 20 devices, we were able to determine two different thresholds for ΔZ% in order to that describe the cell-electrode junction characteristics. Between 52% and 100%, visual inspection confirms that all electrodes actually feature an adherent cell. All electrodes that feature ΔZ % under 21% are totally cell-free. Finally, when ΔZ% is comprised between 21% and 52%, optical inspection shows that only 85% of electrodes have a adherent cell. In other words, some electrodes feature

an impedance increase even if no adherent cell could be detected by optical inspection.

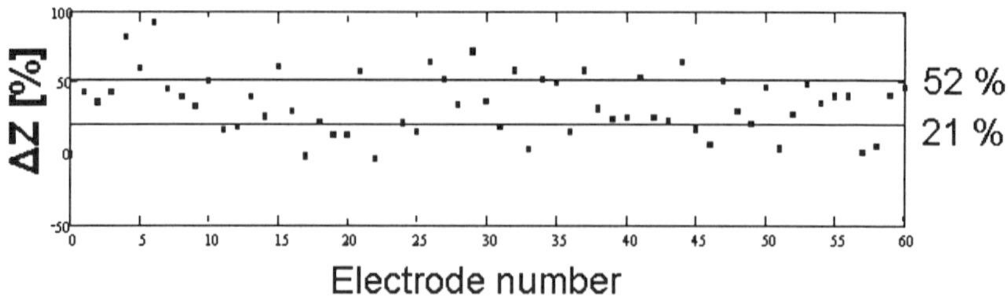

Figure 2 : Comparison for each electrode of measure with and without cell at 66 frequency of analysis signal.

Up to now, we have used the electrical measurements and the optical detection as methods to distinguish electrodes with and without adhering cells. However, optical inspection doesn't give us any information about the quality of the cell-electrode coupling.

On the other side, it is still possible that electrical measurements can be someway complementary to the usual visual inspection. In other words, electrical measurements may be used to gather data on the cell adhesion quality. This is what we are addressing in the following. The main question which arises is how we can interpret EIS data, that is, how we can be confident in translating a EIS datum into a adhesion evaluation. The specific application our device is designed to gives the needed indicator. In fact, electroporation is working with high efficiency only when we have a "good" electrical coupling between the cell and the electrode [7][8]. This way, a macroscopic, "digital" information (stained cell) can be put in relation with a microscopic, "analog" information (cell adhesion).

In detail, after electrical stimulation, the uptake of a small fluorescent dye (Lucifer yellow, LY) appears only for cells in good adhesion with the gold electrodes and not for the whole cell population. By comparing these electroporation data with the correspondent ΔZ% measured on each electrode, and considering 21% as impedance variation threshold, we can obtain an estimation of electrical detection efficiency (see figure 3). In this figure, about 15% of electrodes are false negatives, that is, they feature a negative electrical detection (no cell is detected from EIS measurements), but a) the presence of the cell is found with visual inspection, and b) its good adhesion with electrode is confirmed by the positive electroporation result. On the other side, only 5% of electrodes are false positives, that is, EIS measurements predict the cell presence, but after electroporation we have no LY uptake. However, note that in this case, the cell was actually there, but electroporation yielded a negative result. This is probably due to the low quality of cell-electrode adhesion, insufficient to guarantee a good electrical stimuli transfer from the electrode to cell membrane (see electrodes 36 and 37 in figure 3). Finally, we can derive from Figure 3 that with the EIS measurement setup we developed, it is possible to predict that a cell will be sitting above a given electrode, and that that specific cell will be electroporated, with a 85% confidence.

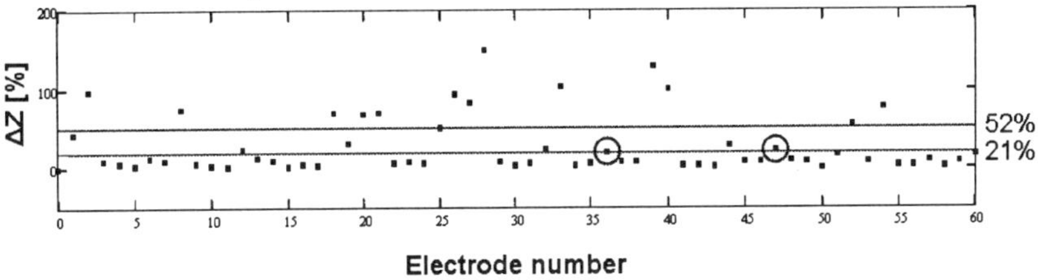

Figure 3 : Variation of impedance for various electrodes. The circled points represent the only electrodes (36, 47) above the borderline of 21% without positive results.

5. Conclusions

In this work, electrochemical impedance spectroscopy measurements are used to detect the presence of a single cell above biochip microelectrode used for single-cell electroporation.

Once we have defined for each electrode ΔZ% as the percent variation of the impedance with and without an adhering cell, it is possible to identify tree different electrode/cell families. When ΔZ% is above 52%, not just a cell is living above the electrode, but the cell/electrode adhesion can be classified as "good". When ΔZ% below 21%, there are no cell above electrode. Finally, when ΔZ% lies in the "grey" region (between 21% and 52%) it becomes difficult to do any good prediction of the adhesion quality. However, by considering these borderline electrodes as "good", one can predict successful electroporation (without visual inspection) with 85% confidence.

References

1. S. Johnstong, D. Tang. Gene gun transfection of animal cells and genetic immunization. [PubMed] Methods Cell Biol. 1994;43 Pt A:353-65.
2. G. N. PROCTOR. Microinjection of DNA into mammalian cells in culture: theory and practice. Methods mol. cell. biol. 1992, vol. 3, no5, pp. 209-231 (5 p.)
3. E. Schenborn, V. Goiffon, Calcium phosphate transfection of mammalian cultured cells. Methods Mol. Biol. 130, 135-145 (2000).
4. R.W. Malone, P. L. Felgner, and I. M. Verma. Cationic Liposome-Mediated RNA Transfection. PNAS, August 15, 1989, vol. 86, no. 16, 6077-6081
5. E.C. Ohki, M.L. Tilkins, V.C. Ciccarone, P.J. Price. Improving the transfection efficiency of post-mitotic neurons. J Neurosci Methods. 2001 Dec 15;112(2):95-9.
6. H. Potter. Electroporation in biology: methods, applications, and instrumentation. Anal Biochem. 1988 Nov 1;174(2):361–373.

7. L. Bandiera, M. Borgo, G. Cellere, A. De Toni, L. Santoni, C. Bersani, and A. Paccagnella. Electrical modeling of a biochip for genetic manipulation of single cells . IEEE IEDM Tech. Digest, 2006, pp.723-726.
8. D.W. Laird, K. Jordan, T. Thomas, H. Qin, P. Fistouris, Q. Shao. Comparative analysis and application of fluorescent protein-tagged connexins. Microscopy Research and Technique. Volume 52, Issue 3 , Pages 263 - 272

CHARACTERIZATION OF A SILICON INTEGRATED MICRO-FLOW CYTOMETER

R. BERNINI, F. BRESCIA, M.R. SCARFÌ
CNR-IREA, 80124 Napoli – Italy

R. PALUMBO
CNR-IBB, 80134 Napoli – Italy

E. DE NUCCIO, A. MINARDO, L. ZENI
Seconda Università di Napoli-DII, 81031 Aversa – Italy

P. M. SARRO
ECTM-DIMES, TUDelft. NL-2600 GB Delft, The Netherland

The flow cytometer is a useful instrument for a selective, fast and accurate analysis of cells or particles [1]. It is widely used in several application fields ranging from cell biology, medicine, toxicology and environmental monitoring [2]. To this aim, a cell suspension is injected in a laminar flow and hydro-dynamically focused in a single cell flow that is illuminated by a monochromatic light. Scattered or emitted fluorescence light is collected and analyzed providing information for cell classification. In the last years several research groups have been involved in the development of micro flow cytometers. Micro flow cytometers have been demonstrated making use of using plastic (PDMS) or glass materials [3-5]. In this paper we propose the characterization of a novel micro flow cytometer based on silicon integrated hollow core AntiResonant Reflecting Optical Waveguides (ARROWs) [6]. ARROW geometry allows one to use the same channel to guide both the sample stream and the fluorescence excitation light, permitting to rearrange the detection system so as to increase the signal-to-noise ratio. The integrated micro flow cytometer has been characterized by using biological samples marked with standard fluorochromes. The experimental investigation confirms the success of the proposed microdevice in the detection of cells.

1. Design

A schematic drawing of the proposed integrated micro flow cytometer and a schematic drawing of a two-dimensional hollow-core ARROW waveguide are shown in Figure 1. The device is based on a hollow core ARROW waveguide., composed of two halves joined together.

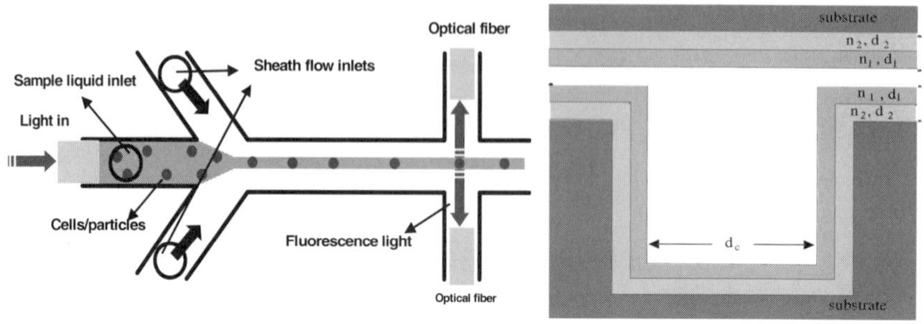

Figure 1. Schematic layout of the integrated micro flow cytometer and transverse section of the hollow core ARROW waveguide

In ARROW waveguides, light is confined inside the core region by two cladding layers designed as antiresonant Fabry–Perot reflectors. It is possible to divide the devicc into two functional blocks. The first block (region A), where the hydrodynamic focusing takes place, is aimed to align single cells or particles. Tuning of the flow rates of the liquids introduced into the channels reduces the cross section of the flowing sample liquid confined between the slabs of sheath fluid, forcing the cells to flow individually through the detection region. As can be observed in Fig. 1, the central channel is used both as a microfluidic channel, where the sample liquid containing the cells is injected, and as an optical waveguide able to confine the fluorescence excitation light. Analysis of the emitted fluorescence is carried out in the second region (region B). In particular, two orthogonal optical fibers are used for guiding the emitted fluorescent light to the photodiodes, located at the exit of the fibers themselves. The finite elements method (FEM) has been used to solve the fluid dynamics equations of the structure, with the aim of finding the optimal ratio between core flow rate and shield flow rate (FRR). This permitted an adequate design of region A of the device.

2. Fabrication

The integrated micro flow cytometer was manufactured from two silicon wafers bonded together by using standard silicon technology. In particular, the hollow waveguide was fabricated by deep silicon dry etching (resulting in a

200×150 μm^2 rectangular-core for the flowing liquid path and in a 140×150 μm^2 core for the orthogonal arms) of the bottom wafer followed by a LPCVD deposition, on both wafers, of the silicon nitride (n_l=2.227) and silicon dioxide (TEOS) (n_2=1.457) cladding layers at a temperature of 850 °C. The cladding layers thicknesses are designed to minimize optical losses [6]: d_1= d_2=266 nm at λ=633 nm and n_c=1.333. After the deposition, micro square holes (700×700 μm^2) were dry etched on the backside of the channel wafer in order to make the inlets holes for liquid injection. Finally, the two halves were joined by silicon nitride wafer bonding. The overall size of the die is 2.5×1 cm^2. The final step is represented by the tubing system, the excitation and collection fibers positioning. The fibers were inserted into the respective locations and sealed by glue. Tygon tubes with internal diameter of 0.7 mm were connected to the inlet holes by using epoxy and successively interconnected to a syringe pumping system. In Figure 2 we show a picture of the bottom wafer and of the assembled device.

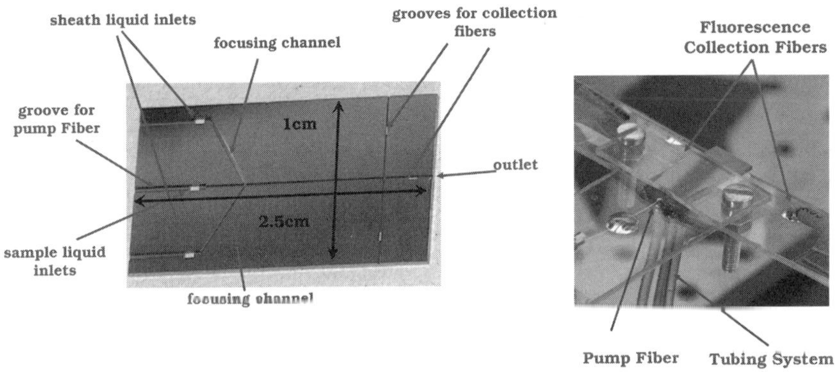

Figure 2. Pictures of the bottom wafer and of the device after assembling

3. Experimental Results

As first analysis, the microfluidics has been characterized to verify the right accomplishment of the hydrodynamic focusing effect. For this purpose the top silicon wafer has been replaced by a glass slide, so as to observe the liquids inside the microchannels. In this case a fluorescent dye solution of resorufin (10ng/ml in de-ionized water) has been used as sample liquid. The experimental results confirms is performed correctly, in accord to the FEM simulation, as shown in Figure 3, in accord to the FEM simulation.

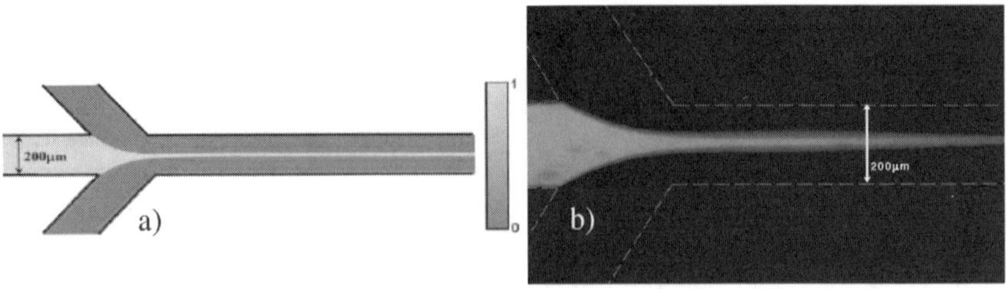

Figure 3. Numerical simulation a) and experimental result b) of the hydro focusing effect

Finally, the integrated micro flow cytometer was fully characterized (microfluidic and optical systems) by using a biological sample. For this purpose, human T leukemia cells (Jurkat) with a diameter of about 20 μm were used. Jurkat cells were grown in RPMI-1640 medium supplemented with 10% heat-inactivated foetal calf serum, 1% L-glutamine, and 1% penicillin–streptomycin in a humidified 5% CO2 atmosphere at 37°C. Cells were maintained at a density of 2–3×105 cells/mL and subcultured three times a week. For the experiments, 10^6 cells were washed and resuspended in diluted propidium iodide (PI) solution (50 mg mL−1 in 0.1% trisodium citrate, 0.1% Triton X100, and 1 mg mL−1 RNAse). PI binds to DNA in a stoichiometric manner so that there is a direct relationship between DNA content and PI fluorescence. In fact, this fluorescent dye is currently used to study the cell cycle. After 15 min incubation in the dark at room temperature, the marked cells, injected by the syringe pump, were focused by the sheath liquids.

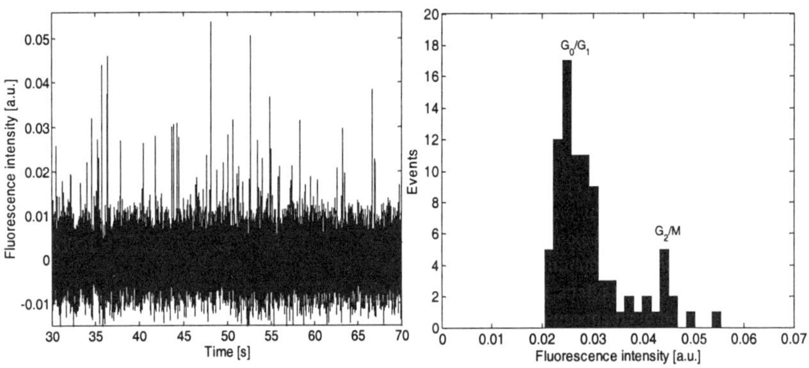

Figure 4. Collected fluorescence intensity of Jurkat cells marked with propidium iodide and Frequency histogram of Jurkat cells, obtained by the data collected by the integrated silicon micro flow cytometer

Due to the low-pass band behaviour of the photodiode (3-dB bandwidth equal to 750 Hz), the absolute velocities of the liquids have been reduced, so as to reduce cells occurrence. The results obtained for a sample flow rate of 1 mL/h and FRR=3 are shown in Figure 4 where the intensity peaks of the collected fluorescence represent the cell transit event and the frequency histogram of the Jurkat cells marked with PI and analyzed with our micro flow cytometer is reported. Two main populations can be observed, namely, G_0/G_1 cells and G_2/M cells, and the fluorescence intensity of G_2/M cells is twice that of G_0/G_1 cells. It should be stressed that the quite low signal-to-noise ratio of our measurements stems from the use of a simple photodiode instead of a photomultiplier.

4. Conclusion

In conclusion, a novel integrated silicon micro flow cytometer has been proposed and characterized. The device is based on a hollow core antiresonant optical waveguide, which permits the simultaneous confinement of both the excitation light and the sample liquid inside the core region. The proposed micro flow cytometer was manufactured using standard silicon technology. The experimental results confirmed the correct behavior of the device. The characterization was carried out by investigating the hydrofocusing effect and by testing the device in the analysis of fluorescent marked Jurkat cells.

References

1. D.Huh, W. Gu, Y. Kamotani, J. B. Groteberg, S. Takayama, *Physiol. Meas.*, **26**, 73-98 (2005)
2. C. Grunde, S. Skerlos, P. Adriaens, *FEMS micorbiology ecology*, **49**, 37-49 (2004)
3. Y. C. Tung, M. Zhang, C. T. Lin, K. Kurabayashi, S.J. Skerlos, (2004), *Sensors and Actuators B*, 98, 356-367.
4. C.H. Lin, G.B. Lee, *J. Micromech. Microeng.*, **13**, 447-453 (2003)
5. J.Kruger, K.Singh, A.O'Neill, C.Jackson, A.Morrison, P.O'Brien, *J. Micromech. Microeng.*, **12**, 486-494 (2002).
6. S. Campopiano, R. Bernini, L. Zeni, P.M. Sarro, *Optics Lett.,* **29**, 1894-1896 (2004).

LASER OXIDATION MICROPATTERNING OF A POROUS SILICON BASED BIOSENSOR FOR MULTIANALYTES MICROARRAYS

L. DE STEFANO[1], L. ROTIROTI[1,3], I. REA[1,2], E. DE TOMMASI[1], M. A. NIGRO[4], F. G. DELLA CORTE[4], AND I. RENDINA[1]

[1]*IMM-CNR - Sezione di Napoli, Via P. Castellino 111, 80131 Napoli, Italy*
[2]*Dept. of Physics, "Federico II" University of Naples, Monte S. Angelo, 80126 Naples, Italy*
[3] *Dept. of Organic Chemistry and Biochemistry, "Federico II" University of Naples, Monte S. Angelo, 80126 Naples, Italy*
[4] *"Mediterranea" University of Reggio Calabria Department of Computer Science, Mathematics, Electronics and Transports, Località Vito di Feo, Reggio Calabria, Italy*

Abstract

In this communication, we present the fabrication and characterization of oxidized porous silicon (PSi) micropatterns, obtained by direct laser writing, and chemically functionalized for biological applications. The laser local oxidation technique is a good alternative to the traditional photolithographic process since this approach allows the development of micropatterned biosensors. In fact, from a technological point of view, the standard masking of PSi by means of photoresist presents remarkable difficulties because of the low resistance to the electrochemical process. The micropatterned PSi oxidized surfaces can be properly functionalized by a specific chemical linker. The modified surface is able to link the carboxylic or aminic groups of biological probes. The binding between the functionalized surface and the biological matter has been tested by directly spotting on the PSi local oxide a fluorescent labelled antibody. The device has been characterized by FT-IR measurements and fluorescence macroscopy.

Keywords

Porous silicon, laser writing, biosensors, microarrays.

INTRODUCTION

Porous silicon based optical biosensors offer several advantages in label-free detection and identification of organic molecules due to the well known properties of this material [1, 2]. PSi has a spongy structure with a large specific surface of the order of 100–500 m^2 cm^{-3} [3], so that a very effective interaction with several adsorbates is assured. When a biological solution penetrates into the PSi pores, it substitutes the air, so that the average dielectric properties of the heterogeneous silicon-air-liquid system change inducing a red-shift in the reflectivity spectrum of the device. The effect depends on the refractive index value of the solution but also on how it penetrates into the pores. Due to this sensing mechanism, PSi optical devices cannot identify the single components of a complex mixture. In order to enhance the sensor selectivity, several biological probes, which exploit very specific interactions only with selected biochemical molecules, can be linked to the PSi surface: most common examples are DNA strands, proteins, enzymes, antibodies and so on. In this work, the fabrication and characterization of oxidized porous silicon micropatterns obtained by direct laser writing [5] and their chemical functionalization for biological applications are reported. In particular we have employed a murine monoclonal antibody (IgG) UN1 previously selected for the specific reactivity with human thymocytes as compared to peripheral blood cells [4]. The antigen recognized by UN1 is a 100–120 kDa transmembrane glycoprotein showing biochemical features of cell membrane-associated mucin-like glycoproteins, a class of macromolecules that are involved in cell-to-cell interactions and cancer progression [6].

EXPERIMENTAL AND RESULTS

The device was a PSi monolayer, 1 μm thick and with a porosity of 70%, produced by electrochemical etch in a HF-based solution, starting from a highly doped p+-silicon, <100> oriented, 0.04 Ω cm resistivity, 400 μm thick. The silicon was etched using a 30 wt. % HF/ethanol solution in dark and at room temperature. After the etching the samples were rinsed in ethanol and dried under a stream of N_2.

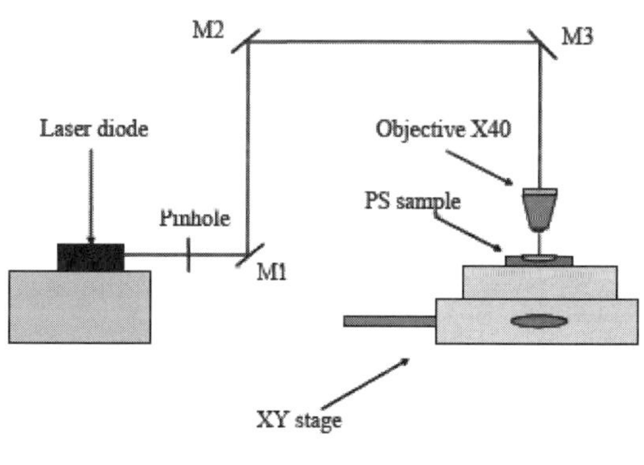

Fig. 1: Scheme of the setup used to write the channel waveguides.

The localized oxidation is obtained by exposing the sample to the beam of a laser diode (μLS Micro Laser Sistem) at 408 nm with an output power of 48mW, using a setup composed of a microscope objective and a motorized xy micrometric stage as shown in Figure 1. The beam dimension is about 1.4/0.32 mm. We have directly written on the sample some oxide strips, 10 μm wide, with a laser fluence of 38 mJ/mm^2.

Starting from our previous experience on chemical modification of the oxidised porous silicon, we have used an organosilane compound such as the aminopropyltriethoxysilane (APTES) and a functional crosslinker such as Glutaraldehyde (GA) which reacts with the amino groups on the silanized surface and coats the micropatterned PSi oxidized surfaces. [7]. The obtained modified surface works as an active substrate for the chemistry of the following attachment of other biological probes, such as DNA single strand, proteins, enzymes, and antibodies.

The functionalization process takes place directly on the chip surface, simply covering it with the chemical solutions. In Fig. 2 is reported the scheme of the functionalisation process. The PSi device has been rinsed by immersion in a 5% solution of APTES and an hydroalcoholic mixture of water and methanol (1:1), for 20 min at room temperature. After the reaction time, we have washed the sample with demi-water and methanol and dried in N_2 stream. The silanized surface was then baked at 100°C for 10 min.

The APTES modified surface is able to link the carboxylic group of a biological probe such as IgG or other antibodies. A further step of functionalization is required to bind the amino group always present in biological matter through the aldehydic group of opportune chemical linker. To this aim, we have treated the chip in a 2.5% glutaraldehyde solution for 30 min. The glutaraldehyde reacts with the amino groups on the silanized surface and coats the internal surface of the pores with another thin layer of molecules.

Infrared spectra were collected by using a Fourier transform spectrometer equipped with a microscope (Nicolet Continuμm XL, Thermo Scientific) and are reported in Fig. 3, where it is well evident that, after the functionalization processes, the Si-OH peaks are completely substituted by the organic linkers characteristics ones.

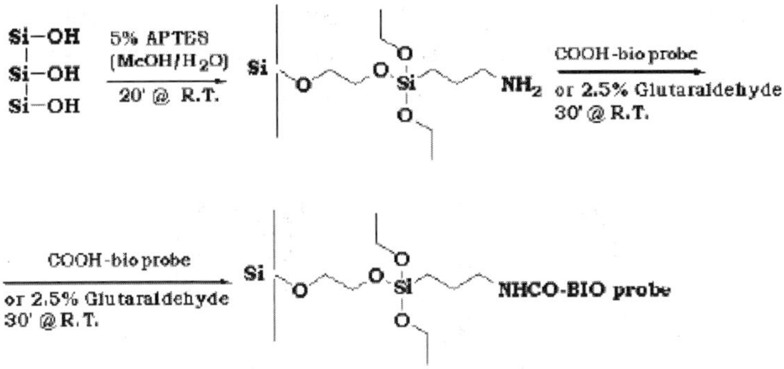

Fig. 2: functionalization steps of PSi surface by means of APTES and Glutaraldehyde and subsequent binding of the biological probe.

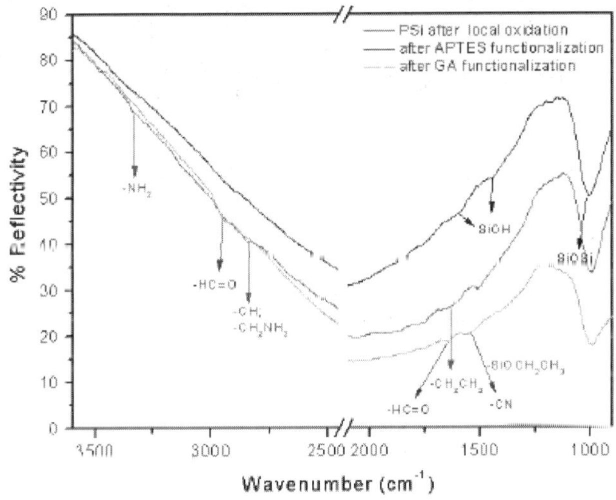

Fig. 3: FT-IR spectra of the micropatterned PSi oxidized surfaces before and after the chemical surface modification treatments.

The binding between the surface and the biological matter has been tested by using fluorescent labelled bioprobes directly spotted on the PSi chemical modified structures. We have spotted on the porous silicon chip a solution containing the rhodamine labelled IgG antibody, 6.8 μM, and incubated overnight at room temperature. The device has been observed by a Leica Z16

APO fluorescence macroscopy system. By 100W high-pressure mercury source we found that the fluorescence is very high and homogeneous on the region locally oxidized by the laser as shown in Fig. 4.

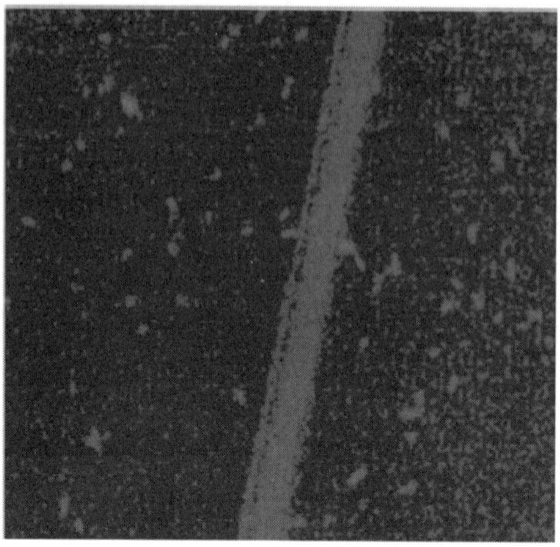

Fig. 4: Post-dialysis fluorescence image of the chip after rhodamine labeled IgG binding

CONCLUSIONS

In this work, we have exploited the possibility to locally functionalize the PSi surface.
A micropattern has been directly written on the porous silicon layer by a laser oxidation process. The pattern, constituted by strips 2 μm wide, has been functionalized by specific chemical linkers such as APTES and glutaraldehyde; then we have spotted on the PSi sample a fluorescent labelled antibody. The interaction between the modified surface and the biological matter has been verified by fluorescence macroscopy; a high fluorescence signal has been observed only in the oxidized region.

REFERENCES

[1] H. Ouyang, C.C. Striemer, P.M. Fauchet, APL 88, 163108, 2006.
[2] C. Pacholski, M. Sartor, M.J. Sailor, F. Cunin, G.M. Miskelly, J. Am. Chem. Soc. 127, 11636-11645, 2005.

[3] R. Herino, G. Bomchil, K. Barla, C. Bertrand, J.L. Ginoux, J. Electrochem. Soc. 134-8, 1994, 1987.

[4] Hilkens J., M.J. Ligtenberg, H.L. Vos, S.G. Litinov, Cell membrane-associated mucins and their adhesion-modulating property, Trends Biochem. Sci. 1992, 17, 359.

[5] A. M. Rossi, S. Borini, L. Boarino, and G. Amato, Phys. Stat. Sol. (a) **197**, No. 1, 284–287 (2003).

[6] Cecco L., H.M. Bond, P. Bonelli, et al., Tissue Antigens 1998, 51, 23.

[7] De Stefano L., A. Vitale, I. Rea, M. Staiano, L. Rotiroti, T. Labella, I. Rendina, V. Aurilia, M. Rossi, S. D'Auria, Extremophiles (2007), DOI 10.1007/s00792-006-0058-6.

FEASIBILITY OF DIRECT CARBON NANOTUBES GROWTH FOR SENSING APPLICATIONS

T. POLICHETTI, O. CALÒ, P. DELLI VENERI, T. DI LUCCIO, E. MASSERA, I. NASTI, P. VACCA, G. DI FRANCIA
ENEA Portici Research Center
Località. Granatello I- 80055 Portici (NA), Italy

The aim of this work is to explore the prospect to grow carbon nanotubes (CNTs) directly on an interdigitated structure to realize a gas sensor. In order to realize a complete device, we start optimizing CNT growth in a Hot wire chemical vapour deposition (HWCVD) reactor on different Ni-coated substrates, finding best pretreatment conditions, and then varying the total pressure, the substrate temperature and the composition of the gases during CNTs synthesis. The material was morphologically and structurally characterized through scanning electron microscopy and Raman spectroscopy, and for some samples by synchrotron X-ray diffraction. Parallelly, with the view to test sensing properties of the CNTs, a gas sensor, obtained dispersing a CNT solution on gold electrode on glass substrate, was prepared. The device was exposed to different analytes. The results are extremely encouraging and make us really confident on the possibility to realize by direct CNTs growth a sensor in perfect workimg order.

1. Introduction

Several synthesis techniques have been developed to grow carbon nanotubes such as laser vaporization and electric arc discharge; the most versatile technique is, however, chemical vapor deposition (or CVD).

Much work have been done these last years to develop CNTs-based electronic devices [1-13]; the main application we are interested in is the field of sensors.

CNTs are utilized as sensing material in pressure, flow, thermal, gas, optical, mass, position, stress, strain, chemical, and biological sensors [8-13].

SWNTs (single-walled carbon nanotubes) and MWNTs (multi-walled carbon nanotubes) have been successfully used as efficient sensors for NH_3, H_2O, CO, CO_2, NO_2 and nitrotoluene at room temperature [12].

Gas sensing at room temperature is of great interest; normally most of the sensors operate at elevated temperatures, except a few types of polymer based gas sensors [14, 15], which unfortunately provide only limited sensitivity;

semiconducting CNTs operate at room temperature with sensitivity as high as 10^3.

Mainly two ways to elaborate CNT-based sensors are reported in literature, one based on the processing and deposition of CNTs dispersion on an electrode network and the second based on the direct synthesis of CNTs on sensor structures. The former technique allows to replace commercially available existing gas sensors due to its simplicity in largescale production, although its sensitivity is limited to the parts per million (ppm) level [12]. The latter solution is more convenient for the possibility to operate an *in situ* purification by burning metallic CNTs with high currents and for the elimination of one process step;tis tecnique has been recentely develped by Wongwiriyapan et al. [13] which fabricated a new type of CNT network sensor with a simple architecture by growing MWNT networks directly on alumina substrate. The sensor exhibited good gas detection performance under room temperature operation.

In this study, we want to explore the practicability to realize a gas sensor by the CNT direct growth. To approach this objective we proceeded on two parallel channels: on one hand we optimized CNTs growth by HWCVD on two type of substrate Si/SiO_2 and Al_2O_3, which are appropriate platform for conventional gas sensor; on the other hand we evaluated the sensing properties of the CNTs, dispersing MWNTs in dimethylformamide and depositing the solution by simple casting on electrodes network. Such device was exposed to different gases and demonstrated good sensing behaviour.

2. Experimental

MWNT growth was performed by HWCVD process. The HWCVD system was designed by IOVAC for the deposition of CNT. The chamber can reach a background pressure of about $5x10^{-7}$ mBar when pumped with a turbo pumping group. The substrate holder is heated via six lamps in order to reach the desired substrate temperature. The actual temperature on the substrate surface can be in some way dissimilar from the measured value due to the distance of the substrates from the measuring point on the substrate holder. A DC power up to 7.5 kW (for a maximum current of 250 A) is applied to three horizontal Ta filaments with the aim of heating and dissociating the gaseous precursors (H_2 and C_2H_2).

We tested the CNTs growth on Si/SiO_2 and Al_2O_3, coated with four different Ni thicknesses deposited by e-beam evaporation at room temperature: 0.1nm, 0.5 nm, 1 nm, 5 nm. After Ni deposition the substrates were loaded into the

HWCVD chamber for thermal reducing pretreatment. The chamber was pumped down to a pressure lower than 10^{-6} mBar, in order to achieve clean conditions.

Keeping fixed at 30 sccm the H_2 flow, the temperature and the duration of the treatment were varied: five substrate temperatures were used: 600°C, 650°C, 700°C, 750°C and 800°C. For each temperature, the duration of the treatment was varied between 10 and 30 minutes. In all these tests the four different Ni thickness substrates were simultaneously mounted on the sample holder.

After pretreatment each sample were examined by Scanning Electron Microscope SEM LEO 1530 in order to establish the dimensions and the distribution of the clusters.

Due to the high roughness of the alumina, only for this kind of substrate the SEM was not able to give information on the clusters structure. So X-ray diffraction experiments were performed at the beamline W1.1 of Hasylab at the Deutsches Elektronen-Synchrotron (DESY). The energy of the x-ray beam was 10.5 keV (λ = 1.18Å). The structural properties of the films were investigated using grazing incidence x-ray diffraction (GID) geometry to reduce the contribution from the Al_2O_3 substrate. The measurements were performed at an incident angle $\alpha_i = 0.25°$. The incidence slits were set at (2x0.2)mm^2 and the detector slits were set at 2mm vertical opening. Using these conditions, detector scans were performed in the direction parallel to the sample surface.

Once the optimal pretreatment conditions were found, CNT growth was carried out varying the substrate temperature between 500 and 800°C, the total pressure between 1.25 and 5.00 mBar, and the H_2/C_2H_2 ratio between 5:1 and 20:1.

A commercial Renishaw inVia Reflex Raman microscope (spot size 1 mm x 1 mm) functioning in backscattering configuration was operated using the 633 nm line of a HeNe laser to establish the quality of the CNT.

The device fabrication process comprises the following steps: the interdigitated electrodes (IDE) pattern is realized using lift-off photolithography after evaporation of 200 nm Au layer on alumina substrate by e-beam technique. The schematic diagram of the IDE pattern is shown in Fig. 1(a): the finger width is 8 μm, the length 2.5 mm and the gap size of 8 μm; pattern total dimensions are 7mm x 5mm. The MWNTs were dispersed in dimethylformamide (DMF) to form a suspension (10 mg/L). The DMF was chosen to debundle the MWNTs ropes because the amide group can easily attach the surface of the nanotubes, providing a uniformly suspended MWNTs solution. This ensures good-quality CNTs deposition on top of the substrate [16]. The suspension was sonicated for 3h at room temperature. The MWNTs-DMF solution was drop-deposited (1μL) onto glass substrates and the DMF was evaporated under vacuum at 60°C for overnight. Figures 1(b) shows the scanning electron microscopy (SEM) images of the magnified view of the electrode region.

The device was placed in the test chamber. Data were collected using a DC voltage of 0.1V. All experiments were conducted at room temperature.

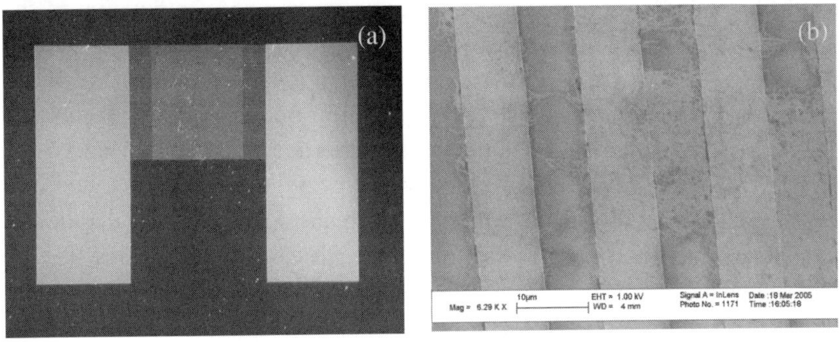

Figure 1. (a) a picture of IDE pattern realized by lift-off pholithography; (b) a detail (SEM photo) of the area where the CNT solution was deposited: it is clearly visible a MWNT bundle between the fingers.

3. Results and Discussions

3.1. *Material optimization*

In fig.2 are reported SEM images of two sample of Ni coated Si/SiO_2 substrates.

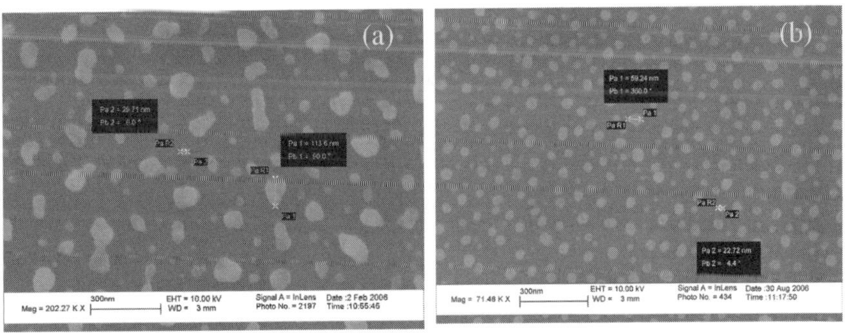

Figure 2. SEM images of catalyst clusters distribution: (a) large clusters (100 nm) not entirely isolated (temperature higher than 700°C); (b) well-defined clusters with size ranging from 20 to 60 nm (deposition parameters in green in table 3). Layer thickness also affect the clusterization; in particular under 5nm, the dimensions of the clusters are below SEM resolution: moreover no CNT growth are observed on samples with Ni layer thinner than 5nm.

The aim of the pre-treatment is to reduce the Ni oxidation by exposure to H_2 and to confer to the metal layer a suited morphology for the CNTs growth, that is the formation of individual and separated clusters with uniform spatial distribution and dimensions between 20 and 50 nm. Catalyst clusters distribution and size were different, depending on the temperature and the duration of the treatment. Temperature lower than 700°C gave rise to large clusters (100 nm) not entirely isolated (see fig.2 (a)). Over the threshold of 700°C , the nickel film resulted in well-defined clusters with size ranging from 20 to 60 nm (see fig.2 (b)). Layer thickness also affect the clusterization; in particular under 5nm, the dimensions of the clusters are below SEM resolution: moreover no CNT growth are observed on samples with Ni layer thinner than 5nm. The duration of the treatment has effect on the definition of the cluster: treatment higher than 10 min. induce the aggregation of the clusters so stuctures larger than 100nm and not well defined are observed.

Due to the small nominal thickness of Ni the SEM analyses did not provide any information about the Ni clusters formation upon the annealing treatment. Alternatively, the GID measurements performed by using the high brilliance x-ray source of the synchrotron radiation confirmed the presence of the clusters and allowed to estimate their mean size. Figure 3 shows the GID measurements on the Ni sample of 5 nm of thickness deposited on Al_2O_3 substrate pretreated at 700°C in 30 sccm of H_2 for 10 min.

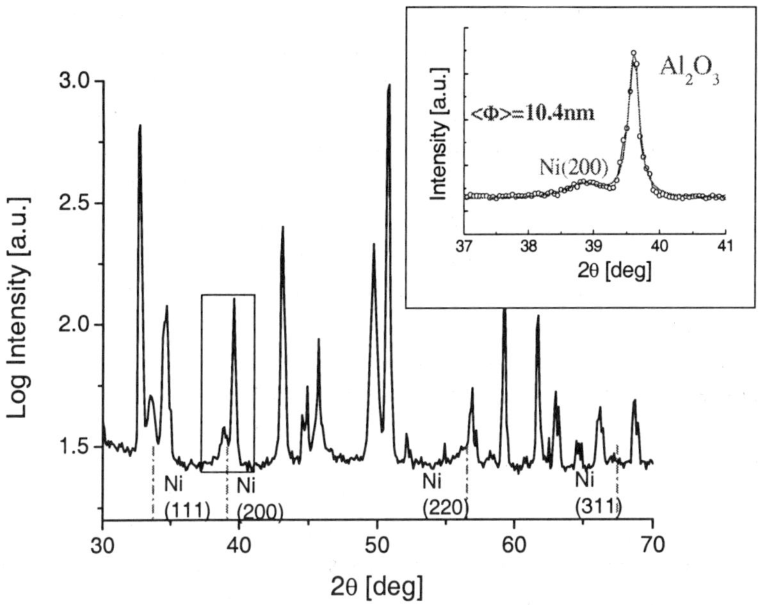

Figure 3. GID measurements on the Ni sample of 5 nm of thickness deposited on Al_2O_3 substrate pretreated at 700°C in 30 sccm of H_2 for 10 min. The inset shows the size of the Ni clusters evaluated by a double-peak Lorentian fitting on the Ni(200) and Al_2O_3(024) peaks.

The diffraction curve was recorded over a wide angular range 2θ between 30 and 70°. A narrower scan in the vicinity of the Ni(200) reflection and the Al_2O_3 (024) peak was performed between 2θ =37 and 41° to get more resolution. The sample exhibits the Bragg peaks due to the cubic phase of Ni [17]. The expected positions of the Bragg peaks at the used wavelength are represented by the vertical grey dashed lines. In the wide range scan most of the peaks is due to the alumina substrate [18] even though their intensity is reduced due to the grazing incidence geometry. The size of the Ni clusters was evaluated by a double-peak Lorentian fitting on the Ni(200) and Al_2O_3(024) peaks of the inset in figure 3 by means of the Scherrer equation. The analysis showed that the mean size is 10.4nm.

The optimal condition for the pretreatment was finally be found to be 700°C in 30 sccm of H_2 for 10 min.

Growth processes were performed on substrates pretreated using the optimal parameters set: gases flows and ratios were changed during growth process.

Below 500°C CNTs growth was not observed, possibly since the temperature was not sufficient for the diffusion of carbon into Ni clusters and thus for the growth; on the other hand at temperature higher than 600°C nickel could diffuse into silicon giving rise to the formation of silicides. As a consequence, less catalyst is available for CNTs growth and carbon reacted with silicon. Then we fixed the substrate temperature at 600°C. The process parameters, varied to achieve a good quality CNTs, were the C_2H_2/H_2 ratio and the total pressure. In fig.4 (a) and fig.4 (b) we report the Raman spectra of carbon nanotubes deposited on Si/SiO_2(70nm)/Ni(5nm) substrate varying the total pressure and the gases ratio respectively.

At constant C_2H_2/H_2 ratio, we expect the presence of two competing phenomena that affect the growth rate and then the CNTs quality: at higher pressure a big amount of material is dissociated near the filament but simultaneosly the mean free path is short so only few atoms can reach the substrate for the growth; at low pressure the situation is reversed. In such cases a maximum condition have to be found. The growth experiments performed at 1.25, 2.5 and 5 mBar, keeping constant the substrate temperature and the carbon concentration, confirmed that too high or too low pressure values affect the purity of CNTs and showed that the maximum condition is reached at 2.5 mBar (see fig.4 (a)).

Keeping constant the total pressure and varying gases ratio, Raman spectra of CNTs grown also in this case show a maximum condition, infact the 10:1 ratio seems to be slightly better than the other (see fig.4 (b)). From Raman analysis we can also deduce that the structures we grew are nanofibers rather than nanotubes.

The optimal growth parameters were found to be: 600°C 10/1 the H_2 /C_2H_2 ratio and 2.5 mBar of the total pressure.

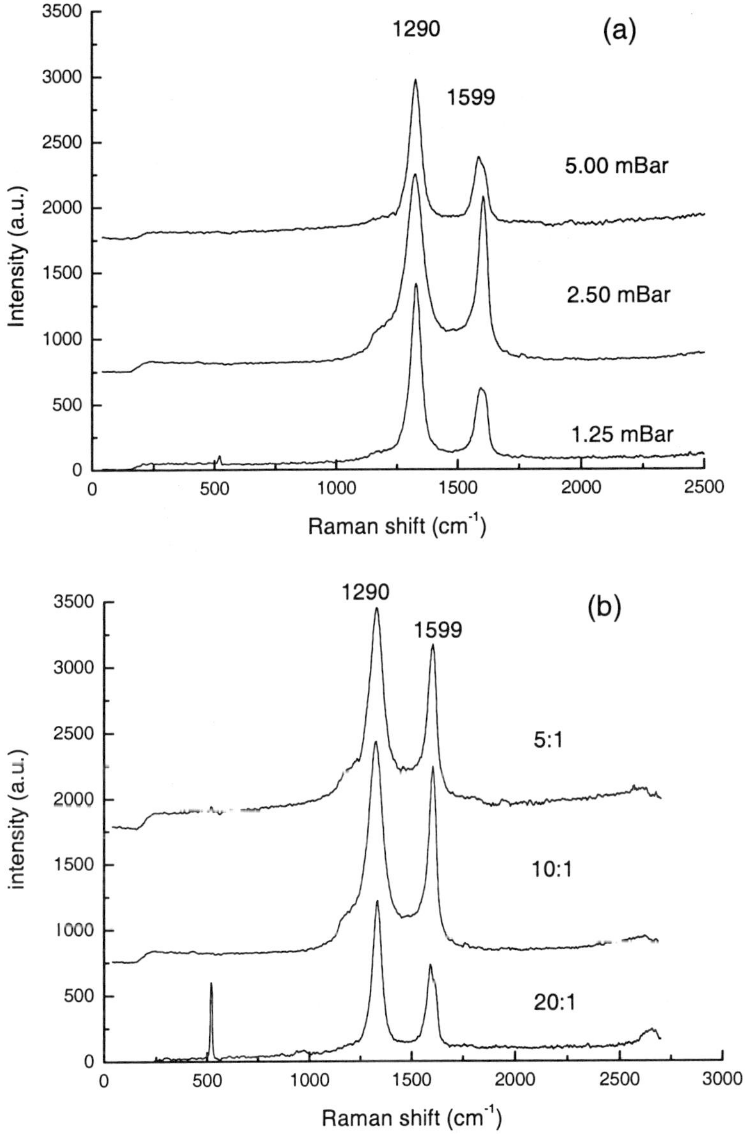

Figure 4. Raman spectra of carbon nanotubes deposited at 600°C on Si/SiO_2(70nm)/Ni(5nm) substrate varying: (a) the total pressure, keeping the C_2H_2/H_2 ratio constant (10:1); (b) and the gases ratio, keeping the total prssure constant (2.5 mBar).

3.2. *Test of the sensors*

The response of device with an initial resistance of 2.8 kW was investigated. The current signal was measured at a constant input voltage of 0.1 V, while different gases with a specified concentration were introduced into the gas flow system at room temperature. The sensor response (see fig.5) S was defined as the ratio of the device conductance after gas exposure to that before gas exposure, G_{gas}/G_0. Nitrogen is used as the carrier gas throughout the work. The total flow rate is set constant at 500 sccm; testing gases are regulated by means of a mass flow controller. The perturbing gases are NO_2, CO_2 diluted with nitrogen and NH_3 diluted with synthetic air. The concentration of gases, during the measurements, were: 5ppm of NO_2, 2% of CO_2, 500ppm of NH_3. During the measurements the sensor was exposed for 10 minutes to each gas; the recovery time was set at 10 minutes in nitrogen. It is found that the electrical conductivity increases when the sensor is exposed to 5 ppm of NO_2. Exposure to NH_3 and synthetic air seems to accelerate the recovery time; the graph also shows a high baseline current drift. We also test the device starting from a step in air and ammonia followed by two step in NO_2 ; this step decreases sensor current baseline while during the steps in CO_2 the device doesn't show any change in response.

For SWNT sensor, it is established that such a long recovery time is due to the higher bonding energy between nanotubes and NO_2 [12,16]. Cho and coworkers have revealed that an NO_2 molecule can bind to a semiconducting SWNT with a binding energy of 0.9 eV (18.4 kcal/mol). This suggests that the nature of the molecule–tube interactions is strong physisorption and approaches the chemisorption regime. The oxidizing NO_2 molecule withdraws about one-tenth of an electron charge from the nanotube. The charge transfer leads to increased hole carriers and enhanced conductivity for the p-type nanotube. The interaction between NH_3 and a SWNT is physisorption in nature. NH_3 is a Lewis base that can donates a small amount of electrons to nanotubes and therefore reduce the hole-carriers, so in turn NO_2 behaves as an acceptor while NH_3 is a donor [12].

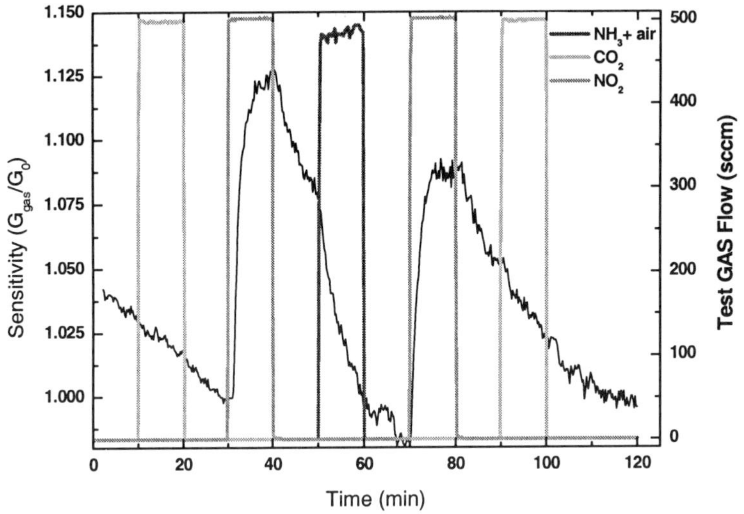

Figure 5. Dynamical response of the device. Data are collected using a DC voltage of 0.1V. All experiments are conducted at room temperature; in particular it is found that the electrical conductivity increases when it is exposed to 5 ppm of NO_2. Exposure to NH_3 and synthetic air seems to accelerate the recovery time , the graph also shows a high baseline current drift. During the steps in CO_2 the device doesn't show any change in response.

4. Conclusions

We have optimized CNT growth in a HWCVD reactor on Si/SiO_2 and Al_2O_3 Ni-coated substrates, finding good pretreatment conditions, and then varying the total pressure, the substrate temperature and the composition of the gases during CNTs synthesis. We verified the quality of the material through scanning electron microscopy and Raman spectroscopy, and for alumina substrate by synchrotron X-ray diffraction. In parallel, with the aim to test sensing properties of the CNTs, a gas sensor, produced by obtained dispersing a CNT solution on gold electrode on glass substrate, was prepared and characterized. The device was exposed to different analytes. The results are extremely encouraging. As the Raman spectrum indicates the material is not optimized yet, we expect that sensing properties of the CNTs could be further improved.

We are really confident on the possibility to realize a sensor in perfect workimg order through CNTs direct growth: interest on this topic is very high since this technique concurs to simplify the flow chart of the sensor fabrication;

infact the CNTs direct synthesis allows to operate an *in situ* purification by burning metallic CNTs with high currents and avoids the step of the dispersion preparation .

References

1. 1.S.J. Tans, A.R.M. Verschueren, and C. Dekker, *Nature* **393**, 49 (1998).
2. R. Martel, T. Schmidt, H.R. Shea, T. Hertel, and Ph. Avouris, *Appl.Phys.Lett.,* **73**, 2447 (1998).
3. A. Bachtold, P. Hadley, T. Nakanishi, and C. Dekker, *Science* **294**, 1317 (2001).
4. P.W. Chiu, G. Gu, G.T. Kim, G. Philipp, S. Roth, S.F. Yang, and S. Yang, *Appl.Phys.Lett.,* **79**, 3845 (2001).
5. P.W. Chiu, G.S. Duesberg, U. Dettlaff-Weglikowska, and S. Roth. *Appl.Phys.Lett.,* **80**, 3811 (2002).
6. Artukovic M. Kaempgen, D. S. Hecht, S. Roth and G. Gruner, *Nano Lett.,* **5** (4), 757 (2005)
7. G.Z. Yue, Q. Qiu, Bo Gao, Y. Cheng, J. Zhang, H. Shimoda, S. Chang, J.P. Lu, and O. Zhou, *Appl.Phys.Lett*., **81**, 355 (2002)
8. Randal J. Grow, Qian Wang, Jien Cao, Dunwei Wang, and Hongjie Dai, *Appl. Phys. Lett.,* **86** (2005)
9. S.Ghosh, A.K.Sood, and N.K umar, *Science,* **299**, 1042 (2003)
10. Sotiropoulou and Chaniotakis, *Anal. Bioanal. Chem.,* **375**, 103 (2003)
11. V.T .S.Wong and W.J.Li, *Proc. IEEE Int. Symp. Circuits Sys.* 4, IV844 (2003)
12. J.Kong, N.R.Franklin, C.Zhou, M.G.Chapline, S.Peng, K.Cho, and H.Dai, *Science,* **287**, 622 (2000).
13. W.Wongwiriyapan, S.Honda, H.Konishi, T.Mizuta, T. Ohmori, T.Ito, T.Maekawa, K.Suzuki, H.Ishikawa, T.Murakami, K.Kisoda, H.Harima, K.Oura and M.Katayama., *Jpn J. Appl. Phys.*, **44**, 8227, (2005)
14. M.J. Madou, S.R. Morrison, *Chemical Sensing with Solid State Devices*, Academic Press, New York (1989)
15. T.Seiyama (Ed.), *Chemical sensors-current state and future outlook, Chemical sensor Technology*, Vols 1/2, Kodnasha and Elsevier, Amsterdam (1998)
16. Jing Li, Yijiang Lu, Qi Ye, Martin Cinke, Jie Han, M. Meyyappan. *Nano Lett*. **3,** 929, (2003)
17. 2001 JCPDS v. 2.2 N. 81-1667
18. 2001 JCPDS v. 2.2 N. 04-0850

OPTICAL SENSORS AND MICROSYSTEMS

METAL-CLADDING LEAKY WAVEGUIDES FOR CHEMICAL AND BIOCHEMICAL SENSING APPLICATIONS

ROMEO BERNINI

I.R.E.A., Consiglio Nazionale delle Ricerche, Via Diocleziano, 328 – 80124, Naples, Italy

M. TONEZZER, G. MAGGIONI, S. CARTURAN , A. QUARANTA, G. DELLA MEA

D.I.M.T.I., University of Trento, via Mesiano 77, 38050 (TN), Italy
I.N.F.N. – Laboratori Nazionali di Legnaro, Viale dell'Università 2, 35020 Legnaro, Italy

F. MOTTOLA, A. MINARDO, L. ZENI

D.I.I. – Seconda Università di Napoli Via Roma, 29, 81031 Aversa, Italy

In this work we present Metal Cladding leaky Waveguides (MCLWs) for chemical and biosensing applications. In particular, we show that MCLW can be used as a basic tool for the realization of liquid, gas and fluorescence sensors. The applications of the device are related to different sensing mechanism, and it is possible to obtain, with a suitable design in according to the application selected, a good sensitivity.

1. Introduction

In the last few decades, there has been increasing interest in (bio-) chemical sensors. In particular, integrated optical sensors play an important role in measurements of different species in gases and liquid. Such sensors offer high sensitivity, fast response, immunity to electromagnetic interference and aggressive environments.

Typically, the basic element of these sensors is a three layers slab waveguide with a high refractive index core. Recently, new optical waveguides have been proposed to improve the sensitivity to various analytes. These devices are similar to conventional slab waveguides, but with an extra metal layer inserted between the guiding layer and the substrate; these devices are conventional named Metal Cladding Leaky Waveguides (MCLWs) [1-3].

In this work, we show that MCLWs, with a suitable design, can be used as a basic tool for the realization of liquid, gas and fluorescence sensors using different sensing mechanism.

In the first case the proposed sensor permits to measure the refractive index of a liquid that acts as cover medium. The sensing mechanism relies on the interaction between the evanescent field of the guided mode, and the analyte that act as cover layer.
In the second case we realize a MCLW sensor for Volatile Organic Compounds (VOCs) detection, in which a free 5,10,15,20 meso-tetraphenylporphyrin (H_2TPP) thin film, forms the guiding layer. In this approach, the sensing mechanism is related with the interaction of the guided mode with the sensing layer, rather than the interaction of the evanescent field. The choice of porphyrin compounds as sensing material is due to their significant solvatochromic effects when they interact with vapours showing distinguishable colorimetric effects [4-7]. Recently, several porphyrin based gas and VOC sensors have been proposed involving optical transducer techniques [8-10].
Finally we show that MCLW can be usefully applied for high efficiency fluorescence detection [11]. In particular we demonstrate that a high directionality of fluorescence emission into glass substrate can be achieved. The first technique used for fluorescence detection is total internal-reflection fluorescence (TIRF) excitation [12]. Recently, several alternative approaches have been proposed, namely supercritical angle fluorescence (SAF) [13], surface plasmon field enhanced fluorescence (SPFS) [14] and surface plasmon-coupled emission (SPCE) [15-16]. The possibility to employ a MCLW as an optical sensor device for fluorescence detection has been proposed the first time in [2], even if the first theoretical study on the emission properties and detection efficiency offered by MCLW has been performed in [11].

2. Theoretical Background

MCLW is a planar four-layered waveguide structure comprising a substrate, a thin metal cladding, a dielectric guiding layer and a cover medium. A scheme of the device is shown in Figure1

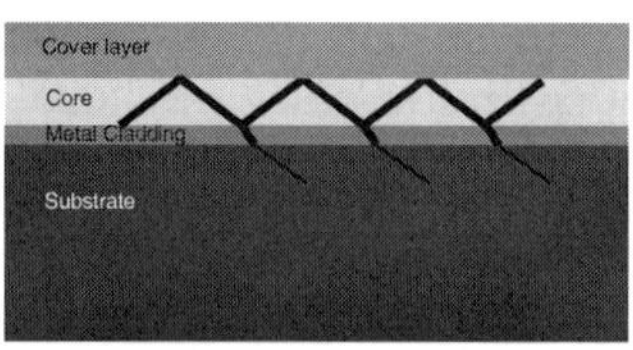

Figure 1. Schematic of a Metal-Cladding Leaky Waveguide.

The structure of the MCLW is similar to the conventional dielectric slab waveguide, but with an extra metal layer introduced between the substrate and the guiding layer.
Light can be guided into the guiding film of the MCLW. However, while guided light experiences total internal reflection at both core region-cover medium and core region-substrate interfaces in traditional slab waveguides, guided light in a

MCLW is only totally internally reflected at the core region-cover medium, while partial reflection occurs at the core region-metal cladding interface. Hence, light is partially transmitted into the metal cladding and partially reflected back into the guiding layer, as it is indicated in Figure 1. As a result, modes propagating into the structure are not truly guided; rather they are referred to as leaky modes [1], and propagation can occur only over a short distance (in the µm-range).

A typical MCLW sensor interrogation scheme consists in illuminating the waveguide from the bottom side by means of a coupling prism, having a proper angle for coupling the light in the core region, similarly to the Kretschmann configuration in surface plasmons resonance experiments [17]. When measuring the reflected intensity as a function of the wavelength, a clear dip appears in the reflected spectrum at the resonant wavelength, i.e. the wavelength corresponding to the coupling of a mode into the guiding layer.

Typically, the sensing mechanism of a MCLW relies upon the interaction between the evanescent field of the guided mode and the cover layer.

By a suitable design, the penetration depth of the evanescent field in the external medium is large, compared with both conventional dielectric waveguides and surface plasmon resonance, resulting in a stronger interaction with the analyte. In this condition is possible to use the device to measure the refractive index of a liquid that acts as cover medium, with a very good sensitivity.

When porphyrin is used as the guiding layer of the MCLW, is possible to use the device as a sensor for Volatile Organic Compounds.

In this case the sensing mechanism is related to changes in the optical properties of the guiding layer, due to the interaction with the VOC. In fact, when exposed to organic vapour, porphyrin films are able to absorb these compounds modifying their optical properties. In particular the porphyrin refractive index, both real and imaginary part, changes owing to VOCs exposure in a measure which depends on the vapour concentration. This change leads to a variation of the effective refractive index of the mode propagating within the MCLW, with a consequent change in the resonant wavelength. Hence, measuring the shift of the dip in the reflected spectrum one can measure the concentration of the vapour under investigation.

In the case of fluorescence detection, the configuration under study is depicted in Figure 2 and consists of a molecule placed in water above the MCLW device.

The theoretical treatment is based on a semi-classical approach, which considers a fluorescing molecule as an ideal dipole emitter. The optical detection properties of interest are calculated by using the emission of dipole emitters in front of a planar surface. The quantities of particular interest are: the angular distribution of radiation (ADR) in the water half space S_w, the ADR in the glass half space S_g, and the total power S_{total} emitted by the molecule. These quantities have been calculated in [11], so here we do not report them for brevity, but we will focus our attention, in the following paragraphs, on the results obtained collecting fluorescence by a MCLW.

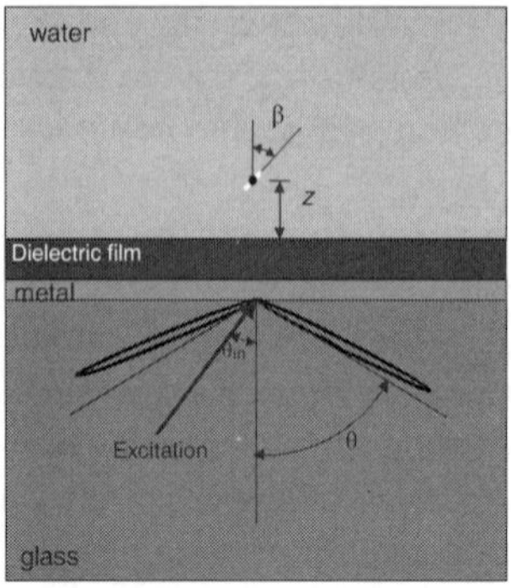

Figure 2. Setup of MCLW-coupled emission: An aqueous solution of fluorescing molecules is placed in top of a MCLW device. Fluorescence excitation is performed from the glass side by a plane wave with incident angle assuring maximum transmittivity. A single molecule is depicted as a dipole emitter with a distance z from the dielectric surface and forming an angle β with the vertical (optical) axis. The angular distribution of radiation into glass is depicted as a bold curve and is a function of angle θ.

3. Evanescent Wave Sensor

For measuring the refractive index of a liquid that acts as cover medium, is useful to design the MCLW to maximize the penetration depth of the evanescent field into the external medium.

The sensor realized for our experiments consists of a glass substrate, whose refractive index is $n_1=1.51$, on which a gold thin film ($n_2=0.12+3.29i$, d2=35nm), acting as a metal-cladding, is sputtered. The guiding layer is deposited by means of the spin coating technique, and consists of a PMMA film ($n_3=1.49$, $d_3=300$nm). All refractive indices are evaluated at a wavelength of 633nm.

In Figure 3, the scheme of the experimental setup is shown. Reflection spectrum measurements were carried out by coupling the light at an angle of 64.5° from the substrate, by means of a coupling prism. A white lamp, followed by a collimator and a polarizer, was used as the source, whereas a spectrophotometer was employed for the acquisition of the reflected spectrum.

In Figure 4 the reflected spectra of the MCLW are shown for three different external refractive indexes ($n_0=1.333$; 1.337; 1.341).

As the refractive index of the external medium changes, so does the wavelength at which light is coupled in the waveguide. Hence, the position of the dip of the spectrum is a measurement of the external medium refractive index, within an opportune range. The dip in reflectivity is sharper, compared with surface plasmon resonance. Because the full width half maximum (FWHM) of the dip is small compared with that of the surface-plasmon resonance (SPR), they offer a high potential for sensitive detection of refractive-index changes in the film [3].

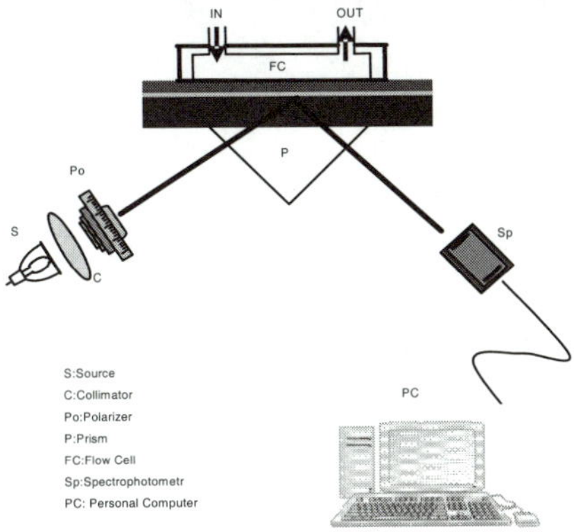

Figure 3. Experimental setup.

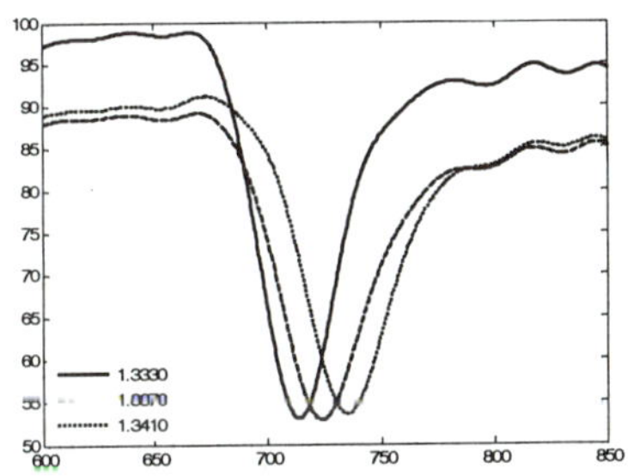

Figure 4. Reflected spectra for the MCLW at three different values of the external refractive indexes (n_0=1.333; 1.337; 1.341).

In Figure 5, the refractive index of the sensed medium n_0, as a function of the resonant wavelength (i.e. the dip position of the reflected spectrum) λ_R is shown for both theoretical and experimental data. As it can be seen, there is a good agreement. The sensitivity $S_n = \partial \lambda_R / \partial n_0$ calculated for a refractive index of the external medium n_0=1.3330 is higher than 2000 nm.

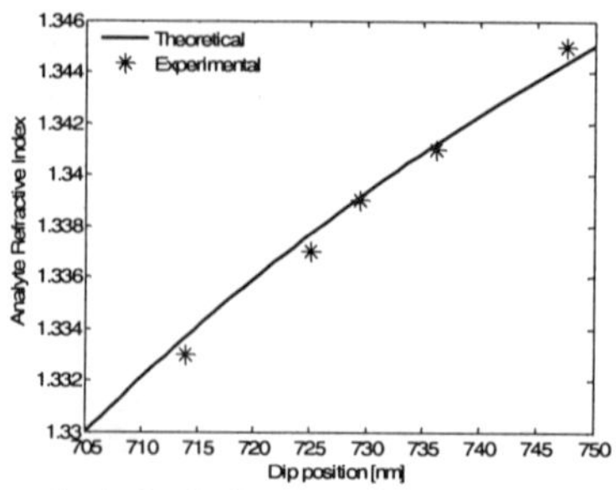

Figure 5. Analyte refractive index in function of the resonant wavelength of the MCLW. Comparison between experimental and theoretical data

4. Guided Wave Sensor

As mentioned above, in the case of VOC sensing, a H_2TPP porphyrin thin film is used as the guiding layer of the MCLW device. The porphyrin choice deals with its property to be easily deposited as thin film and used as sensing material exploiting the change of some physical properties upon analyte exposure. The sensing mechanism of the porphyrin based MCLW for VOC sensing relies upon the interaction between the guided mode of the device and external medium. In fact, when exposed to organic vapour, porphyrin films are able to absorb these compounds modifying their optical properties. In particular the porphyrin refractive index, both real and imaginary part, changes owing to VOCs exposure in a measure which depends on the vapour concentration. This change leads to a variation of the effective refractive index of the mode propagating within the MCLW, with a consequent change in the resonant wavelength. Hence, measuring the shift of the dip in the reflected spectrum one can measure the concentration of the vapour under investigation.

The sensor consists of a glass substrate, whose refractive index is n_l=1.51, on which a gold thin film (n_2=0.12+3.29i, d_2=36nm), acting as metal cladding, is sputtered. The guiding layer consists of a H_2TPP porphyrin thin film, deposited by spinning a solution of 0.1% wt in chloroform ($CHCl_3$). All refractive indices are evaluated at the wavelength of 633nm.

The scheme of the experimental setup is the same used for refractive index measurements, shown in Figure 3.

The normalized reflected spectrum of the porphyrin-based MCLW, measured in air, is shown in Figure 6.

The figure clearly shows that the typical absorbance dips of the porphyrin, one in the Soret band and four in the Q-band, are present in the reflected spectrum of the porphyrin-based MCLW. Note that, as the reflection spectrum is plotted rather than the absorption one, absorbance peaks are presented as dips.

The dip appearing at the right side of the absorbance dips corresponds to coupling of light to the TE mode of the MCLW.

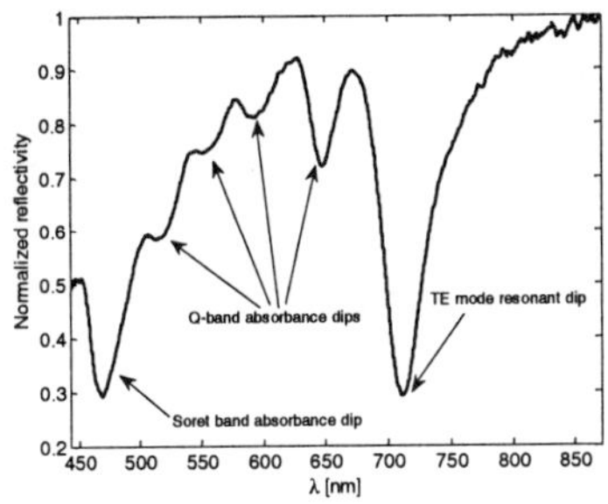

Figure 6. Normalized reflected intensity of the porphyrin based MCLW.

Measurements were carried out by observing the reflected intensity at a fixed wavelength. The wavelength chosen is 695nm and corresponds to the centre of the left edge of the dip associated to the mode coupled into the porphyrin film. The sensor was tested with ethanol vapours at different concentrations in air. The optical response of the sensor was calculated as the normalised reflectivity variation defined as $(I - I_0)/I_0$, where I and I_0 are the reflectivity values during exposure to EtOH vapours and pure air, respectively. Measurements were carried out on two devices, differing each others for the spinning velocity of the porphyrin solution: first device was realized with a spinning velocity of v_{spin}=1200rpm for 60 seconds (device 1); second device was realized with a v_{spin}=500rpm for 60 seconds (device 2).

In both cases the coupling angle has been chosen in order to ensure that the resonant dip falls out of the absorbance dips of the porphyrin.

The sensor response to various concentrations ethanol shows an increase in the reflected intensity for the observed wavelength, that corresponds to a red shift of the resonant wavelength of the MCLW TE mode. The shift can be theoretically explained by an increase of the refractive index of the sensitive layer due to the vapour molecules absorbed in the porphyrin film. The calibration curves for both devices are shown in Fig. 7A.

The increased sensitivity of the sensor having a thicker porphyrin guiding layer is clear. Such an occurrence is in agreement with theory. In fact according Tiefenthaler et al. [20], the refractive index sensitivity of an absorbing waveguide is proportional to the ratio P_f/P, where P is the total power of the guided mode and P_f is the power of the guided mode transported within the guiding film.

When the thickness of the guiding layer increase also P_f increases so the sensitivity increases. For thick waveguides ($N \rightarrow n_f$), all the power of the guided mode P is confined in the core region ($P_f \rightarrow P$) and $\partial n_f / \partial N \rightarrow 1$.

The dynamic response of the devices is fast. In particular the parameter t_{90}, defined as the time taken during the response for the signal intensity to reach the

90% of its final saturated value, is about 30 seconds for both samples. It is moreover worth to note that both the devices are characterized also by very fast recoveries showing parameters t_{10}, defined as the time taken during the recovery phase for the signal intensity to reach the 10% of the saturated value, of about 30 seconds.

Measurements were repeated in the case of the device 2 for a higher wavelength (730nm) in order to investigate the behaviour of the sensor to different wavelength (see Figure 7A). Obviously the coupling angle was adjusted to ensure that this wavelength corresponds to the centre of the left edge of the dip associated to the mode coupled into the porphyrin film.

Results are very similar to those presented above, performed at the wavelength of 695 nm, especially for the parameters t_{90} and t_{10}, but surprisingly the last measurements seems to be more sensitive of that performed at the wavelength of 695 nm. This result can't be explained by the analysis proposed by [20], and suggest that there is a different behaviour of the porphyrin film at different wavelengths: further studies concerning the particular behaviour of the proposed sensors will be the object of future investigations.

In order to investigate prospective selectivity, the device 2 is tested also with different organic vapours. In particular the device is tested with acetone, benzene and methanol vapours. The device shows a good response for both acetone and benzene, whereas the methanol response is very small, so showing a different behaviour to different analytes.

Figure 7B shows the response to various analytes of the device 2.

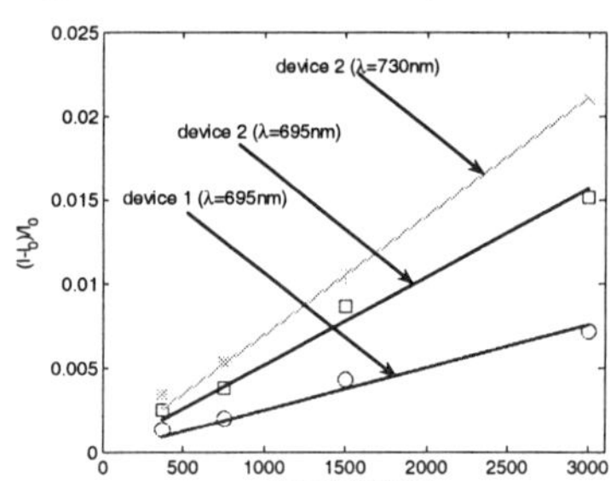

A

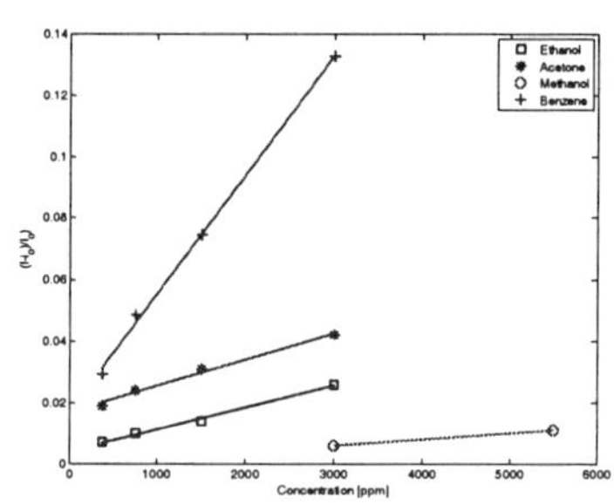

B

Figure 7. (A) Calibration curve for porphyrin based MCLW sensors exposed to ethanol vapours. (B) Calibration curve for porphyrin based MCLW sensor exposed to various analytes

5. Fluorescent Bbased Sensor

For fluorescence based MCLW sensor, numerical calculations were performed for fluorophores that are excited at 532 nm and have their emission maximum at 570 nm. The refractive index of glass substrate was n_g=1.51, whereas an aqueous half space (n_w = 1.33) was considered as upper medium. Silver was chosen as a metal, because of its capability to give narrower angular resonances for the metal cladding waveguide modes. The refractive index adopted for the silver was n_m=0.1195+j3.5278 at 570 nm and n_m=0.1294+j3.2198 at 532 nm. Finally, a refractive index of the dielectric film n_f=1.49 was chosen.

First, we studied the possibility to achieve high directionality of fluorescence emission into glass, through coupling with MCLW modes. To this aim, we performed an optimization on silver film thickness and dielectric film thickness, by considering the maximum of $S_{w,g}(\theta,\phi,z=0,\beta)/S_{total}(z=0,\beta)$ while varying θ and ϕ , as the optimization parameter [11].

Results of simulations carried out for both vertical (β = 0) and horizontal ($\beta=\pi/2$) dipole orientation. Results show that the maximum photon emission probability density increases rapidly when the thickness of the dielectric film raises above ~320 nm, whereas the optimal silver thickness is ~ 46 nm, in the first case, and when the thickness of the dielectric film is ~187 nm and the optima silver thickness is ~40 nm, in the second case.

The dependence of fluorescence emission on dielectric film thickness can be understood by studying the modal properties of the MCLW. As a vertically oriented dipole radiates a p-polarized electric field [18], fluorescence emission can couple in this case only with the waveguide TM mode. When the dielectric film thickness is above the cut-off of the fundamental TM mode (~317 nm), fluorescence emission couples with the leaky waveguide mode, then being re-radiated over a very narrow angular distribution into the glass, the radiated energy being concentrated around the angle corresponding to the MCLW leaky mode de-coupling angle. As horizontally oriented, instead, dipole radiates a s-polarized electric field [18], so fluorescence emission can couple only with the waveguide TE mode, whose cut-off condition is ~187 nm.

In Figure 8 the angular distribution of emission for a molecule is shown, for both horizontally and vertically orientation, is shown.

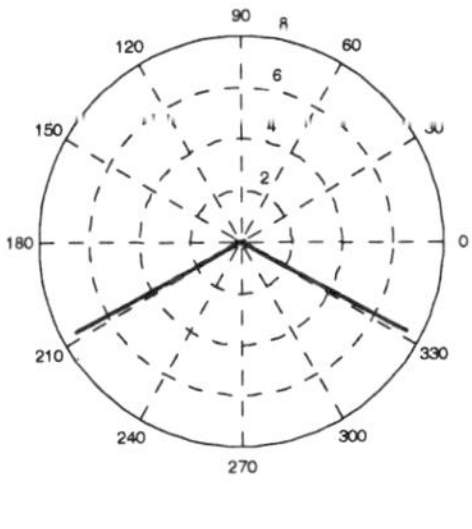

A

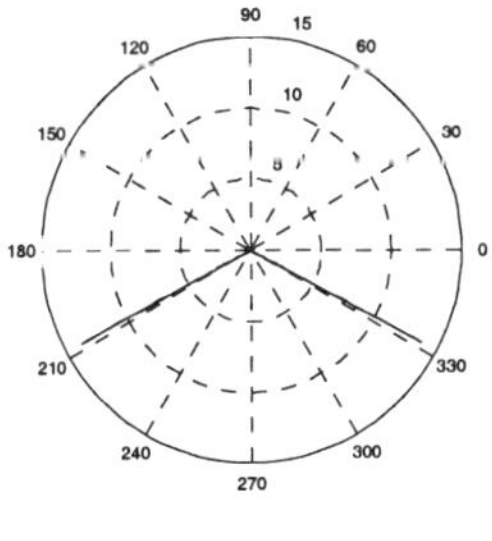

B

Figure. 8 (A) Angular distribution of emission for a vertically oriented molecule directly on the dielectric film surface, calculated for a silver film thickness dm = 46 nm and a dielectric film thickness df = 340 nm. (B) Angular distribution of emission for a horizontally oriented molecule directly on the dielectric film surface, calculated for a silver film thickness dm = 40 nm and a dielectric film thickness df = 200 nm.

The fluorescence brightness was calculated, by considering both fluorescence excitation and collection from the glass side.
We note that illuminating from the glass side may result in an excitation enhancement which can be much higher than the 4-fold improvement reported in [20] for excitation from above. We chose to set the dielectric film thicknesses to the values providing maximum fluorescence collection efficiency, and supposed to employ p-polarized illumination in the case of TM-coupled emission ($d_f = 340$ nm, $d_m = 46$ nm), while using s-polarized illumination in the case of TE-coupled emission ($d_f = 200$ nm, $d_m = 40$ nm). In both cases, we assumed plane wave illumination from the glass side, at the incident angle providing maximum excitation. Random orientation of the fluorescing molecules was also assumed, so that fluorescence collection efficiency was calculated as a weighted sum of efficiency for vertical and parallel dipole orientation (one-third vertical orientation plus two-thirds parallel orientation). Fluorescence brightness, calculated as a function of dipole's distance from MCLW surface, is shown in Figures 9A and 9B. For comparison, we also show in the same figures the results achieved for a pure glass/water interface, both in the case of collection over all the glass half space, and in the case of collection above the supercritical angle (SAF detection). SPCE results are reported only for p-polarized excitation. Both figures indicate that MCLW-coupled emission allows for a stronger fluorescence signal to be detected through the glass, when compared to both SPCE and pure glass/water interface, for dipoles close to the MCLW surface. Remarkably, a 4-fold improvement is observed for the MCLW configuration, with respect to the glass/water configuration, when exciting with an s-polarization and collecting in both cases over the whole glass half space, while an 8-fold improvement is observed with respect to SAF detection.
From Figures 9A and 9B, note also the increased range of dipole's distances for which a sufficient fluorescent signal can be detected in the MCLW configuration, with respect to SPCE and SAF detection. This increased range, which is related, for duality, to the deeper evanescent field of MCLWs exploited in evanescent field sensing, can be an useful property when fluorescence from large emitters has to be detected, such as bacteria, or proteins [1].

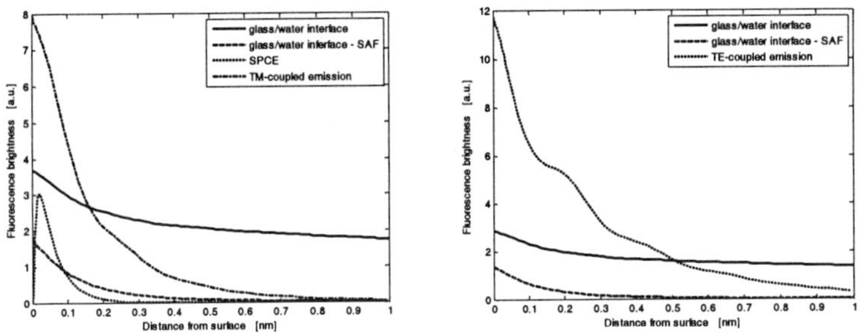

Figure 9. (A) Maximum detectable fluorescence intensity for s-polarized plane wave excitation from the glass side at the incident angle giving maximum electric field intensity at z = 0, when collecting over the whole glass half space, as a function of the molecule's distance from the surface.

Numerical analyses have been carried out so far, by supposing that fluorescence were collected through the glass. Let us now suppose that fluorescence light detection is carried out through a microscope objective immersed into water looking from above. We repeated the calculation of fluorescence brightness, as a function of the silver thickness and the dielectric film thickness, in both cases of p-polarized excitation and s-polarized excitation from the glass side. For each calculation, we still suppose an angle of incidence of the exciting field resulting in the maximum electric field intensity at z = 0. Results show that the fluorescence brightness is more than 70 times the brightness observed for a pure glass/water interface. A so dramatic improvement in signal strength, with respect to fluorescence measurement carried out on a standard glass slide, mostly comes from the strong enhancement of the local exciting field when illuminating from the glass side. However, also collection efficiency is enhanced by the presence of the multilayer. Actually, by using the above determined optimal thicknesses, it turns out that the major part of dipole's emission is directed into the water half space. On the other hand, it must be pointed out that in the case of collection through a microscope objective looking from above, fluorescence emission does not exhibit the property of strong directionality as emission into glass. Simulation performed, not shown here for brevity, prove that radiation emitted in the water half space is spread over a wider angular region. Hence, detecting the fluorescence from above permits on one hand to increase collection efficiency, as a higher fraction of energy can be directed into the water half space, rather than into the glass half space. On the other hand, the strong ADR directional properties characterizing emission into the glass half space are lost in this case. Consequently, this detection configuration reduces the utility of the method, because it may greatly reduce any spatial information imparted via evanescent excitation through the inner filtering mechanism previously shown.

6. Conclusion

In this work we have shown that Metal Cladding Leaky Waveguides are useful tool for sensing applications. In particular, we have shown that MCLW, with a suitable design, can be successfully applied as evanescent field sensor, guided wave sensor, and fluorescence based senor.

References

1. M. Zourob, S. Mohr, B. J. Treves Brown, P. R. Fielden, M. McDonnell, N. J. Goddard, *Sens. Act. B,* **90,** 296 (2003).
2. M. Zourob, S. Mohr, P. R. Fielden, N. J. Goddard, *Sens. Act. B*, **94**, 304 (2003).
3. N. Skivesen, R. Horvath, H.C. Pedersen, *Sens. Act.* B, **106**, 668 (2005).
4. N.A. Rakow and K.S. Suslick, *Nature*, **406**, 710 (2000).
5. M. Tonezzer, A. Quaranta, G. Maggioni, S. Carturan, G. Della Mea, *Sens. Act. B*, **122**, 620 (2007).
6. M. Tonezzer, G. Maggioni, A. Quaranta, S. Carturan, G. Della Mea, Sens.Act. B, **122**, 613 (2007).
7. A. D'Amico, C. Di Natale, R. Paolesse, A. Macagnano, A. Mantini, *Sens. Act. B*, **65**, 209 (2000).
8. T.H. Richardson, C. M. Dooling, O. Worsfold, L. T. Jones, K. Kato, K. Shinbo, F. Kaneko, R. Tregonning, M. O. Vysotsky, C. A. Hunter, *Coll. Surf. A*, **198-200**, 843 (2002).
9. C. Di Natale, D. Salimbeni, R. Paolesse, A. Macagnano, A. D'Amico, *Sens. Act. B*, **65**, 220 (2000).
10. J. Spadavecchia, R. Rella, P. Siciliano, M. G. Manera, A. Alimelli, R. Paolesse, C. Di Natale, A. D'Amico, *Sens. Act. B*, **115**, 12 (2006).
11. A. Minardo, R. Bernini, F. Mottola, L. Zeni, *Opt. Exp.*, **14-8**, 3512 (2006).
12. Hirschfeld, *Can. Spectr.*, **10**, 128 (1965).
13. T. Ruckstuhl, M. Rankl, and S. Seeger, *Biosens. Bioelectron.*, **18**, 1193 (2003).
14. G. Stengel, W. Knoll, *Nucleic Acids Res.*, **33**,e69 (2005).
15. J. R. Lakowicz, *Anal. Biochem.*, **324**, 170 (2004).
16. I. Gryczynski, J. Malicka, Z. Gryczynski, and J. R. Lakowicz, *Anal. Biochem.* **324**, 170 (2004).
17. J. Homola, I. Koudela, S. S. Yee, *Sens. Act. B*, **54**, 16 (1999).
18. K. Tiefenthaler and W. Lukosz, *J. Opt. Soc. Am. B*, **6**, 209 (1989).
19. J. Enderlein, *Chem. Phys. Lett.* **308**, 263 (1999).
20. H. Choumane, N. Ha, C. Nelep, A. Chardon, G. O. Reymond, C. Goutel, G. Cerovic, F. Vallet, C. Weisbuch,, *App. Phys. Lett.*, **87**, 031102 (2005).

STRUCTURED FIBER BRAGG GRATINGS SENSORS: PERSPECTIVES AND CHALLENGES

D. PALADINO, M. PISCO, A. CUTOLO, A. CUSANO

Optoelectronic Division-Engineering Department, University of Sannio, Corso Garibaldi 107, 82100 Benevento, Italy

A. IADICICCO, S. CAMPOPIANO

Department for Technologies, University of Naples Parthenope, Via Medina 40, 80131 Napoli, Italy

M. GIORDANO

Institute of Composite Biomedical Materials, National Research Council, Piazzale Tecchio 80, 80124 Napoli, Italy

Micro-structured fiber Bragg gratings (MSFBGs) – FBGs in which the cladding layer is locally stripped – exhibit attractive spectral features, exalted by their sensitivity to changes in the optical properties of the external medium. In this work, we propose a novel MSFBGs fabrication technique, based on polymeric coatings and UV laser micromachining, as technological assessment to develop tailorable MSFBGs for specific applications. Finally, the technological improvement in the fabrication process enables the development of MSFBG based photonic devices with multifunction properties.

1. Introduction

In the last years, relatively large efforts in the research field were aimed to produce new micro-optical devices as high performance transducers to be integrated with proper materials as sensitive layers and/or micro-fluidics to develop self-contained micro-analysis systems [1-2]. On this line of argument, the authors focused the attention of their research activities on new optical transducers, capable to offer the bases for multifunction integrated optoelectronic systems. The first successful attempt, in this contest, involved fiber Bragg gratings (FBGs), uniformly thinned to monitor the external refractive index [3]. Successively, we found in more complex configurations involving micro-structured FBGs (MSFBGs), a reliable and high performance technological platform to develop advanced photonic devices, well suited for microTAS applications [4-7]. A MSFBG relies on a standard FBG where the selective stripping of the cladding layer in a well-defined region in the middle of the grating acts as a defect along the periodic structure. Its effect on the grating spectral response is the formation of a defect state inside the original FBG stop-band, depending on the geometrical and physical parameters of the etched region [6], leading to appealing features for new active photonic devices. First prototypes of MSFBG were realized by wet chemical etching techniques in hydrofluoric acid (HF) solution combined with a proper holder, involving a Teflon support and epoxy resin. Even if this approach results very inexpensive, it exhibits low control with regards the perturbation geometrical features, especially in terms of longitudinal length. This prevents the realization of MSFBGs with desired geometry and thus spectral features, whereas their potentialities are theoretically and numerically demonstrated [6]. This work deals with new aspects concerning MSFBGs, from a novel fabrication procedure to the analysis of advanced configurations involving multi-defect structures, chirped gratings and the use of high refractive index (HRI) nano-sized polymeric overlays.

2. Novel MSFBG Fabrication Technique

In this section, a novel procedure capable to optimize the MSFBGs fabrication technique, enabling a fine control of their geometric features, is presented. The proposed fabrication approach is based on a new masking procedure involving a polymeric coating layer micro-structured by UV excimer laser micromachining and a successive wet chemical etching based on aqueous HF solution. The adopted laser micromachining system is completely computer assisted and it mainly consists in an argon fluoride (ArF, λ=193 nm) excimer laser equipped with a custom rotating stage able to host and manipulate the optical fiber during the firing process. Finally, a motorized rectangular aperture (MRA) mask, characterized by horizontal and vertical dimensions separately selectable, allows a proper shaping of the laser beam on the target. The first MSFBG prototype was fabricated with a standard 6.0 mm long FBG with resonant wavelength of 1559.10 nm. In light of its good HF resistance, polyamide coating with thickness of 5-8 μm was selected as protective layer.

In this experiment, the MRA mask was selected to form a laser spot size of $180\times250\ \mu m^2$ at the focal point. This mask dimensions, combined with the laser energy value, provides a laser fluence of approximately 300 mJ/cm^2, leading to micro-structuring of the polyamide layer without approach the ablation threshold of silica fibers. Fig. 1 shows a stereo-microscopy image of the coated FBG with the coating layer micro-machined for a length L_{POL}=180 μm +/- 5 μm. Here, the edges of the radiated region are not well defined, probably due to thermal effects during the polymer ablation combined with non perfect shape of the MRA mask. The cladding layer in correspondence of the radiated and thus uncoated part of the grating was removed by wet chemical etching in aqueous HF solution at 24% [3-5]. The double role of the polymeric coating, as protection against external stress and precise mask for the etching process, allowed an easy implementation of the chemical etching procedure: only the adoption of a V-groove channel to sustain the fiber is necessary.

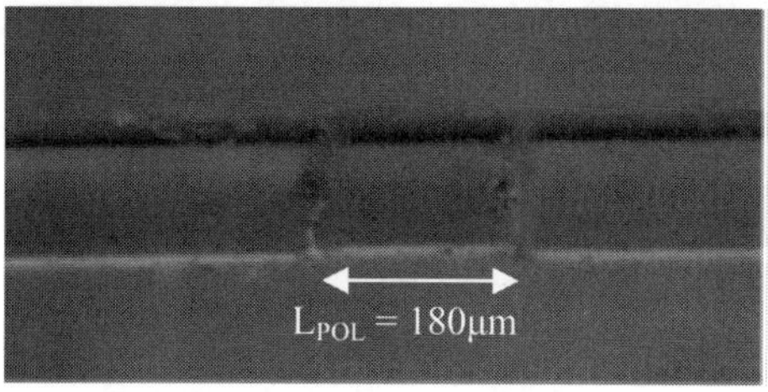

Figure 1 L_{POL}=180 μm structured polyamide layer

Fig. 2.a shows the reflected spectra of the MSFBG during the etching process. After an etching time of approximately 167 minutes, the formation of the defect state within the pristine grating bandwidth is evident. As the etching process goes on, the defect state shifts towards lower wavelengths. After 172 minutes, the acid solution was removed and the holder was washed with pure water and successively filled with calcium oxide (CaO) to neutralize the acid solution. The basic solution quickly slows down the etching process, even if the chemical process completely stops after 34 minutes. In order to analyze the geometric structure obtained in correspondence of the etching region, the MSFBG was characterized by optical microscopy analysis, as shown in Fig. 2.b. As evident, the real profile is different from the ideal one, where the perturbation edges are well defined and sharp. The isotropic nature of the wet etching gets the undercutting of the silica fiber under the polyamide layer, forming tapered regions extending for few hundred of microns. In addition, the uniform thinned region (diameter changes less than 1μm) was found to be approximately L_{Th}=136μm +/- 5μm, and thus it is significantly shorter than the grating region without the polymeric coating (L_{POL}). This effect could be attributed to the dependence of the final profile on the ratio between the mask length and the etching depth. However, this mismatch (between real and ideal profile) can be easily taken into account at the design level [6].

2.1. *Dependence on the SRI*

The main advantage of MSFBGs relies on the possibility to tune the defect state wavelength within the reflected bandwidth of the grating by acting on the surrounding refractive index (SRI) [4-6]. The spectral dependence of the defect state wavelength on the SRI is clearly evident in Fig. 3. As matter of fact, the defect state shifts towards higher wavelengths of 190 pm as the SRI changes from 1.33 to 1.38 or from 1.38 to 1.41. As well known, in fact, the SRI sensitivity increases as well as the SRI [3-6].

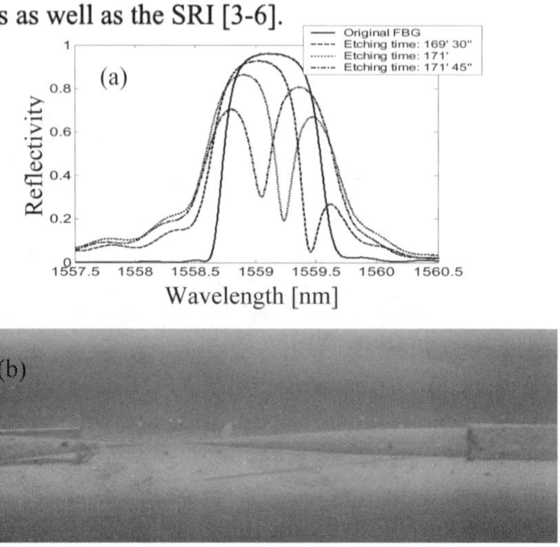

Figure 2 (a) Experimental reflected spectra of the MSFBG during the chemical process; (b) Optical photogram of the etched region forming the MSFBG

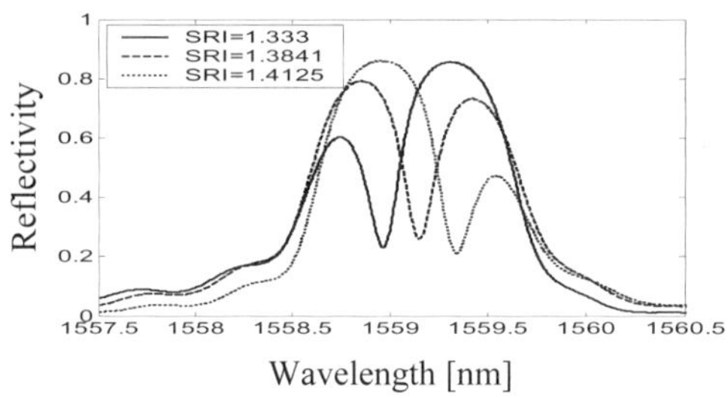

Figure 3 Dependence of the MSFBG spectrum on the SRI.

3. Advanced Structures

The advantages of the laser micromachining based fabrication technique in terms of precise control of the geometrical features of the etching region, can be exploited to develop advanced devices such as multi-defect MSFBG [7]. More perturbations along a periodic structure enable the formation of more defect states inside the band-gap, ruled by multiple interactions of the spectra reflected by the unperturbed grating regions modulated by the phase shifts due to the perturbations (thinned regions). This leads to the possibility to achieve multipoint SRI measurements or multi-channel tunable filter by a single grating element. Also the possibility to develop micro-structured chirped fiber Bragg gratings (CFBGs) with one or more defects is very attractive for sensing and communications applications. The almost periodical structure of CFBGs allows the creation of defect states within the photonic band-gap not dependent on each other, differently from multi-defect MSFBGs. The effect of the local thinnings, if properly exploited, consists in the formation of one or more pass-bands within the pristine grating bandwidth and, consequently, in one or more narrow stop-bands out of the pristine grating bandwidth. The key feature of the proposed structure relies on the strong and exclusive dependence of these defects on the thinned region fabrication features and on the local SRI. Finally, based on previous works regarding HRI coated long-period fiber gratings [1], MSFBGs performances can be enhanced by depositing nano-scale HRI overlays with proper geometrical and physical features along the grating thinned region [7].

4. Conclusions

In this work, we demonstrated the effectiveness of laser assisted fabrication of photonic devices based on MSFBGs. The proposed method involves polymeric coating, UV laser micromachining for polymer stripping and wet chemical etching for cladding removal. A fine control of the structure geometry is achieved. In addition, the technological assessment of the fabrication process enables the development of photonic devices with extended functionalities.

References

1. A. Cusano, A. Iadicicco, P. Pilla, L. Contessa, S. Campopiano, A. Cutolo, M. Giordano, G. Guerra, Journal of Lightwave Technology, Vol. 24, No. 4, pp. 1776-1786, (2006).
2. P. Domachuk, I. C. M. Littler, M. Cronin-Golomb, B. J. Eggleton, Applied Physics Letters 88, 093513, (2006).
3. A. Iadicicco, A. Cusano, A. Cutolo, R. Bernini, M. Giordano, IEEE Photonics Technology Letters, Vol. 16, No. 4, pp. 1149-1151, (2004).
4. A. Iadicicco, S. Campopiano, A. Cutolo, M. Giordano, A. Cusano, IEE Electronic Letters, Vol. 41, No. 8, (2005).
5. A. Iadicicco, S. Campopiano, A. Cutolo, M. Giordano, A. Cusano, IEEE Photonics Technology Letters, Vol. 17, No. 6, pp. 1250-1252, (2005).
6. A. Cusano, A. Iadicicco, D. Paladino, S. Campopiano, A. Cutolo, M. Giordano, "Micro-structured fiber Bragg gratings. Part I: Spectral characteristics", Journal of Optical Fiber Technology (in press).
7. A. Cusano, A. Iadicicco, D. Paladino, S. Campopiano, A. Cutolo, M. Giordano, "Micro-structured fiber Bragg gratings. Part II: Towards advanced photonic devices", Journal of Optical Fiber Technology (in press).

CLASS: AN INNOVATIVE LASER FLOW CYTOMETER FOR THE SIMULTANEOUS MEASUREMENT OF SIZE, REFRACTIVE INDEX, DEPOLARIZATION AND FLUORESCENCE OF CELLS*

LUCA FIORANI, ANTONIO PALUCCI, VALERIA SPIZZICHINO

ENEA, FIM-FISLAS

Via Enrico Fermi 45, 00044 Frascati RM, Italy

A new laser scanning flow cytometer system (CLASS) has been here employed for the characterization of real marine one-celled organisms. Our system has been able to measure simultaneously size, refractive index, depolarization and fluorescence of phytoplankton cells, giving both morphological and biological information. Hence, CLASS has proved to be a valuable non-destructive tool for the *in-vivo* recognition and classification of different microorganisms.

1. Introduction

A new laser scanning flow cytometer system (CLASS) [1] has been developed in the frame of the project MIAO (microsensor systems for extreme and hostile applications) [2]. It has been designed to characterize marine phytoplankton morphology and composition from the simultaneous observation of Mie scattering, light depolarization and fluorescence. Size and refractive index are retrieved by a solution of the inverse light-scattering problem.

Classic flow cytometers (FC) are usually equipped with two or more channels that detect fluorescence and light scattered at determined angles (typically side and forward scattering) [3]. The main difference of a scanning FC (SFC) is the capability of performing light scattering measurements in a very wide-angle range (from 5° to 100° typically). This feature leads to a better morphological characterization of the particles under study. In the following sections some tests performed by CLASS on aqueous samples of phytoplankton cells will be presented. The results demonstrate that the system is able to retrieve size and refractive index with a good accuracy and that the depolarization and fluorescence measurements allow the classification of particles otherwise indistinguishable.

---

* This work has been supported by MIUR FIRB MIAO project.

2. Experimental Set-up

The experimental apparatus of CLASS consists of three main subsystems: hydrodynamics, optics and electronics. Hydrodynamics allows us to produce the hydrofocusing of the sample fluid and to regulate its speed and diameter. Such subsystem is quite conventional except that it embeds two closed loop electronic pressure regulators, the first one for the sheath fluid and the second one for the sample fluid. The optical layout is shown in Figure 1. Its heart is the cuvette where particles flow in an on axis capillary (diameter: 250 μm) from the bottom. The beam delivered by the diode laser emitting at 405 nm is focused by the lens L_1 in the capillary where it is scattered by the particle after going through the mirror with hole. The beam diameter in the interaction region is nearly constant (FWHM: 25 μm). The forward scattered light is reflected by the spherical mirror constituting the bottom wall of the cuvette. The light coming out from the cuvette is reflected by the mirror with hole M, filtered spatially by the iris I_1, filtered spectrally by an interference filter and detected by the photomultiplier PM_1 measuring the scattering as a function of angle (indicatrix). The scattering angle is related to the particle position along the capillary axis, and thus to the detection time. A reference particle position is fixed by the side scattered light focused by the lens L_2, spatially filtered by the pinhole I_2 and detected by the photomultiplier PM_2 measuring the trigger. Recently, depolarization and fluorescence channels have been added. The first one, introduced by splitting of the indicatrix line, measures the indicatrix after a polarizer. In this way the depolarization induced by non-spherical particles can be inferred by the comparison between indicatrix before and after the polarizer. The second one detects the fluorescence emitted by the particle at a wavelength selected by a suitable interference filters. The main parts of electronics are four preamplifiers (one after each photomultiplier) and a four-channel analog-to-digital converter with 14 bit resolution and 2 MS/s sampling.

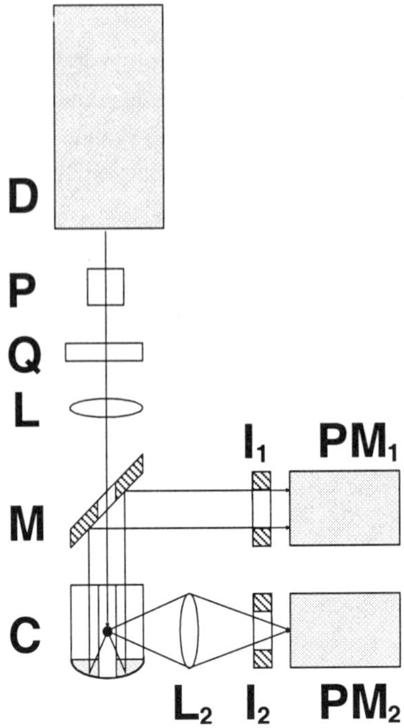

Figure 1. Scheme of the experimental apparatus.

3. Materials and Method

Two different kinds of marine microorganisms have been studied by CLASS: *Chlamydomonas reinhardtii* and *Synechocystis*. The first one is an eukaryote single celled green alga, about 10 μm long, which moves using two flagella, about 10 μm long. That cell offers an ideal test for the system having elliptic shape and containing chlorophyll-a that imply, respectively, non zero depolarization and fluorescence around 680 nm. *Synechocystis* is a photosynthetic cyanobacterium that has become a model cyanobacterium used by scientists to study pigment synthesis and its regulation, respiration, photosynthesis and several other processes. Moreover, in order to evaluate the functionality of the experimental apparatus and to perform an initial system calibration, non-fluorescent and fluorescent labeled polystyrene beads have been studied by means of CLASS. Thus, suspensions of a variety of specified sizes of

carboxylate-modified microspheres coated with a hydrophilic polymer (Molecular Probes® F8888) have been analyzed.
Experiments on *Chlamydomonas* solutions mixed with fluorescent and non-fluorescent polystyrene spheres (6 μm diameter) have been performed too.
For every sample described above trigger, indicatrix (before and after polarizer) and fluorescence signals (at 680 nm) have been collected. From such data, information on size and diameter of particles has been obtained and a classification, based also on fluorescence and depolarization, has been performed. The light scattering inversion scheme here used to retrieve size and refractive index of spherical particles is a parametric solution of the inverse light scattering problem. This scheme is an evolution of the flying light scattering indicatrix method [4], based on the spectral decomposition of the detected angular light scattering. In fact, the features of the Fourier spectrum of experimental signals can be related to the particle characteristics. The inversion scheme, that was proved to be more robust in case of noisy data, is thoroughly described elsewhere [5] and the reader is referred to that reference for algorithm details.

4. Results and Discussion

At the beginning a calibration of the system has been carried out with samples of microspheres with diameter equal to 0.5, 1.0, 2.0 μm. An example of indicatrix signal for such particle is given in Figure 2, where the theoretical fit of the experimental curve is also shown.

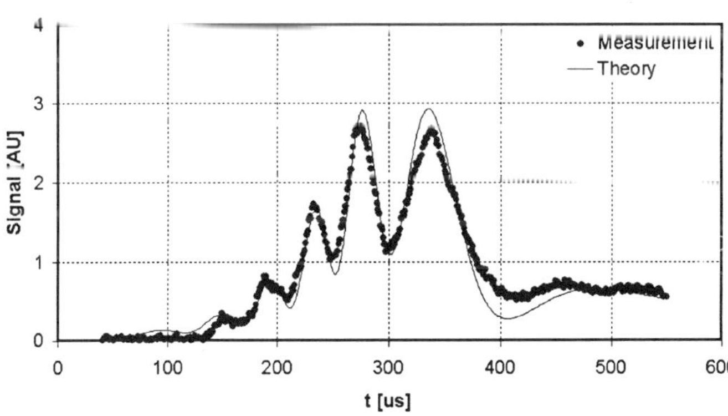

Figure 2. Indicatrix signal and its theoretical fit for 2 μm diameter spheres.

Size and refractive index have been then calculated by a parametric solution of the inverse light scattering problem both for microspheres, as a test of the method used, and for biological particles. In Figure 3 size and refraction index distributions obtained for 2.0 μm beads are shown.

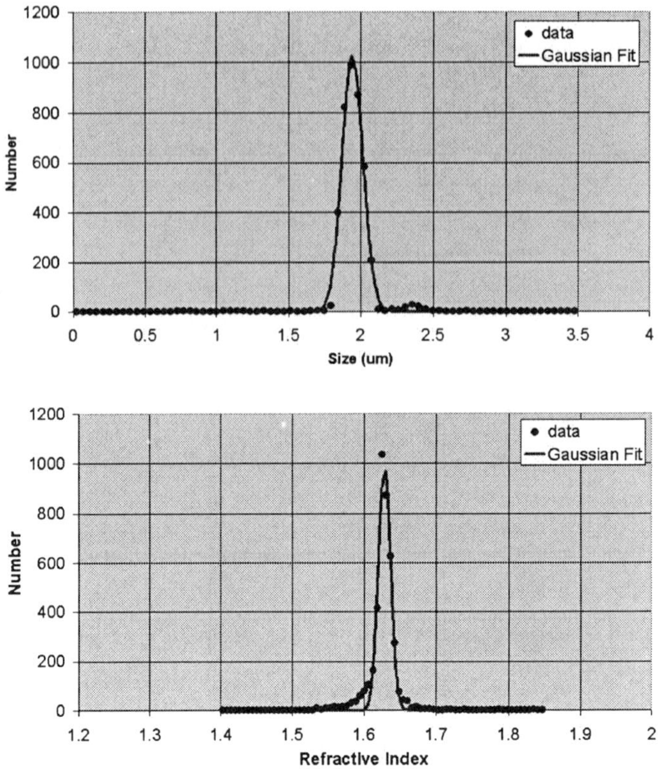

Figure 3. Size (above) and refractive index (below) distribution measured for 2.0 μm spheres.

It is possible to note how the system measures correctly size and refractive index of the spheres. In fact, the expected values: 2.0 μm and 1.60 [6], respectively, are compatible with the measured values: 1.95±0.10 μm and 1.63±0.02, corresponding to relative errors around 1% in both cases.

Results on *Synechocystis* are given in Figure 4. For size distribution of *Synechocystis* two different Gaussian fits have been performed and two peaks, at 1.0 and 1.7 μm can be seen. They correspond to single cells and two-cell aggregates, respectively. This result shows the capability of CLASS to distinguish, count and classify very similar particles that can be found mixed in a real marine environment.

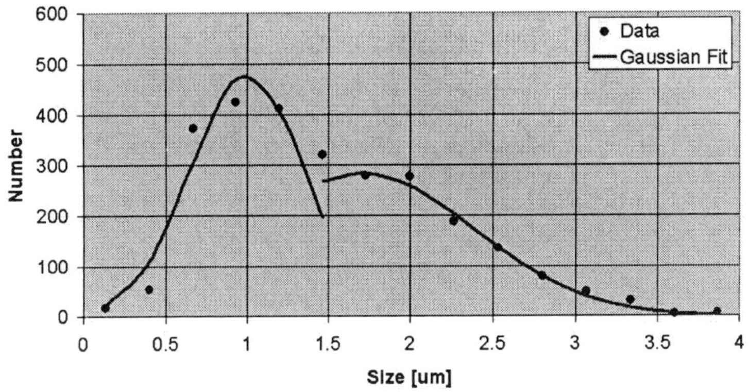

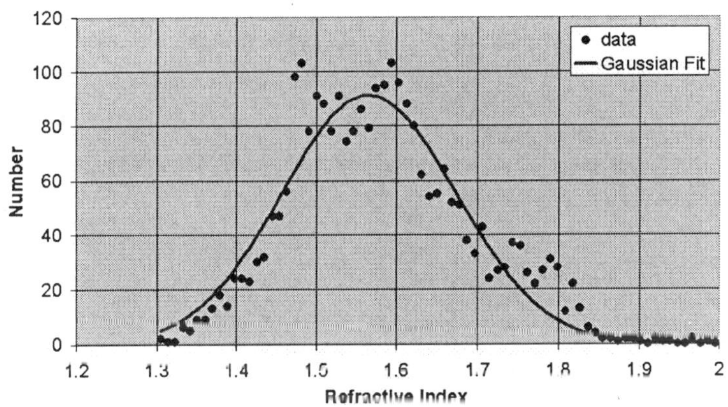

Figure 4. Size (above) and refractive index (below) measurements of *Synechocystis* cells by means of CLASS.

Measurements on solutions of *Chlamydomonas* mixed with fluorescent and non-fluorescent polystyrene spheres, characterized by sizes comparable to the *Chlamydomonas* ones, have been also performed.

Due to the large natural variability in size of the cells, their experimental size distribution is rather wide and so such kind of data is not the suitable tool to distinguish cells from spheres. For the refractive index, as it is possible to see in Figure 5, the distribution is narrower and, because the significant difference between its value for *Chlamydomonas* and polystyrene, such quantity can be used to discriminate the two kinds of particles. However, a non-negligible overlap in the histograms exists and mistakes in the identification can be done.

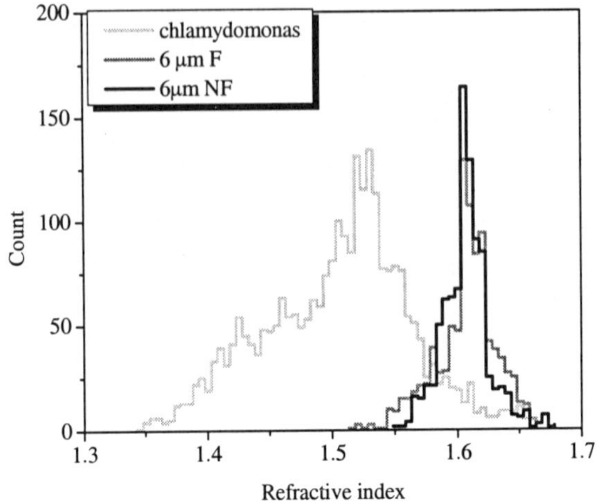

Figure 5. Refractive index distributions for *Chlamydomonas* cells, fluorescent (F) and non-fluorescent (NF) microspheres with diameter equal to 6 μm.

Instead, Figure 6 shows how, if the depolarization is taken into account, it is practically impossible to mistake cells and spheres.

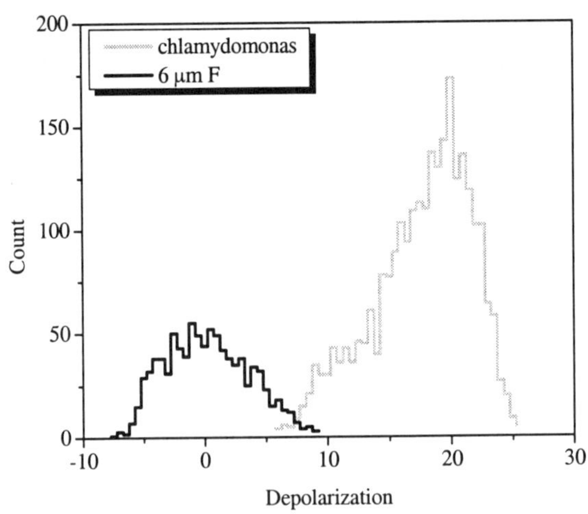

Figure 6. Depolarization distributions for *Chlamydomonas* and non fluorescent spheres.

5. Conclusion

The new laser scanning flow cytometer CLASS has proved to be a very powerful tool for measurements and classification of phytoplankton cells, thanks to its capability to measure simultaneously size, refractive index, fluorescence and depolarization.

Acknowledgments

The authors are deeply grateful to F. Barnaba, K. Semyanov and P. Tarasov for their contribution to this work and to F. Colao and R. Fantoni for their involvement.

References

1. 1.F.Barnaba, L.Fiorani, A.Palucci, P.Tarasov, *Optics of Biological Particles*, V.Maltsev, A.Hoekstra, G.Videen (Editors), NATO Novosibirsk (Russia) **34** (2005).
2. F.Colao, R.Fantoni, L.Fiorani, A.Palucci, P.Tarasov, RT/2004/60/FIS ENEA (2005).
3. H.M.Shapiro, Practical flow cytometry, 4th ed. Hoboken, Wiley (2003).
4. K.A.Semyanov, P.A.Tarasov, A.E.Zharinov, A.V.Chernyshev, A.G.Hoekstra, V.P.Maltsev, *Applied Optics* **43**, 5110 (2004).
5. F.Barnaba, L.Fiorani, A.Palucci, P.Tarasov, *Journal of Quantitative Spectroscopy and Radiation Transfer* **102**, 11 (2006).
6. X.Ma, J.Q.Lu, R.S.Brock, K.M.Jacobs, P.Yang and X.H.Hu, *Phys. Med. Biol.*, **48**, 4165 (2003).

OPTICAL PROBE FOR THE TURBINE INLET TEMPERATURE MEASUREMENT IN GAS TURBINE PLANTS*

I. GIANINONI, E. GOLINELLI, U. PERINI

CESI RICERCA, Via Rubattino 54, 20134 Milano, Italy

We present an optical probe for the measurement of the turbine inlet temperature (TIT) in gas turbine plants. The probe carries out spectroscopic photometric measurements of the IR radiation emitted in a selected wavelength band by the CO_2 molecules present in the combustion gases. It is mechanically robust and potentially compatible with operating gas turbines. The system has been installed and tested on a full scale combustor test bed.

1. Introduction

The Turbine Inlet Temperature (TIT) is a critical parameter of Gas Turbine systems influencing both material and coating lifetime of turbines as well as their efficiency. Therefore on line TIT monitoring systems are important since they allow to maximize the gas temperature (although it should remain within the limits foreseen for materials and coatings operation) thus increasing the Gas Turbine overall efficiency.

The measuring system we present here is a temperature probe based on spectroscopic photometric measurements of the IR radiation emitted in a selected wavelength band by the CO_2 molecules in the combustion gases. It is mechanically robust and thermally resistant so to withstand typical operating conditions of industrial gas turbines. To minimize installation requirements, the probe has been designed in such a way it can operate through a single optical access.

2. Principle of Operation

The measuring technique is a sort of emission - absorption pyrometry ("Planck – Kirchhoff method") like the one utilized in non reacting gas mixtures [1]. The measured quantity is the grey body spectral irradiance $H(\lambda,T)$ defined by the product of the blackbody spectral irradiance $W_B(\lambda,T)$ (over 2π solid angle) and the grey body absorption $Abs(\lambda,T)$:

$$H(\lambda,T) = W_B(\lambda,T) \cdot Abs(\lambda,T) \tag{1}$$

the blackbody spectral irradiance being defined by the Planck's law:

* This work has been financed by the Ministry of Economic Development with the Research Fund for the Italian Electrical System under the Contract Agreement established with the Ministry Decree of March 23, 2006.

$$W_B(\lambda,T)\cdot\Delta\lambda = \frac{1}{\lambda^5}\cdot\frac{C_1}{\exp\left(\frac{C_2}{\lambda T}\right)-1}\cdot\Delta\lambda \quad [\mathrm{W/m^2}] \qquad (2)$$

where $C_1 = 3.74\mathrm{x}10^{-16}$ [J m^2/s], $C_2 = 1.44\mathrm{x}10^{-2}$ [m K] and $\Delta\lambda$ is the selected bandwidth. Under the hypothesis that the combustion gas is strongly absorbing in the selected bandwidth and therefore can be considered as a blackbody, the only dependence on temperature is given by the blackbody irradiance, so that the gas temperature can be determined. In order to achieve such conditions an operational spectral range has to be selected (e.g. by means of the HITRAN database [2]), which satisfies the requirement of having the carbon dioxide molecules in the hot gases strongly absorbing IR radiation along the test optical path (to avoid contributions from the opposite hot wall) and, at the same time, transparent enough to the emitted radiation to allow it to cross a significant portion of the test region. By taking into account that the hot gases going out of the combustor are at pressures of 15–18 bars with temperatures exceeding 1600 kelvins and that the measuring volume can vary from 10 to 40 cm in depth, the recommended IR bandwidth turns out to be 8–10 nm wiae (~ 4-5 cm^{-1}), centered around 4461 nm.

3. The Measuring System

As shown in Fig. 1, the measuring system is made by three units: an optical probe (interfaced with the combustor), a detection unit (connected via an optical fiber to the optical probe) and a data acquisition and analysis unit.

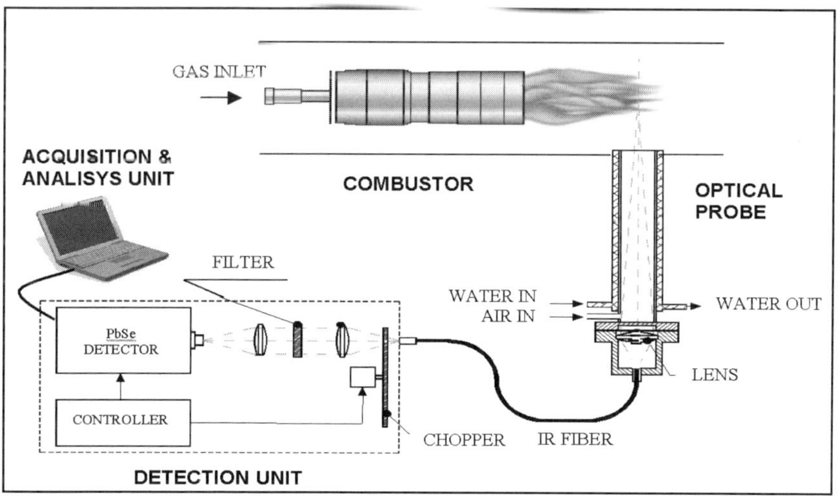

Figure 1. TIT measuring system.

The optical probe is cooled by water and is purged with air to keep the optical window clean. On the outer part of the probe a ZnSe lens collects the emitted radiation onto the input face of a Hg halide optical fiber supported by a micro-positioner. In the detection unit (Fig. 2) the radiation emerging from the fiber is first collimated, then filtered by means of an IR interference filter (centered at λ_c = 4461 nm, 10 nm FWHM) and finally focused onto the PbSe photoconductive detector (0.5x0.5 mm^2 active area). The IR detector, which is mounted on a heat sink, is provided with an active cooling system that controls the sensor temperature. A mechanical chopper is placed at the input close to the end of the optical fiber. The signal out from the detector is first amplified and then fed to the acquisition board (16 bit, 20 kS/s). Data acquisition and analysis are carried out via a PC based analysis unit. All the system is controlled by means of a Labview dedicated software.

Figure 2. Detection unit.

4. Experimental Results

Following a set of preliminary laboratory tests aimed at checking the dependence of the detected IR signal on the gas temperature as well as the repeatability and sensitivity of the technique [3], the TIT measuring system has been installed on a full scale combustor test bed (see Fig. 3). The probe is correctly positioned through the outer combustor case by means of a protecting metallic tube and properly designed mechanical interfaces. Since the probe consists of passive components, it only requires purging air and cooling water. The measuring system can be remotely operated from the control room.

Figure 3. TIT probe on a full scale combustor.

An example of experimental results recorded during combustor operation is shown in Fig. 4. Plot 1 is the averaged signal (mV) obtained from the TIT optical probe, whereas the other plots correspond to parameters available from the test bed control system, i.e.: the computed adiabatic temperature (2), the CO_2 concentration (3) and the pressure inside the combustion chamber (4).

The trend of the computed adiabatic temperature (2) can be compared with the experimental TIT measurements (1) in the range where the pressure inside the combustion chamber remains quite stable, i.e., within the two vertical dashed lines. As it can be noticed, measured and computed data are well correlated. Preliminary estimates indicate a measurement sensitivity of the order of ±10 K, which is high enough to follow typical fluctuations of gas turbine combustors. Based on these results as well as on its manufacturing simplicity and capability of being adapted to existing inspection ports, the measuring systems turns out to be promising for real plant applications. The foreseen activity will investigate the possibility of achieving absolute temperature values from the measured signals by extending the measurement to multiple wavelengths. The validation of the measuring system will also be performed in the frame of a FP6 European Project ("HEATTOP"), which is devoted to development and testing of high temperature instrumentation for gas turbine performance and condition monitoring.

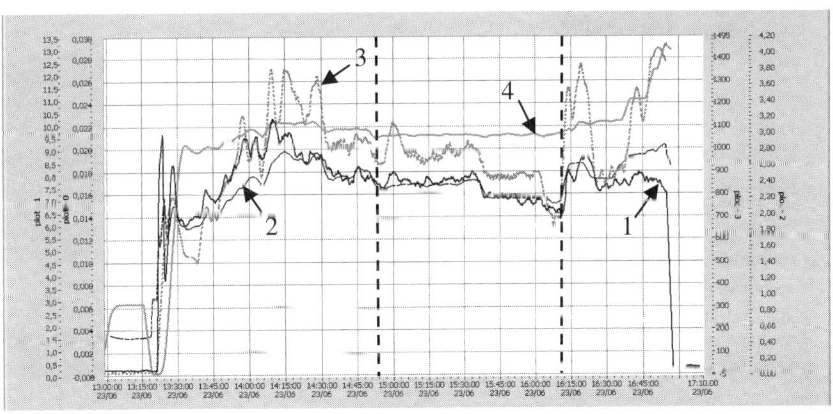

Figure 4. Experimental results. Plot 1: TIT measurements; Plot 2: computed adiabatic temperature; Plot 3: CO_2 concentration; Plot 4: pressure inside the combustion chamber.

References

1. G.A. Hornbeck, Optical methods of temperature measurement. *Appl. Opt.* **Vol. 5, No. 2**, 179 (1966).
2. University of South Florida, HITRAN Database, Copyright 1997.
3. I. Gianinoni, E. Golinelli, U. Perini and L. Fiorina, *Proc. 9th Nat. Conf. "Strumentazione e metodi di misura elettroottici"*, Frascati, 165 (2006).

HOLLOW-CORE OPTICAL FIBERS INTEGRATED WITH SINGLE WALLED CARBON NANOTUBES AS VOCS SENSORS

MARCO PISCO[†], MARCO CONSALES, ANTONELLO CUTOLO, ANDREA CUSANO

Optoelectronic Division - Engineering Department, University of Sannio, Corso Garibaldi 107, 82100 Benevento, Italy

MICHELE GIORDANO

Institute of Composite and Biomedical Materials, CNR, 80124, Naples, Italy

PATRIZIA AVERSA, MICHELE PENZA

ENEA, C.R. Brindisi, Materials and New Technologies Unit, 72100 Brindisi, Italy

STEFANIA CAMPOPIANO

Department for Technologies, University of Naples Parthenope, Via Medina 40, 80131 Napoli, Italy

In this work, Hollow-core Optical Fibers (HOF) functionalized with Single Walled Carbon NanoTubes (SWCNTs) are proposed for volatile organic compounds (VOCs) detection. The sensing probe is composed by a piece of HOF with a termination coated and partially filled by SWCNTs. The infiltration of the SWCNTs inside the HOF holes has been accomplished by means of the Langmuir-Blodgett technique. Far field transmission characteristics have been carried out within the HOF bandwidth. Finally the sensing capability of the proposed sensors has been investigated by exposure in a proper designed test chamber to traces of toluene. The experimental results obtained demonstrate the success of the SWCNTs partial filling within the HOF holes and the sensor capability to perform VOCs detection with a good sensitivity and fast response times.

1. Introduction

Hollow-core Optical Fibers (HOFs) are wavelength scale microstructured optical fibers, able to guide light within the air core over a wide bandwidth by means of a two dimensional Photonic Bandgap (PBG) formed by the periodic structure of the cladding [1]. Since the HOFs are composed by an array of air-holes, they also constitute a promising platform to develop new classes of

[†] Corresponding author: pisco@unisannio.it

optical devices. The microstructuration of the cladding in fact allows to fill the HOF holes with materials which are able to supply to the final device advanced properties in terms of tunability [2-3] and sensitivity [4]. In this work, we present an all fiber opto-chemical sensor useful for volatile organic compounds (VOCs) detection based on HOFs covered and partially filled with carbon nanotubes. Single Walled Carbon NanoTubes (SWCNTs), which yet widely demonstrated their capability to work for chemical sensing applications [5], have been employed as sensitive materials and hence injected within the HOF holes. The sensing probes, fabricated with different fabrication parameters, have been characterized by far field transmittance. Finally the realized sensors, employed in a single wavelength reflectometric configuration, have been exposed in a test chamber to several toluene impulses in order to demonstrate their capability to work for VOC detection and to show the impact of the fabrication parameters on the sensing performances.

2. Methodology

2.1. *Sensor Fabrication*

The sensing probe is composed by a piece of HOF, spliced at one end with a single mode standard optical fiber (SOF) and covered and partially filled with SWCNTs at the other termination. Before the HOF-SOF splicing, the deposition of SWCNTs was performed at atmospheric pressure by means of the Langmuir-Blodgett (LB) technique [5]. This deposition technique involves the dipping of the HOF through the LB solution, perpendicularly to it, in order to pick up carbon nanotubes monolayer by monolayer. In this way the injection of the LB solution inside the HOF holes is guaranteed and the adhesion of carbon nanotubes onto the air-holes sides is provided. In the following, two fabricated sensing probes are analyzed. The former one, labeled sensor 1, is constituted by a 13.3 cm long HOF with 10 monolayers of SWCNTs, the latter, labeled sensor 2, by a 4cm long HOF and 20 SWCNTs monolayers. The deposition parameters were the same used to coat SOF with SWCNTs and can be found elsewhere [5].

2.2. *Far field Characterization*

The far field emerging from the sensors has been collected by means of an infrared vidicon camera (Hamamatsu C2741-03) while a narrowband laser source at 1550nm lights the sensing probes. The end face of the samples has been positioned in the nearby of the receiving lens. In figure 1.a and 1.b the far field transmittance of sensor 1 and sensor 2 have been reported, respectively. The far field emerging from the sensor 1 presents the typical quasi-Gaussian shape of the field associated to the HOF fundamental mode, shown in fig. 1.c. At the bottom of the Gaussian waveform, little peaks of circular shape are observable, corresponding to the cladding modes also observed in the far field plot of fig. 1.c. The presence of the SWCNTs hence is not able to strongly modify the field distribution of the propagating mode while it confers the double interferometric behavior evidenced in the previous section for sensor 1. In addition, a slightly higher light content can be observed in correspondence of the HOF cladding in the coated case. This effect demonstrates a slightly worse confinement of the filled HOF consistent with refractive index contrast reduction induced by a partial filling of the cladding holes.

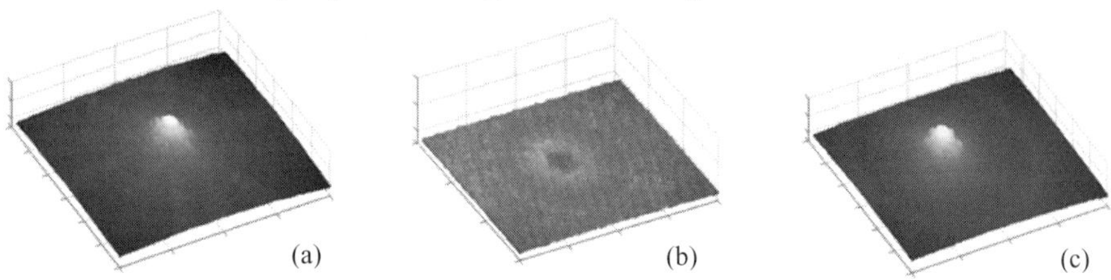

Fig. 1. Distribution of the far field transmittance of the HOF sensor 1 (a), of the HOF sensor 2 (b) and of an HOF (c)

With regards sensor 2, inversely, a strong modification of the field distribution occurs and the amount of power transmitted is widely reduced. The emerging field presents a circular crown shape and the core mode is not more visible. In other words, the larger amount of SWCNTs used for sensor 2 penetrated within the holes are able to suppress the fundamental core mode thus leading to a strong diminution of the confinement power. This circumstance is attributed to the modification of the PBG occurring in consequence of the cladding and core holes filling which yields the functionalized HOF not more able to guide meaningfully the light. The PBG modification affects particularly the sensor 2 which was fabricated with an higher number of monolayers with respect to the sensor 1. As a consequence an overall deeper infiltration of the carbon nanotubes and also their higher density within the HOF holes occur. The penetration depth of the SWCNTs within the holes and their spatial distribution is currently under investigation as well as the effects of the deposition parameters on the resulting functionalized HOF and it will be the object of further work. In summary, the far field characterizations reveal the success of the SWCNTs deposition onto the distal end of the HOF [4] and demonstrate the partial filling of the nanotubes within the microstructured optical fibers. Also, the obtained results show how by a proper control of the filling procedure, especially in terms of spatial distribution, it is possible to modify the PBG and thus the propagation properties of HOFs for specific applications.

3. Experimental Results

In order to investigate the sensing capability of the fabricated HOF sensors, they have been placed in a test chamber [5] and exposed to traces of VOCs. At the same time, a reflectometric system, which allows reflectance measurements at single wavelength has been employed. In order to light the sensing probes a Superluminescent Light Emitting Diode with 40nm bandwidth centered at 1550nm has been used [5]. During the measurements, the temperature within the chamber has been monitored by using a commercial thermocouple. The sensors sensitivity to temperature variations was previously characterized and used in order to properly compensate the sensors output. Here a comparison among the performances exploited by sensors 1 and 2 is carried out by considering their relative reflectance change due to toluene exposure. To this aim, both sensors have been exposed to three toluene pulses of 45 minutes with increasing concentrations in the ppm range. Fig. 2.a reports the time responses of both sensors. Upon exposures, the reflectance of the sensor 1 decreases linearly with the toluene concentration as confirmed by further exposure measurements resumed in fig. 2.b. Differently, sensor 2 reflectance increases upon exposure presenting a significantly lower sensitivity, opposite in sign. Moreover, sensor 2 is not able to reach a steady state within 45 minutes especially for high concentration exposures, this also limits the maximum reflectance change observed during the exposure tests. The higher response times and the lower sensitivity of the sensor 2 can be explained on the basis of the higher extension of the SWCNTs region (higher diffusion times and higher losses in the filled region). In fact, the VOCs, able to interact with the SWCNTs, lead to a change of the SWCNTs agglomerate dielectric function and the induced variations are detected as fast as short is the SWCNTs region. In addition, as demonstrated by the far field characterization, the SWCNTs penetration depth in the sensor 2 yields the functionalized HOF not more able to guide the light and thus it reduces the capability of the light to interact with the sensitive material. Further investigation is required to assess the sensitivity dependence on the SWCNTs distribution within the HOF structure [4] and also to identify characterization features able to predict the correct functioning and the performance of the final device.

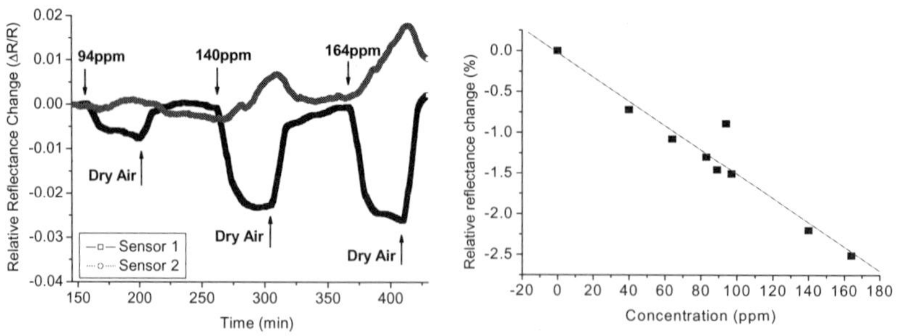

Fig. 2. Time responses of HOF sensors 1 and 2 to 3 Toluene impulses (a). Relative Reflectance Change ($\Delta R / R$) of the sensor 1 corresponding to the steady states reached upon exposures to different concentration pulses of toluene (b)

In conclusion, here, HOFs have been functionalized by means of the infiltration of SWCNTs within the HOF holes. The integration between the carbon nanotubes and the HOF yields the final device capable to work for chemical sensing applications. The presented experimental results also demonstrate that the fabrication parameters affect strongly the sensors performances. In such a way a suitable control of the fabrication process allows the functionalized HOF to perform chemical detection of VOCs with a good sensitivity and fast response times. In addition, by a selective filling procedure and by a proper control of the SWCNTs spatial distribution, the propagation properties of HOFs and the same PBG offered by the functionalized HOF can be modified for specific applications.

References

1. J. C. Knight, "Photonic crystal fibres", Nature, vol. 424, (2003), pp 847-851
2. Y. Huang, Y. Xu and A. Yariv, "Fabrication of functional microstructured optical fibers through a selective-filling technique", Applied Physics Letters, Vol. 85, No. 22, 29 November 2004, pp.5182-5184
3. A. Cerqueira S. Jr., F. Luan, C. M. B. Cordeiro, A. K. George and J. C. Knight, "Hybrid photonic crystal fiber", Optics Express, Vol. 14, No. 2, 23 January 2006, pp926-931
4. A. Cusano, M. Pisco, M. Consales, A. Cutolo, M. Giordano, M. Penza, P. Aversa, L. Capodieci, S. Campopiano, "Novel Opto-Chemical Sensors Based On Hollow Fibers And Single Walled Carbon Nanotubes", Photonics Technology Letters, Volume 18, Issue 22, Nov.15, 2006 pp2431 - 2433
5. M. Penza, G. Cassano, P. Aversa, A. Cusano, A. Cutolo, M. Giordano, L. Nicolais, "Carbon nanotube acoustic and optical sensors for volatile organic compound detection", Nanotechnology, 16 (2005) pp2536-2547.

MULTI-SPECTRAL EXTINCTION BASED OPTICAL SYSTEM FOR THE CHARACTERISATION OF PARTICLES AND GASES IN THERMOELECTRIC POWER PLANTS EXHAUSTS *

E. GOLINELLI, S. MUSAZZI, U. PERINI

CESI RICERCA, Via Rubattino 54, 20134 Milano, Italy

We present a novel non intrusive optical system for the on-line characterization of coal fired power plants flue gases. The system carries out both real time particle sizing of solid particles (in the range 0.1 – 10 μm) and concentration measurements of chemical species (as small as few ppm) downstream the electrostatic precipitators. The principle of operation relies on a multi-wavelength extinction technique based on the measurement of the extinction coefficient as a function of the illuminating wavelength.

1. Introduction

Coal/oil fired boiler systems generate the largest fraction of the total worldwide available electric power. In addition, because of the larger availability of coal with respect to other fossil fuels, coal fired plants will probably continue to increase their production in the future. As a consequence, methods for monitoring the exhausts content of thermoelectric power plants are assuming increasingly importance since they can be used both to monitor/feedback the combustion process and, at the same time, to maintain under control the release of environmental pollutants. Among others, optical diagnostic tools seem to be the most promising ones because of their poor intrusiveness and the capability of performing real time on-line measurements. Although optical systems for monitoring particulate emissions from power plants already exist, they suffer the limitation of requiring periodic calibration (opacimeters) or, as it is the case of very sophisticated instruments, a further development phase is required before their installation in power plants become feasible. It should be noticed, in fact, that because of the very critical operative conditions (i.e. high temperatures, corrosive gases, mechanical vibrations) optical instrumentation can hardly be

* This work has been financed by the Ministry of Economic Development with the Research Fund for the Italian Electrical System under the Contract Agreement established with the Ministry Decree of March 23, 2006.

used in industrial plants. It is therefore mandatory to develop a new class of innovative optical tools capable of operating in such hostile environments. In this paper we present a new optical system properly conceived for the real time characterization of particles (in the range 0.1 - 10 μm) and chemical species (with a sensitivity of few ppm) in coal/oil fired boilers exhausts.

2. Principle of Operation

2.1. *The optical particle sizer*

The optical particle sizer is based on the detection of the extinction coefficient measured as a function of the illuminating wavelength. This measurement is carried out by illuminating the test region by means of a white light source and by detecting the spectrum of the transmitted radiation after propagation. By comparing this spectrum with the one obtained in absence of particles, it is possible to determine the spectral extinction coefficient due to the presence of particulate material in the optical path. The particle size distribution and concentration are determined via a properly developed inversion algorithm. The principle of operation of this method is briefly described in the following.

The extinction coefficient $\alpha(\lambda)$ is defined by the Lambert-Beer law [1]

$$P_T = P_0 e^{-\alpha(\lambda)L} \tag{1}$$

which describes the attenuation suffered by a monochromatic light beam (of wavelength λ) propagating through a test region of length L. P_0 and P_T represent the incident and transmitted power respectively. In the case of a dilute suspension of particles of different sizes (which is the case of flue gases downstream the electrostatic precipitators), the spectral extinction coefficient $\alpha(\lambda)$ is given by:

$$\alpha(\lambda) = \int \pi r^2 Q_{ext}(r, \lambda', m) N(r) dr \tag{2}$$

where N(r) is the number concentration of particles with radius ranging between r and r + dr while Q_{ext} is the extinction efficiency (provided by the Mie theory) [1]. Here λ' is the wavelength of the radiation in the medium, and m is the refractive index of the particles relative to the medium. Eq. (2) is a typical first-kind Fredholm integral equation where $\alpha(\lambda)$ is obtained from the experiment, $\pi r^2 Q_{ext}$ is the known kernel and N(r) is the unknown particle size distribution to be recovered. Unfortunately, this is an ill-posed problem, i.e. when noisy input data are used, different distributions N(r) can fit the same data set $\alpha(\lambda)$ with the same level of accuracy [2]. As a consequence, the inversion of Eq. (2) becomes a difficult task that can be fulfilled only by using stable and reliable inversion algorithms. The inversion of Eq. (2) is carried out by means of a non linear iterative algorithm we have developed and properly tested in the past (see Ref. [3,4]).

2.2. *The DOAS system*

Chemical species are detected via DOAS (Differential Optical Absorption Spectroscopy) measurements. Since this is a well known technique [5], we simply remember here that the concentration c of a given chemical specie is provided by the equation:

$$c = \frac{\log\left(\frac{P_0(\lambda)}{P_T(\lambda)}\right)}{\alpha(\lambda)L} \tag{3}$$

Where, referring to Eq. 1, $\alpha(\lambda)$ represents in this case the absorption cross-section of any species at a given wavelength.

3. The Measuring System

The measuring system is schematically shown in Fig.1. It is made by two separate units (the transmitting/receiving unit and the retroreflecting unit) that have to be positioned on the opposite sides of the test region.

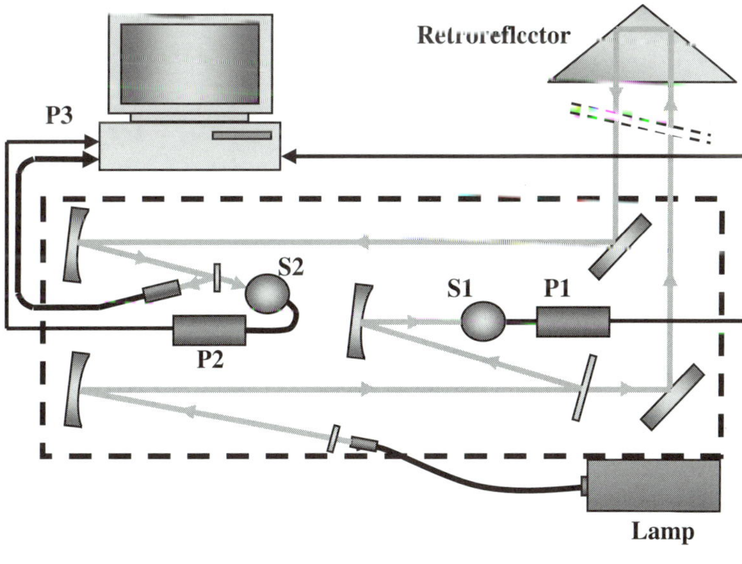

Figure 1. Schematic description of the measuring system.

The transmitting unit consists of a white light source and a projection/receiving optical system. To get the required spectral region two different light sources are needed: a Xenon lamp has to be used for particle sizing while a Deuterium lamp is required for DOAS measurements. The lamp output (carried by a 1mm core optical fiber) is fed to the retroreflector via a collimation-reflection scheme based on the use of an off axis parabola and a 45° tilted mirror. A fraction of this light is also collected by the integrating sphere S1 and analyzed by a polychromator P1 for background measurements. After crossing the test region, the light coming from the retroreflector is collected by the integrating sphere S2 and fed to a polychromator P2 whose output is properly analyzed for particle size measurements.

The fraction of the light to be used for DOAS measurements is collected at the entrance of the integrating sphere S2 by means of a beam splitter and brought to a polychromator P3 (a PC board inside the desktop that controls the overall system) via a properly coupled optical fiber.

4. Experimental Results

Examples of measurements carried out in a 600 MW coal fired power plant are presented in Fig. 2 and 3. In Fig. 2 we show a two hours recording of the particle concentration measured by the multispectral extinction based instrument (crosses) compared to the one obtained during the same period by a conventional opacimeter installed on the stack (dots).

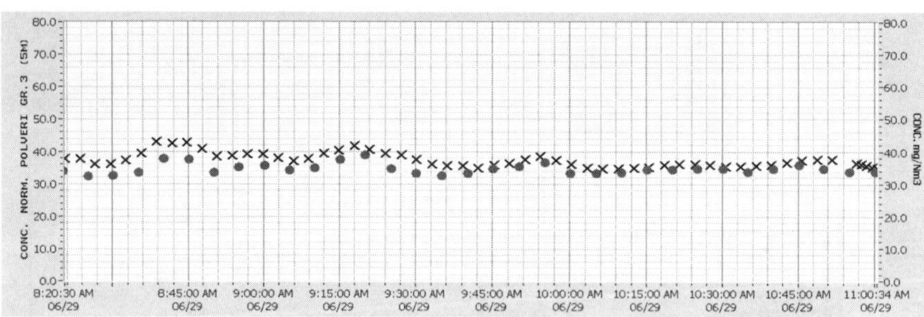

Figure 2. Particle concentration measurements

Fig. 3 shows concentration measurement curves of NO and SO_2 carried out by our instrument (crosses) compared to those obtained by conventional instrumentation (dots) during the same period. As one can notice both particle and gas concentration measurements carried out by the multispectral extinction based instrument are in good agreement with those obtained with conventional instruments.

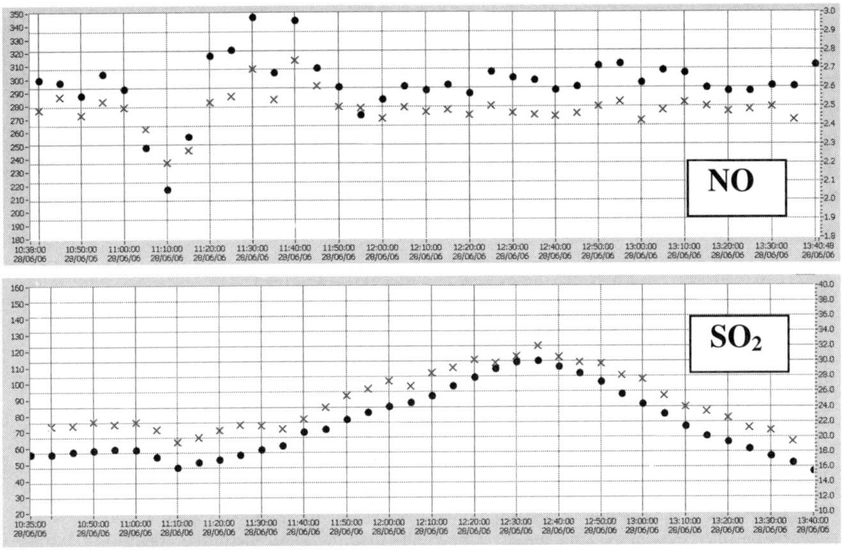

Figure 3. Chemical species concentration measurements

References

C. Van de Hulst, "Light Scattering by Small Particles", Dover Publications (New York), 127 and 388 (1981).

S. Twomey, "Introduction to the Mathematics of Inversion in Remote Sensing and Indirect Measurements", Elsevier, Amsterdam (1977).

A. Bassini, F. Ferri, E. Paganini, " Modified version of Chahine algorithm to invert spectral extinction data for particle sizing", Appl. Opt., 34, 5829-5839 (1995).

A. Bassini, F. Ferri, M. Giglio, E. Paganini, S. Musazzi, U. Perini "Optical particle sizer based on the Chaine inversion scheme", Opt. Eng., 31, 1112-1117 (1992).

U. Platt, "Modern methods of the measurement of atmospheric trace gases", Phys. Chem. Chem. Phys, 1, 5409-5415 (1999).

OPTICAL DEVICE FOR INTEGRITY ASSESSMENT OF THERMAL BARRIER COATINGS

LETIZIA DE MARIA, CLAUDIA RINALDI
CESIRICERCA, via Rubattino 54
Milan, 20134, Italy

A portable optical device, for an in-shop integrity assessment of gas turbines ceramic coatings, is described. The main features of the optical scheme are reported together with the description of the miniaturised probe shaped to extend the PLPS detection also at the leading edge. Measurements performed both on specimens aged in laboratory conditions and on operated components are presented and compared with the initial condition. The important meaning of the obtained results is shown with the help of metallographic analyses performed on sections of specimens and components. Finally limits and future perspective of the technique are evidenced.

1. Introduction

Ceramic thermal barrier coatings (TBCs) are applied to hot parts in gas turbines to increase resistance to high temperatures. Their use onto rotating blades allowed to increase inlet temperatures and thus significantly improve turbine efficiency. The endurance of such coatings is still a key issue as very variable lifetimes are registered. As a consequence it is important both to assess the condition of the coating before component mounting onto the rotor and to monitor the actual degradation level of the TBCs after operation, during inspections.

For these reasons there is an increasing interest towards reliable non-destructive techniques able to detect incipient failure of TBCs before spallation, to avoid irreversible component damage. Several techniques are under development and were recently reviewed by the authors in [1]. Among the several methods under study, the optical ones are particularly interesting, due to their non-contacting nature. For example thermography is already widely used as quality control technique to detect coating delamination of some millimeters on new components, before mounting them on gas turbine rotors. This technique was also recently developed to be applied in situ, for on line monitoring of TBC adhesion on rotating blades during operation [2].

Among the more innovative techniques to detect incipient damage phenomena before delamination, the Photoluminescence Piezospectroscopy [3-4] is very promising for its sensitivity to slight modifications happening at the interface under the ceramic layer. Actually one of the main damage mechanisms leading to TBC failure is the growth of a thin alumina layer which develops at the interface between the ceramic topcoat and the metallic layer beneath (named bond-coat (BC) and giving protection from oxidation and hot corrosion phenomena). This alumina layer is called Thermally Grown Oxide (TGO) and grows subjected to very high compressive residual stresses, due to both its volume increase inside a solid body and to the thermal expansion mismatch between bond-coat/TGO/top coat.

The Photo-Stimulated Luminescence Spectroscopy (PLPS) is a technique demonstrated to be able to provide information on the stress state of the TGO by monitoring the frequency shift of some suitable spectroscopic bands with respect to a stress-free reference [3]. A significant decrease of such stress level is a useful indicator of incipient local detachment of the TBC. A Round Robin showed the reproducibility and the reliability of this type of measurements [5].

On the market there are only instruments to perform measurements of the PLPS spectra with large size and sensor configuration which can be used only in the laboratory [5].

In this paper a new sensor with the relative portable PLPS instrument is presented, developed to allow the in situ detection of PLPS signals on new and service operated blades.

2. Principle of the PLPS technique

The Chromium ions present inside the TGO alumina layer, if excited by a green laser light, can produce a red fluorescence peak, influenced by the presence of residual stresses. A typical TGO spectrum registered beneath the ceramic topcoat is shown in Fig.1. In this figure, the solid line refers to a compressed TGO layer; residual stress values higher than 2.5 GPa (in this example 2.8 GPa) are typical of the initial condition of the coating on a new blade. Analysing the shape, the position and width of this luminescence spectrum, the magnitude of the residual stress of TGO can be measured. A very interesting aspect of the technique [3] is that this measurement can be done through the ceramic thermal barrier coating of Zirconia partially stabilised with Ittria, as this material is transparent to visible radiation. In Fig. 2 a metallographic section of the two layer thermal barrier coating is shown, for the sake of clarity. On the base material, a Ni based superalloy; a metallic layer of a MCrAlY (land based

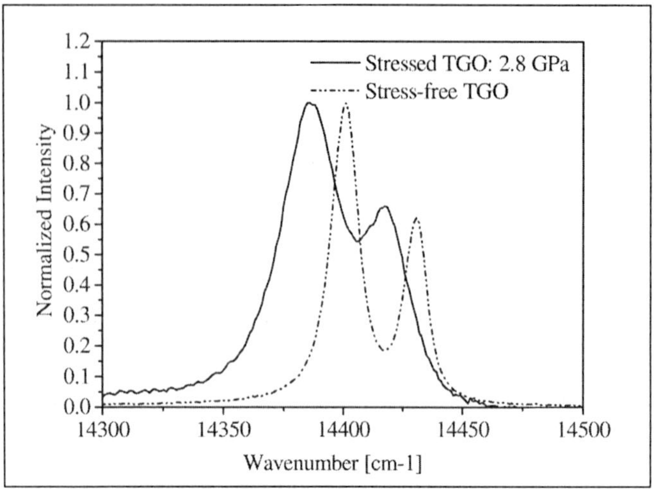

Fig.1 : Typical TGO spectrum beneath the ceramic topcoat

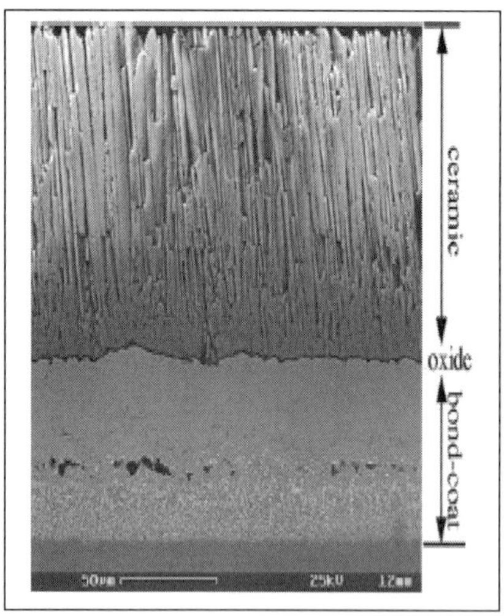

Fig. 2 Typical section of a thermal barrier coating for aeronautic blade

turbines) or Pt-Aluminide (aeronautic motors) is present with a Zirconia-Ittria ceramic layer on it. One important comment has to be done about the microstructure of the ceramic layer: the columnar structure, shown in this example, is ideal to obtain the necessary transparency also with TBC thickness up to 300 μm, as happens on real blades, coated with electron beam assisted physical vapour deposition (both on the aeronautic motors and on the first stage rotating blades of the last generation land based gas turbines for energy production).

3. Sketch of the optical device

The exciting source is a Nd:YAG laser @ 532nm wavelength. Both the exciting and the backscattered signal are conveyed and collected by the removable fibre optic sensor (Fig.3) [2]. The sensor is a compact side-view probe of reduced size (*Φ*13mm x 50 mm) (Fig.4), to extend the measure also to the leading edge of the blade. The TGO luminescence spectra is acquired by a compact spectrometer, specifically designed to match the TGO luminescence spectral range, whose resolution is approximately 1 cm^{-1} in the frequency range of the Cr^{3+} peaks. The spectrometer is controlled and calibrated by means of a proprietary software interface developed under Labview environment.

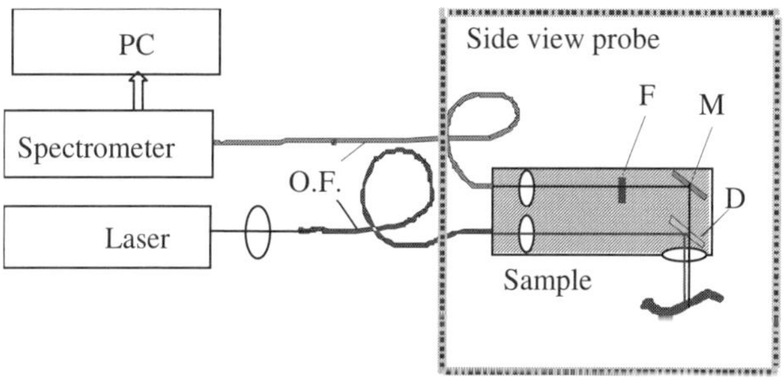

Fig.3: Sketch of the device: L= Laser Source; S=spectrometer; O.F.=optical fibre; F=Filter; M=Mirror; D= Dichroic mirror

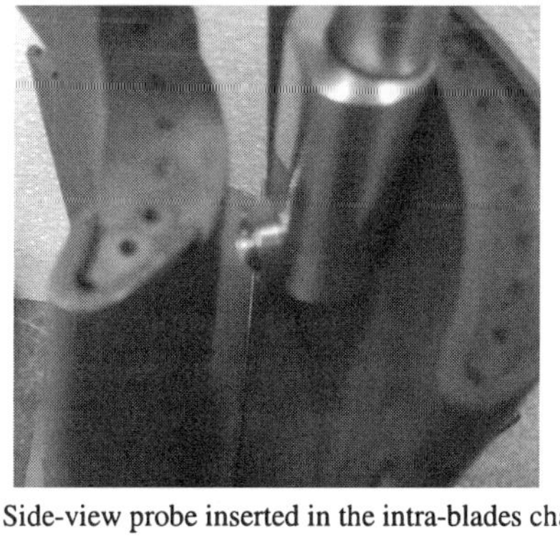

Fig.4: Side-view probe inserted in the intra-blades channel

4. Results

Instrument validation was performed through tests on a series of TBC coated specimens aged in cyclic and isothermal oxidation at different temperatures. The comparison with results of measurements with a commercial instrument was successful (Fig.5). The differences between the results obtained on the same specimen are much smaller than the standard deviation of the single measure.This is the case both for new coatings (high stresses) and for damaged coatings (stresses around 1GPa), with various degradation mechanisms (edge delamination, buckling).

The new portable system was used to assess the coating integrity on the following samples:

- Laboratory aged specimens: (fig.6). The measurements of TGO stresses by PLPS of samples to thermal cycling were compared with microstructural analysis performed on the corresponding metallographic sections. The results confirm the system capability of prematurely detect and localise initial damage process on cycled samples, before EB-PVD TBC spallation.

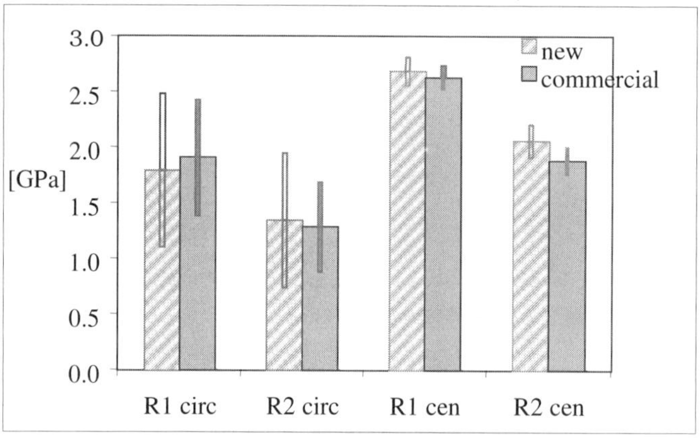

Fig. 5 Comparison between TGO stress measurements performed with the new optical device and commercial one

- New components assessment in laboratory: measurements were taken in regions of both suction and pressure sides; high stress levels, typically higher than 2.5 GPa, were generally found independently from the curvature of the blade, and in case of blades with good quality coatings.
- Serviced components characterization in laboratory (Fig.7): The technique was used to assess the spent life fraction of TBC coated first stage blades of the last generation gas turbine, from a combined cycle plant. The TGO

residual stresses, beneath the thermal barrier, were mapped on different areas of the blade. The system identified regions of low stresses (high damage) towards the trailing edge and of high stresses typically after cooling holes (lower temperature and lower damage).

- Service operated components also in an in-shop condition: with its long fibre optic and lateral view/small size sensor, the system was easily applied to blades also in shop operation. Differences of damage levels between blades having different load histories (base load and cyclic operation) could be evidenced.
- Results of PLPS measures are particularly useful in diagnostic activities on multilayer coatings, with TBC, integrated with the results of other innovative non destructive methods: the acquired data on the topcoat/metallic interface stress level are complementary to those of the metallic coating residual life, as shown previously in [6].

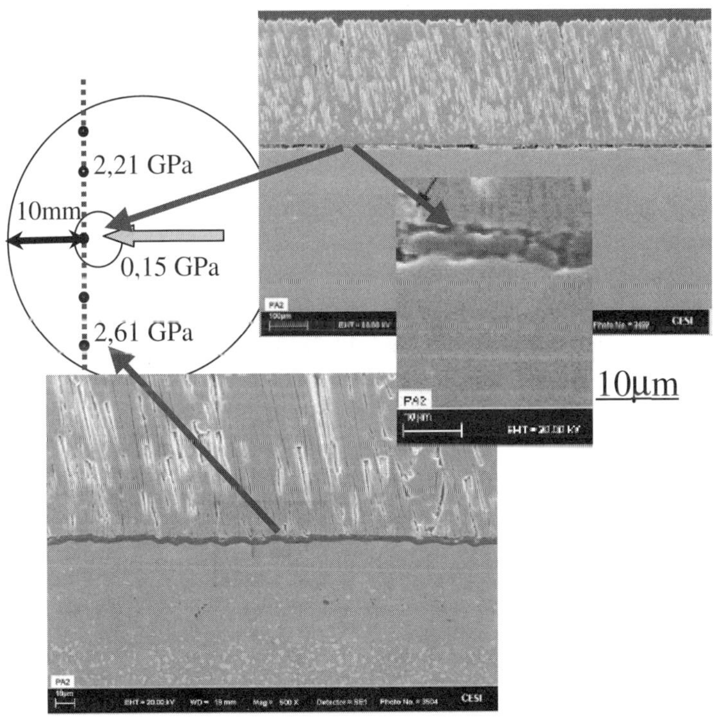

Fig. 6 Map of TGO stresses beneath the TBC measured on specimen coated with Aluminide bondcoat after 143 cycles at 1080°C

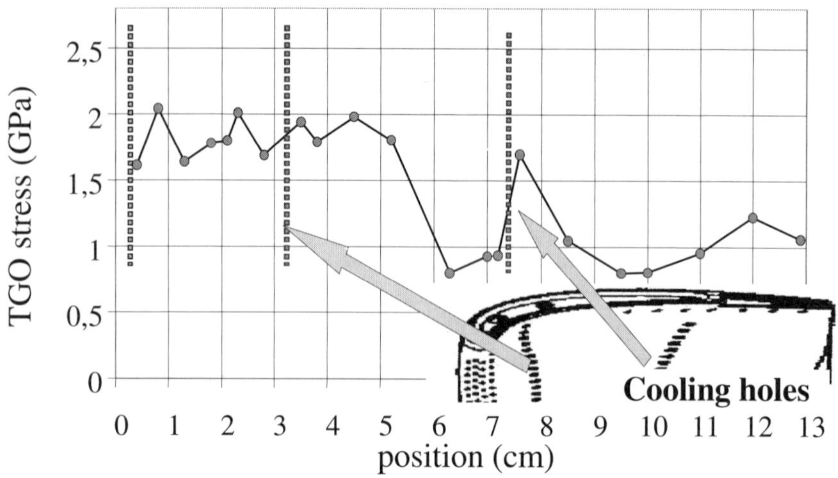

Fig. 7 TGO stress values[GPa] along the pressure side blade airfoil. The scan started at leading edge

The results obtained till now are encouraging: the system can be easily implemented to be used also during endoscopic hot parts inspections, to collect information about TBC coating condition along the turbine life, not only without dismounting blades but even in closed casing conditions.

At present, the technique has been successfully applied to the columnar thermal barriers only; in fact the cheaper (and less resistant) thermal barriers obtained by air plasma spray (APS) have a disordered microstructure rich in porosity and splat boundaries, which induce a too high light scatter. So the PLPS signal cannot pass through the ceramic layer with its normal thickness and come back to be detected out of the external surface of the coating.

5. Conclusion

In this paper a new portable device for PLPS non destructive technique was described. Its performances were successfully validated on aged specimens with different damage levels and on new and service operated blades. The recent development of a side-view prototype probe increased the ease-of-use of the device and a better component accessibility. A further miniaturisation of the probe and the development of a system for tip-insertion and positioning are under development for an endoscopic application during gas turbine hot parts inspections in "closed casing" conditions.

6. Acknowledgments

This work has been financially supported by the Research Fund for the Italian Electrical System established with Ministry of Industry Decree DM 26/1/2000 and 2006.

References

1. S. Osgerby, C. Rinaldi, L.De Maria, in **Proc. of Liege Conference: Materials for Advanced Power Engineering**, vol.1 (2006) 217
2. A.Zombo presentation at the Int. Conf. on Gas turbine technology, Bruxelles, July (2003), www.Siemens.org
3. Christensen R.J, Lipkin D.M, Clarke D.R., Murphy K.S., *Appl.Phys. Lett.*, **69 [24**] (1996) 3754-3756
4. L. Xie, E.H. Jordan, M. Gell, K.S. Murphy, *Material Sci. & Eng.*, **A393** (2005) 51-62
5. A Nychka, D R Clarke, S Sridharan, E Jordan, M. Gell, M J Lance, C J Chunnilall, I M Smith, S R J Saunders, R. Pillan, V. Sergo, A. Selçuk, A. Atkinson, and K S Murphy, *Surf. Coat. Tech.*, **163-164** (2003) 87-94
6. C. Rinaldi, L. De Maria, F. Cernuschi and G. Antonelli, **Proc. ASME ESDA 2006,** Torino, July 4-7 (2006), paper ESDA2006-95555.

SILICON RESONANT CAVITY ENHANCED PHOTODETECTOR BASED ON THE INTERNAL PHOTOEMISSION EFFECT

M. CASALINO

Università "Mediterranea" di Reggio Calabria, Località Feo di Vito
Reggio Calabria, 89060, Italy

L. SIRLETO

Consiglio Nazionale delle Ricerchey, Via P. Castellino, 111
Napoli, 80131, Italy

L. MORETTI

Università "Mediterranea" di Reggio Calabria, Località Feo di Vito
Reggio Calabria, 89060, Italy

F. DELLA CORTE

Università "Mediterranea" di Reggio Calabria, Località Feo di Vito
Reggio Calabria, 89060, Italy

I. RENDINA

Consiglio Nazionale delle Ricerchey, Via P. Castellino, 111
Napoli, 80131, Italy

In this paper, a methodology for the analysis of a resonant cavity enhanced (RCE) photodetector, based on internal photoemission effect and working at 1.55 μm, is reported. In order to quantify the performance of photodetector, quantum efficiency including the image force effect, and its dependence from the inverse voltage applied, are calculated. We propose a comparison among three different Schottky barrier Silicon photodetectors, having as metal layers gold, silver or copper respectively. We obtain that the highest efficiency (0.2%) is obtained with metal having the lowest barrier (Copper). The device fabrication is completely compatible with standard silicon technology.

1. Introduction

Silicon photodetectors have already found wide acceptance for visible light (0.400-0.700 micron) applications [1], while for applications in optical communications in the near-IR wavelength range between 800–900 nm they suffer from low bandwidth-efficiency products due to the long absorption length necessitated by the small absorption coefficient. In silicon (Si), considering the interband transition, a cut off wavelength of about 1.1 micron is obtained, therefore, in order to obtain photodetector working at 1.3-1.55 micron fiber optic communication wavelength range, we have two possible option. The former is to use a semiconductor which is sensitive around the 1300–1550-nm wavelength range. Germanium (Ge) is a good candidate, given its smaller direct energy band gap of 0.8 eV, but unfortunately bulk Ge is still a relatively weak absorbing material at 1550 nm. As a result, a thick Ge active region would be required to obtain a certain level of quantum efficiency, resulting in a slow device. Besides the growth of this compound on silicon is still a challenge in terms of cost and complexity [2]. Therefore the direct monolithic integration of photodetectors in a chip should be a more attractive solution to integrate receivers with electronics. The latter option is the exploitation of the internal photoemission effect over the metal-semiconductor Schottky barrier [3]. Schottky photodiodes based on the internal photoemission effect are very attractive because of their simple material structure and fabrication process. The main advantage of these devices resides in their extremely high switching speed, but due to the leakage photon flux within the metallic layer, their quantum efficiency is small. In order to enhance the quantum efficiency, the Resonant-cavity-enhanced (RCE) detection scheme is particularly attractive for Schottky-type photodetectors, since the semitransparent metal contact can also function as the top reflector.

In Resonant-cavity-enhanced photodetectors (RCE-PD) the enhancement of quantum efficiency η is obtained by placing the active layer inside a Fabry-Perot cavity. The optical field enhancement in the cavity allows the use of thin absorbing layers, which minimizes the transit time of the photogenerated carriers without hampering the quantum efficiency [4]. RCE-PD have been successfully demonstrated for a range of operating wavelengths, including Si-based detectors optimized for 850 nm [5] and Ge-based detectors designed for operation around 1550 nm [6].

In this paper, the design of a RCE metal Schottky barrier Si photodetector, based on internal photoemission effect and operating at 1.55 μm, is described. A comparison among three different photodetectors, having as Schottky metal: gold, silver or copper respectively is proposed and analysed. In order to estimate the theoretical quantum efficiency, we take advantage of analytical formulation of the internal photoemission effect (Fowler theory) including the image force effect, and its extension for thin films, while for the optical analysis of device a numerical method, based on the Transfer Matrix Method, has been implemented.

2. Design and optical analysis

The sketch of device is shown in fig. 1. The RCE-PD is based on internal photoemission effect over a Schottky junction metal-Si, top illuminated and operating at 1.55 micron. The resonant cavity is a Fabry-Perot vertical-to-the-surface structure. It is formed by a buried reflector, a mirror top interface and in between a λ/2-silicon-layer. The buried reflector is a Bragg mirror formed by alternating layers of Si-SiO_2 structures.

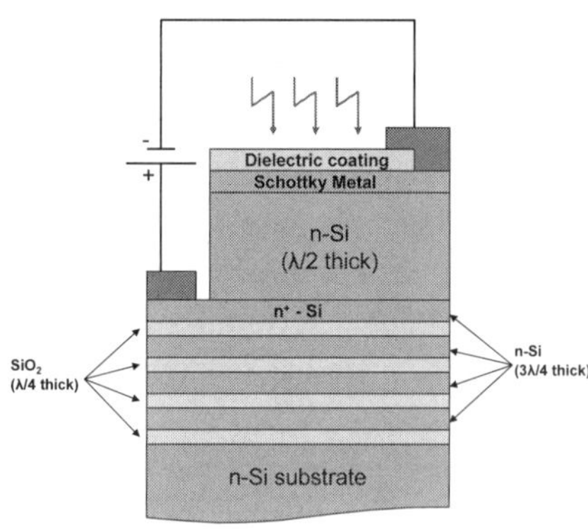

Figure 1. Schematic cross section of our RCE Schottky photodetector.

Deposited on λ/2-silicon-layer is a semitransparent Schottky metal and a dielectric coating layer, working as the top reflector of the resonant cavity. We underline that our structure is different from the RCE Schottky PD's in which the Schottky contact is only an electric contact and not the active layer, whereas in our device the metal layer works as top contact and as active (absorbing) layer at the same time.

One of the many benefits of silicon is the large index contrast provided by $Si\text{-}SiO_2$ structures, allowing the realization of high-reflectivity, wide spectral stop-band Distributed Bragg Reflector (DBR) made of few periods [7]. Anyway, limitations in fabrication process usually do not allow for layer thickness as thin as ($\lambda/4n$), for this reason ($3\lambda/4n$) layers for Si were used [7]. Starting from this results, in our design we propose a DBR centered at 1.55μm. The DBR could be formed by alternate layers of Si and SiO_2 having refractive index 3.45 and 1.45, and thickness of 340 nm and 270 nm, respectively. In our design a bottom mirror formed by 4 periods of Si/SiO_2 is considered. In order to get ohmic contact, the top layer of the DSOI is supposed to be realized by a very thin but heavily doped $10^{19}cm^{-3}$ silicon layer. Regarding the top reflector of the resonant cavity, we consider three metals: gold, silver and copper, whose optical and electrical properties are summarized in table 1 [8]-[11]:

Table 1. Optical and electrical properties for three metals: gold, silver and copper

Metal	Complex refractive index (N)	Mean free path (L_e) [μm]	Fermi level (E_F) [eV]	Potential barrier (Φ_B) [eV]
Au	0.174-j9.960	0.055	5.530	0.780
Ag	0.450-j9.290	0.057	5.480	0.780
Cu	0.145-j9.830	0.045	7.050	0.580

The quantum efficiency of a RCE-PD, based on internal photoemission effect, is given by the formula [12]:

$$\eta = A_T F_e P_E \eta_c \tag{1}$$

where A_T is the total optical absorbance of the metal, F_e is the fraction of absorbed photon which produce photoelectrons with appropriate energy to contribute to the photocurrent, P_E is the total accumulated probability that one of these photoexcited electrons will be able to overcome the Schottky barrier after scattering with cold electrons and with boundary surface and η_c is the barrier collection efficiency, which is bias dependent due to the image force effect.

2.1. *Theory of internal photoemission*

Internal photoemission is the optical excitation of electrons into the metal to an energy above the Sckotty barrier and then transport of these electrons to the conduction band of the semiconductor. The standard theory of photoemission from a metal into the vacuum is due to Fowler[13]. In a gas of electrons obeying the Fermi-Dirac statistic, if energy photon is close to potential barrier ($h\nu \approx \Phi_B$), the fraction (F_e) of the absorbed photons, which produce photoelectrons with the appropriate energy and momenta before scattering to contribute to the photocurrent, is given by:

$$F_e = \frac{\left[\left(h\nu - (\phi_{B0} - \Delta\phi_B)\right)^2 + \frac{(kT\pi)^2}{3} - 2(kT)^2 e^{-\frac{h\nu - (\phi_{B0} - \Delta\phi_B)}{kT}} \right]}{8kTE_F \log\left[1 + e^{\frac{h\nu - (\phi_{B0} - \Delta\phi_B)}{kT}}\right]} \tag{2}$$

where hv is photons energy, Φ_{B0} is the potential barrier at zero bias, $\Delta\Phi_B$ is the lowering due to image force effect (as we will see later) and E_F is the metal fermi level. In fig. 2, Fe as a function of bias voltage for three different metals, at room temperature, is reported. As metal we choose gold (Φ_{B0}=0.78eV), silver (Φ_{B0}=0.78eV) and copper (Φ_{B0}=0.58eV). We obtain that increasing the bias voltage, F_e increases.

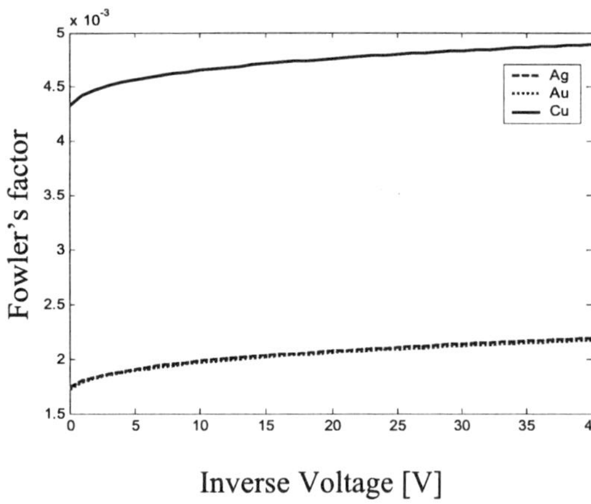

Figure 2. Fowler's factor (F_e) versus bias voltage applied.

The equation (2) was obtained without taking into account the thickness of the Schottky metal layer. In order to study the quantum efficiency for thin metal films, the theory must be further extended, taking into account multiple reflections of the excited electrons from the surfaces of the metals film, in addition to collisions with phonons, imperfections and cold electrons. Assuming a thin metal film, a phenomenological, semiclassical, ballistic transport model for the effects of the scattering mechanisms resulting in a multiplicative factor for quantum efficiency was developed by Vickers [12]. According to this model the accumulated probability P_E that the electrons will have sufficient normal kinetic energy to overcome potential barrier is given by:

$$P_E \cong \left[1-e^{-\frac{d}{L_e}}\right]^{\frac{1}{2}} \tag{3}$$

where d is the metal thickness and a L_e the mean free path.
Moreover in order to estimate the quantum efficiency, the calculation of the absorptance (A_T) is numerically carried out by the Transfer Matrix Method (TMM) [14].

2.2. *Image-force lowering of the barrier*

In order to estimate quantum efficiency, we must take into account the image force between an electron and the surface of metal. The effect of the image force is that the barrier which an electron has to surmount in passing from the metal into the semiconductor is lowered by an amount ($\Delta\Phi_B$) while the maximum potential energy occurs at a position x_m. The barrier lowering ($\Delta\Phi_B$) and the barrier maximum position (x_m) are given by [10]:

$$x_m = \sqrt{\frac{q}{16\pi\varepsilon_{Si}}\frac{W}{|V_{Bias}|}} \qquad \Delta\phi_B = \sqrt{\frac{q}{4\pi\varepsilon_{Si}}\frac{|V_{Bias}|}{W}} \tag{4}$$

where ε_{Si} is the permittivity of silicon (10^{-12} C/cmV), W is the depletion width and V_{Bias} the applied bias voltage (fig.3).

The probability that an electron travels from the metal-semiconductor interface to the Schottky barrier maximum without scattering in the Si is taking into account by the barrier collection efficiency η_c, which is given by:

$$\eta_c = e^{-\frac{x_m}{L_s}} \tag{5}$$

where L_s is the electron scattering length in the silicon. It is worth noting that increasing the bias voltage, a shift of Schottky barrier closer to metal/semiconductor interface is obtained. Therefore, the barrier collection efficiency increases.

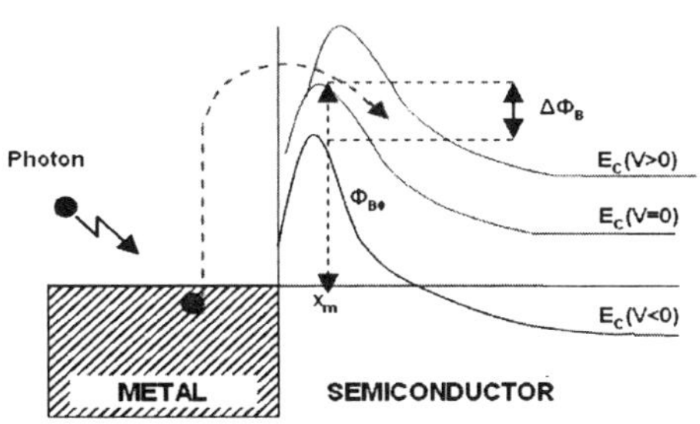

Figure 3. Energy band diagram for a metal/n-semiconductor junction.

3. Quantum efficiency

An analytical formulation of the quantum efficiency for a simplified RCE-PD structure with lossless mirrors was given by Kishino et. al. [15]. In the case of an absorptive mirror, such as the semitransparent metallic top mirror of RCE Schottky PD's the previous formulation is no longer valid. Therefore in order to calculate the maximum quantum efficiency of a RCE metal Schottky barrier Si photodetector, based on internal photoemission effect and operating at 1.55 μm, the following methodology has been adopted:

1. The calculated values of bottom mirror reflectivity and phase (R_2, ψ_2) are 0.990 and 3.11 rad, respectively.
2. for metal thickness into the range (0-50 nm) top mirror reflectivity and phase (R_1, ψ_1), the value of silicon cavity thickness in resonance condition and the absorbance is calculated using the TMM method. We obtain a curve of absorbance as a function of metal thickness and we consider the maximum.
3. Dielectric coating thickness, chosen in order to avoid perturbation of resonance condition, is a Si_3N_4 layer, having refractive index 2.0 and thickness of 390 nm.
4. Obtained the maximum absorbance, the parameters of optimized cavity are fixed and the quantum efficiency as a function of wavelength in the range of interest is calculated using Eq. (1).

The methodology takes into account the dependency of efficiency on reflectivity of the top and buried mirrors and the normalised absorption coefficient αd of the metal, while the peak wavelength is a function of cavity length (L) and of the phases shift due to the top and bottom mirror (Ψ_1, Ψ_2). The parameters calculated are summarized in the table 2.

In our device, we suppose that the metal-semiconductor junction is polarized in opposite way. It is simple to prove that for an inverse voltage of -0.85 V the depletion layer is greater than the $\lambda/2$-layer of cavity (i.e. the path that carriers generated into metal have to cross before being collected by ohmic contact).

Table 2. Cavity parameter coming out from our simulations

Metal	Cavity thick-ness (L) [μm]	Metal thick-ness (d) [nm]	R_1	Ψ_1 [rad]	A_T	Q value
Au	0.420	30	0.920	-2.140	0.780	525
Ag	0.410	20	0.730	-2.110	0.930	153
Cu	0.420	32	0.930	-2.360	0.740	585

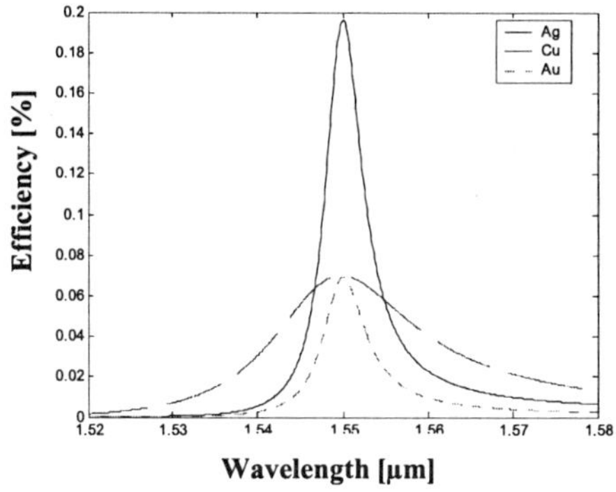

Figure 4. Calculated quantum efficiency versus wavelength for 0.85 V of inverse voltage applied.

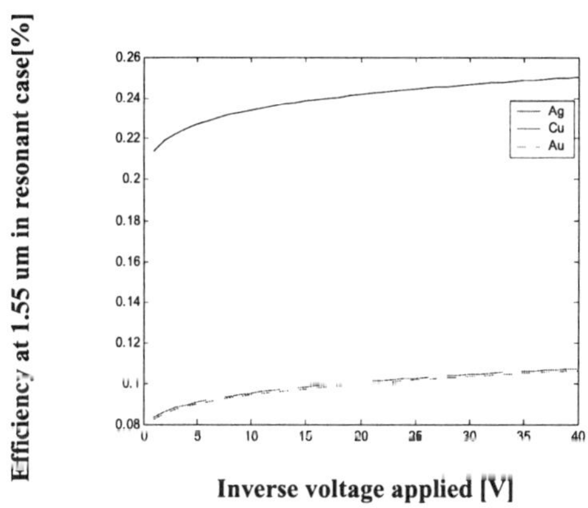

Figure 5. Calculated quantum efficiency at 1.55µm versus the inverse voltage applied.

Considering the range of bias voltage [1V-40V], the previous condition is always satisfied, therefore we can assume that the depleted region is just the cavity length (W=L).

Efficiency versus wavelength (fig. 4) and maximum efficiency at the resonant wavelength versus inverse voltage applied (fig. 5), for the three different metals, are reported. We obtain that a significant enhancement in quantum efficiency can be achieved by using a resonant cavity structure. Among the different metals, the quantum efficiency increases, lowering the barrier. It is worth noting that copper has the best quantum efficiency (about 0.2%) and selectivity due to its lower potential barrier and to its higher reflectivity, respectively. It is interesting to compare also gold and silver, due to the same value of barrier we get the same order of efficiency, but in the case of gold we get a better selectivity, due to the higher reflectivity. Finally, we prove that for a given metal, increasing the inverse applied voltage, the quantum efficiency at 1.55 micron increases (fig. 5).

4. Conclusions

In this paper, the design of a silicon RCE photodetector based on the internal photoemission effect and working at 1.55 micron, was proposed. Quantum efficiency, including the image force effect, as a function of bias voltage is calculated for three different metals as gold, silver and copper. Results prove that photodetectors having copper as top mirror have best results The main advantage of this structure is that the fabrication process is simple and completely compatible with standard silicon technology. We believe that these results open the way to investigate more complex structure having a higher Q value (for example ring resonator) which could provide a significant improvement of quantum efficiency.

References

1 H. Zimmermann, Silicon Photo-Receivers. Heidelberg : Springer Berlin, 2004.

2 S. Famà, L. Colace, G. Masini and G. Assanto, “High performance germanium-on-silicon detectors for optical communications,” Applied Physics Letters, vol. 81, pp. 586, 2002.

3 Kimukin, E. Ozbay, N. Biyikli, T. Kartaloglu, O. Aytur, S. Unlu, G. Tuttle, “High-speed GaAs-based resonant-cavity-enhanced 1.3 m photodetector,” Applied physics Letter, vol. 77, pp. 3890, 2000.

4 M. S. Unlu and S. Strite, Appl. Phys. Rev., “Resonant cavity enhanced (RCE) photonic devices,” vol. 78, pp. 607, 1995.

5 M. K. Emsley, O. I. Dosunmu and M. S. Unlu, “High-speed resonant-cavity-enhanced silicon photodetectors onreflecting silicon-on-insulator substrates,” IEEE Photonics Technology Letters, vol. 14, pp. 519, 2002.

6 O. I. Dosunmu, D. D. Cannon, M. K. Emsley, L. C. Kimerling and M. S. Unlu, “High-Speed Resonant Cavity Enhanced Ge Photodetectors on Reflecting Si Substrates for 1550-nm Operation,” IEEE Photonic Technology Letters, vol. 17, pp. 175, 2005.

7 M. K. Emsley, O. Dosunmu, M. S. Unlu, “Silicon substrates with buried distributed Bragg reflectors for resonant cavity-enhanced optoelectronics,” IEEE Journal of Selected Topics in Quantum Electronics, vol. 8, pp. 948, 2002.

8 E. Y. Chan, H. C. Card, "Near IR interband transitions and optical parameters of metal-germanium contacts", Applied Optics, vol. 19, pp. 1309, 1980.
9 E. Y. Chan, H. C. Card, M. C. Teich, "Internal photoemission mechanisms at interfaces between germanium and thin metal films," IEEE Journal of Quantum Electronics, vol. QE-16, pp. 373, 1980.
10 S. M. Sze, Physics of Semiconductor Devices. New York: John Wiley & Sons, 1981, ch. 5.
11 P. Yeh, Optical Waves in Layerer Media. New York: A Wiley Interscience Publication, 1988.
12 V. E. Vickers, "Model of Schottky Barrier Hot-electron-Mode Photodetection," Applied Optics, vol. 10, pp. 2190, 1971.
13 R. H. Fowler, "The analysis of photoelectric sensitivity curves for clean metals at various temperatures," Physical Review, vol. 38, pp. 45, 1931.
14 M. A. Muriel, A. Carballar, "Internal field distributions in fiber Bragg gratings," IEEE Photonics technology letters, vol. 9, pp. 955, 1997.
15 K. Kishino, M. S. Unlu, J. Chyi, J. Reed, L. Arsenault, and H. Morkoc, "Resonant cavity-enhanced (RCE) photodetectors," vol. 27, pp. 2025, 1991.

ELECTRIC AND OPTICAL SENSING IN NIR ENVIRONMENTAL MONITORING

MEDUGNO MARIO

Consiglio Nazionale delle Ricerche, Institute for Microelectronics & Microsystems, Via P. Castellino 111, Napoli, 80131, Italia

Non Ionizing Radiation monitoring in wide areas or built environments is a challenging task, as the electromagnetic fields are largely variable over space and time. In this paper we review several NIR sensing instruments and some standard procedures adopted in environmental monitoring activities for the Province of Naples. Moreover we propose electronic and optical sensors which can be massively adopted for fast detection of electromagnetic fields, and a no-perturbation inducing device suitable to locally monitor the telecommunication infrastructures.

1. Introduction

Monitoring the Non-Ionizing (or ionising) electromagnetic Radiation (NIR) in wide areas or built environments is a challenging task, as the electromagnetic fields are largely variable over space and time depending on the infrastructure source type, the state of electrical power supply and the physics properties of matter where field propagation occurs in the real world.

Under known and standard conditions, hard to obtain and/or realize, the fields produced by sources on wide areas could be in theory determined from electrical and physical parameters, but the field measurement in real world is a local and time-consuming process which involves a wide set of unknown factors on wide area.

NIR acronym in other contexts also stands for near infrared; the broad frequency spectrum of the electromagnetic field ranging from ELF to UV bands is called NIR because the field has enough energy to move atoms in a molecule around or cause them to vibrate, but not enough to remove electrons (ionisation). Examples of frequently used NIR are extremely low frequency (ELF), radio frequency waves (RF), microwaves (MW), infrared (IR) and the light in the visible band (VIS). The interaction of NIR fields with the matter therefore generally promotes vibration and translation motions of the molecules, depending on the field frequency and the structural properties of the matter. For example MW can induce strong currents and thermal effects in organic matter, so power tubes radiating several hundreds of watt are used as field source in ovens, or power HF sources radiating several hundreds of watt can be used for industrial soldering and etching processes.

While all kinds of ionising radiation can cause cancer and other acute, stochastic or long-term health effects (these are out of the scope of the present paper), NIR under defined range of fields and/or power density values are not considered dangerous for human health by World Health Organization. Several research works on the matter are therefore in progress, and they are frequently financed by private funds. The only pathway of concern with NIR prevention is the direct or external exposure of people to electromagnetic field; several national authorities [1] makes the conservative (cautious) assumption that any increase in radiation exposure is accompanied by an increased risk of effects on the human health. So under the cautious assumption several field threshold values and mechanisms of exposition of the population are defined by European, national, regional and local laws. The population is aware of the presence of a NIR source when a macroscopic infrastructure is visible over a wide territorial area or a built environment, or when is object of a study or census [2,3]; thus people tends to under/overrate the risk on the base of personal sensibility to risk perception. Recently human exposure at direct-energy of NIR, inducing superficial thermal effects on the skin, are also been proposed for defence and military applications (Active Denial Systems) as mobile non lethal denial systems [2].

2. Environmental NIR Monitoring

The environmental electromagnetic field is the sum of a lot of fields produced by time dependent analog and digital modulated sources. The field measurement procedure returns a signal which is temporally and spatially averaged according to the administrative law technical specifications [3]. Minimal local field modification is introduced with optical probes, but no such requirement is specified in any law; it's only a recommended practice for the electromagnetic field measurement.

A measurement campaign in wide territorial area requires large funds and time, thus occasionally occurs in a spot area when a lawsuit is filed or a new equipment is installed. Several monitoring activities of NIR are described in the project "Censimento delle emissioni elettromagnetiche e redazione del Piano Provinciale delle installazioni in Provincia di Napoli" [3]. Data of existing infrastructures in the 2003 year, high resolution aerial orthophoto, measures and administrative limits were collected in a Geographical Information System. Figure 1 shows several data on features as the NIR infrastructures and electromagnetic field measurements performed in several "hot spots" with standard instruments as the PMM 8053 and the related triaxial probes as the EP300 for RF and EHP50 for ELF band. Measures were performed according to the Italian and regional laws illustrated in [3].

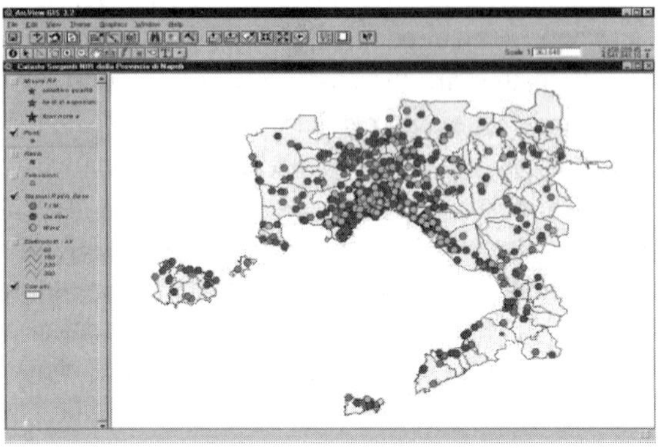

Fig.1. An Arcview GIS map of NIR infrastructures and measurements in the Province of Naples.

In connection with the GIS, a Microsoft Access database for infrastructures and monitoring procedures was proposed for standard data management, and for further software simulations of the NIR. It contains electric data for RF power systems and the technical specifications of the radiating elements.

An analysis of telecommunication infrastructure on the communal territories was developed by a Microsoft Access procedure, in order to provide an useful information system for the authorization of new infrastructure and to implement the communal installation plans. It provides statistical indicators as RF infrastructure densities (per surface unit and number of inhabitants).About eight hundreds radio-TV infrastructures and eight hundreds Base Transceiver Stations were georeferenced; data were classified as "sensible", therefore only a statistical summary map is shown in Figure 2. Indicators of NIR infrastructures in Naples are comparable with ones of the Milan metropolitan area. NIR monitoring was performed at same time in France[4]; here differently from Italy, the data of each infrastructure were classified as "non sensible" and published on the web together with the cartographic data.

NIR in the broad frequency range of 0-300GHz are generated by a huge set of electronic devices and infrastructures, mainly power lines (aerial or underground) and radio stations for broadcasting and cellular networks. In a general way guidelines for measuring fields are given in [5,6] while methods for estimating field are known from literature when a dataset for the infrastructure is given in [3]. Here the suggested readings include the standard radio-communication bibliography for RF estimates.

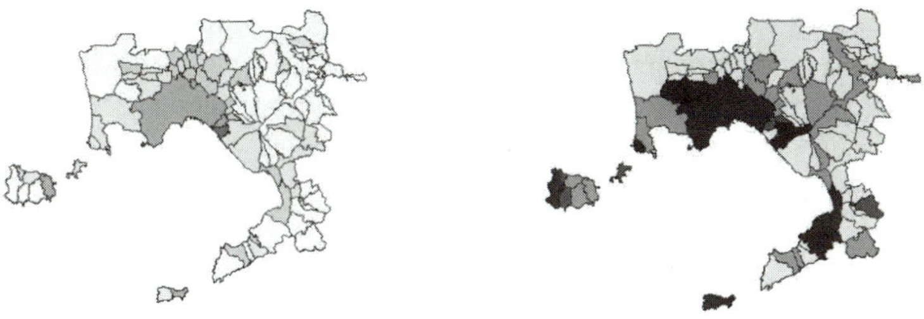

Fig. 2. Density maps for NIR infrastructures: SRB in green and radio-TV in blue.

The normalized densities of broadcast and cellular network plants and "potential risk" indicators were computed on an administrative base. Figure 2 shows the 2003 statistic data on Neapolitan wide area generated by ESRI GIS: in green the map for SRB infrastructures, and in blue for radio-TV. As can be seen the former have a typical local coverage proportional to residential people density while the latter shows a slight different dynamic due to economical influence of radio-TV media.

3. Measuring Magnetic Fields by a "dowser rod" Electric Sensors

Specific guidelines for ELF field measurement are given in [7]; underground filled power lines in particular are not visible, and a precise measure in wide territorial area is impossible to perform with a huge number of local magnetic field measurement. Therefore an electronic sensor for fast detection of magnetic field hot spots in ELF-LF band has been designed and realized; it is used as a "dowser rod". Figure 3 shows the schematic of the ELF magnetic sensor.

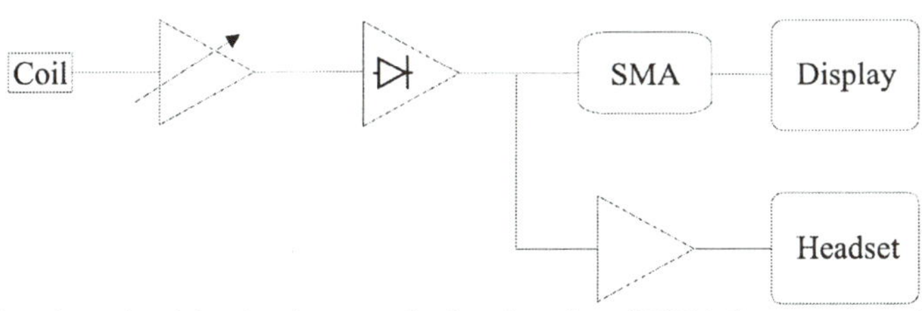

Fig. 3. The schematic of the electric sensor for fast detection of ELF infrastructures and magnetic field hot spots.

Such magnetic field sensor consists of a magnetic field sensor with a variable gain amplifier, an AM detector, a Simple Moving Average (SMA) with a meter for detecting space-temporal magnetic field variation. The SMA signal in a fixed 200 ms temporal window is displayed by a 4 LED array.

The AM detector output signal is amplified and issued to an audio headphone for spectral and amplitude aural sensing, in fact the mono and three-phase fields yield different harmonic contents.

Fast prototyping of the invisible electric infrastructure and field hot spot detection also in wide communal areas are easily allowed by a trained human using such sensor, and the measurement process could be automated by a specialized moving robot.

A precise measurement by the PMM EPH50 triaxial standard probe with optical lines has ever been performed according to the Italian specifications [7] after an hot spot detection by the "dowser rod" electronic sensor.

The AM detector is also able to detect modulated magnetic fields in the LF band generated by electric switch arcs or envelopes generated by inverters, dimmers and other power electronic switching devices. Unexpectedly strong FM radio broadcasting in the VHF band can be heard when the rod is pointed to 2 meters long metallic targets, as it could occur with a Foster-Seeley discriminator for FM detection.

4. Measuring Electric Fields by an Optical Sensor

Standard metallic probes for electric and magnetic field measurements, also in the Fresnel region, are often equipped with optical transmission lines to avoid coupling with the measuring field.

The heads of standard probes generally need both power source and electronics for i) conditioning and processing of the signals received by an external metal antenna, ii) electro-optical (EO) conversion to eventually drive an optical link with the measure equipment. The mutual coupling between EMF and a standard metal probe is the main source of precision loss, which can strongly be reduced by a completely optical probe. Recently have been developed new EMF sensors elements (we use the terms probe and sensor as synonyms) made by dielectric materials to overcame this problem.

We propose an optical electromagnetic field sensor, made only of $LiNbO_3$ and a very tiny Indium Tin Oxide (ITO) tapered antenna [7]. The probe is an optical driven device with input-output optical fibers connecting it to a laser source and a photodetector, as shown in Figure 4.

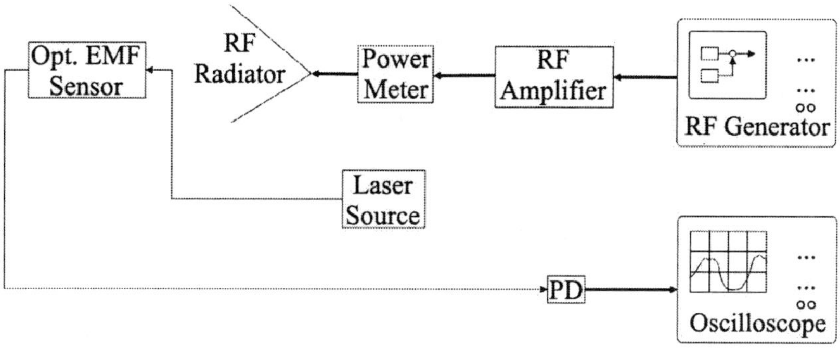

Fig. 4. The experimental setup of the optical electromagnetic field sensor.

A very tiny resistively loaded antenna, with a flat response in a wide band, yields the electric input for the integrated Mach-Zehnder interferometer. The optical probe is fabricated on a $LiNbO_3$ substrate, as sketched in the Figure 5, with the antenna superimposed on the modulator. The waveguide is underlying the substrate as described in the following section, and a phase shift in the arms of the interferometer is yielded by the electro-optic effect of the $LiNbO_3$. This phase shift produces an amplitude modulation in the optical output, where in particular the potential difference $V\pi$ applied to electrodes is such to induce an optical phase shift $\Gamma=\pi$ between the modulator arms.

Optical probes result in small dimension devices mainly made by non metallic parts, thereby dramatically less invasive then standard EMF probes. The dimensions of electrical and optical gaps G_{EL} e G_{OPT}, as well the antenna widths are in the order of microns, and the arm length is 80 mm.

When a metallic probe is applied in proximity of a directive RF radiator at 2.4 GHz, the radiator impedance changes and introduces a power loss due to impedance mismatch with the power line. The optical probe applied in proximity of the same directive RF radiator did not introduce any appreciable power loss, confirming that is non-metallic structure does not induce any perturbation on RF transceiver infrastructures.

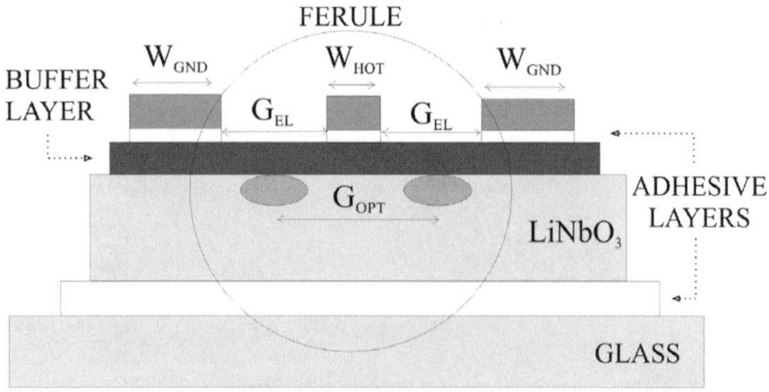

Fig. 5. The structure of the optical electromagnetic field sensor.

As can be seen in Figure 5, a section of the optical device where the layer thickness are not in scale, the ceramic ferules of the input-output fibers are aligned with the optical waveguides; the only upper conductive layer measures 1,4 μm.

The $LiNbO_3$ probe provided the electro-optical response expected by theoretical tapered antenna framework, optical power budget and photodetector conversion gain in the VHF band [7].

The monitoring activities described in the previous section showed that the principal NIR radiating infrastructures are the antennas of radio broadcasting. The optical probe seems a suitable add-on device for permanently monitor NIR infrastructures instead of a power meter insertion in the RF chain (cited as a black-box) in the Naples metropolitan area regulation law.

5. Conclusions

We described NIR sensing instruments and some standard procedures adopted in previous experiences of environmental monitoring for the Province of Naples, and in particular the sensors adopted for measurement electromagnetic field.

A new electric sensor for characterization and fast detection of strong magnetic fields in the ELF/LF band generated by indoor and outdoor hidden infrastructures is discussed.

The optical electromagnetic field sensor proposed in this paper seems an optimal probe for local monitoring the sources of NIR in wide area network as is dramatically less invasive then standard EMF probes and can be added in the proximity of existing radiating infrastructures as a no-perturbation inducing device.

References

1. International Commission on Non-Ionizing Radiation Protection: Guidelines for limiting exposure to time-varying electric, magnetic and electromagnetic fields (up to 300 GHz), Health Physics, vol.74, n.4, pp.494-522,1998.
2. http://news.bbc.co.uk/1/hi/world/americas/6297149.stm
3. M. Medugno and M.Fontana, Radiazioni Non Ionizzanti ed Esposizioine Umana: dai Fondamenti Fisici alla Gestione Territoriale degli Impianti Radianti, Provincia di Napoli, 2004.
4. Stations radioélectriques et points de mesure de champs électromagnétiques, www.cartoradio.fr
5. IEEE Standard C95.3 – 2002, "IEEE Recommended Practice for Measurements and Computations of Radio Frequency Electromagnetic Fields With Respect to Human Exposure to Such Fields, 100 kHz–300 GHz", ISBN 0-7381-3520-8, Dicembre 2002, pp. 1-133.
6. Comitato Elettrotecnico Italiano, Guida per la misura e per la valutazione dei campi elettromagnetici nell'intervallo di frequenza 10 kHz - 300 GHz, con riferimento all'esposizione umana. Appendice A: Centraline di monitoraggio dei campi elettromagnetici a radiofrequenza: procedure e finalità di utilizzo, CEI 211-7, 2006.
7. Comitato Elettrotecnico Italiano, Guida per la misura e per la valutazione dei campi elettrici e magnetici nell'intervallo di frequenza 0 Hz - 10 kHz, con riferimento all'esposizione umana, CEI 211-6, 2001.
8. L. Ciccarelli e M. Medugno, Electric Field Measurements by a LiNbO3 Probe, Proc. Aisem 2006, World Scientific, in press.

MULTI-SPECTRAL INFRARED SYSTEM FOR TOXIC GAS DETECTION

C. CORSI, N. LIBERATORE, S. MENGALI, A. MERCURI, R. VIOLA, D. ZINTU

Consorzio CREO, via Pile 60,
67100 L'Aquila, Italy

M. SEVERI, G. CARDINALI, I. ELMI, M. PASSINI

CNR-IMM Sezione di Bologna, via P.Gobetti, 101,
40129 Bologna, Italy

Abstract: Advanced IR emitters and sensors are under development for high detection probability, low false alarm rate and identification capability of toxic gases. One of the most reliable techniques for gas identification is Absorption Spectroscopy, especially in the Mid- and Far Infra-Red, where most molecules exhibit their strongest roto-vibrational absorption bands. A compact non dispersive infrared multi-spectral system is here presented, designed and developed for security and environmental monitoring applications. It utilizes a few square millimeters thermal source, a novel design multi-pass cell, and a smart architecture micro-bolometric sensor array coupled to a linearly variable spectral filter to perform toxic gases detection and identification. Preliminary tests for sensitivity and selectivity are undergoing using mixtures of ammonia and ethylene. Detection capability down to tens of ppm has been demonstrated. Possible improvements through the implementation of open path and hollow-fiber based sensors are discussed.

Keywords: IR micro-bolometers, Smart sensors, Gas sensing..

1. Introduction

Accidental or intentional release in the air of Toxic Industrial Compounds TICs or other toxic chemicals (including Chemical Warfare Agents CWAs) represents a serious threat to life and goods. Such threat generates a demand of toxic gas sensors, which may be divided into two main classes, depending on the scenario and application for which they are intended.

A first class of very high sensitivity sensors should be used to detect very low concentration levels of toxic compounds being continuously released ex.

from an industrial plant, or to 'sniff' toxic compounds being hidden ex. in a luggage at check points and security portals.

A second class of more rugged sensors is intended for scenarios in which release of the toxic gas is in the form of a local, high concentration burst, as due to an industrial accident, a leak of chemicals from a cargo container, or a terroristic attack. Sensors of the second class should therefore be implemented into a sensor-network and used for area-surveillance and early warning in a number of public and private settings, these including ex. port and airport civil- and cargo-terminals, rail- and metro-stations, sport arenas, etc..

Sensors of the two classes are quite different, as they have to fulfill pretty different requirements.

In particular, for sensors of the second class, which are the focus of this paper, high sensitivity is not the most important issue, as such sensors have first to guarantee: a) good identification capability (ability to distinguish toxic gases per classes and from interferents); b) fast response (within tens of seconds); c) cost effectiveness (this including easy installability and maintainability); d) continuous operability.

Mid- and far-Infra-Red absorption spectroscopy has long been considered a very attractive gas sensing technique, because of its identification capability (most toxic gases show strong absorption fingerprints in the spectral range 3-15 micron) and continuous operability (there is no chemical interaction between the sensing head and the gas). However, standard equipment (ex. Fourier Transform FTIR systems) is too bulky and expensive for the application that we are discussing.

An innovative smart sensor[1] is here presented, based on Non Dispersive Infrared Spectroscopy (NDIR), which tries to combine good identification capability with compact size and cost-effectiveness.

2. Principle of Operation

A thermal source of infrared radiation is optically coupled to a linear detector array through a multi-pass cell[2]. Presence of a gas within the cell causes absorption of IR radiation, which can be measured by the detectors. Each detector in the array selects a specific band in the IR spectrum. At low gas concentration, absorption is proportional to optical path, gas concentration, and average absorption strength of the specific gas in the specific band. Spectral separation is accomplished by a linearly variable filter mounted in front of the detector array. The linear filter is a band-pass filter in which the position of max transmittance changes linearly along the main axis of the detector array. Spectral resolution depends therefore on both the punctual bandwidth of the filter and the size of the pixel.

3. System Description

The system was conceived to operate in the IR spectral range between 2 and 20 um, and a first prototype has been designed and set up to operate in the long wavelength infrared (LWIR) from 8 micron to 12 micron, where the main rotational and vibrational absorption bands of several very toxic compounds are located.

A blackbody type thermal source at 800K temperature and of 5mm x 5mm emitting area has been utilized to fulfill the whole spectral range of interest. The radiation emitted from the source is collected and propagated through a multipass cell to increase the absorption path length. The radiation is finally focused onto a micro bolometric sensor array by means of the same optics constituting the cell.

The multi-pass cell gives a total path length of about 1 meter within a volume of 10 x 10 x 10 cm^3, through a 10X path length enhancement respect to its linear size. The multi-pass cell was specifically designed to have a larger throughput in comparison with existing more conventional schemes (e.g. White type or Herriot type). Although starting from a blackbody type source, which suffers of lower spectral density power and brilliance with respect to typical spectroscopic laser sources, the novel cell design allows to collect a useful amount of radiation at each spectral band of interest, even though considering a narrow bandwidth of 0.1 micron. Owing to the implementation of high luminosity f/2 optics in the multipass cell, an overall geometrical extent (entendue) of 0.037 sr cm^2 has been achieved. The infrared source is imaged on the detector, which whole sensitive area is the same of the emitting area of the source[2].

Aiming to the development of an unattended, even expendable, system it is strategic to develop a room temperature operating system using a micro bolometric sensor array specifically designed and developed to make the spectral analysis of the incoming radiation. The spectral selection is obtained by optically coupling the sensor array with a spectrally linear variable filter, which is positioned behind the sensor array very close to its sensing area. This configuration allows to acquire the whole spectrum simultaneously and continuously so enabling fast processing and early warning. Spectral resolution of this first prototype is around 0.3 micron.

Starting from a single suspended membrane of silicon nitride the sensitive area of the detector is separated in multiple sub-elements of asymmetric shape. This allows to increase the number of suitable spectral bands within the useful sensing area and to maximize the useful collecting area for each band at the same time. The final result is shown in figure 1.

Figure 1: 1D array of 16 micro-bolometric sensors on a single SiN membrane. The array is optically coupled to a linearly variable filter, and each pixel in the array becomes sensitive to a specific band in the IR spectrum

The sensor array exhibits characteristics at the state of the art. The responsivity (R) is up to 3×10^4V/W, the detectivity (D*) is up to 1×10^9cmHz$^{1/2}$/W, and the response time is down to 20ms, depending on the final shape and area of the single pixel. Single frame NETDs (f/1, 300 K, 8-12 μm) better than 100 mK can be achieved. In a first prototype the whole sensitive area of 5 mm x 5 mm was separated in sixteen elements (pixels). Each element is a microbolometer strip sensor of 0.25 mm x 5 mm shape. This design allows to obtain a large collecting area, owing to the large bi-dimensional geometry of the whole sensitive area, and to operate the sensor as a linear array at the same time. When the sensor array is coupled with the spectrally linear variable filter, each sensing element will collect the spectral content selected by the corresponding portion of the filter only. The sensor and the filter are assembled together and enclosed in a commercially available package for sealing and connecting the sensor with the read-out electronics. The opto-mechanical layout of the system prototype is shown in figure 2

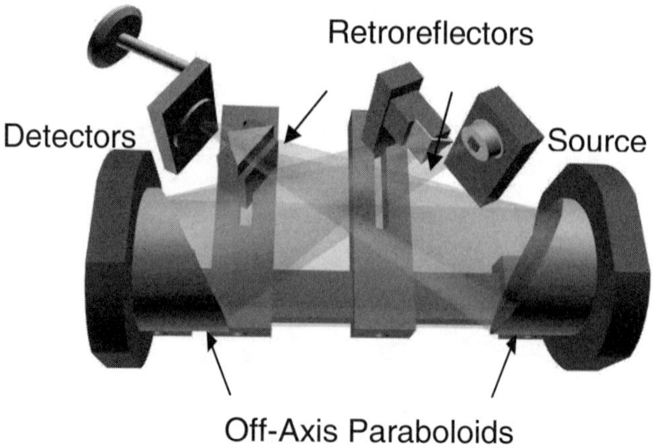

Figure 2: Opto-mechanical lay-out of the sensor prototype

In order to improve the system sensitivity, the source is modulated and the sensor array signals are acquired using a synchronous detection technique by means of parallel multi-channel lock-in amplifier based electronics. The electronics has a modular design in order to match with the useful number of spectral channels and it was set up to be contained in a compact portable unit. The block diagram of the single channel synchronous detection scheme is illustrated in figure 3

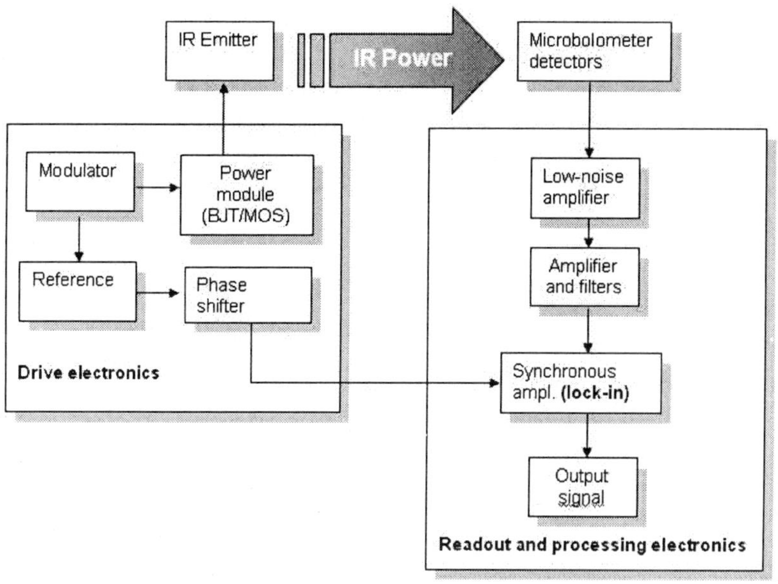

Figure 3: electronic scheme

Although the sensor should be operated in vacuum to achieve higher sensitivity, the overall sensitivity of the system is slightly degraded when sealing the sensor at room pressure. This result is explained considering the behavior of the sensor responsivity at different operative pressures and as a function of the frequency modulation of the source. This is shown in the figure 4. The sensor exhibits higher responsivity, even though with higher time constant, in vacuum condition respect to room pressure operation, as expected owing to the higher thermal insulation of the sensor in vacuum. Moving toward faster modulation of the source signal the responsivity in vacuum asymptotically overlies the value at room pressure.

The overall efficiency of the system, in terms of signal to noise ratio, is determined by the sensor responsivity and noise at the frequency of modulation, and also by noise from the synchronous detection electronics. The sensor responsivity increases at lower frequency. In particular, vacuum sealing gives a noteworthy better performance at modulation frequencies of the source lower than few Hz, up to a factor of ten in DC operation. The responsivity drops rapidly at frequency higher than about 10 Hz, both in vacuum and at room pressure, owing to the intrinsic response time constant of the sensor that is tens of milliseconds. On the other hand, the main component of the sensor noise which is the one proportional to 1/f, (where f is the frequency of modulation), and the lock-in synchronous detection electronics efficiency (e.g. noise vs. frequency of modulation), push toward higher frequencies to increase the system signal to noise ratio. As a result, the room pressure operation at about 10 Hz has been identified to be a good trade-off. This is especially true if additional cost-benefit consideration are included in the trade-off analysis, considering reliability and the additional cost and complexity associated to vacuum sealing

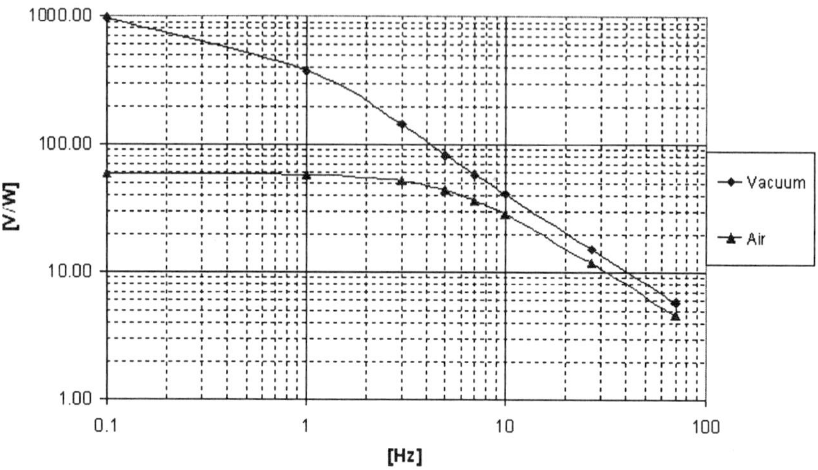

Figure 4: Microbolometer responsivity vs. frequency in vacuum and at ambient pressure

4. Results

A prototype system has been set up accordingly to the previous described design in order to evaluate its effectiveness. A test bench has been settled to make preliminary tests of sensitivity and selectivity. For this purpose a prototype of the system optical head was assembled and enclosed into a sealed test chamber connected to a vacuum system in order to operate in controlled temperature and pressure conditions. The prototype was connected to an external control and read-out electronics by means of sealed connectors to allow safe measurements when using known concentration of toxic samples. In figure 5 the system prototype inside the test chamber is shown.

Figure 5: sensor prototype mounted inside the testing chamber

Ammonia and ethylene were selected as appropriate target gases, owing to their spectral absorption features that are shown in figure 6. Ammonia shows several absorption peaks within the 8-12 micron spectral range and in particular it has two main absorption peaks around 10.35 micron and 10.75 micron. Ethylene exhibits weak absorption within the 9-12 micron spectral range but has a strong absorption peak at about 10.55 micron, just between the two main peaks of ammonia. As above mentioned the spectral resolution achievable by our system is of about 0.3 micron. The absorption bands of ammonia and ethylene are well contained within the system spectral range of operation, moreover their separation is near to match with the spectral resolution of the system. Therefore, ammonia and ethylene have been utilized to test both the system sensitivity and selectivity.

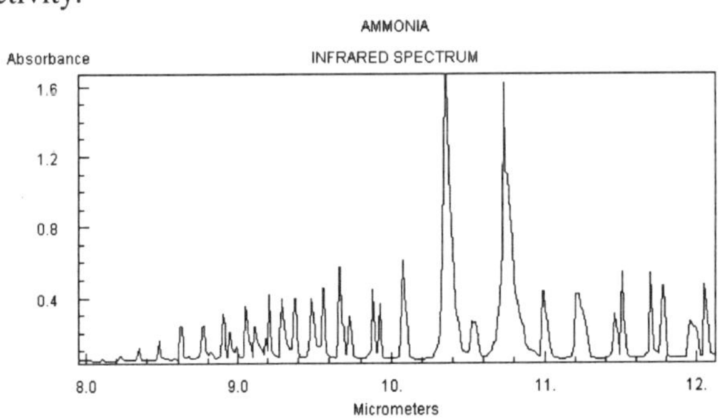

Figure 6: high resolution spectra of ammonia and ethylene

Figure 7 to 10 show results obtained using eight contiguous spectral channels corresponding to the eight neighbouring pixels of the sensor array spanning a spectral range from 10.2 micron to 11.1 micron using a linear variable filter of about 0.4 um/mm linear spectral variation. Figure 7 reports the time evolution of the measured output signals from the lock-in amplifier multi-channel electronics as recorded in real time. They appears quite stable owing to their noticeable signal to noise ratio (S/N better than 10^3) at utilized frequency of modulation of about 9 Hz.

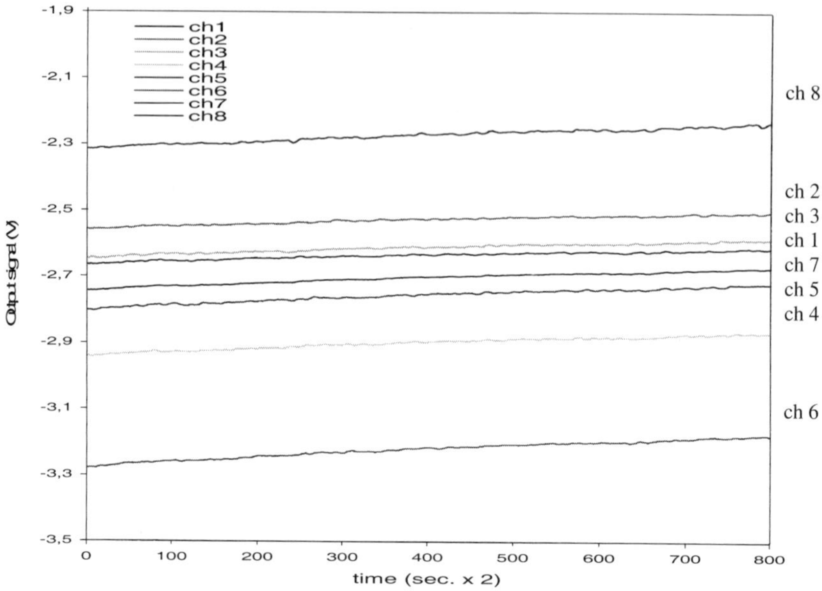

Figure 7: time evolution of output signals for the eight channels of the sensor prototype in pure nitrogen at standard T and P.

Considering the ammonia and ethylene absorption cross sections, a sensitivity in the order of tens ppm should be expected for these gases assuming this S/N figure. This was confirmed by simple post processing of the recorded signals. The percentage signal variation after introducing a small concentration of ammonia/ethylene in the test chamber corresponding to a few hundreds ppm was computed from measured data for each spectral channel.

Figure 8 below shows the response of our sensor when tested with 200 ppm ammonia. The ratio 'off' / 'on' is calculated for eight active channels which cover the spectrum 10 ÷ 11.3 micron. 'off' is the signal measured in an atmosphere of pure nitrogen at standard pressure and temperature, while 'on' is the signal measured after the introduction of 200 ppm of ammonia. Both 'off' and 'on' are averaged over 1 minute. The high resolution absorption spectrum of ammonia is shown in the lower part of the figure for comparison. Our sensor reproduces correctly the main features of the absorption pattern. Values of 'off' / 'on' below 1 for channels 1, 7 and 8 are due to small drifts of source and detector during the measurement.

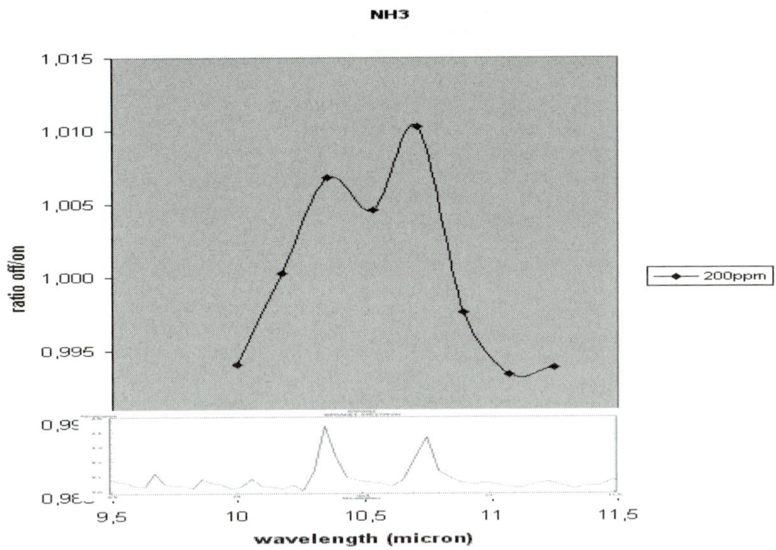

Figure 8: ratio 'off' / 'on' of signals measured before and after the introduction of 200 ppm ammonia in the test chamber.

Figures 9 and 10 show results of experiments with ethylene and ammonia at different concentrations. Both the two absorption peaks of ammonia and the single absorption peak of ethylene are identified, though they appear smoothed, owing to the relative low spectral resolution of the measurement. Furthermore, looking at the plot for ammonia, where also the concentrations are reported, it can be noted how the signal variation is proportional to the gas concentration in correspondence of the absorption bands. This is in agreement with the linear approximation of the Lambert-Beer absorption law for absorbing gas at small concentration (e.g absorption less than a few percent), and demonstrates quantitative concentration measurement capability of the system.

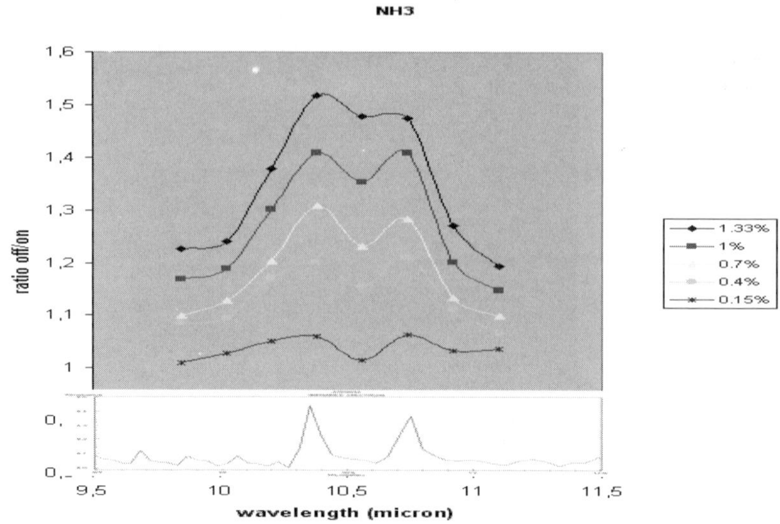

Figure 9: ratio ‘off’ / ‘on’ for ammonia at different concentrations

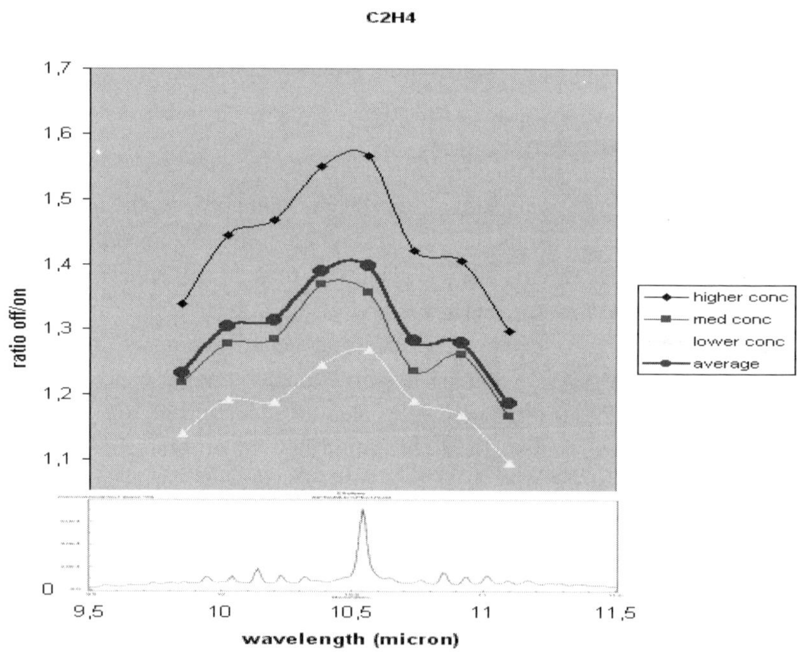

Figure10: ratio ‘off’ / ‘on’ for ethylene at different concentrations

5. Plans for the Future

This work was focussed on compact point-sensors with optical path of 1 m and sensitivities in the order of 10s of ppms.

CREO is now considering also line-sensors with extended optical path, and suitable to achieve sensitivities of 1 ppm or better.

In particular, CREO is planning to develop rugged and cost-effective open path systems, and hollow-fibre-based line-sensors.

An open-path configuration that appears particularly interesting consists of large area bolometric detectors, filters and diffractive lenses in a planar array, coupled to a thermal source at a distance of 10s meters. Advanced spectral separation and gas identification capabilities might be achieved through the use of a number of passive channels, with interferential filters of b.w. ≈ 0.1 micron to cover most of the spectrum, complemented with a few active channels, which use acousto-optic tunable filters to provide high-resolution fast random access to critical regions of the spectrum.

In a hollow-fibre-based line-sensor, the fibre acts both as a gas cell and as a bendable optical coupler between source and detector. Hollow-fibre sensors appear particularly attractive if combined with quantum cascade laser sources. These sensors therefore, when implemented with proper air sampling systems, may represent the flexible and general purpose device which, spread to ground or fixed to walls, can monitor streams of people or luggage, air ducts, or any other site with strong spatial constraints.

Open path and hollow-fibre line sensors altogether represent a very interesting solution as building blocks of a flexible sensor-network for surveillance of small heavily-crowded areas such as ports, airports, rail and metro-stations.

6. Conclusions

A compact room temperature infrared system for gas detection and identification has been originally designed and set up. It operates with a thermal source, a multipass cell and a room temperature sensor array coupled to a spectrally linear variable filter to detect the presence of toxic gases by means of the multi-spectral measurements of their absorption spectra. The system can work completely unattended in a automatic way without any operator control. Moreover gas detection sensitivity down to a few tens ppm has been demonstrated for sample gases like ammonia and ethylene. Gas identification capability even though using a small number of selected spectral bands has been verified.

These early results though obtained with a limited number of spectral channels are very promising. Also, when dealing with the challenging task of low concentrated gas detection, the system could allow to quantify the gas concentration once detected. Further work will be dedicated to assess the overall effectiveness of the system in presence of more composite gas mixtures and interfering species.

References

1. Carlo Corsi Proceedings SPIE Vol. 6297 Infrared Spaceborne Remote Sensing XIV, Marija Strojnik, Editor, 62970W (Sep. 7, 2006)
2. R. Viola, "High luminosity multipass cell for infrared imaging spectroscopy", Applied Optics, vol.45, No.12, pp.2805-2809, April 2006.

CHARACTERIZATION OF A PIEZOELECTRIC CAP USING A FIBER BRAGG GRATING

SANDRO RAO, FRANCESCO G. DELLA CORTE

Department of Information Science, Mathematics, Electronics and Transportations (DIMET), Mediterranea University, Via Graziella Località Feo di Vito, I-89060 Reggio Calabria, Italy

Optical fiber sensor technology based on intra-core Bragg Gratings can be used to measure many different parameters including strain, temperature, pressure, displacement etc. In this work the mechanical deformation of a piezoelectric cap is characterized by means a Fiber Bragg Grating (FBG) at frequencies up to 200kHz. We applied to the piezoelectric material a sinusoidal voltage and we calculated the consequent longitudinal cap's deformation. At the resonance frequency of f_{RES} = 1.5kHz the measured strain $\Delta L/L$ is $2.87 \cdot 10^{-4}$.

1. Introduction

Piezoelectric materials are very attractive for many applications in Micro-Electro-Mechanical-Systems (MEMS). Lead zirconate titanate (PTZ) is widely used for its high piezoelectric properties, as sensor or as actuator, for the generation of ultrasonic and acoustic vibrations (buzzers). The reverse piezoelectric effect is used in micro positioning, where an electric field is applied to produce precise motion. Examples of applications are: fibre optic alignment, machine tool alignment. Actually different techniques based on laser interferometry are used for the characterization of devices made with these materials. In this paper we propose the optical fiber sensor technology based on intra-core Bragg Gratings for a complete characterization of a piezoelectric cap.

2. Experimental setup

A light source, from a tunable laser diode, is coupled into the optical fiber with a FBG written in it. The grating is perfectly stuck on the piezoelectric cap and the transmitted light is detected by an InGaAs high speed photodiode. Figure 1 shows the schematic cap structure: a circle of piezoelectric material, specifically PZT, is located between two metal contacts. When an electric field is applied across the electrodes, the piezoelectric layer elongates horizontally creating a net bending motion.

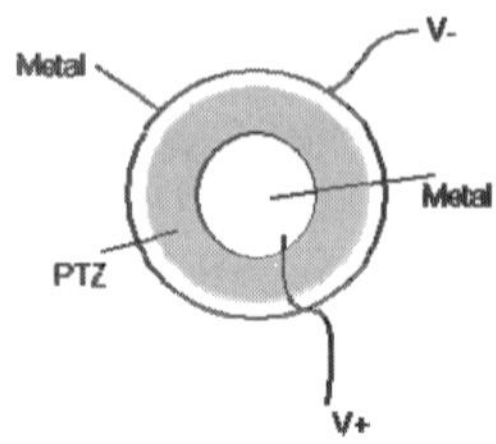

Figure 1. Piezoelectric cap's structure.

The output detector voltage signal is displayed, finally, on an oscilloscope. The schematic setup is shown in Figure 2.

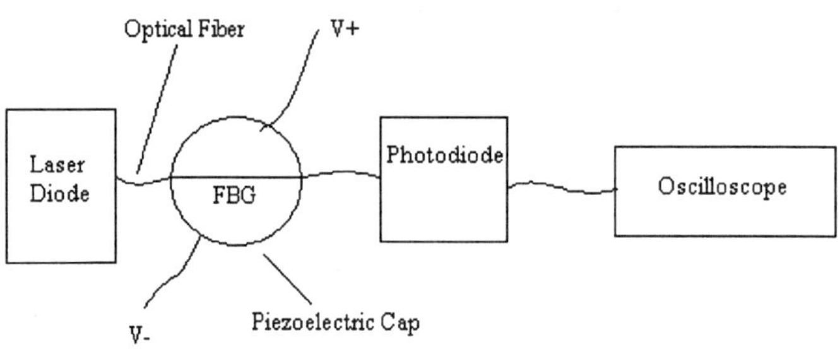

Figure 2. Experimental Setup of the system.

3. Principle of Operation

We applied to the piezoelectric cap a sinusoidal voltage ($V_{peak\text{-}peak} = 22$Volt, f = 1.5kHz), in this condition a system vibration is generated. The FBG central wavelength, $\lambda_B = 2n\Lambda$, varies as a consequence of both the grating pitch, Λ, and the refractive index, n, change: when we apply $V_{MAX} = +11$V, on the oscilloscope a minimum value is displayed if the laser diode lases at $\lambda_{LD} = \lambda_{B1} = 1550.078$ nm, in fact this wavelength is reflected by the grating.

For the same reason if we apply V_{MIN} = -11V, a minimum light intensity at $\lambda_{LD} = \lambda_{B2}$ = 1550.425 nm is transmitted. So we calculated the $\lambda_{\Delta B} = \lambda_{B2}-\lambda_{B1}$ quantity. The corresponding translated transmission spectrums are reported in Figure 3.

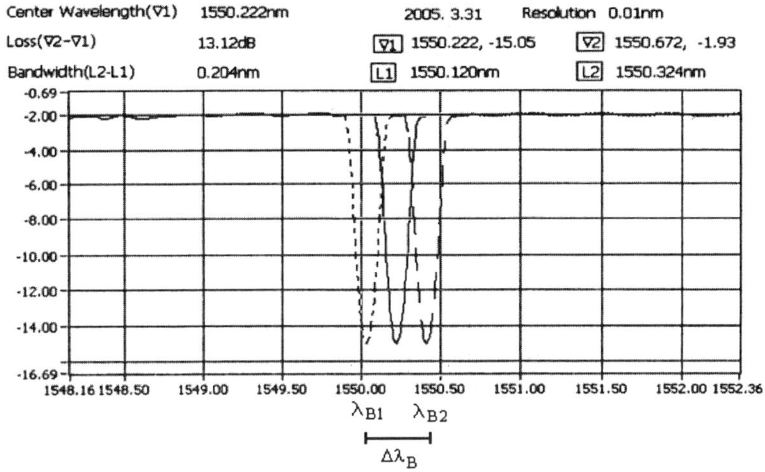

Figure 3. FBG spectrum shift λ_B = 1550.222nm (V = 0V); λ_{B1} = 1550.078nm (V = +11V); λ_{B2} = 1550.425nm (V = -11V).

The continue curve represents the FBG spectrum when no voltage is applied, the dotted and the broken lines represent respectively the transmission spectrum when V_{MAX} = +11V and V_{MIN} = -11V is applied.
We can deduct that positive voltage values cause a grating compression, vice versa happens when the sinusoidal voltage covers its negative half cycle.

4. Experimental Results

The wavelength shift, $\Delta\lambda_B$, for an applied longitudinal strain ε is given by the following relation: $\Delta\lambda_B = \lambda_B(1-\rho_e)\varepsilon$, where ρ_e is the photoelastic coefficient of the fiber optic core (for silica fibre at λ = 1.55μm ρ_e~0.22 [1]). Therefore we obtained: $\varepsilon = 2.87\cdot10^{-4}$.
The Bragg grating length is ~ 1.5cm, consequently in that part the piezoelectric cap deformation is ~ ∓ 2.15μm.
In Figure 4 is reported the signal voltage applied to the cap and the correspondent transmitted light source, at λ_{LD} = 1550.425 nm, modulated at f = 1.5kHz (characteristic resonance frequency).

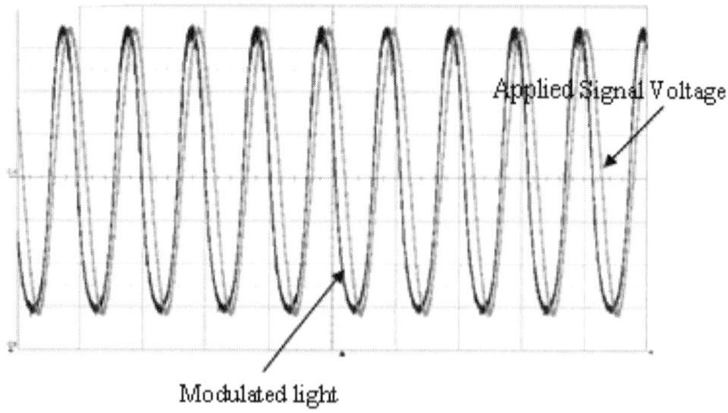

Figure 4. Signal voltage (Vpp = 22 Volt) applied to the cap and the correspondent transmitted light source modulated at f = 1.5kHz .

In correspondence of the resonance frequencies the piezoelectric cap deformation is at its maximum, and the modulated light signal has a Fourier spectrum with only one spectral line at the same frequency of the sinusoidal voltage signal impressed to the cap. In fact, in these conditions the harmonics introduced by the mechanic deformation of the piezoelectric material are negligible.

Therefore the proposed system allows to determine all cap's resonance frequencies, the highest value calculated is $f_{Max} = 196kHz$.

5. Conclusions

In this work we presented the characterization of a piezoelectric cap using a Fiber Bragg Grating. We applied to the device a sinusoidal voltage in order to deform it and we calculated the consequent cap deformation by analysing the FBG transmission spectrum.

This system allows to calculate all cap's resonance frequencies through the analysis of the Fourier spectrum of the modulated light signal. The highest observed resonance is at $f_{Max} = 196kHz$.

References

1. K. O. Hill and G. Meltz (Member IEEE) "Fiber Bragg Grating Technology Fundamentals and Overview" Journal of Lightwave Technology, Vol. 15, No. 8, August 1997

OPTICAL FLUORESCENCE ENHANCERS FOR TRACE DETECTION OF M1 AFLATOXIN

C. CUCCI, A.G. MIGNANI

CNR-IFAC, Via Madonna del Piano, 10 – 50019, Sesto Fiorentino (FI), Italy

c.cucci@ifac.cnr.it

C. DALL'ASTA, G. GALAVERNA, A. DOSSENA, R. MARCHELLI

University of Parma, Dept. of Organic and Industrial Chemistry,

Parco Area delle Scienze,17 – 43100 Parma, Italy

Aflatoxin M1 (AFM1) is classified as hazardous food contaminant occurring in several dairy products. A portable and easy-to-handle fluorometer for the AFM1 rapid detection is presented. The system is intended to be used as an "early warning system" so as to quickly single out risk or alarm situations. The addition of cyclodextrin (CD) to the AFM1 solution is investigated, as a tool to enhance the fluorescence signal, in order to increase the system sensitivity. Preliminary results are reported which show that succinyl-β-CD can be successfully used to improve the overall sensitivity.

Introduction

Aflatoxin M1 (AFM1) belongs to the class of mycotoxins, naturally occurring contaminants produced by widespread moulds which can colonise food and feeds, (cereals, oilseeds, spices, etc.) under particular micro-climate conditions. Aflatoxins are classified as toxic and carcinogens, and are strictly regulated at international level [1, 2]. AFM1, also known as "milk toxin", can be found in dairy products of animals that have ingested contaminated feed with Aflatoxin B1 [3-6]. Due to the fundamental role of milk in the human diet, in particular as infant nourishment, the presence of AFM1 in milk is regarded as highly hazardous for human health. According to the EC legislation, the maximum legal limit for AFM1 in milk is set at 0.05 μg/Kg (50 ppt) for all EU Member States, and further guidelines suggest to adopt lower limits (25 ppt) for baby food [7]. Due to the strict regulations established, in the past decades an increasing demand for new rapid screening and low cost methodologies, that would complement the official analysis techniques (HPLC, ELISA, etc.) for AFM1 detection, has been registered [8, 9].

The present work reports on advances achieved in development of a novel device, previously presented [10], for the rapid detection of low concentrations of AFM1 in liquid solutions. The prototype is a portable fluorometer specifically tailored for AFM1 detection. It was designed as an "early warning system", so as to rapidly single out risk/alarm situations.

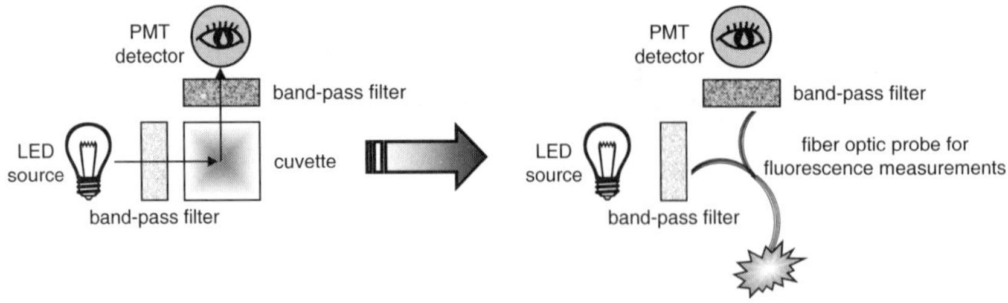

Figure 1. Schematic diagram of the fluorometer (left) and its possible implementation using fiber optics (right).

Based on the AFM1 titration curve, three levels of contamination were identified: 1) "alarm" 2) "alert for baby-food" and 3) "safe". This approach appeared promising in a perspective of large-scale milk controls, since it would allow to identify contaminated milk stocks and to limit the deepest analytical controls only to these cases.

The preliminary results showed that AFM1 below 50 ppt could be detected in non pre-concentrated aqueous solutions, provided that the fluorometer was equipped with a quartz ultra-micro cell, that allowed minimisation of the signal-to-noise ratio. Nevertheless the use of a quartz ultra-micro cell involved a time-consuming procedure of cell cleaning. Therefore, further possibilities to improve the overall system sensitivity and to shorten the measurement procedure have been here investigated. The use of disposable and low costs plastic cells was considered. Indeed, such a modification would allow to skip the cleaning cell step, with a great advantage in terms of time and costs.

In this study a new methodology based on addition of cyclodextrins (CD) to the AFM1 sample was investigated. The idea was to use CD as fluorescence enhancers in order to increase the overall system sensitivity, and to avoid a mandatory use of quartz ultra-micro cells for detection of AFM1 traces. Results of new tests carried out are reported, with a comparative study on the response of the fluorometer to AFM1 synthetic solutions with and without additional succinyl-β-CD.

Instrumentation

A schematic view of the instrumentation is reported in Figure 1 The fluorometer was a commercially available model (PMT-Fialab Instr.) [11], modified and optimised for the selective detection of AFM1 in liquid samples. A LED was used as light source (UVLED365-10, λ=365 nm), whereas the detection module was a highly sensitive PMT detector. Detailed technical features of the instrumentation are already reported in [10].

Materials and methods

Different types of cuvettes, both in quartz and disposable were tested. Quartz cuvettes were micro-cell quartz SUPRASIL® (45 μl Vol.). Disposable cells were PMMA macro-cell, certified for fluorescence, (Ocean Optics CVD-UV1U; 1.5-3.0 mL Vol.). Aqueous solutions of AFM1 were prepared at different concentrations by diluting the AFM1 standard solution (1 ppm, Biopure, Tulln, Austria). The samples covered the 0-125 ppt concentrations range. Samples of AFM1 with additional CD were prepared using succinyl-β-CD randomly substituted, substitution degree 3.5 (Sigma-Aldrich Steinheim, Germany). The solutions were prepared with a fixed concentration value of AFM1 (50 ppt), and by varying the succinyl-β-CD concentration in the 3-15 mM range.

Experimental results

Six sets of four samples of AFM1 in the 0-125 ppt concentration range were measured, and the corresponding titration curve is reported in Figure 2.

With a view to practical applications, the main interest was to provide a reliable screening device. Hence, starting from the titration curve, a simplified reference scale based on three main contamination levels (thresholds) was obtained (Figure 2):

- level 1: "safe" - [AFM1< 25 ppt]
- level 2: "alert for baby food" [25 ppt < AFM1 < 50 ppt]
- level 3: "alarm" - [AFM1 > 50 ppt]

Using this scale, the fluorometer could be used as starting point for "tree decisions" when measuring the fluorescence of unknown samples. If the result is level 1, the sample is "safe" and the control chain ends at this point; if the fluorescence value falls in level 2, a potential contamination is present and deeper analyses are needed; if it falls in level 3, the sample is contaminated over the admitted limit, and the corresponding stock should be discarded. The previous results were obtained without any need of pre-concentration of the sample, thanks to the use of a quartz ultra-micro cell, which allowed to optimise the signal-to-noise ratio, by reducing stray-light and scattering effects.

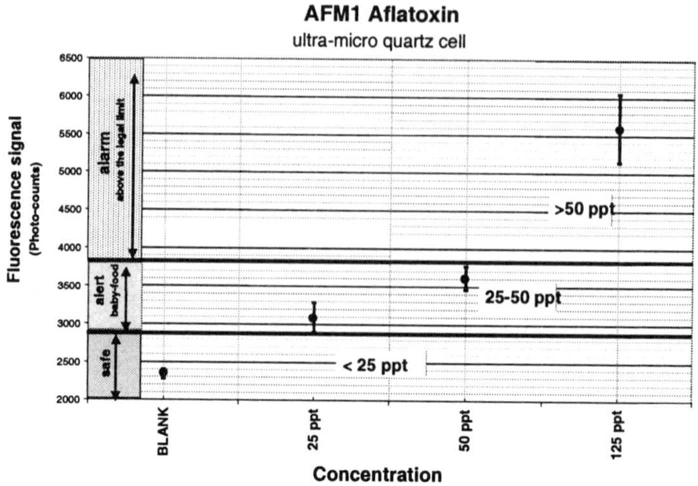

Figure 2. Titration curve for AFM1 aqueous solutions in the 0-125 ppt concentration range.

Nevertheless, this configuration required a careful and time-consuming experimental procedure for cell cleaning. Therefore, a study on alternative solutions to improve the overall sensitivity of the system was conducted. The possibility of increasing the fluorescence signal by adding suitable CD concentrations to the AFM1 sample was therefore investigated. The use of CD as fluorescence enhancers for aflatoxin detection is widely reported in the literature [12-14]. Among the different CDs, the succinyl-β-CD was selected, since it was reported to be the most efficient fluorescence enhancer for AFM1 [15, 16].

Solutions with a fixed concentration of AFM1 at 50 ppt were prepared by adding succinyl-β-CD at different molar ratios in the 3-15 mM range; the best results were obtained by using succinyl-β-CD at 5 mM ratio.

A new titration curve of AFM1 aqueous solutions with additional succinyl-β-CD (5 mM), was obtained in 0-125 ppt AFM1 concentration range. The two AFM1 titrations curves obtained with and without CD respectively, are reported in Figure 3. The increased sensitivity is evident from a comparison between the two curves. In particular, the value for the lowest detectable limit (25 ppt) increased by about a factor of 3 in presence of succinyl-β-CD, although an increased error bar affects measurements due to the CD scattering effects. Moreover, the improvement in sensitivity is testified by the higher slope in the linear best fit, which increased about 1.5 times in the presence of the fluorescence enhancer.

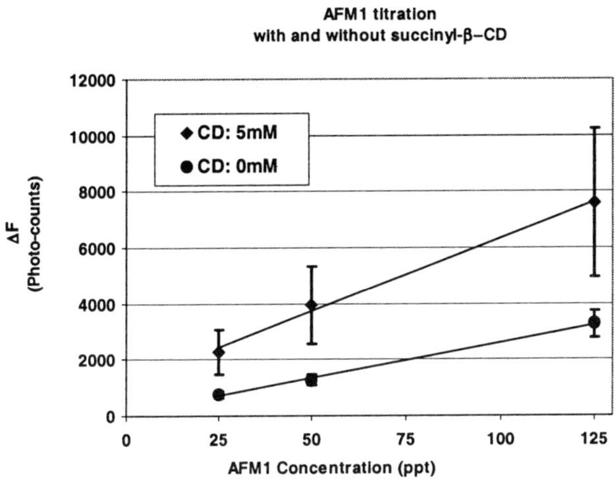

Figure 3. Comparison between the titration curves obtained with and without addition of succinyl–β- CD to the AFM1 aqueous solutions.

This result appeared promising in view of the development of an easy-to-handle system, suitable for non-trained users. Indeed, a possible modification of the original prototype could consist on the substitution of the quartz ultra-micro cuvette by a disposable plastic (PMMA) cell. This change would make it possible to skip the cleaning procedure, with a great advantage in terms of timing and easiness of the procedure. On the other hand, due to poorer optical quality of PMMA with respect to the quartz, a lack of sensitivity is expected in the case PMMA cells would be used. In fact, preliminary tests were performed with aqueous AFM1 solutions in PMMA cells, that showed that, due to the worsening of the signal-to-noise ratio, these cells were not suitable for a direct detection of AFM1 below 50 ppt. Nevertheless, the successful results obtained by using additional succinyl-β-CD as fluorescence enhancer, encouraged to consider this strategy as a possible tool to increase the sensitivity of system equipped with disposable cells.

Conclusions

A portable fluorometer was proposed for the rapid and straightforward detection of low concentrations of AFM1, without any need of pre-concentration of the sample. The technical features of the fluorometer were presented, with a special focus on its potentialities as an "early warning system", which makes it possible to distinguish between safe and potentially contaminated samples, according to the legal definition. The addition of a suitable succinyl-β-CD concentration to the AFM1 sample solution was also investigated, as a possible tool to improve the overall system sensitivity. The titration of the system with and without addition of succinyl-β-CD to AFM1 synthetic solutions was performed, in the 0-125 ppt concentration range. The results obtained showed that the use of CD as fluorescence enhancers can be considered a promising strategy to improve the overall system sensitivity.

Acknowledgments

Project MIUR-FIRB #RBNE01KZZM 'Biosens' is gratefully acknowledged for the partial financial support. The authors would like to thank Mr. Franco Cosi for the technical assistance.

References

1. FAO, Worldwide regulations for mycotoxins 1995 - A compendium. *FAO Food and Nutrition Paper,* 64 Rome, ISSN0254-4725 (1997).
2. IARC Monographs on the Evaluation of Carcinogenic Risks to Humans. *International Agency for Research on Cancer*, Lyon, **56** (1993).
3. R. Alleroft, R.B.A. Carnaghan, *Vet. Rec.*, **65**, (1963), 259.
4. H.P. Van Egmond, *Elsevier Applied Science*, London (1998).
5. P.M. Scott, *J. Assoc. Off. Anal. Chem.*, **70**, (1987), 276.
6. J. Leitao, G. De Saint Balnquat, J.R. Bailly, Ch. Paillas: *J. Chromatogr. A*, **253**, (1998), 229-234.
7. European Committee Regulation No. 1525, Bruxelles, 16 July 1998.
8. J. Gilbert, E.A. Vargas, *J. Toxicol. Toxin Rev.*, **22** (2003), 381–422.
9. C.M. Maragos, *J. Toxicol. Toxin Rev.* **23** (2004), 317–344.
10. C. Cucci, A.G. Mignani,C. Dall'Asta, R. Pela, A. Dossena, *Sens. Act. B: Chem.* (2007), *In press* doi:10.1016/j.snb.2007.03.036
11. PMT-FL. Fialab Instruments Inc., Bellewue WA, http://www.fialab.com./
12. M.I. Vasquez, C.M. Franco, A. Cepeda, P. Prognon, G. Mahuzier, *Anal. Chim. Acta,* **269** (1992) 239–247.
13. A. Cepeda, C.M. Franco, C.A. Fente, B.I. Vazquez, J.L. Rodriguez, P. Prognon, G.J. Mahuzier, *J. Chromatogr. A,* **721 (1)** (1996) 69–74.
14. C. Dall'Asta, G. Ingletto, R. Corradini, G. Galaverna, R. Marchelli, *J. Inclus. Phenom. Macrocyc. Chem.*, **45,** (2003), 257–263.
15. E. Chiavaro, C. Dall'Asta, G. Galaverna, A. Biancardi, E. Gambarelli, A. Dossena, R. Marchelli, *J. Chromatogr. A*, **937** (2001), 257–263.
16. C.M. Franco, C.A. Fente, B.I. Vazquez, A. Cepeda, G. Mahauzier, P. Prognon, *J. Chromatogr. A*, **815 (1)** (1998), 21–29.

MAGNETIC FIELD SENSOR
AT DIFFERENT PRE-STRESS LEVEL

C. AMBROSINO, D. DAVINO, C. VISONE, A. CUTOLO, A. CUSANO
Eng. Dept. of the Universit. of Sannio, C.so Garibaldi 107 82100 Benevento ITALY

S. CAMPOPIANO
Dept. for Techn. of the Univ. of Naples Parthenope Naples ITALY

M. GIORDANO
Inst. for Composite and Biomedical Materials, Nat. Res. Council, Naples ITALY

In this work, magnetic field sensors based on the integration of a Fiber Bragg Grating (FBG) with giant magnetostrictive materials (Terfenol-D rod and Magnetic Shape Memory Alloy, MSMA) are presented. Moreover , the influence of the pre-stress in the magnetic responses have been investigated. This dependence can be exploit to optimize the design of sensor for a successive use in specific application.

1. Introduction

Magnetic sensors are useful to control and analyze a large number of devices and physical processes. Therefore, they are widely spread in several fields of science and technology. For all the applications optical current sensors and, in particular, Fiber Bragg Gratings (FBGs) [1] are an excellent choice, due to their many advantages compared to conventional iron-core current transformers, including their immunity to electromagnetic interference, high dynamic range, easily integration within structures, competitive costs [2] and good features in terms of sensitivity, resolution and bandwidth [1]. In general, the common mechanism used to detect magnetic field with FBG sensors is Faraday effect, but due to the low value of the Verdet coefficient, they showed very low sensitivity. Therefore, different transducers have been employed in order to improve performances. In fact, recently, novel fiber optic sensors exploiting magnetostriction to sense magnetic field have been proposed [3,4]. They are based on the integration of FBG devices and magnetostrictive materials [5]. These materials show interesting properties like high energy densities and large magnetostriction being able to strongly deform itself in response to a magnetic field. In particular, Terfenol-D [5] shows strains up to thousands ppm; conversely, MSMAs [6] exhibit extremely large strains, up to 10%, similar to those of SMA but faster responses, being them actuated by magnetic fields and not by temperature and higher magnetic range of operation. So, the integration of such materials to a FBG-based sensor yield to passive, small size, and high sensitivity magnetic field sensors [3,7].

Along this line of argument, the authors have investigated two different FBG magnetic sensor configurations, employing alternatively Terfenol-D and MSMA as the active material and FBG as the transducer. An important element to be considered, when magnetostrictive materials are considered, is the dependence of their characteristic on applied mechanical loads (i.e. the applied pre-stress). In fact in dependence of the applied load, the magnetostrictive branches change their shape [5] and their slope, modifying the sensitivity for the sensor and the measurable range for magnetic field. Such dependence is usually disregarded, as the device is characterized to work at constant pre-stress level. In this work this dependence has been investigated in order to exploit it to optimize the design of sensor toward specific applications.

2. Experimental setup

After a suitable bonding protocol choice, two FBGs (central wavelength, λ_B at 1550 nm and 1550.14 nm) have been bonded one on a 20mm rod with a 4.7mm diameter of Terfenol and the other one on a 3x5x20 mm^3 stick of MSMA in

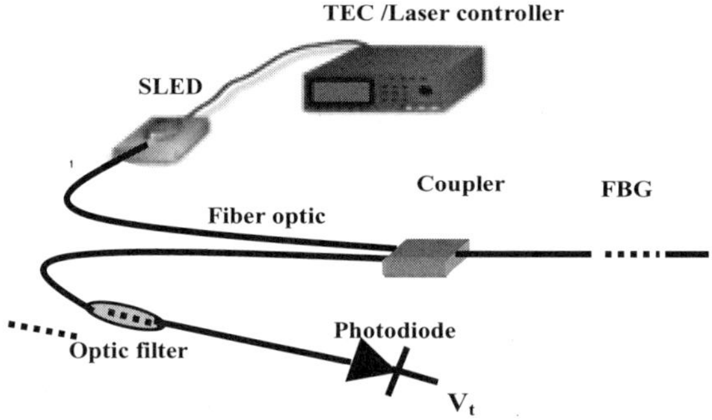

Fig. 1 Interrogation FBG sensor system block diagram

martensitic state. The reversibility of magnetic-field-induced strain of each material is obtained by pre-stressing it. In fact, each sample was placed in a tension-compression testing system by Lloyd between two rods. The compressive stress was applied along the [100] crystallographic direction of

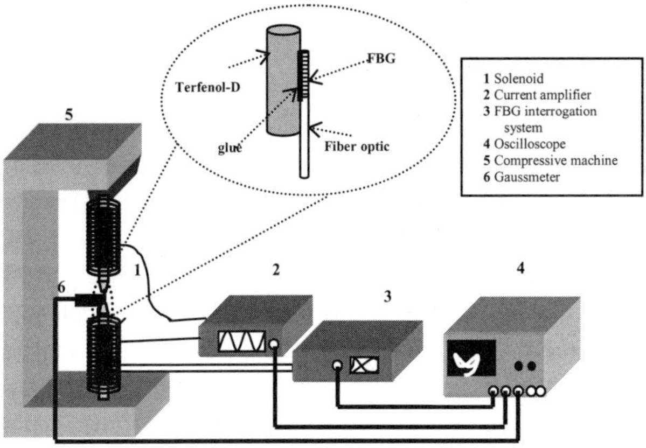

Fig. 2 Schematic diagram of the whole setup

each sample and the magnetic field was successively applied using a solenoid for the Terfenol based one and an electromagnet for the MSMA one. Magnetization was measured in the direction of the applied magnetic field by using a DSP-based Hall effect gaussmeter. When a magnetic field is applied to the samples, magnetic domains in materials' crystal rotate, providing proportional and positive expansion. This deformation actuates the FBG bonded on the active material, and consequently modifies its Bragg wavelength [2]. The wavelength shift is converted in intensity variation by using the FBG interrogation system described in [8] and depicted in Fig. 1. When a FBG is irradiated by a broadband source, it reflects a narrowband signal centered at Bragg wavelength called λ_B. The key element of the FBG interrogation technique is the in-fiber optic filter responsible for the wavelength-intensity conversion. It consists of an optical device with a transmittance and reflectance linearly varying with the optical wavelength and can be easily realized by using the well assessed Bragg technology [9]. Thus, when the Bragg wavelength shifts, for instance due to an applied strain, the optic filter will convert this wavelength shifts into amplitude ones. In this way, the photo-detector output signal variation ΔV_T is a good measure of Bragg's wavelength shift and with a proper calibration procedure, this voltage signal can be directly expressed in terms of the strain applied to the FBG. In particular, the optoelectronic setup relies on a super luminescent diode (40nm FWHM bandwidth centered at 1550nm) two FBGs previously characterized and an optical fiber filter (nominal linearity range of 7nm, centered at 1550nm). The whole adopted setup for Terfenol one is depicted in Fig.2.

3. Results and discussion

Two magnetic field sensors characterizations have been carried out. In order to characterize the sensing systems and the materials several tests at monitored temperature have been performed for each configuration. The tests have been performed also with different level of pre-stress because the magnetostrictive relation of the material is strongly dependent on the applied pre-stress. In fact in dependence of the applied load, the magnetostrictive branches change their shape [8] and their slope, modifying the sensitivity for the sensor and the measurable range for magnetic field. In Fig.3, regarding to the Terfenol based configuration, the FBG sensor output voltage vs. time is plotted for two different loads. In this case a sinusoidal magnetic field with amplitude up to 30 kA/m and a frequency of 0.05 Hz has been applied to the system. In Fig.4, regarding to the Terfenol based configuration, the system strain characteristic for two different load has been plotted. In this case the maximum value for the magnetic field

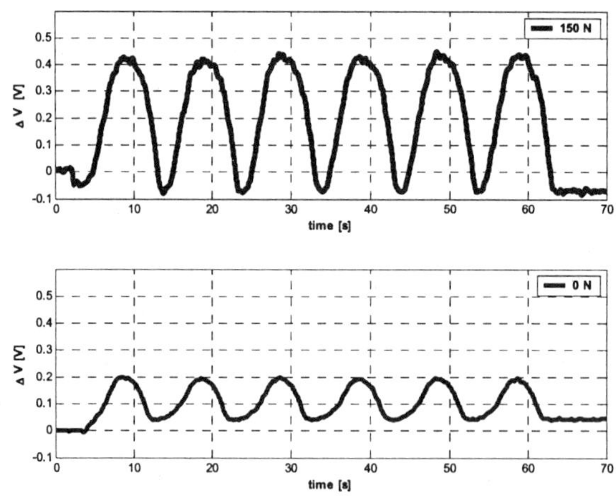

Fig.3 Measured output voltage of the FBG strain sensor versus time at two different pre-stress load condition (150N and 0N) for the Terfenol based configuration of the sensor. The strain is proportional to the measured voltage. The materials act always in tensile stress state.

applied is 97 kA/m. Contemplating these two typical butterfly closed loop shapes, minor is the load higher is the slope of characteristic and so the sensitivity but shorter is the magnetic measurable range. Whatever is the material, the magnetostrictive responses show strongly hysteresis. This effect cannot allow the device to correctly reconstruct the field. In order to take in account this rate independent memory effects , algorithms for hysteresis compensation based on a Preisach could be employed [10]. Such algorithms will allow to improve the performances of the device in terms of accuracy and linearity range compensating hysteresis and non linearity.

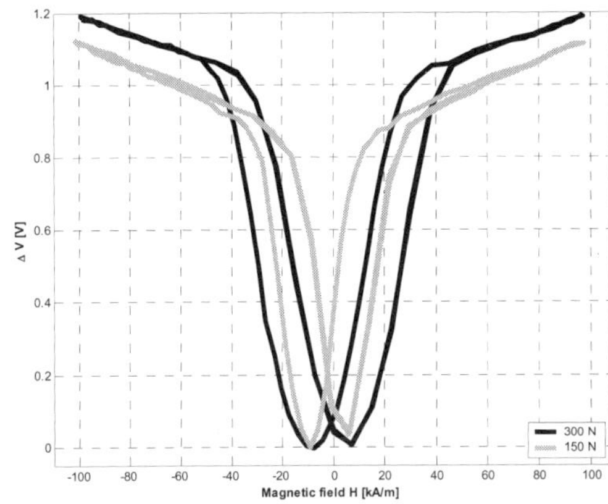

Fig.4 System strain characteristic at two different load (150 N and 300 N)

References

1 K. Bonhert, P. Gabus, H. Brandle, P. Guggenbach,, 17th Int. Conf. on Optical Fibre Sensors, Proc. of SPIE Vol 5855 pp 210-213,2005.
2 B. Culshaw and J. Dakin, "Optical Fiber Sensors: Applications, analysis, and future Trends", Artech House inc., Norwood, 1997
3 C. Ambrosino *et al.* *IEEE Sens. Lett.* Vol. 7, Issue 2, Feb. 2007 pp.228-229
4 J. Mora, A. Diez, J. L. Cruz, M.V. Andres, *IEEE Phot. Techn. Lett.*, vol. 12, No. 12, December 2000, 1680-1682
5 G. Engdhal (editor), Handbook of Giant Magnetostrictive Materials, Academic Press, 2000
6 J. Tellinen *et al.* “Basic Properties of Magnetic Shape Memory Actuators”, 8th. ACTUATOR ‘02, Germany, 2002
7 D. Davino *et al.* “Compensation of hysteresis in magnetic field sensors employing Fiber Bragg Grating and magneto-elastic materials”, Sensors and Actuators (submitted)
8 Cusano A., Breglio G., Giordano M., Nicolais L., Cutolo A., IEEE/ASME Transactions on Mechatronics, 9(1), 40-50, 2004
9 R.W. Fallon, L. Zhang, L. A. Everall, J. A. R.Williams, and I. Bennion, Meas. Sci. Techn., vol. 9, pp. 1969–1973, 1998.
10 I. D. Mayergoyz, Mathematical Models of Hysteresis,Springer, 1991

THE HYPER-SPECTRAL OPTICAL SIGNATURE OF EXTRA VIRGIN OLIVE OIL

L. CIACCHERI, A.G. MIGNANI
CNR-IFAC, Via Madonna del Piano, 10 – 50019 Sesto Fiorentino (FI), Italy
Email: l.ciaccheri@ifac.cnr.it

H. THIENPONT, H. OTTEVAERE
Vrije Universiteit Brussel, Department of Applied Physics and Photonics, Belgium

A. CIMATO, C. ATTILIO
CNR-IVALSA, Via Madonna del Piano, 10 – 50019 Sesto Fiorentino (FI), Italy

Italian extra virgin olive oils bearing labels of certified area of origin were considered. Their multispectral digital signature was measured by means of absorption spectroscopy in the 200-1700 nm spectral range. The instrumentation was a fiber optic-based, cheap, and compact device. The spectral data were processed by means of multivariate analysis and plotted on a 2D classification map. The map showed sharp clusters according to the geographic origin of the oils, thus demonstrating the potentials of UV-VIS-NIR spectroscopy for optical fingerprinting. Then, the spectral data were correlated to the content of the most important fatty acids. The good fitting achieved, demonstrated that the optical fingerprinting can be used also for predicting nutritional and chemical parameters.

Introduction

Many industrial process controls depend increasingly on diagnostic systems capable of distinguishing products and identifying them through reliable methods. In the food sector, inexpensive diagnostic systems are needed to distinguish high-value products in order to protect their brands and to ensure that food peculiarities and safety requisites are met. Italian extra virgin olive oils bearing labels of certified area of origin are characterized by distinctive taste and exceptional nutritional properties. These peculiarities derive from the kind of olives used and the traditional, low-mechanization production means. New methods and technologies are envisaged for qualifying and certifying these oils, to protect both producers and consumers [1, 2].

This work presents a method for measuring the multispectral digital identity of extra virgin olive oils bearing labels of certified area of origin. The method complements conventional laboratory procedures that monitor key production chemical parameters. It was applied to 85 samples of extra virgin olive oil produced in the regions of Sicilia, Calabria, and Toscana during the 2005-2006 season. The differences in the taste and nutritional properties of the oils derived from variation in geographic origin, cultivar, climate, and harvest time.

The samples were fingerprinted by UV, VIS, and NIR absorption spectroscopy covering the 200-1700 nm spectral range. A fiber optic-based, low-cost, compact spectrometric device, precursor to an online diagnostic system, was used. The spectral data were processed by means of Principal Component Analysis (PCA) and Linear Discriminant Analysis (LDA) and plotted on a 2D classification map. The map showed that the oil samples were sharply clustered according to geographic origin, thus demonstrating the potentials of UV-VIS-NIR absorption spectroscopy for optical fingerprinting. Then, an attempt was made to demonstrate the possibility of implementing an authentication device by using the results of optical fingerprinting. The spectral data were processed by means of Partial Least Square (PLS) analysis looking for a correlation with the content of the most important fatty acids. A good fitting was achieved, demonstrating that cheap micro-optic technologies and multivariate analysis can provide a tool for low-cost traceability.

Instrumentation and data processing

Absorption spectroscopy in the 200-1700 nm spectral range was performed by means of the experimental set-up sketched in Figure 1. It consisted of a deuterium/halogen lamp (Micropack Inc., DH-2000-BAL) coupled to a fiber optic bundle, which provided illumination to a quartz cuvette containing the oil sample. Another bifurcated fiber optic bundle was used to detect the transmitted light intensity of the oil and to split it into two spectrometers for UV-VIS (Ocean Optics Inc., HR4000) and NIR (Ocean Optics Inc., NIR512) spectroscopy, respectively.
The working ranges of the spectrometers were slightly overlapping, with different spectral resolution. The 200-1100 nm UV-VIS range offered the best resolution of 0.23 nm, while the 900-1700 nm NIR range provided a lower resolution of 1.7 nm. Therefore, each oil sample was characterized by 4160 spectral data.

The spectral data were processed by means of multivariate data analysis techniques. PCA was used for dimensionality reduction, while LDA was used for oil classification. The data processing protocol considered the following steps [3-5]:

1. Two matrices were created by means of the UV-VIS and NIR transmission spectra, respectively, each row representing the spectrum of a different oil sample.
2. The two matrices were separately processed by means of PCA, to compress the relevant information into a limited numbers of features (scores), thus creating two score-matrices, for UV-VIS and NIR PCA-data, respectively.
3. The two score-matrices were concatenated along the 2nd dimension, thus operating the UV-VIS and NIR data fusion.
4. LDA was applied to the joint matrix, extracting two Linear Discriminating Functions, DF1 and DF2, for oil classification.

Finally, DF1 and DF2 were used to build a 2D map, each point of which representing an oil with its own peculiar spectral characteristics.

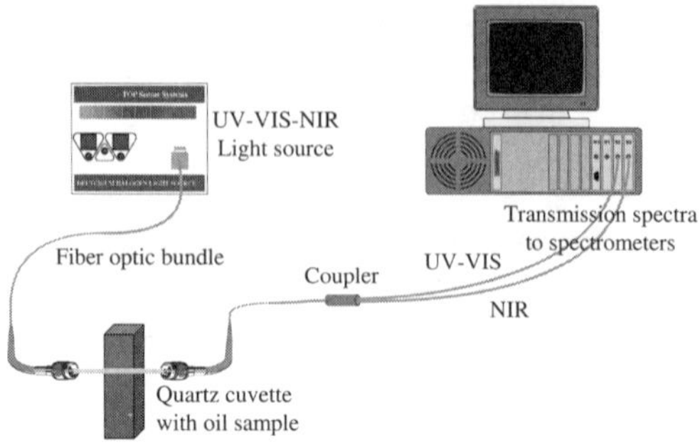

Figure 1. Experimental set-up for spectroscopy of the oil samples.

Discrimination of extra virgin olive oils from Sicilia, Calabria, and Toscana

A library of 85 oil samples from the 2005-2006 harvesting was considered. They were produced in three different Italian regions, Sicilia (deep South of Italy), Calabria (South of Italy), and Toscana (Central-North part of Italy), respectively. These oils have distinctive tastes not only because of the different cultivar, but also because of the different environmental conditions and agronomic practices.
The transmission spectra of the full library (Figure 2) show at a glance different spectral signatures for the various oils. The main differences in the UV-VIS region are due to the diverse content of phenol-derivatives, carotenoids and chlorophylls, while the little differences in the NIR region are mainly caused by the diverse fatty acid content.

Three PCAs were sufficient to model the NIR spectra, explaining the 99% of the data variance. The UV-VIS spectra showed a more complex structure requiring four PCAs to describe the data, with an explained variance of 93%. Therefore, the data matrix for LDA was made of 7 columns. Figure 3-left shows the result of PCA/LDA processing that is the 2D map in the DF1-2 space. A very good clustering is achieved according to the geographic region of origin of the oils.

Although the library was made of 85 samples only, an attempt of training the system to predict the geographic origin was made. The training set was made of 65 samples, which were used to create the 95%-confidence ellipses of each cluster. The validation set was made of 20 samples. The result of prediction is shown in Figure 3-right: the standard-error-prediction was less than 3%, which can be considered a successful result.

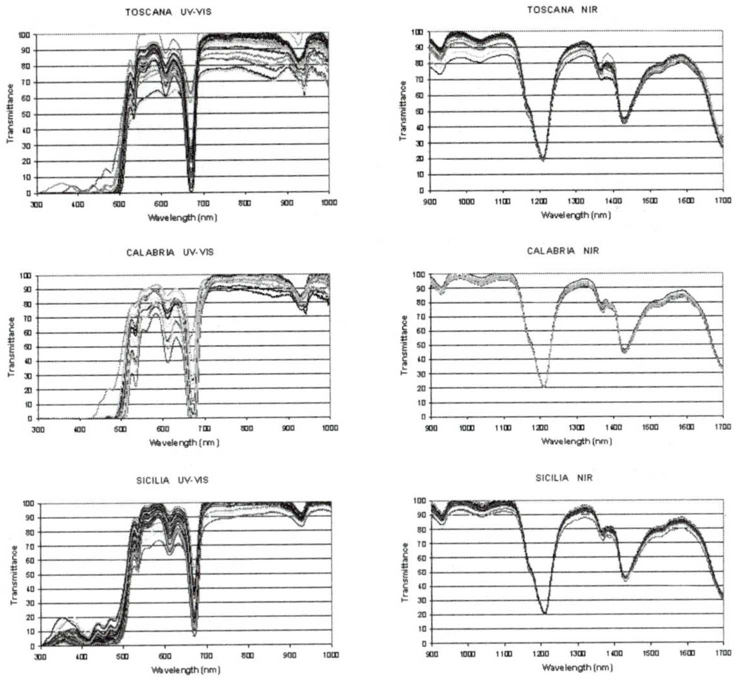

Figure 2. UV-VIS-NIR spectral signatures of the full library of the extra virgin olive oils.

More thorough analysis of extra virgin olive oils from Sicilia

Among the 85 samples of Italian extra virgin olive oils constituting the analyzed library, those from the region of Sicilia were in the majority, with 42 samples. This number was sufficient to run multivariate data analyses in order to make a more thorough discrimination as well as to test which kind of correlation might exist between the optical data and the nutritional and chemical characteristics of the oils.

Discrimination according to PDO

The oils from Sicilia were selected from five legally-certified areas, usually referred to Protected Designation of Origin (PDO) regions. Three PDOs were populated by a sufficient number of samples to run a multivariate analysis for the purpose of attempting a classification according to the PDOs. The results achieved by LDA are shown in Figure 4. Although each PDO was populated by only 9 samples, a clustering according to PDO-Valli Trapanesi, PDO-Val di Mazara, and PDO-Monte Etna was achieved, with a Standard Error of Cross-Validation (SECV) of better than 37%. In spite of the high error, which might be lowered by analyzing a higher number of samples, the clustering according to

PDOs can be considered a satisfactory result, since all oils have the same region of origin.

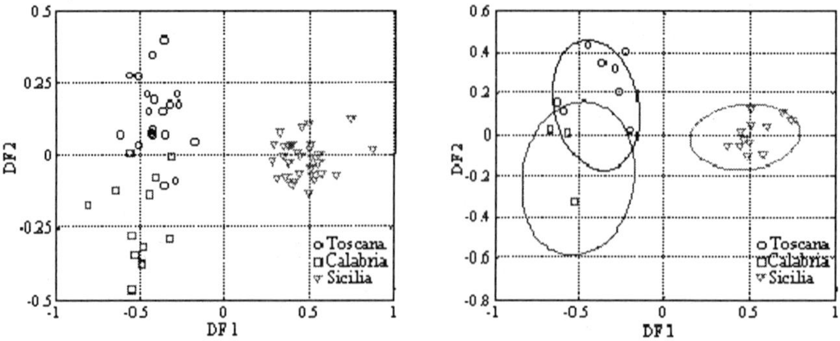

Figure 3. 2D maps in the DF1-2 space, clustering the extra virgin olive oils. Left: full library clustered according to the geographic region of origin. Right: 95%-confidence ellipses and validation set points.

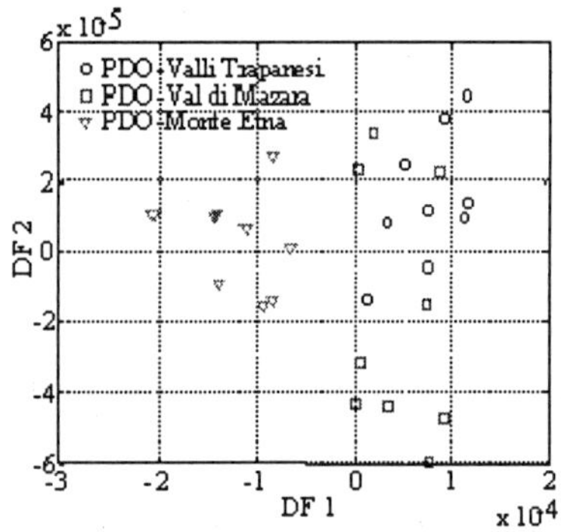

Figure 4. 2D maps in the DF1-2 space, clustering the PDOs of extra virgin olive oils from Sicilia.

Correlation of the spectral data with the most important nutritional and chemical parameters of the oils

An attempt was made to verify what kind of correlation existed between the spectral data and some of the most important nutritional and chemical parameters of the oils. The spectral data were processed by means of PLS analysis to find the degree of correlation between the spectral and chemical data. Both UV-VIS and NIR bands were tested as predictor matrices, and the most convenient one was then chosen. The olive oil is a complex compound made principally of fatty acids, these representing 99% of its composition; the related percentage of fatty acid is able to reveal the nutritional value of the oil itself. For these reasons, the most important fatty acids were considered for the correlation with the spectral data. The set of fatty acids was obtained by means of conventional analytical techniques. Table 1 lists the fatty acids giving the best fitting by processing the spectral data of the NIR spectral band, together with the fatty acid average value, the SECV and the correlation coefficient, R, between the analytically measured and predicted values. Significant correlation occurred with the Oleic acid, the main monounsaturated fatty acid of olive oil, as well as with Palmitic acid, and for the total content of Palmitic+Stearic acids.

Table 1. Best fitting of spectral data and the most important fatty acids of the oils.

Parameter	Average content	SECV	R
% Oleic Acid	74.4 %	2.5 %	0.91
% Palmitic Acid	11.9 %	1.7 %	0.87
% Palmitoleic Acid	0.8 %	0.4 %	0.81
% Palmitic + Stearic Acids	14.1 %	1.4 %	0.97
% Eptadecanoic Acid	0.1 %	0.05 %	0.70
% Eptadecenoic Acid	0.2 %	0.07 %	0.80
% Linoleic + Linolenic Acids	9.6 %	1.8 %	0.66

Perspectives

A low-cost spectrometric device and a suitable multivariate data processing allowed to clustering a group of Italian extra virgin olive oils according to the geographic region of origin, that were Sicilia, Calabria, and Toscana, respectively. The oils from Sicilia, which were the majority, were also clustered according to three PDO. Then, an attempt was made to correlate the optical fingerprinting with the most important fatty acids. The results were satisfactory and demonstrate the possibility of implementing an authentication device based on cheap micro-optic technologies. Of note is also the use of optical fibers, which is precursor to an online diagnostic system. New measurements are scheduled after the next harvest, by taking into account a larger number of oil samples as well as looking for the correlation with other important chemical parameters.

Acknowledgments

The European Network of Excellence 'NEMO', the CNR Short Term Mobility Programme, and the Regional Board of Sicilia (Assessorato Agricoltura e Foreste, Servizio IX°, Palermo) are acknowledged for partial financial support. For the technical support, acknowledgements are also due to the Customs Laboratory of Palermo and the UOs of the Extension Service of the Regional Board of Sicilia.

References

1. J. Harwood et alii, 1999, *Handbook of Olive Oil*, Aspen Pbl, USA, 1st edition.
2. M. Jae, 2002, *Oils and Fats Authentication*, Blackwell Publ., CRC Press, Boca Raton FL.
3. M.A. Adams, 1995, *Chemometrics in Analytical Spectroscopy*, The Royal Society of Chemistry, Letchworth UK.
4. H. Martens et alii, 1989. *Multivariate Calibration*, John Wiley&Sons, Chichester UK.
5. B.G.M. Vandeginste et alii, 1998, *Handbook of Chemometrics and Qualimetrics*, Elsevier Science BV, Oaklawn AcademyAmsterdam.

PHYSICAL SENSORS

FAST SCINTILLATION READOUT BY MULTI-PIXEL PHOTON COUNTING *

R.SCAFÈ ♦, F.PISACANE, P.G.GABRIELLI, G.ALONGE, D.DELLA SALA
Casaccia Research Center, ENEA, Rome, Italy

M.SALMI, R.SALMI
RTS Instruments Srl, Rome, Italy

R.PANI
INFN and Dept. of Experimental Medicine, Univ. of Rome "La Sapienza", Rome, Italy

S.SALVATORI, F.ZOCCOLI, G.CONTE AND F.DE NOTARISTEFANI
Dept. of Electronic Engineering, Univ. of Rome "Roma Tre", Rome, Italy

This paper presents a preliminary evaluation of the new Multi-Pixel Photon Counter (MPPC) by Hamamatsu Photonics for single-event fast scintillation readout in the nanoseconds range. The MPPC is a segmented high-gain avalanche photodiode realized using cutting edge semiconductor technology. Each pixel integrates the resistance for avalanche quenching allowing the operation in Geiger-mode at pixel level. Because of the adequate time resolution of the S10361-11 series, $LaBr_3$, $LaCl_3$, GSO and YAP Ce-doped fast scintillators were selected for investigating the respective scintillation decays. Single-event voltage signals were analyzed by a LeCroy digital sampling oscilloscope in single-shot mode. Both the MPPC and the oscilloscope are fast enough to perform good sampling of the signal along the exponential decay of light emission.

1. Introduction

The Hamamatsu MPPC (Multi-Pixel Photon Counter) [1], also referred to as SiPM (Silicon Photo-Multiplier), is a solid state Si-APD [2] (Avalanche Photo-Diode) suitable for Geiger-mode operation with a superior gain for photon counting. The most remarkable feature is segmented active area. Each pixel integrates its quenching resistance, so that the Geiger discharge is

* This work was supported in part by the Italian National Institute of Nuclear Physics (INFN) under SCINTIRAD Project and in part by Italian National Council for New Technologies, Energy and the Environment (ENEA) under P466 Project.

♦ Corresponding author: scafe@casaccia.enea.it

independently controlled pixel by pixel. As a result, the MPPC's output is in proportion with the number of pixels, which are excited by the incident light. Due to the limited number of pixels illuminated, the photon counting can be considered digital. The MPPC is a new generation of segmented Si-APD realized by Hamamatsu with cutting-edge semiconductor technology [3]. The high performances will bring new photon counting applications mainly in nuclear medicine, radiation monitoring, high-energy physics and medical diagnosis.

2. Materials and methods

2.1. *The multi-pixel photon counter*

Generally speaking, the MPPC is a low cost device characterized by (a) high quantum efficiency for photons in the visible, (b) single photon counting capability, (c) high time resolution, (d) low-voltage operation and (e) insensitivity to magnetic fields.

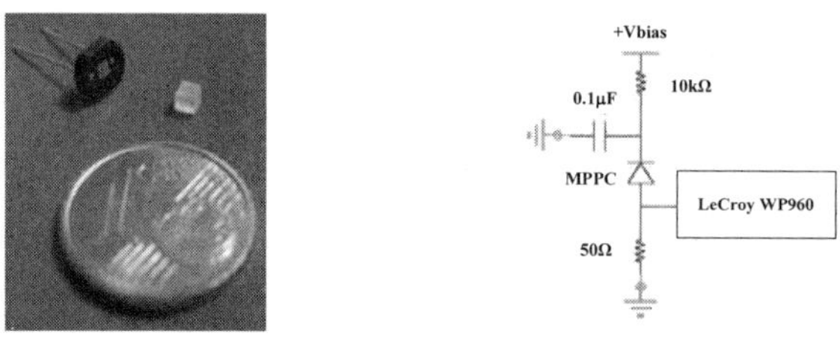

Fig.1 (Left) Hamamatsu S10361-11-050C Multi-Pixel Photon Counter and 2x2x2 mm^3 GSO:Ce crystal shown with a 1 Euro Cent coin Ø 16.2mm; (Right) Connection diagram for a single-shot operation.

The new class of Hamamatsu solid state photon counter improves such features. In particular these detectors show enough charge multiplication to allow the monitoring of signals by oscilloscope in single-shot mode. Two versions of MPPC were available: S10361-11-025C and -050C (fig.1), with the same chip size of 1.5x1.5mm^2 and 1.0x1.0mm^2 active area but with different theoretical pixel sizes 25x25μm^2 and 50x50μm^2 respectively. Stable operation is performed with low voltage polarization (77V typically). From the sensitivity point of view, they operate at the same 400 nm peak wavelength with a quantum efficiency of 70% but, due to different geometric factors, with a photon detection efficiency of 40% for the 50x50 model and 20% for the 25x25 model respectively [1].

As a consequence of high-gain values these devices show high rate of measurable dark events but, when used in pulse-height analysis modality, the latter can be easily discarded using a low-voltage threshold. In particular, at 77V and 25°C, the 50x50 model shows typical values of gain = 7.50×10^5 and dark-events rate = 470kcps, while the 25x25 version is characterized by 2.75×10^5 and 220kcps respectively.

2.2. *Electronics and scintillators*

Measurements were performed in the range from 20°C to 25°C using the connection diagram (fig.1) proposed by Hamamatsu for MPPC inverse polarization and signals readout [1]. Voltage signals across the 50Ω resistor connecting the MPPC to ground were digitized and stored by LeCroy WP960 2GHz digital sampling oscilloscope operating with 50Ω input impedance in single-shot mode.

To evaluate the MPPC response in the range of nanoseconds, this work analyzed the scintillation decay time characterizing some fast scintillators. Due to the larger MPPC pixel-size, that implies a reduced dead area and a better event visibility, the -050C model was selected.

Measurements have been done with a series of Ce-doped fast scintillators ($LaBr_3$, $LaCl_3$, GSO and YAP). Decay time, light output and geometries of the employed crystals are shown in tab.1. It is worth to note that, depending on the light output values, the Lanthanum-based scintillators gave enough response in planar geometry, while the others were used as pixels to concentrate the light-photon spot toward the MPPC active area.

Table 1 : Some properties of Ce-doped scintillators

Scintillator	$LaBr_3$ $^{+}$	$LaCl_3$ *	GSO [4]	YAP [4]
Decay time (ns)	16	27	56 , 400	27
Light output for γ-rays (ph/keV)	63	40	9	18
Crystal geometry	planar	planar	pixel	pixel

$^{+}$ BrilLanCe ® 380; * BrilLanCe ® 350, both by Saint-Gobain Crystals, France.

Each scintillator was coupled by optical grease to the MPPC and the assembly was darkened in a box. The device was biased at the appropriate value and measurements were performed, for dark and scintillation events investigations respectively, without or with gamma-ray irradiation by ^{22}Na radioisotopic source to induce scintillations in the crystal.

The dark events' analysis was preliminarily performed using the assembly without scintillator and without irradiation. In this condition the exclusive observation of MPPC intrinsic dark events was reasonably guaranteed.

The MPPC transient responses to individual scintillation flashes, produced inside the scintillation crystal by photoelectric interactions of 511keV or 1274keV gamma-rays from ^{22}Na irradiation, were studied on a time scale extended to the complete photocurrent decay.

Furthermore the transient responses to scintillations produced inside the crystal by α-decays of ^{227}Ac impurities, contaminating the natural Lanthanum at levels of fractions of parts per billion [5], were also analyzed.

3. Results and discussion

Voltage traces show the effect of lightening and quenching of individual MPPC pixel(s) overlapping the scintillation decay. Fig.2 (top-left) shows a weak dark signal recorded without scintillator and without irradiation and (top-right) a strong scintillation signal produced in a $LaBr_3$ crystal by ^{227}Ac α-rays.

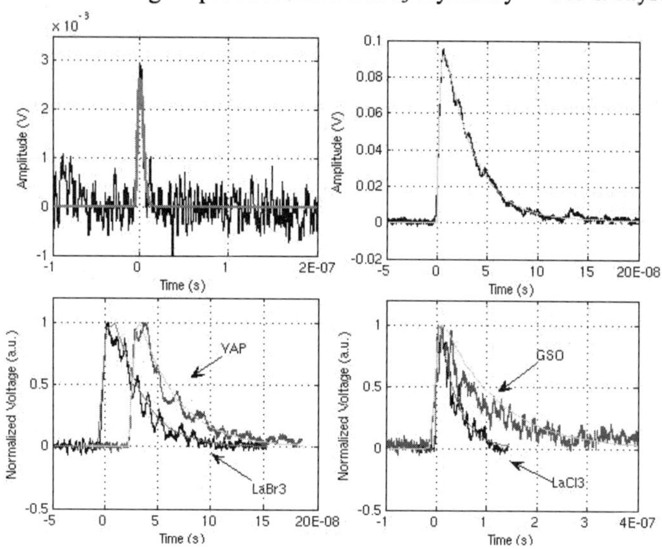

Fig.2 : (Top-left) Dark signal from Hamamatsu S10361-11-050C w/o scintillator and w/o gamma-ray irradiation; (top-right) α signals of ^{227}Ac naturally contaminating the La from the same device with $LaBr_3$ scintillator and w/o source irradiation. (Bottom) Normalized responses to 1274keV gamma-ray from Na^{22} source, concerning the scintillators coupled to the MPPC. The YAP time-shift (left) was applied for better visibility.

Fig.2 (bottom) reports some signals corresponding to scintillations induced by ^{22}Na 1274keV gamma-rays inside the considered crystals.

All traces were preliminarily analyzed by non-linear fitting using the *pulse* model of Eq.(1), where τ_1 and τ_2 are the rise and decay constants, respectively.

$$v_{(t)} = A\left(1 - Exp\left(-\left(t - t_0\right)/\tau_1\right)\right)^P Exp\left(-\left(t - t_0\right)/\tau_2\right) \qquad (1)$$

Numerical results are reported in tab.2. The data overview suggests the following: **(a)** the dark signals, as detected by the present readout system, were about 6ns in total length; **(b)** the simplified model produces values of scintillator decay time approximating the literature data within normal uncertainties with the exception of GSO due to the presence of the relevant (10%) slow component (400ns) [4]; **(c)** the α-rays scintillations from ^{227}Ac impurities of $LaBr_3$ crystal seem quite slower than events excited by gamma-rays; **(d)** the rise time values for the examined cases appear somewhat different, particularly for GSO.

Table 2 : Time constants (ns ± 1σ) of responses measured by S10361-11-050C MPPC

Scintillation From:	Gamma-rays*		Alpha-rays	
	Rise time	Decay time	Rise time	Decay time
$LaBr_3$:Ce	3.8 ± 2.0	21 ± 6	2.5 ± 0.8	29 ± 2
$LaCl_3$:Ce	1.9 ± 1.2	24 ± 9	-	-
GSO:Ce	5.6 ± 0.2	105 ± 5	-	-
YAP:Ce	2.2 ± 0.7	34 ± 4	-	-
None (dark)	3.9 ± 0.2	2.1 ± 0.3	-	-

* Averaged out values over 511KeV and 1274KeV γ rays of ^{22}Na.

Better results are expected improving the model by considering the signals oscillations due to effect of lightening and quenching of individual MPPC pixel(s). The *sine-damp* model for residuals fitting is presently under investigation by subtracting the oscillations of the experimental data from the model of Eq.(1).

Acknowledgments

The MPPC samples were kindly provided by Massimo Aversa from Hamamatsu Photonics Italia Srl, Rome Office.

References

1. Hamamatsu Photonics K.K.: Preliminary - MPPC S10361-11-025, S10361-11-050, Technical information; K00-150000 (2006), Japan
2. D.Renker, Nucl. Instr. and Meth. in Phys. Res. A, 567 (2006) 48
3. S.Gomi *et al.*, IEEE 2006 Nuclear Science Symposium, Conference Records, San Diego, California, USA
4. G.F.Knoll, Radiation Detection and Measurement, 3rd Edition, John Wiley & Sons, Inc., ISBN 0-471-07338-5; (2000) 235
5. B.D.Milbrath *et al.*, Nucl. Instr. and Meth. in Phys. Res. A, 547 (2005) 504

BUILT-IN STRAIN MEASUREMENTS IN POROUS SILICON BY RAMAN SCATTERING

M. A. FERRARA[a),b)], L. SIRLETO[a)], G. MESSINA[c)], M. G. DONATO[c)], S. SANTANGELO[c)], AND I. RENDINA[a)]

[a)] *Istituto per la Microelettronica e Microsistemi – CNR*
Via P. Castellino 111 - 80131 Napoli
[b)] *DIMET - Università "Mediterranea" Reggio Calabria*
[c)] *MECMAT - Università "Mediterranea" Reggio Calabria*
antonella.ferrara@na.imm.cnr.it

ABSTRACT

In this paper, porous silicon layers of different porosity and thickness have been investigated by Raman spectroscopy. The spectral characteristics of the first order Raman scattering (line shape and peak position) were analyzed according to the phonon confinement model in order to estimate the size of the crystallite and evaluate the built-in strain in porous silicon.

1. Introduction

Studies on porous silicon (PS) were first concentrated on the applications of this material in the field of silicon-on-insulator technology[1,2]. Recently, the discovery of the optical properties[3-5] of PS has stimulated renewed research on this material. Visible photoluminescence[3] and electroluminescence[4,5] have been obtained at room temperature on thin PS silicon layers with a porosity of 70% or more. The interpretation of these properties is based mainly on the quantum confinement of electrons and holes in nanometric-size silicon crystallites[3,6,7]. Thus, the determination of the size and the shape of the silicon crystallites in PS is an important issue. Direct pictures of the PS structure can be obtained by electron microscopy, but sample preparation is destructive and can modify the observed structures. X-ray diffraction[8] and Raman scattering experiments[9] have been performed in order to obtain a better knowledge of the structural properties of PS.

For an as-formed PS sample, strain depends on various parameters, the most important being the porosity of the PS layer, the type and the level of doping of the silicon substrate[10,11]. The PS lattice parameter is slightly expanded in direction perpendicular to the surface[11], while in direction parallel to the surface, the porous layer has the same lattice parameter as the substrate. Due to lattice mismatch with bulk Si, a porous silicon layer is expected to be under compressive stress[12] when attached to the Si substrate.

In this paper, Raman spectroscopy has been applied to the study of the distribution of strains in porous silicon layers having different porosity and thickness.

2. Theoretical Background

The Raman effect in crystals is due to the scattering of light by the lattice vibrations[13], leading to annihilation or creation of crystal phonons. The energy lost or gained by the lattice is compensated by the increase (anti-Stokes part of the Raman spectrum) or the decrease (Stokes part of the Raman spectrum) of the energy of the scattered light. Moreover, as optical photons have very small wave vectors (approx. 10^5 cm^{-1}) if compared with lattice vibrations (up to 10^8 cm^{-1}), due to the conservation of the crystal momentum in the scattering process only phonons at the centre of the Brillouin zone (namely, having $\mathbf{q}=0$ wavevector) can be involved in the scattering process. In crystalline silicon, the scattering of optical photons from the threefold degenerate optical phonon at the centre of the Brillouin zone gives rise to a strong Stokes peak at approximately 521 cm^{-1} from the laser line, having a full width at half maximum (FWHM) of approx. 4 cm^{-1} at room temperature.

Porous silicon is composed of wires and/or dots of not uniform dimensions. When the size of the particles reduces to the order of nanometers, the wave function of optical phonons is no more a plane wave[14]. The localization of the phonon wavefunction (phonon confinement effect) leads to the relaxation of the wave vector conservation in the Raman scattering process in a crystal. Thus, not only the phonons with zero wave vector, but also those with $\mathbf{q}\neq0$ take part in the Raman scattering process, resulting in the red shift and broadening of the PS Raman line[14,15]. Campbell and Fauchet[15] developed a quantitative model that calculates the Raman spectrum of porous silicon as depending on the size L and on the shape of the porous silicon crystallites. If PS is modelled as an assembly of quantum wires, the phonon confinement is assumed to be two dimensional, while if the PS is modelled as an assembly of quantum dots, the confinement is three dimensional. The localization of the phonon wavefunction is imposed by means of a weighting factor W and the resulting localized wavefunction is expanded in a Fourier series with coefficients[9,14,15] $C(0,q)$. The Raman spectrum is given by:

$$I(\omega) \cong \int \frac{d^3q\left|C(0,q)\right|^2}{\left(\omega-\omega(q)\right)^2+\left(\Gamma_0/2\right)^2} \tag{1}$$

Campbell and Fauchet[15] showed that the weighting function and the Fourier coefficients that best fit the experimental data are:

$$W(r,L)=\exp\left(-\frac{8\pi^2r^2}{L^2}\right) \qquad \left|C(0,q)\right|^2=\exp\left(-\frac{q^2L^2}{16\pi^2}\right) \tag{2}$$

The first-order Raman spectrum $I(\omega)$ is thus given by[9]:

$$I(\omega)=\int_0^{\frac{2\pi}{a_0}} \frac{4\pi q^2 \exp\left(-\frac{q^2L^2}{16\pi^2}\right)dq}{\left[\omega-\omega(q)\right]^2+\left(\frac{\Gamma}{2}\right)^2} \tag{3}$$

where a=0.54 nm is the lattice constant of silicon, Γ is the natural linewidth for c-Si at room temperature and $\omega(q)$ is the dispersion relation for optical phonons in c-Si[16] that, according to Sui *et al.* [17], can be taken as:

$$\omega(q)=\omega_0-120\left(\frac{q}{q_0}\right)^2 \tag{4}$$

Here, ω_0 is the position of the c-Si Raman peak.

However, Yang *et al.*[9] have shown that the experimental results, obtained by Raman scattering study of PS, could only be explained by adding a "built-in" strain to the phonon confinement effect, having the effect of further decreasing the PS peak position with respect the c-Si value. Thus, the built-in strain may be qualitatively estimated by means of the difference between the experimental and theoretical PS Raman peak position at the same linewidth.

3. Experimental results

In these experiments, porous silicon monolayers have been obtained by electro-chemical etching on p^+ type (ρ=8-12 mΩ cm) standard silicon wafers (001). In order to study the effect of both porosity and thickness on the Raman spectrum of porous silicon, two series of samples having different thickness (10 μm - 20 μm) have been prepared. At fixed thickness, the porosity of the samples ranged between 50% and 80% (step 10%).

Unpolarized Raman spectra have been detected, at room temperature, in backscattering geometry using a Jobin Yvon Ramanor U-1000 double monochromator, equipped with a microscope Olympus BX40 for micro-Raman sampling and an electrically cooled Hamamatsu R943-02 photomultiplier for photon-counting detection. The excitation source was a Coherent Innova 70 Ar^+ laser, operating at 514.5 nm wavelength. In order to prevent laser-annealing effects, the laser power was about 2 mW at the sample surface. Using a 50X objective having long focal distance, the laser beam was focused to a diameter of few microns. Its position on the sample surface was monitored with a video camera. All components of the micro-Raman spectrometer were fixed on a vibration damped optical table.

As a preliminary measurement, the Raman spectrum of c-Si has been recorded. A lorentzian line, centred at 521.4 cm^{-1} and having FWHM of approximately 4.4 cm^{-1} was obtained.

In Fig.1, the Raman spectra of 20 μm thick PS samples having different porosity are shown. As expected on the basis of the phonon confinement model[14,15], the red shift and the asymmetry of the PS Raman peak increase with increasing porosity.

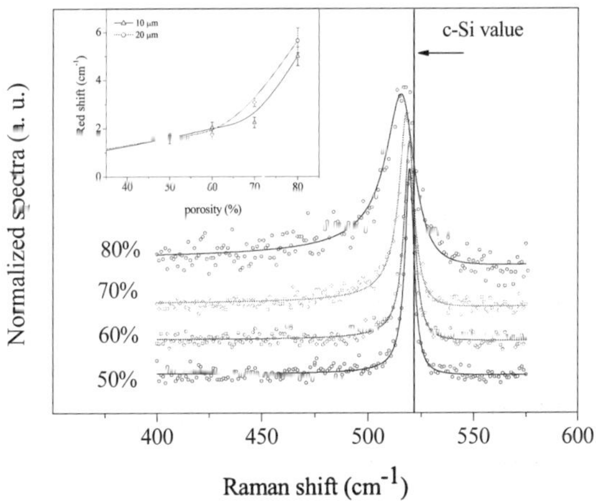

Figure 1. Raman spectra (circles) of PS samples (20 μm) at different porosity. In the inset, the positions of the PS Raman peaks registered on the samples with different thickness as a function of the layer porosity are shown. The solid lines are drawn as guides for the eye.

In the inset, the position of the PS Raman peak is shown as a function of the layer porosity for the two series of samples at increasing thickness. A greater red shift with respect the c-Si value is observed in samples of higher porosity and thickness. Thus, according to the data reported by Papadimitriou *et al.*[18], we may hypothesize gradients of porosity and/or intrinsic stress along the thickness of the films. The evaluation (Fig.2) of the difference $\omega_{PS}-\omega_{th}$ between experimental and theoretical Raman peak shift at the same linewidth shows that built-in strain is more effective in thicker samples. From the figure, an approximately constant built-in strain in thicker samples can be deduced.

By fitting the measured Raman spectra with the model proposed by Campbell and Fauchet (eq. (3)), information about the size of the silicon crystallite have been obtained. In Tab. I are listed, for each sample at a given porosity and thickness, the estimated size of the crystallite. The size of the crystallite decreases with increasing porosity. Moreover, at fixed porosity, a lower size of the crystallite is obtained in samples of greater thickness.

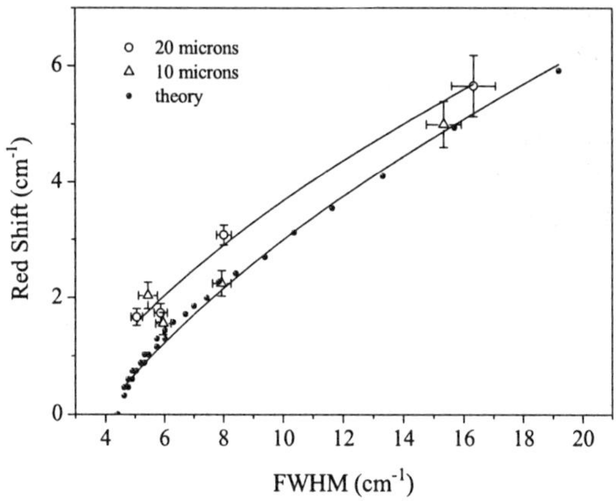

Figure 2. Relationship between the linewidth and the peak shift of experimental and theoretical PS Raman peak with that of c-Si. The solid lines are drawn as guides to the eye.

Porosity (%)	Thickness 10 μm	Thickness 20 μm
	L [nm]	
50	8.7	8.3
60	8.0	7.7
70	6.0	6.0
80	4.7	3.9

Table I. Size L of the silicon crystallite calculated by means of the confinement model for the PS samples with different porosity and thickness.

4. Discussion and Conclusions

In this paper, Raman spectroscopy has been applied to the study of porous silicon layers having different thickness and porosity. As expected from previous theoretical works[14,15], the shift of the PS Raman peak towards lower wavenumbers and the broadening of the line is greater in samples of higher porosity (Fig.1). In addition, the red shift is more sensitively observed in samples having higher thickness (Fig.2). As already noted, this effect may be associated to an inhomogeneous distribution of porosity and/or strain along the thickness of the samples. In fact, the position of the PS Raman peak is the result of a competition[18] between confinement effects[15] and built-in strain[9], which cause a red shift of the peak position with respect to bulk Si, and compressive stress[12], induced by lattice mismatch with Si substrate, which causes a blue shift of the peak. In this regard, it is reasonable to hypothesize that in thicker samples the effect due to the c-Si substrate is weaker than in thinner samples, resulting in a greater contribute of confinement effect and built-in strain, and, thus, in a greater red shift of the PS Raman peak with respect crystalline silicon.

References

1. M. I. J. Beale, N. G. Chew, A. G. Cullino, D. B. Gasson, R. W. Hardeman, D. J. Robbins, and I. M. Young, J. Vac. Sci. Technol. B **3**, 732 (1985).
2. R. Herino, G. Bomchil, K. Barla, C. Bertrand, and J. L. Ginoux, J. Electrochem. Soc. **134**, 1994 (1987).
3. L. T. Canham, Appl. Phys. Lett. **57**, 1046 (1990).
4. A. Halimaoui, C. Oules, G. Bomchil, A. Bsiesy, F. Gaspard, R. Herino, M. Ligeon, and F. Muller, Appl. Phys. Lett. **59**, 304 (1991).
5. N. Koshida and H. Koyama, Appl. Phys. Lett. **60**, 347 (1992)
6. J. C. Vial, A. Bsiesy, F. Gaspard, R. Herino, M. Ligeon, F. Muller, R. Romesain, and R. M. Macfarlane, Phys. Rev. B **45**, 14171 (1992).
7. A. Bsiesy, J. C. Vial, F. Gaspard, R. Herino, M. Ligeon, F. Muller, R. Romesain, A. Wasiela, A. Halimaoui, and G. Bomchil, Surf. Sci. **254**, 195 (1991).
8. D. Bellet, G. Dolino, M. Ligeon, P. Blanc, and M. Krisch, J. Appl. Phys. **71**, 145 (1992).
9. M. Yang, D. Huang, P. Hao. F. Zhang, X. Hou, X. Wang, J. Appl. Phys. **75**, 651 (1994).
10. D. Bellet, G. Dolino, Thin Solid Films **276**, 1 (1996).

11. I. M. Young, M. I. J. Beale, J. D. Benjamin, Appl Phys. Lett **46**, 1133 (1985).
12. U. Gruning, A. Yelon, Thin Solid Films **255**, 135 (1995).
13. R. Loudon, Advances in Physics **13**, 423 (1964).
14. H. Ritcher, Z. P. Wang and L. Ley, Solid State Communications **39**, 625 (1981).
15. I. H. Campbell and P. M. Fauchet, Solid State Communications **58**, 739 (1986).
16. F. Kozlowski, W. Lang, J. Appl. Phys. **72,** 5401 (1992).
17. Z. Sui, P.P. Leong, I.P. Herman, G.S. Higashi, H. Temkin, Appl. Phys. Lett. **60**, 2086 (1992).
18. D. Papadimitriou, J. Bitsakis, J.M. Lopez-Villegas, J. Samitier, J.R. Morante, Thin Solid Films **349**, 293 (1999).

DIAMOND DETECTORS FOR X-RAY BEAM MONITORING

G. CONTE*, M. GIROLAMI, S. SALVATORI
Dept. of Electronic Engineering, University of Rome "Roma Tre"
Via della Vasca Navale, 84 – 00146, Rome, Italy

R. SCAFÈ, F. PISACANE, D. DELLA SALA
Casaccia Research Centre, ENEA
Via Anguillarese 125 - 00123 S.M. di Galeria, Rome, Italy

Metal-diamond-metal structures were realized and tested under continuous X-ray irradiation. Photocurrent versus applied voltage curves were used to evaluate the mobility-lifetime product of collected charges at different beam intensities. $\mu\tau$ values in the order of 6×10^{-6} cm^2/V were estimated. As the X-ray source is not monochromatic, Bremsstrahlung contribution was removed and the detector response to $K\alpha$ copper radiation was evaluated.

1. Introduction

Polycrystalline diamond deposited by CVD techniques is a material with 5.5 eV of band gap. Due to improvements in growth technology, the electronic quality of this material has improved in the last few years, and is now better than natural diamond in some respects. Recently, the promise and expectation for this material have been demonstrated through the realization of high frequency electronic devices [1]. On the other hand, as a photoconductor, polycrystalline diamond shows a high dark resistance and a large breakdown electric field. It is a robust wide-gap semiconductor material which can withstand high temperatures, a wide range of corrosive environments and exhibits high radiation hardness This combination of properties makes it extremely attractive for use as photon and particle detector. Devices based on such a material have potential application in very high rate radiation fields, such as radiotherapy [2] and nuclear beams, high voltage X-ray metrology [3], X-ray microscopy, astrophysics, solar physics, medical imaging as well as homeland security. Aimed to contribute to overcome limitations observed in the past, mainly due to un-reproducible material quality, we report on the realization and testing of detectors based on high quality polycrystalline diamond, mechanically polished on both the surfaces, to be used for continuous and pulsed X-ray sources analysis and monitoring. The collection efficiency of charge carriers under irradiation was studied to infer the transport mechanisms as a function of beam

* Corresponding author: gconte@uniroma3.it

energy and intensity. The evolution of the device sensitivity to the ionising radiation has been analysed and a correlation with the mobility-lifetime product has been proposed.

2. Experimental

Thick diamond deposits were realized on silicon starting from a mixture of methane and hydrogen by using a microwave plasma enhanced CVD technique. Silicon substrate was then removed by wet etching, and both surfaces treated to remove about 100 μm of material. After this polishing, the average roughness was estimated to be 2-8 nm. A 0.8×0.8 cm^2 specimen, 200 μm thick, was selected and cleaned by dipping in sulphochromic solution and aqua regia. In order to provide electric contacts, Silver deposits were made on both the faces of the diamond specimens by thermal evaporation, then multi-finger structures 40 μm apart were defined by photolithography on one face to perform measurements both in a planar or sandwich configuration. I-V characteristics are symmetric, i.e. independent of polarity, indicating that the front and back contact are of the same type.

Studies in continuous mode X-ray irradiation were carried out by using a commercial source equipped with a Nickel filtered Copper microfocus tube (10x1 mm focal spot) powered at 40 kV and 30 mA. A collimator 1 mm in diameter was used to deliver the radiation on the sample. A revolver with 15 filters of increasing Aluminum thickness was interposed between the source and the device, aimed to reduce in a controlled way the beam intensity. Current versus voltage (I-V) measurements were conducted up to 100 V by using a Keithley 617 electrometer.

3. The operational principle of a diamond detector

The detector has a metal-diamond-metal structure made of contacts realized on both the surfaces of diamond. When ionizing radiation impinges on the device, electron-hole pairs can be excited, transferred to electrodes in the electric field and collected as signal by electrodes. To evaluate the collection efficiency we assume that the electric contact between the Silver and the polycrystalline diamond is ohmic (as addressed by the symmetry of I-V characteristic in the dark). On this basis, we consider a uniform electric field within the entire sample thickness. Moreover, with the assumption that all the active material is bedewed by the X-ray beam, neglecting diffusion transport, the collection distance of photo generated carriers is simply given by the drift length $l_c=\mu\tau F$, where $\mu\tau$ is the mobility-lifetime product and F the electric field. Assuming an exponential dependence of the probability to collect charge carriers generated at the distance x from the impinging surface, the collection efficiency, η_{coll}, is given by [4]

$$\eta_{coll} = \frac{\mu\tau F}{d}\left[1-\exp\left(-\frac{d}{\mu\tau F}\right)\right] \tag{1}$$

being *d* the sample thickness. The beam interaction with the material is mainly determined by absorption of atoms associated to the photoelectric effect and the Thomson elastic scattering which only deviates part of the X-ray intensity without absorption.

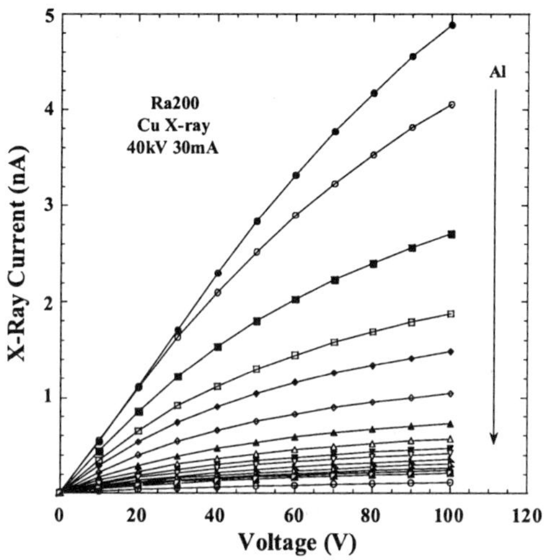

Figure 1. Photocurrent versus voltage at different beam intensities. Aluminum thickness changes from top (naked beam) to bottom in the range from 0 to 0.22 cm.

4 Results and discussion

Typical photocurrent response under 8.06 keV X-ray irradiation is reported in Figure 1 at different irradiation intensities obtained by Aluminum foils attenuation. The photocurrent increases proportionally to the electric field at low voltages (i.e. 100 V/cm), while a further increase of bias leads to a saturation trend following the expression

$$I_{ph} = \eta_{coll} I_{ph}^{sat} \tag{2}$$

where I_{ph}^{sat} is the saturation value of X-ray photocurrent density. As the source is not monochromatic, increasing the Al thickness, Cu K_α radiation and average Bremsstrahlung are attenuated with a different linear absorption coefficient: 135 cm^{-1} and 6 cm^{-1} respectively. This effect is shown in Figure 2 where the photocurrent as a function of Al thickness is reported. The superposition of two exponential regimes is apparent. Absorption coefficients were evaluated by curve fitting with expression

$$I_{ph} = I_0 + I_{K\alpha} \exp(-\mu_{K\alpha} d) + I_B \exp(-\mu_B d) \tag{3}$$

where I_0 is a contribution due to un-filtered radiation components, $I_{K\alpha}$, $\mu_{K\alpha}$ and I_B, μ_B are the photocurrent and linear absorption coefficient of 8.06 keV and average Bremsstrahlung (i.e. 22 keV) respectively.

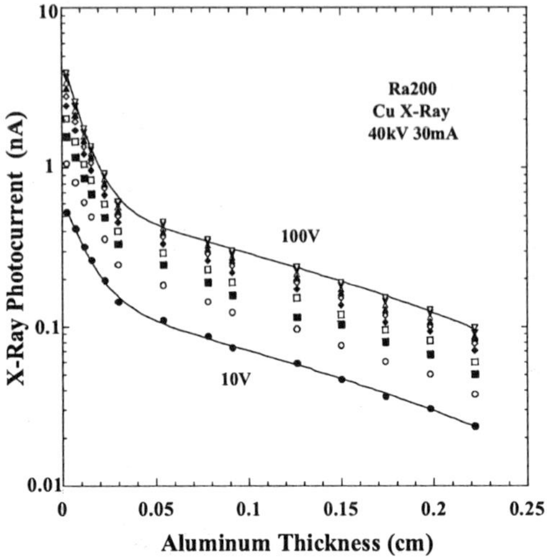

Figure 2. Detector response at different beam intensities obtained by Aluminum filtering. Continuous lines are the fitting with expression (3).

While μ_B stays constant around 6±1 cm^{-1}, $\mu_{K\alpha}$ changes with source intensity between 74 and 108 cm^{-1}. Such a behavior addresses a non-linear response to Cu Kα radiation. Assuming the photocurrent dependent on the beam intensity as stated by Rose's theory [5], device response to X-ray flux, Φ, can be evaluated from

$$I_{ph} \propto \Phi^{\beta} \tag{4}$$

where $0.5<\beta<1$ being indicative of the active recombination regime: β=0.5 is observed when a bimolecular mechanism is working, addressing a pure semiconductor, whereas β=1 is obtained for a monomolecular process, when states inside the bandgap participate to the recombination mechanism. On such basis, β corresponds to the ratio between $\mu_{K\alpha}$ found and 135 cm^{-1}, the theoretical value. Once isolated, K_α contribution to the photocurrent has been used to evaluate $\mu\tau$ values by using expression (1). Figure 3 reports the result of the fit versus the normalized X-ray intensity. At low generation rate, $\mu\tau$ stays constant around 6×10^{-6} cm^2/V whereas increasing the beam flux an exponential decrease is observed. Such a behavior can be interpreted on the basis of carrier-carrier scattering [6] addressing the need of a stronger electric field to improve charge collection.

Realized devices demonstrated the high quality of the diamond used. Notwithstanding the mechanical treatment, the defect density is low and a response proportional to the beam intensity over three order of magnitude was observed. A higher defect density appears needed to obtain a linear response whereas a higher electric field seems to be needed at the higher intensities.

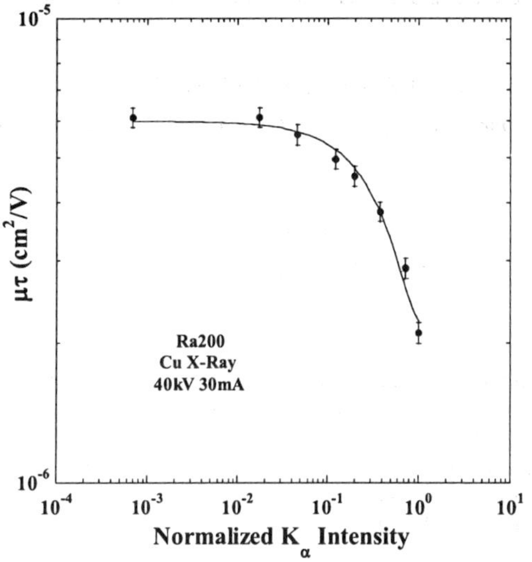

Figure 3. $\mu\tau$ values versus X-ray intensity.

Acknowledgements

V. Ralchenko is gratefully acknowledged for providing the diamond sample.

References

1.K. Ueda, M. Kasu, Y. Yamauchi, T. Makimoto, M. Schwitters, D. J. Twitchen, G. A. Scarsbrook and S. E. Coe, *IEEE EDL* **27,** 570 (2006).
2.M. J. Guerrero, D. Tromson, C. Descamps and P. Bergonzo, *Diamond Relat. Mater.*, **15,** 811 (2006).
3.H. Kagan, *Nucl. Instr. Meth.* **A541,** 221 (2005).
4.G. Conte, M. C. Rossi, S. Salvatori, P. Ascarelli and D. Trucchi, *J. Appl. Phys.* **96**, 6415 (2004) and references therein.
5.A. Rose, *Concepts in photoconductivity and allied problems*, R.E. Krieger Publishing Co., Huntington, New York, 1978.
6.L. S. Pan, D. R. Kania, *Diamond: electronic properties and applications,* Kluwer Academic, 1995.

A FINITE ELEMENT 2-DIMENSIONAL MODEL FOR THE PREDICTION OF THE FREQUENCY RESPONSE OF THERMAL GAS VELOCITY DETECTORS

P. BRUSCHI, M. SCHIPANI, N. BACCI
Dipartimento di Ingegneria dell'Informazione, via Caruso 16
I-56122 Pisa, Italy

M. PIOTTO
IEIIT-Pisa, CNR, via Caruso 16
I-56122 Pisa, Italy

The frequency response of integrated thermal gas velocity detectors is estimated by means of 2-Dimensional finite element simulations. The study is devoted to examine the possibility to extend the application field of simple integrated flow meters to the detection of acoustic waves. An original approach for taking into account non negligible heat conduction paths along the third axis is described. The role of the thermal mass of the active elements is highlighted.

1. Introduction

Acoustical Velocity Field (AVF) sensors are devices capable of detecting the local velocity produced by acoustic waves in a gas. Recently, AVF sensors based on miniaturized thermal flow sensors have been proposed and commercial products denominated microflown™ are currently available [1]. These devices are used in professional applications to fully characterize the acoustical field in closed environments, where the knowledge of the sole pressure field is not sufficient[2,3]. The fragility of the structure involved in the microflown™ and the complex fabrication procedure make this device unsuitable to replace traditional microphones in common applications.

On the other hand, very robust and CMOS compatible thermal gas velocity detector are widely used as core elements in flow meters [4]. With a proper design, aimed to improve sensitivity and frequency response, such structures might be used as inexpensive, extremely miniaturized integrated microphones. Unfortunately, complicated phenomena contributing to the velocity detector response, such as formation of the mechanical and thermal boundary layers, hinder the development of analytical or numerical design procedures [5]. In this work we describe the use of 2-dimensional finite element simulations for the prediction of the frequency response of velocity sensors produced by post-processing of integrated circuits produced with standard microelectronic processes. Two structures, differing mainly for the thickness of the oxide layers covering their active parts, are compared.

2. Device Description

The structure of the devices considered in this work is shown in Fig. 1, where the top view (a) and cross section are represented (b). The devices are differential thermal flow meters based on a well known principle, briefly recalled in this section [5]. A heater, consisting in a polysilicon resistor driven with constant power, is placed in symmetric position between two thermopiles. The heater and the thermopile hot junctions are placed over dielectric membranes, suspended on a cavity etched onto the silicon substrate. In this way, the three mentioned elements are thermally insulated from the substrate. The thermopile cold junctions, placed over the substrate, are separated from the latter by a thin oxide layer so that their temperature can be considered equal to the substrate temperature. Each thermopile produces an output voltage proportional to the difference between its hot junction and the substrate. In condition of still air, the heater produces equal overheating of the two thermopiles. Conversely, a component of the gas velocity along the direction shown with x in the figure (direction of maximum sensitivity of the structure) induces a temperature difference between the thermopiles. The corresponding difference of the thermopile output voltages constitutes the sensor output signal.

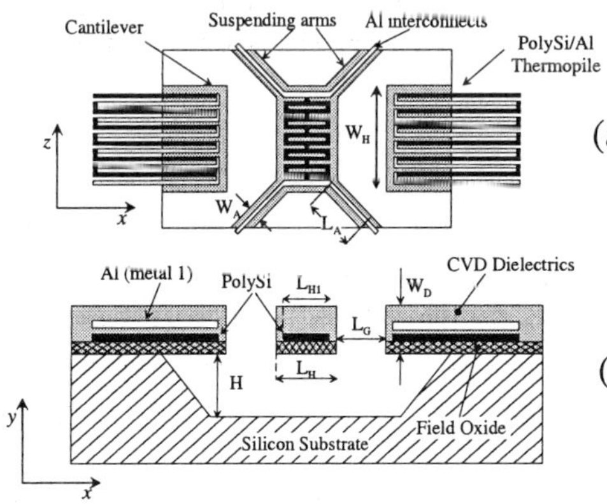

Figure 1. Top view (a) and cross section (b) of the thermal gas velocity detectors.

An acoustic wave produces pressure and velocity oscillations. Pressure oscillations may cause small variations of the thermal conductances [6] but, due to the symmetrical structure, the output signal is unaffected. Instead, velocity oscillations produces corresponding oscillations in the output signal, provided that their frequency is below the upper band limit of the devices. Currently, a convincing model to predict the frequency response of these devices is not available in the literature.

3. Device Modeling

The model has been obtained by improving a previously introduced approach, used only for static simulation. Here we will briefly summarize the innovative aspect of the previous model; more details are reported in Ref. [7].

Since 3-Dimensional simulations are time consuming and often present convergence problems, we have simulated the 2-D cross-section shown in Fig. 1(b). This is a common choice [8], dictated by the difficulty of solving the Navier-Stokes equations in a domain with features dimensions ranging from several centimeters (external environment, such as a pipe portion) to a few microns (e.g. membrane thickness). Our original contribution is that of introducing *ad hoc* terms in the 2-dimensional equations in order to take into account also non negligible heat conduction paths along the third axis. In particular, the thermal power produced into the heater flows to the external environment mainly through the four suspending arms, which are not represented in the simulated cross-section. To take into account these contributions we have calculated the total conductivity from the heater middle section to the substrate. Calling G this conductance we have modified the power generation term in the polysilicon heater according to this heat drainage mechanism, obtaining:

$$\frac{dP}{dA} = \left[I_H^2 R_H - G(T - T_{sub})\right] \frac{Z_1}{t_{poly} W_H L_{H1}} \tag{1}$$

where dP/dA is the local heater power generation source, T the local heater temperature, T_{sub} the substrate temperature, I_H and R_H the heater current and resistance, respectively, t_{poly} the polysilicon thickness, and Z_l the unity length along the third axis (1 m) considered by the simulator. Note that a 2-D model is equivalent to a 3 D model where all the parameters and variables are uniform along the whole third axis. We have chosen to consider a uniform condition only across the heater /thermopile width (W_H) and to neglect what occurs outside this domain. Even over this limited domain, the material properties are not uniform, since, for example, the thermopile and heater layer are not continuous. For this reason we have introduced effective properties (e.g. effective thermal conductivities and specific heat) obtained by a weighted average of the materials alternating along the z axis.

The model was adapted for transient simulations by introducing a few simplifying conditions that increases the computational efficiency and significantly improve convergence. One simplification was that to consider the air as an uncompressible fluid. In fact, we are interested to consider only the effect of the gas velocity which, in the audible frequency range can be considered uniform over the whole sensing chip. The geometry of the simulated configurations is shown in Fig. 2. The chip is included into a 2-dimensional channel with ideally rigid upper and lower walls. At one end of the channel, a sinusoidal velocity is imposed, while at the other extreme a constant pressure is assigned.

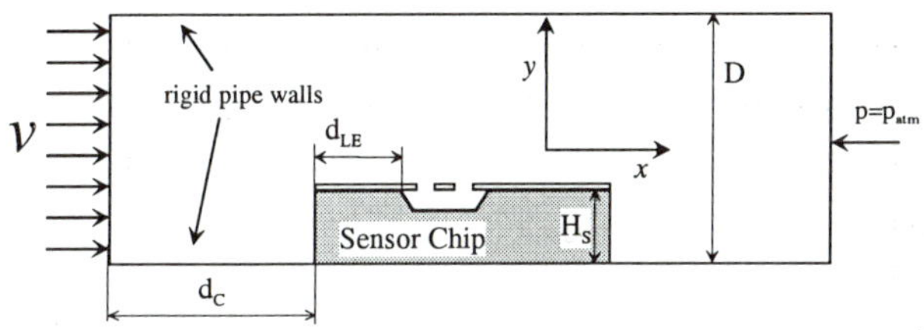

Figure 2. Complete configuration of the system used in the simulations.

The dimensions indicated in Fig. 1 and 2 for the sensing structure and test channel, respectively, are reported in Table 1.

Table I. Parameters of the simulated structure and test channel

Parameter	value (μm)	Parameter	value (μm)
W_H	80	L_A	74
L_G	61	d_C	5000
H	80	D	3000
L_{H1}	34	H_S	400
L_H	42	d_{LE}	500
W_A	25	Heater power: 7 mW.	

The simulations have been devoted to study the frequency response of two different real velocity detectors, fabricated by means of post-processing steps applied to standard microelectronic chips. The two structures differ for the process employed and post-processing procedure. The first structure [6] was fabricated using the 1 μm, Bipolar-CMOS-DMOS process BCD3s of

STMicroelectronics. The particular post-processing procedure adopted to etch the cavity into the substrate resulted also in removal of the upper dielectric layers. This did not occur for the second structure, fabricated using the 0.35 μm process BCD6 (STMicroelectronics). As a result, the two structures were very similar for all the dimensions except for dimension W_D, which was 2.5 μm for the BCD3s structure and 5.2 μm for the BCD6 one.

4. Simulation Results

The model was used to build a 2-D description of the system, suitable to be solved by the finite element environment FEMLAB™. The standard fluid-thermo-dynamic equations were modified in selected domains (e.g. inside the polysilicon heater volume) in order to take into account heat exchange paths along the third axis, according to Eq. (1). A series of transient simulations were performed varying the frequency of the sinusoidal velocity *v*.

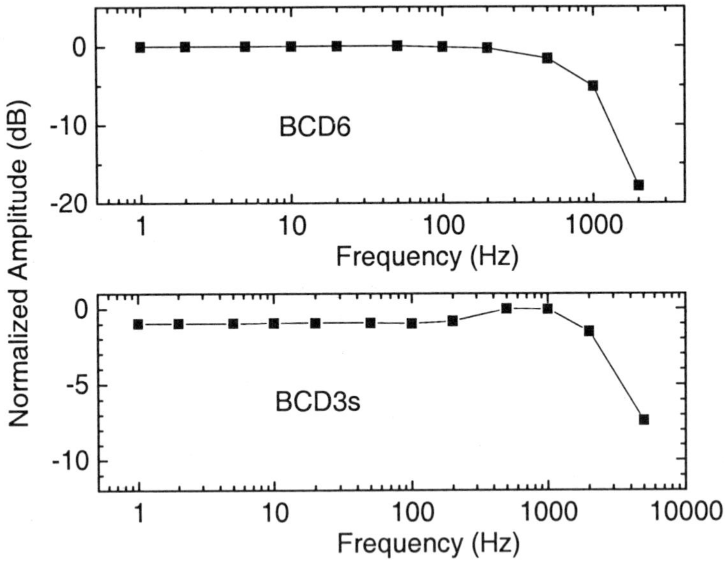

Figure 3. Normalized frequency response of a BCD6 and BCD3s structure.

The amplitude of the velocity oscillations were set to 4.5×10^{-4}m/s, equivalent to an acoustic intensity of 80 dB. Such a high velocity level was required to achieve a sufficient resolution on the relevant signals (thermopile temperatures) with no need of excessively small tolerances for the numerical solver. The temperature difference between the hot junctions of the two thermopiles has been considered as the output signal. From this signal we have extracted the magnitude of the oscillations at the same frequency of the excitation signal (input velocity). Fig. 3 shows the resulting frequency response for the two different structures, identified by the technological process used for their fabrication. The curves have been normalized to the maximum value. It can be observed that the structure obtained with the BCD3s process has a wider bandwidth. In particular the -3 dB cut-off frequency is equal to 2.5 kHz for the BCD3s structure and to 800 Hz for the BCD6 one. This can be a consequence of the smaller thermal mass of the suspended structures in the BCD3s, due to the reduced dielectric layer thickness. Another noticeable difference is the peak visible in the BCD3s structure. This can be due to the fact that boundary layer formation is less efficient at high frequencies. This phenomenon is probably present also for the BCD6 structure, though the lower cut-off frequency of the latter compensates the related amplitude increase.

Acknowledgments

The Authors wish to thank the STMicroelectronic R&D group of Cornaredo (Milan, Italy), for fabricating the prototypes that has been modeled in this work.

References

1. H.E. de Bree, *Acta Acustica*, **89,** 163 (2003).
2 F.J.M van der Eerden, H.-E de Bree, H Tijdeman,. *Sensors and Actuators A*, **69**, 126 (1998).
3 F. Jacobsen, Y. Liu, *Journal of the Acoustical Society of America*, **118**, 3139 (2005).
4 H. Baltes; O. Paul, O. Brand, *Proc. of IEEE*, **86** 1660 (1998).
5 M. Elwenspoek and R. Wiegerink, *Mechanical Microsensors*, Springer-Verlag, Berlin, 2001.
6 P. Bruschi, M.Piotto, G. Barillaro, *Sensors and Actuators A*, **132,** 182 (2006).
7 P. Bruschi, A. Ciomei, M. Piotto, *proc of EUROSENSORS XX*, Goteborg, 17-20 September 2006.
8 U. Schmid, *Sensors and Actuators A*, **97-98,** 253 (2002).

DEVELOPMENT OF A MULTISENSOR LAYOUT FOR ROBOTS

MARCO SANTORO, CLAUDIO MORICONI

Department of Robotics, ENEA Research Center "Casaccia", Via Anguillarese 301 Rome, 00123, Italy

Robots operating in harsh and hostile environments need reliable multisensorial systems able to automatically acquire as much information as possible. Sensory data from a range of disparate and independent multiple sensors have to produce an improved model or estimate of the domain of interest. Therefore the automated intelligent combination of data from multiple sensors can derive less ambiguous/uncertain information about the desired state. ENEA, the *Italian National Agency for New Technologies, Energy and the Environment*, is therefore developing a multisensorial layout for robots operating in hostile environments.

1. Multisensors

Multisensors are widely present in nature, in human beings, animals, vegetables. Robots can fruitfully harness such a kind of devices [1] for better interpreting their operating environment and improve their dependability, i.e. their ability of operate in complex, non-structured, environments, like airports, streets, crowded spaces . A multisensorial distributed unit can generally contain several discrete sensors over, inside and beneath a definite surface area/volume.

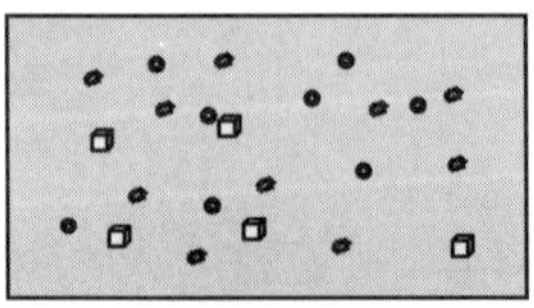

Figure1 A multisensor unit. Discrete sensors are distributed over and inside the sensor's volume

1.1. *Our multisensor*

A Pressure Pad based on a foam material, whereby light introduced to the foam is scattered in a manner dependent on the force applied to the surface of the foam, is our pressure detecting device.

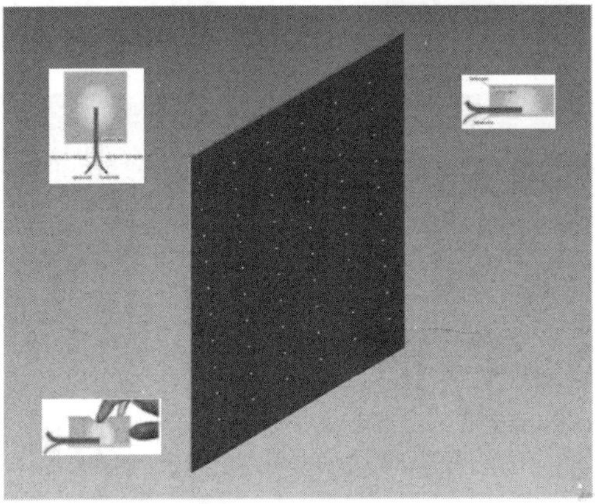

Figure 2 The CAD layout of the pressure pad and its operating principle

60 pressure sensors are spaced evenly throughout the area of the touchpad (210mm X 350mm). Each individual pressure sensing element consists of a tiny light source and a tiny detector imbedded in a thin cellular elastomer. As pressure is applied to the elastomer in the region of the sensor, the cell size distribution is altered along with the bulk optical scattering properties of the material in that area. The light detector measures the change in scattered light intensity at the sensor, and interprets this as a change in pressure.

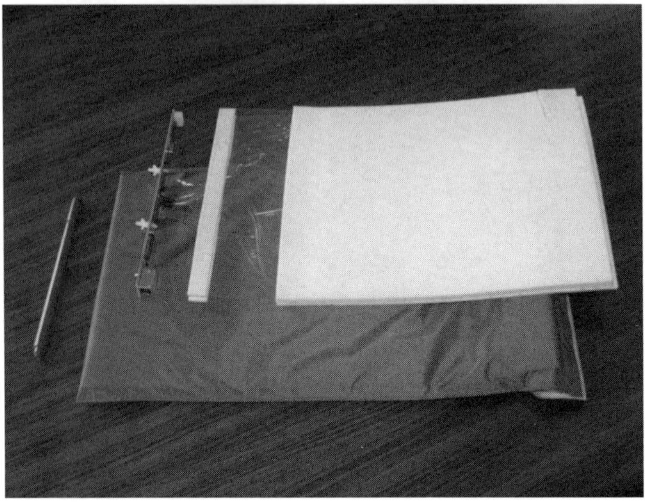

Figures 3 - 4 - 5 The pressure sensing device. A nylon cover, beneath the pressure pad in Fig. 5, covers completely the device when pressures are being detected

The light source is a light emitting diode (LED) connected to a bundle of fibers, and the light detector is an array of photodiodes connected to a second set of fibers. A microcontroller converts the signals from the detectors to digital signals, conditions those signals, and computes the required pressure information or decision data using a set of algorithms implemented in firmware.

Optical fibers laminated into the structure are used as the "nervous system", to deliver transmitted light to the location of the sensors, and to deliver received light back from the sensor. It is convenient and less expensive to use these fibers rather than individual light sources and detectors actually located within the material. This pressure system is thus unaffected by electromagnetic interference, since light intensity is used for measuring the pressure applied.

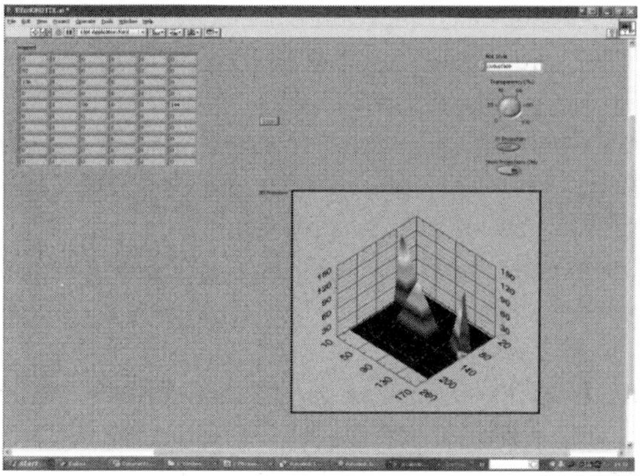

Figure 6 A pressure figure is acquired from the pad

Support Vector Machines [2] is a kind of pattern recognition algorithm able to classify non-separable data set too. An implementation of such an algorithm to pressure figures acquired from the pad has allowed a first raw classification of data. By this way we can e.g. discriminate among round and polygonal items lying over the pad.

In order to improve the quality of information available from the device, others sensors are being added to the pad.

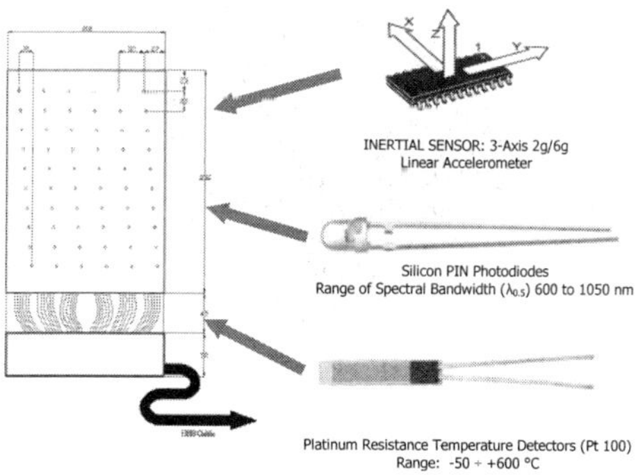

Figure 7 Heterogeneous sensors are added to the pressure pad

30 Platinum Resistance Temperature Detectors with 100Ω@0°C nominal resistance (PT100) are evenly placed over the pad cover for detecting temperature figures, e.g. a hand pattern figure.

30 Silicon PIN Photodiodes, with a range of Spectral Bandwidth ($\lambda_{0.5}$) from 600nm to 1050nm, are evenly placed over the pad. They are able to detect shadows and lights when an item is approaching the pad. This information is integrated with the temperature figure for accurate sensor fusion.

4 three-axis inertial sensors measure low-frequency accelerations of the pad.

Analog data from these sensors is scanned from a 32-bit ARM7-based microcontroller and provided to a PC in serial mode. This data set is continuously classified finally by an SVM algorithm.

Figure 8 32-bit acquisition board

References

1. Dario, P., Laschi, C., Micera, S., Vecchi, F., Zecca, M., Menciassi, A., Mazzolai, B., and Carrozza, M.C., "Biologically - inspired microfabricated force and position mechano-sensors", in "Sensors and Sensing in Biology and Engineering", by Friedrich G. Barth, Timothy W. Secomb, Joseph A., C. Humphrey, eds, Springer Verlag; 1st edition, March 28, 2003

2. Bernhard E. Boser, Isabelle M.Guyon, Vladimir N. Vapnik, *"A Training Algorithm for Optimal Margin Classifiers"*, Fifth Annual Workshop on Computational Learning Theory, Pittsburgh, 1992, ACM

OPTICAL STRAIN SENSOR BASED ON POLYMERIC DIFFRACTION GRATINGS

VERA LA FERRARA, IVANA NASTI, ETTORE MASSERA AND GIROLAMO DI FRANCIA

Research Center ENEA, Loc. Granatello, 80055 Portici (NA)

We look at an innovative method for the strain, stress measurement on surface. Our project is a simple imaging technique illuminating a polymer (PDMS) grid, attached on specimen, with a large spectra light source, and recording, with a ccd camera, the grid image obtained by diffracting light. Image acquired under stress is subtracted from one without stress giving a colour map of the surface correlated to the stress intensity. This technique arises from asserted optical techniques. Moreover, taking in account the technological progress (polymer, ccd, LEDs), we make the hardware setup very simple because it doesn't need coherent light source and the detector may be a commercial colour ccd.

1. Introduction

Strain measurement is very important in mechanics, material science and engineering. A number of optical techniques such as moiré, speckle and holography [1] have been developed and are routinely used in these areas. Among these optical techniques, diffraction methods provide strain information directly. In other words, the grating diffraction technique avoids the difficulty in fringe pattern interpretation associated with most optical techniques. This technique also possesses several advantages, such as the fact that no master grid is necessary, the system is simple and easily integrated, and no special high resolution recording system is required. The use of diffraction grating is widely studied since 1956 [2,3]. Perhaps only today is possible to fabricate diffraction gratings on polymer replica [4,5], opening new perspective on strain measurement with this technique. Polymeric diffraction gratings are high flexible and can be attached easily on a wide spread of specimen.

2. Experimental

In our experiment we have fabricated a polydimethylsiloxane (PDMS) replica using a silicon template with regular inverted pyramids (Fig. 1a). Starting from n-type silicon substrate, with 10 Ωcm resistivity, we have first deposited, by e-beam evaporation, 100 nm of SiO_2. In order to realize inverted pyramids we have used a photolithographic process where the mask consisted in a series of 2 μm in diameter holes, 4 μm center to center separated each other. The whole patterned area was about 2.25 cm^2 with a total number of about 2 millions of holes. After photolithography, SiO_2 has been etched using a BHF etching for about 1' at room temperature. An anisotropic etching, 50:50 solution of KOH and IPA at 80°C, has, then, permitted to realize the inverted pyramids (Fig. 1a). After PDMS solution preparation (base:curing agent=10:1), we have dropped it onto silicon pyramids substrate and peeled off polymeric replica after drying (Fig. 1 b).

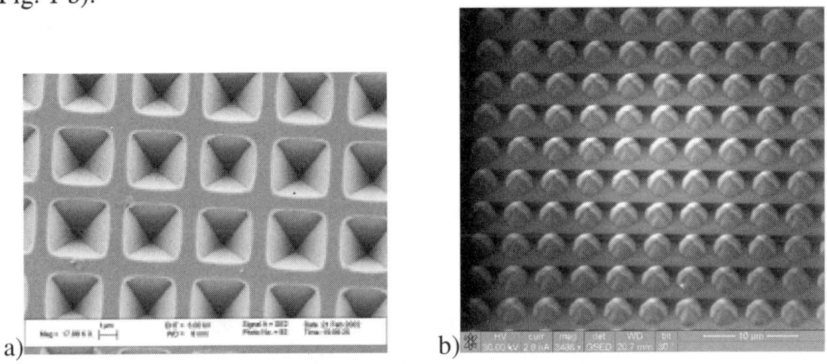
a) b)

Fig. 1: a) SEM image of Silicon Pyramids; dots are 2 μm large and space between two dots is about 1 μm. b) SEM image of PDMS peeled off from silicon pyramidal grooves

The PDMS grating is put on a specimen. Without glue, the grating naturally adheres. For the strain measurement we use a simple imaging technique illuminating the grid with a large spectra light source, and recording the grid image obtained by diffracting light with a ccd camera. Image acquired under stress is subtracted from one without stress giving a colour map of the surface correlated to the stress intensity. This technique simplifies the hardware setup because don't need coherent light source and the detector may be a commercial colour ccd.

3. Results and Discussion

To discover the better experimental conditions, strain detection tests with PDMS grids were made on a spread variety of material like aluminium, silica, plastic; with different optical surface (glossy, rough, opaque, etc..); under several light conditions (diffuse light, spot light, alogen, LED, crown LED). We find that PDMS grid on glossy surface, illuminated by a white LEDs crown around the lens camera gives the best S/N rate, and the best homogeneity in the illumination and diffraction of light. In Fig. 2 an image is illustrated of the diffraction grid under this condition.

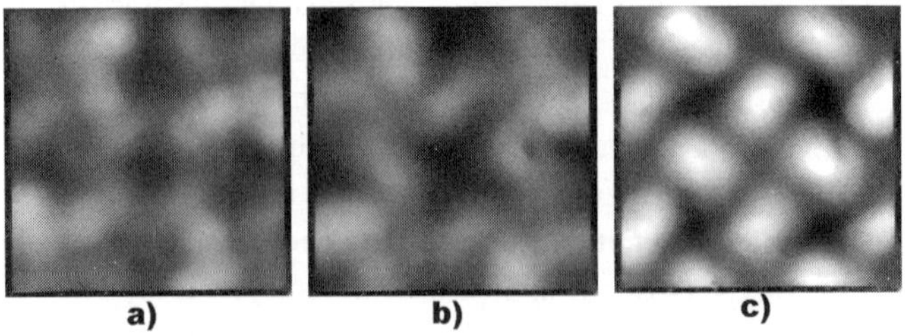

Fig. 2: ccd image of Polymeric grid, posted on glossy silica, illuminated by crown LEDs around ccd camera. Grid dimension is 15 X 15 mm2. a) channel RED; b) channel GREEN; c) channel BLUE.

As we can see the three principal colour component are diffracted with a different pattern as predicted by the classical optical diffraction theory.
For the strain measurement, 4'' Glossy silica wafer is put on a central anchor bolt while border screws give the desired strain to the wafer as is shown in Fig. 3. It is possible to give a spherical strain focused on grid with a resolution of about 1000 μstrain and a max stress of 200.000 μstrain.

Fig. 3: experimental setup for strain measurement on glossy silica wafer: 1) grating attached on silica wafer; 2) screw for strain regulation; 3) LEDs crown around the ccd camera lens.

Measurements show that ccd images are able to detect stress under 1000 μstrain. Grid image under strain, subtracted from the reference one, appears as reported in Fig 4. Blue Channel shows black areas suggesting that in this areas intensity of blue radiation didn't change under stress. Perhaps this can due to the saturation in this area of the blue channel of the ccd. All three primary colour channels show black lines that edge areas where the camera records a change of intensity. Although the stress applied on PDMS is uniform, images are complex. We are studying correlation between images of grid under stress and how diffraction changes when the grid change its path due to a deformation. Black lines can be linked to the force-lines in the PDMS highlighting non homogeneity in the material. Red, Blue or green radiation should indicate with different images the same information on the grid deformation.

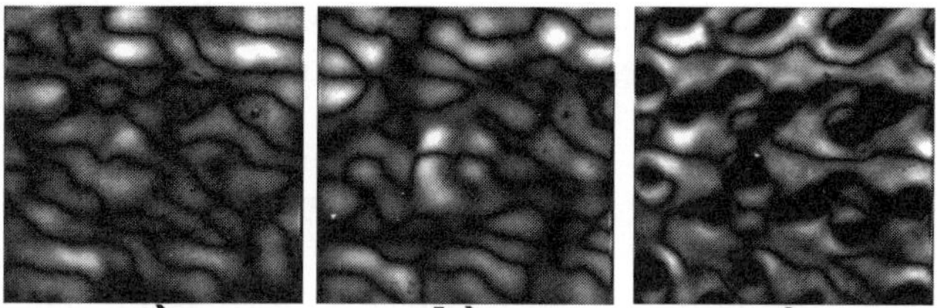

a) b) c)

Fig. 4: ccd image subtraction from reference of Polymeric grid, posted on glossy silica **under uniform stress** of about 200.000 μstrain. a) channel RED; b) channel GREEN; c) channel BLUE.

Measurements varying stress intensity are made by recording 16 shots in 4 second while screw for strain regulation change stress intensity on silica Wafer between 1.000 μstrain and 200.000 μstrain.
Shots reported that there is only a change on the intensity of each pixel but not on the picture, i.e., correlation between pixel is unaffected by stress. This is reported in Fig. 5 where, for example, we compare two image: a) channel green (linearly amplified to make visible the image of PDMS grid under about 2.000 μstrain and b) channel green (no amplification) same grid with 200.000 μstrain applied on the silica wafer.

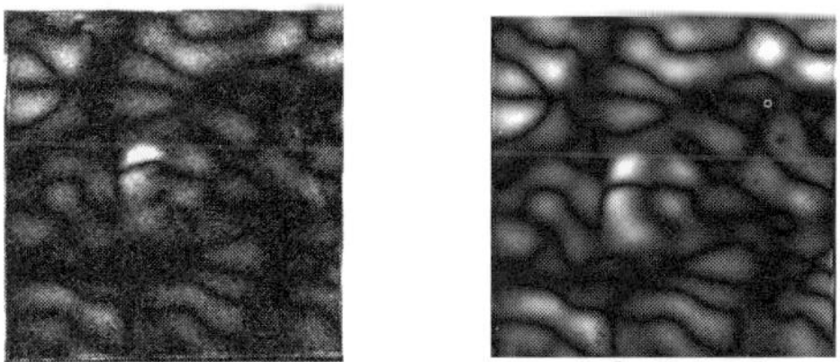

a) low stress b) high stress

Fig. 5: comparison between two diffraction image (channel green) of PDMS grid applied on glossy silica wafer under two strain strenght: a) about 2000 μstrain; b) about 200.000 μstrain.

Picture remains the same while S/N rate is better for the high stress figure (b).

Work is in progress for studying diffraction PDMS grid images with non uniform stress, non linear distortions like stretching and torsion applied on the specimen.

4. Conclusions

This study demonstrates the possibility of develop an optical strain sensor using PDMS diffracting grids, realized by replica technique, and a commercial optical system for the local strain detection. We show how diffraction grid images can give information on uniform strain applied on glossy silica wafer. Work is in progress to understand the optical behaviour of this diffractive grid in more complex strain scenarios.

References

1. A. Asundi, ''Developments in the moire´ interferometric strain sensor,'' Holographic Interferometry and Speckle Metrology, Proc. SPIE **1163**, 63–68 (1989).
2. J. F. Bell, ''Determination of dynamic plastic strain through the use of diffraction gratings,'' J. Appl. Phys. **27**, 1109–1113 (1956).
3. Y. Ma and M. Kurita, ''Strain measurement using high-frequency diffraction grating,'' JSME Int. J. A, 36 (3), 309–313 (1993).
4. K. Hosokawa, K. Hanada, R. Maeda, "A plydimethylsiloxane (PDMS) deformable diffraction grating for monitoring of local pressure in microfluidic devices,'' J. Micromech. Microeng. **12**, 21057-8 (2002)
5. W.C. Chuang, C. T. Ho, Y.R. Luan, C.K. Chao, R.F. Shyu, W.C. Wang, "transducing Mechanical Forces Using a Polymer Optical Grating Sensor", Materilas Cience Forum, 505-507, 91-96, (2006)

SYSTEMS AND ELECTRONIC INTERFACES

UNCALIBRATED HIGH-DYNAMIC RANGE RESISTIVE SENSOR FRONT-END WITH PARALLEL CAPACITANCE ESTIMATION

ANDREA DE MARCELLIS, GIUSEPPE FERRI, VINCENZO STORNELLI

Department of Electrical and Information Engineering, University of L'Aquila, Monteluco di Roio, 67100 L'Aquila, Italy

ALESSANDRO DEPARI, ALESSANDRA FLAMMINI

Department of Electronic for Automation, University of Brescia, via Branze 38, 25123 Brescia, Italy

In this paper we present an uncalibrated CMOS fully-integrable interface, based on oscillating topology, for wide-range resistive gas sensor applications. The proposed circuit is a low-voltage, low-cost and very simple front-end, able to reveal with good linearity and precision more than six decades of resistance variation and, at the same time, to estimate the sensor parasitic capacitance up to 50 pF. Moreover, the interface has also been designed, at transistor level, in a standard CMOS technology (AMS 0.35μm). In this case, no external resistors or capacitors are required, since passive component values are all integrable. This integrated solution has been developed so to be powered with a low supply voltage (2V) and to have characteristics independent both from supply variations and temperature drifts. This makes the proposed circuit suitable for portable applications. Preliminary experimental results, performed through a discrete element board fabricated with commercial components, have demonstrated the validity of the proposed interface and a good agreement between experimental results and theoretical values.

1. Introduction

In microsensor applications and electronic noses, resistive gas sensors can show either base-lines varying in a very high range or resistive values heavily dependent on reagent concentrations. This wide range can include very high resistances, in the order of tens of GΩ, owed to the use of new materials and new fabrication processes and to reduced power consumption. For these applications, the implementation of oscillating circuits, performing a resistive-to-period (R-T) conversion, seem the best solution as first electronic front-end, because output signal is very easy to manage and especially because other kind of interfaces cannot give a wide output range without the use of either scaling factors or high-resolution pico-ammeters [1-4]. Moreover, sensor signal conditioners with frequency output offer a number of benefits, when compared to voltage output circuits, such as good noise immunity and easiness in data processing. Unfortunately, sensor behavior cannot be modeled by a pure resistance, but the equivalent circuit of the sensing element appears as the parallel between a resistance and a capacitance, whose value is quite low (few pF) [5]. This is due either to the miniaturization processes employed in sensor fabrication or to the presence of the heating element. Therefore, the parasitic capacitances must be estimated so not to affect the measurement accuracy. In this work we present a CMOS fully-integrable interface for wide-range resistive gas sensors based on oscillating topology, which does not need any calibration, able to reveal more than six decades of resistance variation and, at the same time, to estimate the sensor parasitic capacitance.

2. The Proposed Front-End

Figure 1 shows the block scheme of the proposed front-end: it is a low-voltage, low-cost solution and presents a very simple circuit based on an oscillating topology. This interface has been designed at transistor level in a standard CMOS integrated technology (AMS 0.35μm), implementing a completely integrable solution, powered with a low supply voltage and showing characteristics independent from both supply variations and temperature drifts. In the integrated solution, no external resistors or capacitors are required, since passive component values are all integrable, so it is suitable for very compact portable applications.

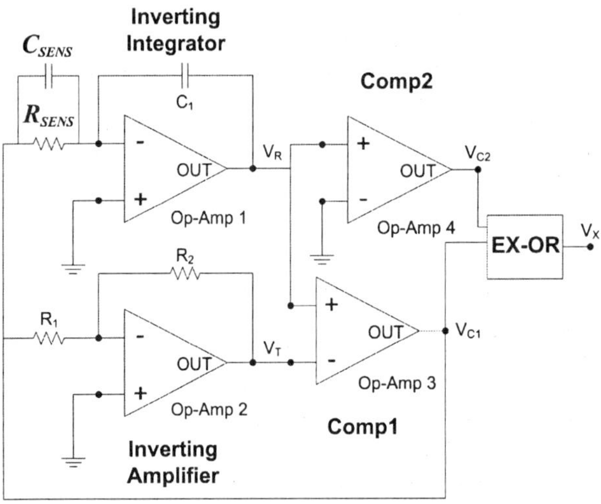

Figure 1: Block scheme of the proposed interface.

Interface design has been performed so to minimize the following error sources: finite Op-Amp Slew-Rate, finite open loop voltage gain, asymmetrical sensor excitation voltage, inverting integrator input bias current and offset voltage, comparator input offset voltage and time delays. In this sense, the active blocks used as comparators and amplifiers have been implemented simply by Operational Transconductance Amplifiers (OTA), whose internal topology, at transistor level, is shown in Figure 2. Considering Figure 1, Comp1 generates a square wave voltage signal (V_{C1}), as it shown in Figure 3, whose period is proportional to sensor resistance R_{SENS} and, under the hypothesis of R_{SENS} and C_{SENS} constant during the measuring time, it has the following expression:

$$T = 4GC_1R_{SENS}\left(1-\left(\frac{C_{SENS}}{GC_1}\right)\right) \tag{1}$$

being G the ratio between R_2 and R_1. From this equation, it comes that the proposed topology shows two degrees of freedom, in particular, C_1 and G, to select interface sensitivity. The EX-OR logic block, together with Comp2, allows to estimate the value of the sensor parasitic capacitance.

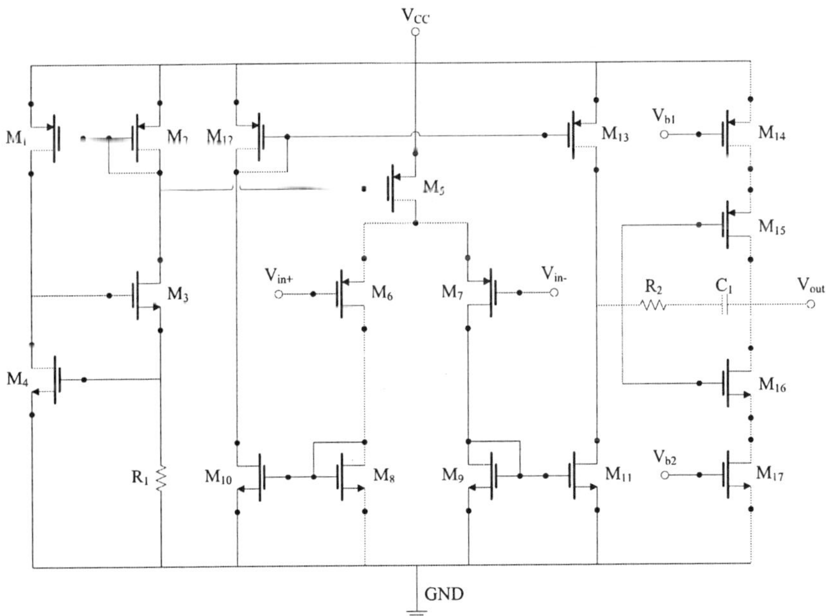

Figure 2. Internal topology of the operational amplifier at transistor level.

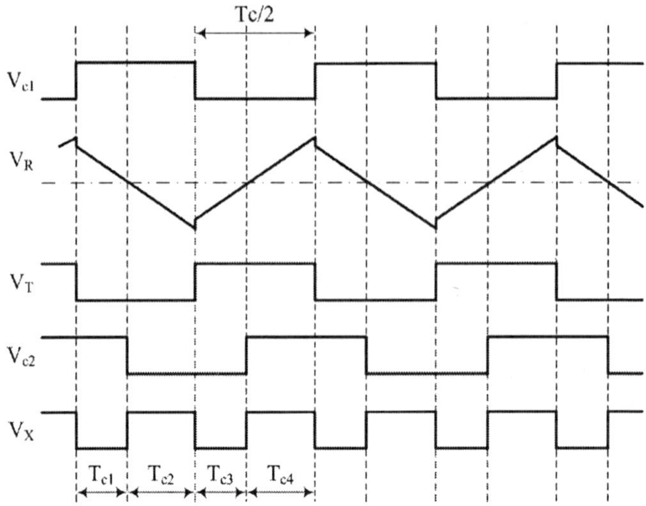

Figure 3: Voltage levels generated by each active block.

More in detail, considering Figure 3, the EX-OR generates a square wave signal, whose duty-cycle depends on sensing element capacitance, which allows to estimate both C_{SENS} and R_{SENS} values, according to the following expressions:

$$R_{SENS} = \frac{T_{C2} + T_{C4}}{2GC_1} \tag{2}$$

$$C_{SENS} = GC_1 \frac{T_{C2} + T_{C4} - T_{C1} - T_{C3}}{2T_{C2} + 2T_{C4}} \tag{3}$$

3. Experimental Results

Preliminary experimental measurements, performed through a discrete element board fabricated with commercial components, shown in Figure 4, have demonstrated the validity of the proposed interface and a good agreement between experimental results and theoretical expectations. In fact, the proposed front-end has been proved to reveal, with good linearity and precision, both more than six orders of magnitude of sensor resistive variations and a parallel capacitance up to 50 pF.

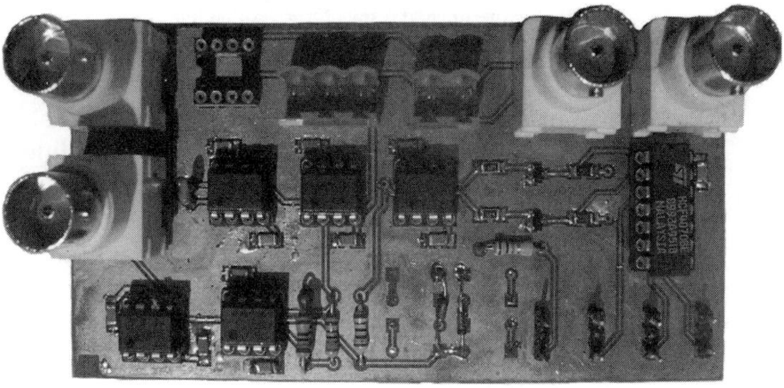

Figure 4: The fabricated prototype.

Experimental results show a relative standard deviation of R measurement always below 0.7%, while standard deviation of C is always lower than 0.05pF, as reported in Table 1 and Table 2, respectively, so confirming the validity of the proposed interface.

Table 1. Experimental results for Cs estimation.

C_S [pF] (expected) (RS=10GΩ)	$<C_S>$ [pF]	**Absolute error [pF] (LMS line)**
0	1.03	-0.067
1	2.02	-0.068
2.2	3.08	0.060
4.7	5.65	-0.049
10	10.76	0.077
15	15.54	0.234
22	22.87	-0.187

Table 2. Experimental results for Rs estimation.

R_S [MΩ] (expected)	<C_S> [pF] (expected 0pF)	<R_S> [MΩ]	Relative error [%] (WLMS line)
0.1	1.95	0.10	-0.095
1	-0.24	0.99	0.860
10	0.08	9.93	0.619
100	0.36	99.44	0.443
1000	0.59	1000.71	-0.193
10000	1.03	10050.48	-0.627
100000	0.48	100942.40	-1.065

References

1. L. Fasoli, F. Riedijk, J. Huijsing, IEEE Transactions on Instr. and Meas., Vol. 46, No. 4, 954, (August 1997).
2. M. Grassi, P. Malcovati, A. Baschirotto, Proceedings of Eurosensors XVIII Rome, 513, (September 2004).
3. A. Flammini, D. Marioli, A. Taroni, IEEE Transactions on Instr. and Meas., Vol. 53, N. 4, 1052, (August 2004).
4. M. Grassi, P. Malcovati, A. Baschirotto, Proc.ESSCIRC Montreaux, (September 2006).
5. G. Ferri, V. Stornelli, A. De Marcellis, A. Flammini, A. Depari, Proc.11th IMCS, Brescia, (July 2006).

A 77 HZ LOCK-IN AMPLIFIER FOR SENSOR APPLICATIONS

GIUSEPPE FERRI, ANDREA DE MARCELLIS, VINCENZO STORNELLI
Department of Electrical and Information Engineering, University of L'Aquila, Monteluco di Roio, 67100 L'Aquila, Italy

ARNALDO D'AMICO, CORRADO DI NATALE, CHRISTIAN FALCONI, EUGENIO MARTINELLI
Department of Electronic Engineering, University of Tor Vergata,00060 Roma, Italy

In this paper we propose a low supply voltage integrated lock-in amplifier, completely designed at transistor level and suitable for those sensor applications where the signal to be measured is buried into noise. The proposed system has been designed for operating with a 77 Hz reference signal and all its blocks have been internally designed, at transistor level and supplied at 2V, in a standard CMOS technology (AMS 0.35μm), so to have a completely integrated solution. Waiting for the chip fabrication, we have also developed and fabricated a lock-in system through a discrete element board with commercial components, for preliminary analysis. It has been proved that the system is able to recover with success signal from noise till to 0.1 *S/N* ratios without performance degradation: this corresponds to the detection of very small quantities of gas.

1. Introduction

In some sensor applications the signal to be measured has a small amplitude and sometime is buried into noise. Generally in these cases a linear filtering operation is not sufficient to extract the signal information, so different strategies have to be adopted, such as the use of: lock-in amplifiers (analog and digital), waveform averages (as signal averages and box-car integrators), auto- and cross-correlators. All of these solutions, even if different, have the goal to reduce the noise bandwidth [1]. Among them, lock-in amplifier is a signal recovery instrument, utilized mainly in optics but also in those applications where the signal buried into noise has a small amplitude at a fixed and well-known frequency [2].

Commercial lock-in amplifiers [3-5] are large-sized and absolutely not suitable for portable applications. For this reasons we have designed a low supply voltage lock-in amplifier, at transistor level, completely integrable in a standard CMOS technology (AMS 0.35μm).
In the meantime, waiting for the chip fabrication, we have developed a lock-in system through a discrete element board with commercial components for preliminary measurements.

2. The Lock-In Amplifier

The proposed system has been designed for operating with a 77 Hz reference signal so avoiding any kind of interference with the 50 Hz net supply oscillation frequency and its harmonics. In its basic implementation, a lock-in amplifier uses a phase-sensitive detection strategy which measures the amplitude and phase of a noisy signal. In the architecture here proposed, depicted in Figure 1, the low amplitude signal embedded in noise, coming from the sensor, is amplified though an AC differential amplifier, whose block scheme is shown in Figure 2, and combined with a reference signal (having, as mentioned, the same frequency) into a Mixer (or PSD), whose block scheme is shown in Figure 3. The Mixer is followed by a low pass filter, with a very low cut-off frequency, which reduces the noise through a DC extraction whose characteristic time is related to the desired *S/N* output ratio. The low-pass filter has been implemented through the cascade of two Gm-C cells, as it shown in Figure 4.

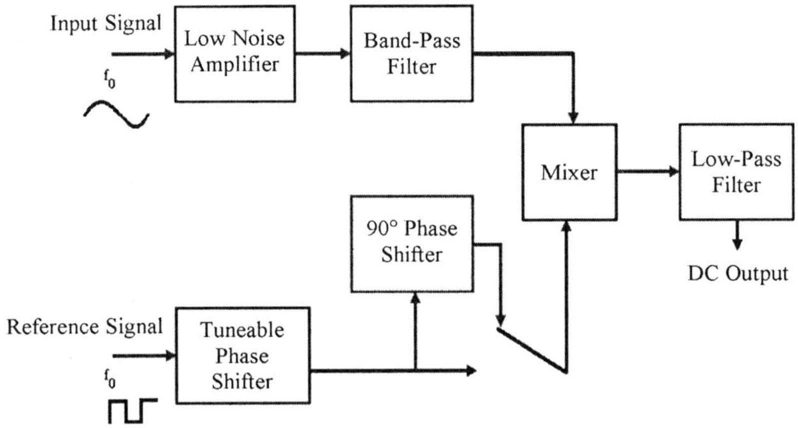

Figure 1. Block scheme of the proposed lock-in amplifier architecture. Upper path: signal channel. Lower path: reference channel.

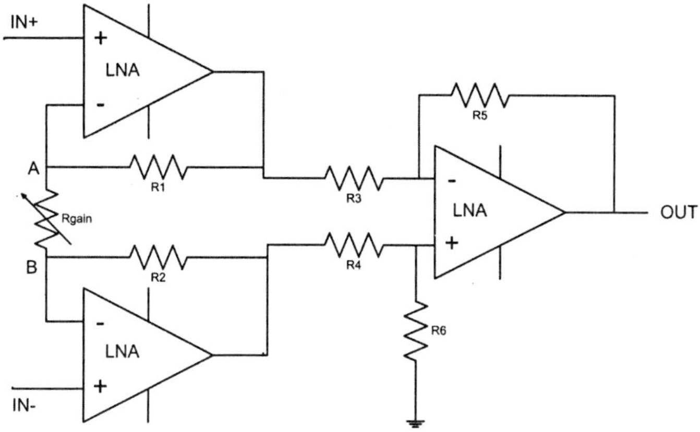

Figure 2. Differential instrumentation amplifier.

The filter output, in this manner, is a DC voltage signal whose level is proportional to the amplitude of the input signal (depending on the reference signal phase). All the lock-in blocks have been internally designed, at transistor level, in a standard CMOS technology, supplied at 2 V, achieving a complete integrated solution. In particular, it has been designed a Low Noise Amplifier (LNA), whose schematic circuit is shown in Figure 5, so to implement a more accurate differential instrumentation amplifier. Referring to Figure 2, each LNA, at 77 Hz, shows a 23nV/Hz½ equivalent input noise, while the complete differential instrumentation amplifier has a 34nV/Hz½ equivalent input noise, for a 92dB DC gain.

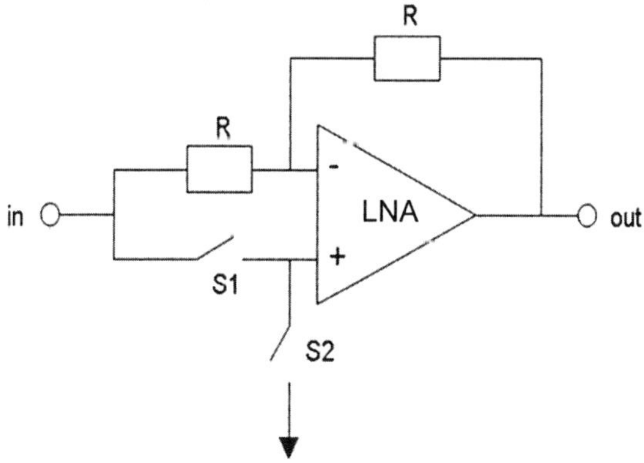

Figure 3. Block scheme of the Mixer (PSD).

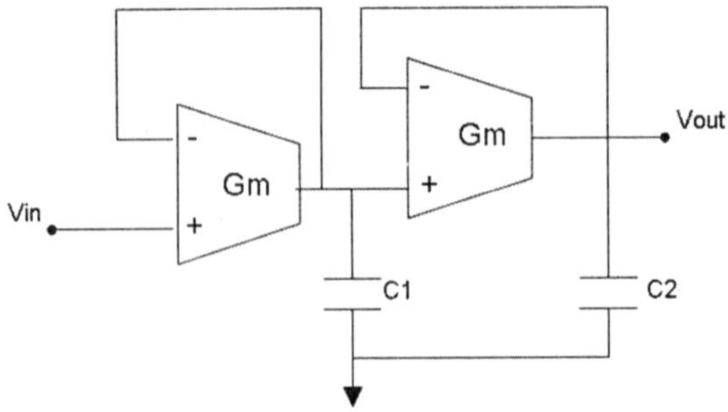

Figure 4. Block scheme of the active Low-Pass Filter.

3. Preliminary Experimental Results

Figure 6 shows the prototype board, fabricated for preliminary analysis, while Figure 7 depicts the measured mixer output signal, to be filtered by the final low-pass module, when the input sensor signal and the reference are "in phase". The designed lock-in system has been proved to recover with success signal from noise till to 0.1 S/N ratios without performance degradation. In Figure 8 the measurement system, comprehensive of the input conditioning passive circuit with the "equivalent" sensor resistance (*Rs*), is shown, being V_{REF} a 77 Hz sinusoidal wave, the same frequency of the voltage excitation signal, V_{ECC}.

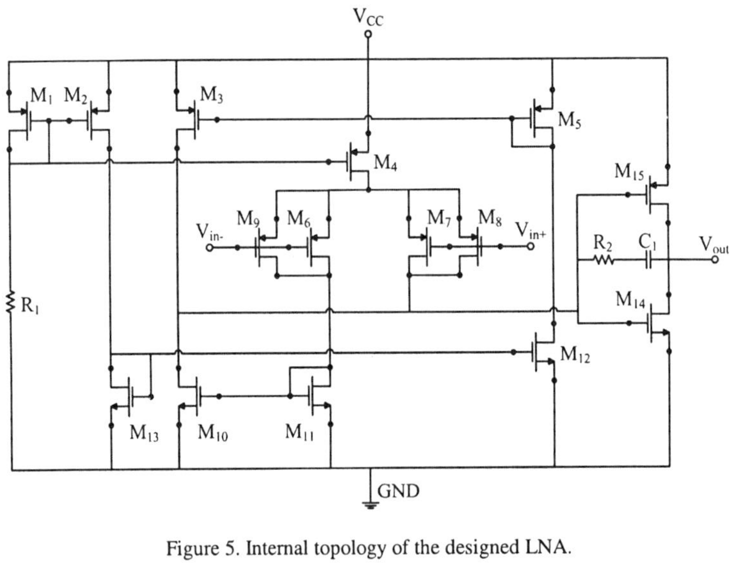

Figure 5. Internal topology of the designed LNA.

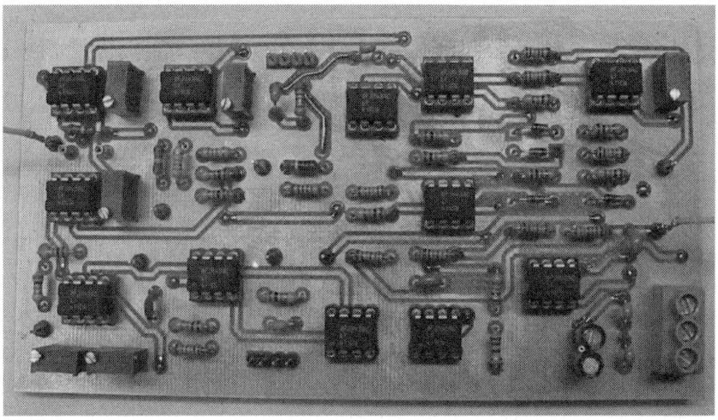

Figure 6. The fabricated prototype lock-in board.

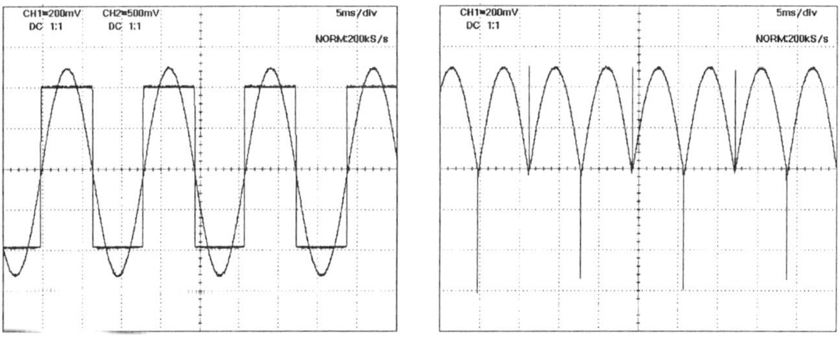

Figure 7. Measurement results: voltage at mixer output (right), with in-phase input and reference signals (left).

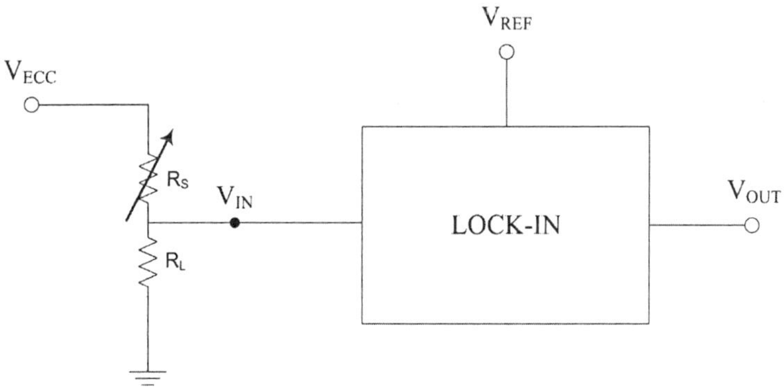

Figure 8. Measurement scheme for the proposed lock-in amplifier.

Through this scheme, we are aimed to detect very small *Rs* variations. More in details, the measurement results operated by prototype board are reported in Table 1: they are in a good agreement with theoretical values.

R_L [Ω]	R_S [Ω]	$V_{ECC\,p\text{-}p}$ [V]	AC Amplifier Gain	$V_{IN\,p\text{-}p}$ [V]	Measured DC V_{OUT} [V]	Theoretical DC V_{OUT} [V]
9.86 K	5.66 M	30 m	≈ 5691	52.17 μ	98.4 m	94.5 m
9.86 K	5.66 M	10 m	≈ 5691	17.39 μ	28.6 m	31.5 m
9.86 K	5.66 M	2 m	≈ 5691	3.48 μ	7.2 m	6.3 m
0.99 K	5.66 M	2 m	≈ 34142	0.350 μ	2.8 m	3.8 m

Table 1. Measurement results.

Acknowledgement

This work has been supported by PRIN Project N° 2005092937.

References

1. C.Falconi, E.Martinelli, C.Di Natale, A.D'Amico, F.Maloberti, P.Malcovati, A.Baschirotto, V.Stornelli, G.Ferri, "Electronic Interfaces", Accepted for publication on Sensors and Actuators B.
2. G.Ferri, P.De Laurentiis, C.Di Natale, A.D'Amico, Sensors and Actuators A, vol.92, pp. 263, (2001).
3. M.L. Meade, Peter Peregrinus Ltd (1983).
4. Lock-in amplifiers and pre-amplifiers, Princeton Appl. Res.Corp., data sheets (1971).
5. Lock-in amplifiers, application notes, Stanford Res.Syst., data sheets, (1999).

CCII-BASED OSCILLATOR FOR SENSOR INTERFACE

VINCENZO STORNELLI, GIUSEPPE FERRI, ANDREA DE MARCELLIS

Department of Electrical and Information Engineering,
University of L'Aquila, Monteluco di Roio, 67100 L'Aquila, Italy

In this paper we propose a low-voltage low-power application of a quasi-ideal second generation current conveyor (CCII) as resistive sensor interface. The proposed architecture, which has been implemented using a standard CMOS 0.35μm process, consists of a single block oscillating circuit performing a resistance to period conversion. The same topology could be also used for capacitive sensor interfacing.

1. Introduction

Sensor interfaces have recently assumed a fundamental importance in sensor systems. In fact, it is mandatory to design, close to the sensor itself, the electronics for controlling and reading the sensor answer which has to be processed and elaborated for a suitable data display. In sensor system research, there is actually a great demand in micro-systems and technologies which combine a low power consumption with excellent general performance. The main fields of applications are the following: environment, medicine, biotechnology, automotive, consumer electronics, etc.. The interfacing of the sensitive element with a suitable integrated circuit implements the so-called *smart sensors* [1-2]. In this sense, CMOS technology is widely used, because it allows to match the reduction of costs of the silicon technology with the possibility of designing new low voltage low power interface circuits [3-4]. In the scientific world, several gas sensor systems have been already developed, also in complete system environments, but many of them are characterized by large dimensions, large power consumption and high costs. Moreover, these interfaces are not always optimised for the specific sensor especially when the measure of high-value resistances is a target. This is, in fact, a crucial problem in many research and industrial applications. In this case, a R-T conversion is necessary because other kind of interfaces cannot give a wide output range without the use of either scaling factors or high-resolution pico-ammeters [5-6].

Current-mode design is actually a new challenge in sensor interface [7-8]. The basic current-mode analog block is the Second Generation Current Conveyor (CCII), which operates as both a current and a voltage buffer.
In this paper, we propose a low-voltage low-power application of a quasi-ideal CCII as resistive sensor interface in a single block oscillating circuit performing the R-T conversion. Simulated results are presented showing an excellent agreement with the theoretical expectations for one-two decades of resistive variation.

2. Second Generation Current Conveyor (CCII)

CCII is the current-mode basic block [7] which, in numerous applications, can favourably replace operational amplifiers, both in linear and nonlinear contexts. In an ideal CCII device (see Figure 1), if a voltage is applied at Y node, an equal voltage is produced at X node and the current flowing into X node is equal or opposite to the current flowing into Z node. Moreover, Z and Y nodes show infinite impedances and X node shows zero impedance. Positive (CCII+) and negative (CCII–) current conveyors are respectively obtained for $I_Z = I_X$ and $I_Z = - I_X$.

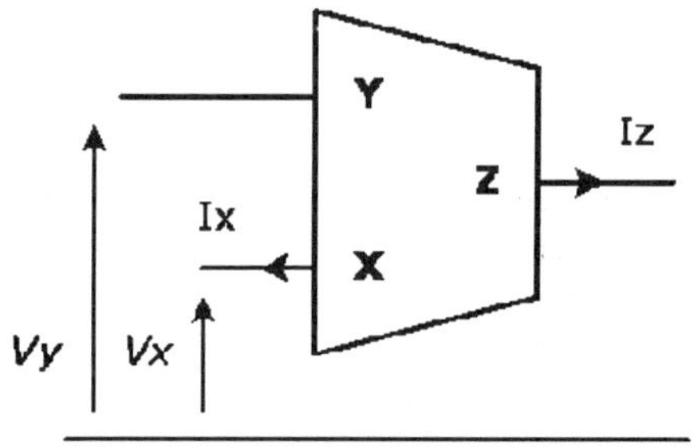

Fig. 1. Ideal CCII block scheme.

Typically, in CCIIs, a unity voltage transfer function V_X/V_Y is ensured by implementing a differential transistor input pair, but a residual offset due to components mismatch affects the follower operation.
Recently, we have designed, at transistor level, a quasi-ideal CCII showing negligible parasitic impedances and unitary voltage and current gains for a very large bandwidth [9]. The circuit, shown in Figure 2 and implemented using a standard CMOS 0.35μm process, is also able to operate at a reduced supply voltage (± 1.5V). In particular, it is formed by a differential input stage (M1-M7; Rc3), a class AB output stage (M8-M11; Rc1,Rc2; M16-M17) and a Wilson current mirror (M12-M21).

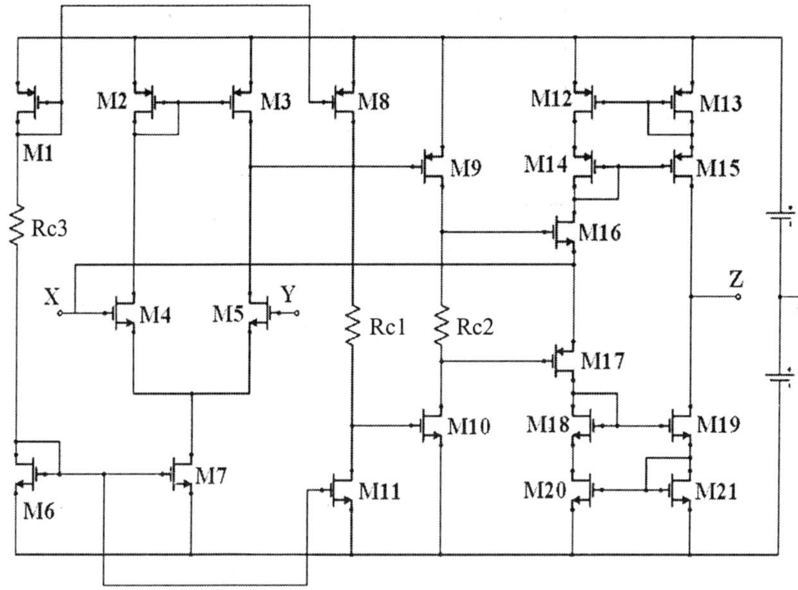

Fig. 2. Quasi ideal CCII schematic at transistor level.

3. Proposed CCII Oscillator

In Figure 3 the proposed CCII-based oscillator circuit is presented, where CCII parasitic effects have been highlighted at X and Z nodes. Routine analysis of Figure 3 circuit gives the following expression for the oscillation frequency:

$$f_o = \frac{1}{2\pi\sqrt{C_1\left(C_2 + C_Z\right)R_1R_2}}$$

From this equation, we can notice that the circuit can work as both resistive and capacitive sensor interface.

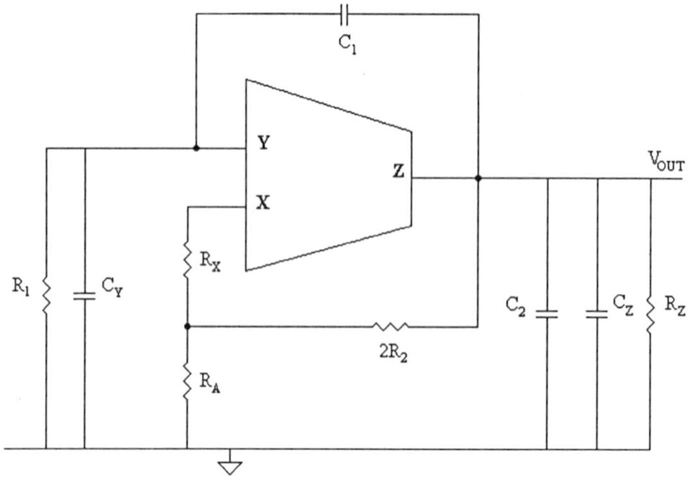

Fig. 3. Proposed CCII based oscillator.

In Figure 4 the simulated oscillation frequency vs. R_1 variation (dashed line: simulated, continued line: theoretical) is depicted, considering all integrable values for passive components. The simulated results are in an excellent agreement with the theoretical expectations. Since the circuit is a sinusoidal oscillator, the interface works in a relatively reduced range of resistance variation (one-two decades).

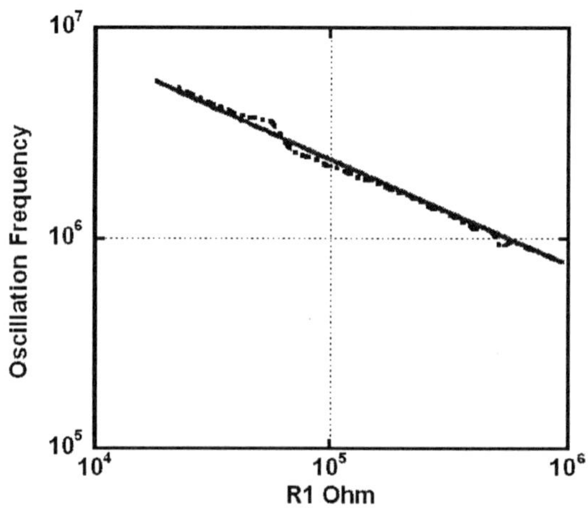

Fig. 4. Output frequency vs. R_1 resistance.

4. Conclusions

In this paper we have presented a low-voltage low-power quasi-ideal CCII-based resistive (or eventually capacitive) sensor interface circuit implemented through a single block oscillating circuit performing the R-T (or C-T) conversion. The topology, designed in a standard CMOS technology, shows a relatively simple implementation at transistor level.

References

1. S.Middeloeck, S.A.Audet, P.French, *Silicon Sensors*, Acad.Press, London, 2000.
2. R.Pallas-Areny, J.G.Webster, *Sensors and signal conditioning*, Wiley Ed., 2001.
3. A.Bakker, J.H.Huijsing, IEEE Journal of Solid State Circ. 31 (7) 1996, pp.933-937.
4. E.Wouters, M. De Cooman, R.Puers, IEEE Journal of Solid State Circ. 29(8) 1994, pp.952-956.
5. V.Ferrari, C.Ghidini, D.Marioli, A. Taroni, Proc.IMTC 1999, vol.2 pp.1233-1238.
6. G.Ferri, V.Stornelli, A. De Marcellis, A. Flammini, A. Depari, Proc. IMCS 2006.
7. G.Ferri, N.Guerrini, *Low voltage low power current-conveyors*, Kluwer Ac.Publ., Boston, 2003.
8. C.Cantalini, G.Ferri, N.Guerrini, S.Santucci, Proc. ICM 2004.
9. G.Ferri, V.Stornelli, M.Fragnoli, Analog Integr. Circuits and Signal Processing, vol.49 (3), Sept. 2006, pp.247-250.

INTEGRATED WIRELESS TEMPERATURE SENSOR WITH ON-CHIP ANTENNA

F. AQUILINO, M. MERENDA, F.G. DELLA CORTE

DIMET, University "Mediterranea" of Reggio Calabria, Via Graziella, Loc. Feo di Vito, 89060 Reggio Calabria, Italy

The widest deployment of wireless sensors may take advantage from the integration of an antenna realized directly on chip [1], [2].

In this paper a new wireless temperature sensor made with a standard 0.8 μm VLSI process, with an on-chip antenna, is presented. An on-chip antenna allows the elimination of external components, which are usually placed on an additional PCB. As a result, extreme system miniaturization is obtained. On the other hand, system efficiency is degraded by the antenna low Q factor in the practical transmission frequency ranges.

The realised chip includes a 3-stage ring oscillator structure which transforms the silicon substrate temperature variation into a frequency modulation. The signal is transmitted by a small loop antenna structure which is realized by aluminium deposition on the top surface of the chip. The antenna is connected between the input and output pins of the ring oscillator.

1. Introduction

Compared to traditional temperature sensors such as platinum resistors and thermocouples, integrated temperature sensors have the advantage that signal conditioning and interface electronics can be integrated on the sensor chip. Furthermore, integrating traditional sensors into a chip requires extra fabrication steps and materials, while an integrated sensor can be made in a standard high volume IC process, so production costs can be minimized. Sometimes such smart sensors provide a standardized digital output signal, which makes it easier to incorporate them in a measurement system.

In contrast to the classic wired configuration, a wireless communication channel might pave the way to new unexplored applications. In this case the output signal can be transmitted using an antenna realized directly on chip, eliminating the need for external transmission line connections and sophisticated packaging, which can radically reduce cost of IC systems.

During the last ten years the feasibility of an integrated antenna has been investigated [1], [2]. This technology can potentially be applied to the implementation of a true single-chip radio for general purpose communication, on-chip and inter-chip data communication systems, RFID tags, RF sensors/radars, and others.

A transceiver with on-chip antenna forms a radio-chip that can provide a communication link for sensor network nodes. The nodes can have a size of the order of mm^3 and are sufficiently inexpensive. Such nodes could help accelerating the realization of the Smart Dust vision [3]. An on-chip antenna has also been used as a transformer in radio frequency identifiers (RFIDs) [4]. For this application, the use of on-chip antennas completely eliminates the need for bondwire connections.

In this paper a new wireless temperature sensor made with a standard 0.8 μm VLSI process, with an on-chip antenna, is proposed. The chip actual area is 7.9 mm^2, while the active area is 0.63 mm^2. The circuit meeting the design goals consists of a 3-stage ring oscillator made of CMOS inverters which transforms the silicon substrate temperature variation into a frequency modulation. The signal is transmitted by a small loop antenna structure which is realized by aluminium deposition on the top surface of the chip. The antenna is connected between the input and output nodes of the ring oscillator.

2. Temperature Dependence of the Ring Oscillator frequency

The ring oscillator is inherently a temperature-sensitive circuit. Silicon substrate temperature variations cause oscillator's frequency modifications. The critical parameters that vary with temperature in MOS transistors are the charge carriers mobility and the threshold voltage. The relationships governing the variation can be approximately given by [5], [6]:

$$\mu \propto T^{-3/2}, \tag{1}$$

$$V_T(T) = V_{T_0} - K(T - T_0) = V_{T_0}(1 + \alpha_{V_T} T), \tag{2}$$

where the temperature coefficient α_{V_T} is negative. A similar linear dependence with a negative temperature coefficient can also be used for the oxide capacitance:

$$C_{ox_T}(T) = C_{ox_{T0}}(1 + \alpha_{C_{oxT}} T). \tag{3}$$

The time delay introduced by a single stage CMOS inverter is given by:

$$t_d \approx \frac{C(V_H - V_L)}{I_d}, \tag{4}$$

where C is the total capacitance seen at the output of the considered stage ($C \approx C_{oxT}$), I_d is the bias current of the circuit:

$$I_d \approx \frac{1}{2}\mu C_{ox}\frac{W}{L}(V_{GS}-V_T)^2, \tag{5}$$

and $(V_H - V_L)$ is the output voltage swing.

The oscillation frequency is:

$$f = \frac{1}{Nt_d}, \tag{6}$$

where N is the number of delay stages and t_d the time delay of each element.

In (2) a temperature variation, for example an increase, causes a lowering of V_T. However this does not produce an increase in current, as expected according to (5). In fact the drain current of a MOS device in saturation is directly proportional to the mobility and to the gate capacitance, which decrease according to (1) and (3). Furthermore the effects of mobility and gate capacitance are dominant in (5), so the total effect of a temperature increase is a current reduction.

According to (4), this induces a time delay increase, hence a frequency downshift according to (6).

3. Chip Design

3.1. *Circuit Design*

Following a design approach similar to that described in [7], the circuit has been accurately designed to obtain a linear dependence of the frequency with temperature. The circuit has been simulated using Virtuoso Spectre Circuit Simulator. The p-MOS to n-MOS channel width ratio was set to $W_p/W_n=4$, which determined the oscillation frequency to be about 750 MHz @ 30 °C. Figure 1 shows the harmonic contents of the output periodic signal at different temperatures obtained simulating the circuit: the shifting of the frequencies with temperature is substantial linear.

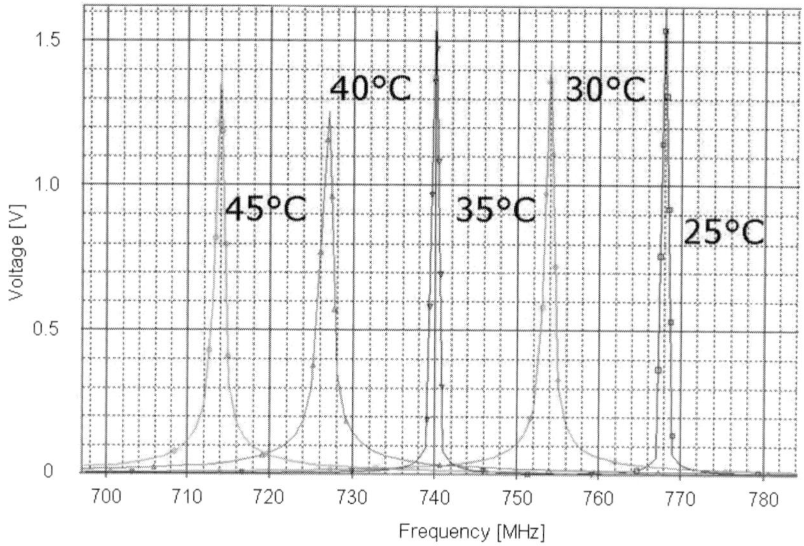

Figure 1. FFT of the output signal obtained simulating the circuit.

3.2. *Antenna Design*

The feasibility of an on-chip antenna integration has been fully demonstrated only recently with a wide use of dipole and patch antennas in RF interfaces. For example, on chip integration of a conventional resonating antenna for wireless communication among different sub circuits integrated on the same chip has been demonstrated [2].

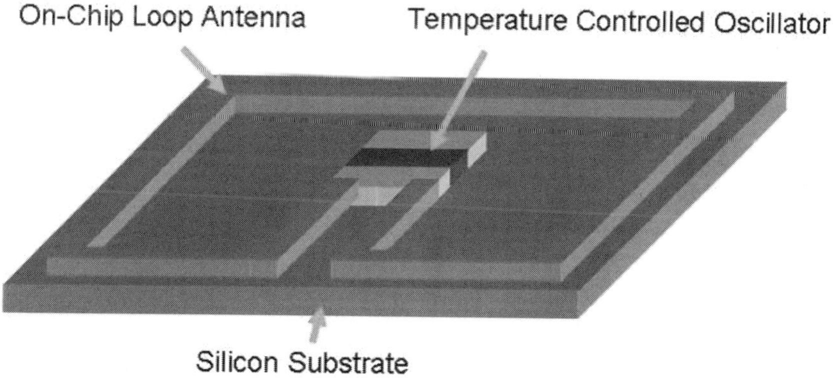

Figure 2. The integrated antenna model.

Figure 2 shows the model of the antenna, which is in fact the feedback lined of the ring oscillator.

Two different sources of losses exists, affecting the radiation efficiency: losses due to the metal conductivity (these losses are neglected) and losses in substrate volume. A lossy substrate decreases the radiation efficiency, which depends on the conductivity of the substrate material.

Comsol Multiphysics has been used to simulate the radiating element, with the aim of estimating the emitted electromagnetic field and the effect of the Si substrate on it. In addition, the 2D EM simulator Designer has been used for comparison purpose.

The simulations shows that the substrate is responsible for a reduction of the electric field of about 2 or 3 orders of magnitude when compared to a loop antenna positioned in the free space. In the former case, the electric field intensity is of the order of tens μV/m at a distance of 10 cm (figure 3).

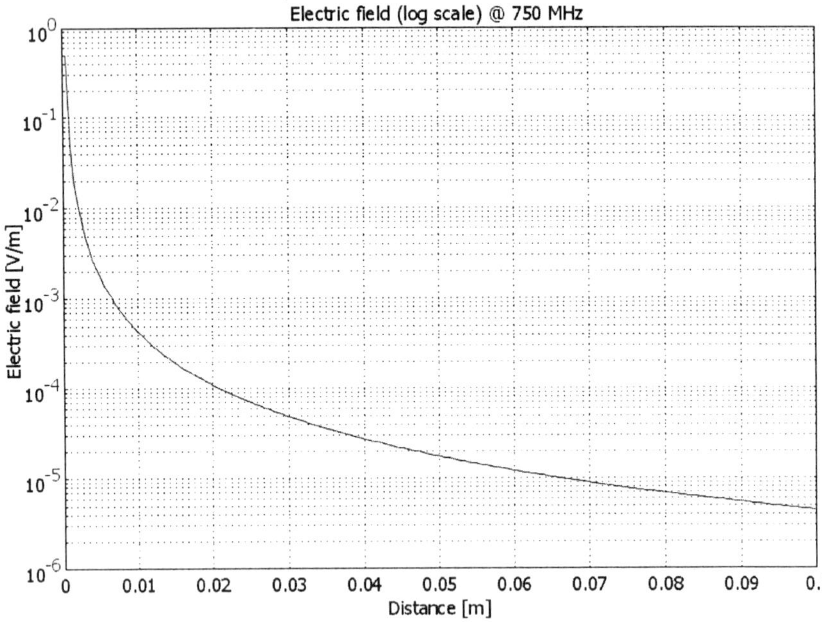

Figure 3. Electric field intensity versus distance along the axis of the loop antenna.

The sensor has been fabricated on a 300 μm thick Si wafer with a resistivity of 10 Ω·cm. The oxide layer thickness is 1 μm and its resistivity is 10^9 Ω·cm. Over this layer a 1 μm thick aluminum layer for antenna formation has been deposited. Antenna width is 10 μm. There is no ground plane in the bottom of the chip.

In figure 4 a microphotograph of the realized chip is shown; antenna is highlighted with a white line depicted on the core.

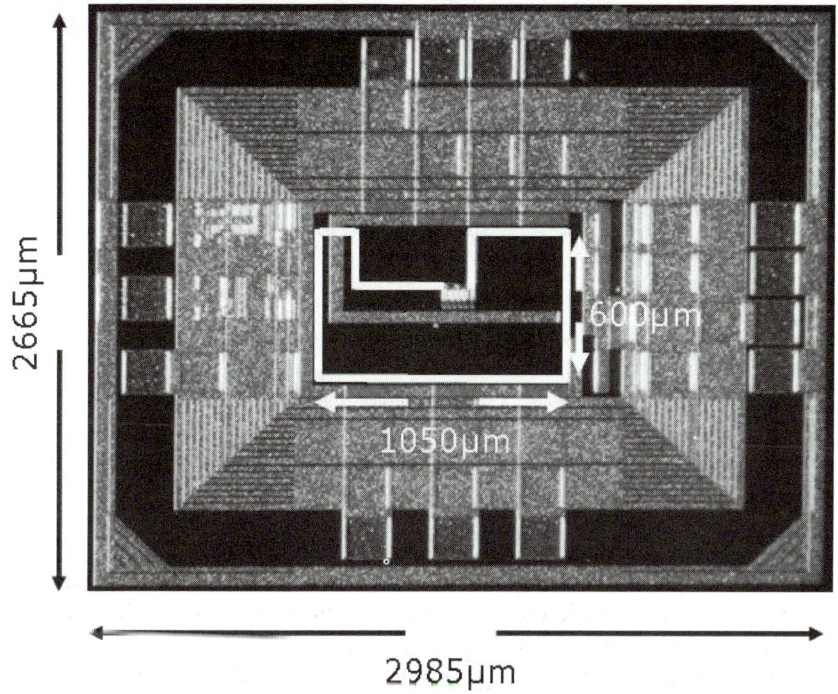

Figure 4. Chip microphotograph: antenna is highlighted by a white line depicted on the core.

4. Experimental Results and discussion

RF power measures performed in an anechoic chamber using a wide band dipole as the receiving antenna and a spectrum analyzer (Advantest R3131) have shown a power level of about -61.5 dBm at a temperature of 30 °C at a distance of about 10 cm, in good agreement with simulations.

The test chip has been characterized using a PID temperature-controlled oven. Frequency vs. temperature data curve at 3.0 V bias voltage is shown in figure 5. The sensor shows a sensitivity of 1.8 MHz/°C and a resolution of ± 0.025 °C with the used setup measure. The theoretical f(T) curve is shown for comparison on the same plot.

The frequency stability over time has been tested over a time period of 12 hours. The maximum observed deviation from the centre frequency has been of 0.189 MHz @ 30 °C and 3.0 V bias voltage.

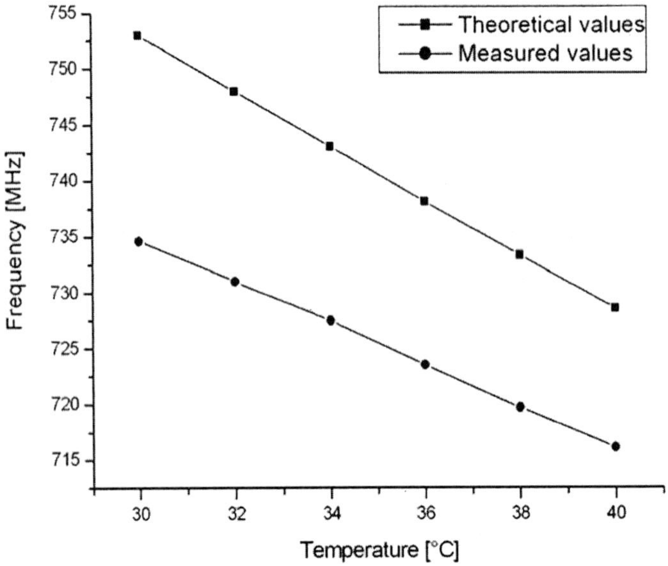

Figure 5. Frequency vs. temperature data curves of theoretical and measured values @ 3.0 V bias voltage.

The chip nominal voltage is 3.0 V and the power dissipated is 3.0 mW. To decrease the power consumption and avoid the consequent moderate self heating phenomenon, in future the oscillator might have a duty cycle of 1/1000, so the power consumption would be only 3.0 μW.

In consequence of its dimensions, antenna is far from resonance conditions and therefore its gain and radiation diagram are of minor relevance for this applications.

The design of the circuit elements was targeted at obtaining a linear dependence on temperature. However, another important effect to be taken into account concerns the frequency dependence on the circuit bias voltage. The measures has been done biasing the sensor with a constant voltage, but an autonomous chip may suffer of bias voltage oscillations. It is not simply to take into account this effect. A possible solution regards a voltage stabilizer circuit integrated on the same chip.

5. Conclusion

A new wireless temperature sensor made with a standard 0.8 μm VLSI process, with an on-chip antenna, has been performed. Temperature is sensed by a 3-

stage ring oscillator of CMOS inverters. The sensed data are transmitted by a small loop antenna structure realized by aluminium deposition on the top surface of the chip.

Measurement results show linear frequency dependence with temperature, a sensitivity of 1.8 MHz/°C and a resolution of ± 0.025 °C with a -61.5 dBm RF power level at a distance of about 10 cm.

A potential improvement of the transmitted power level signal is expected from the tuning of the on-chip antenna gain.

Additionally, the chip compatibility with VLSI processes allows to optimize the system with respect to an autonomous power supply, as a solar cell, a micro-battery or both of them.

The system can be seen as the core element of an intelligent active RFID tag, with a great improvement represented by on-chip antenna. Furthermore, with respect to a RFID system, there are some advantages in terms of packaging costs reduction.

References

1. K. Kim, H. Yoon, K. O Kenneth, "On-Chip Wireless Interconnection With Integrated Antennas", *IEEE-Electron Devices Meeting*, 2000, IEDM Technical Digest, International.
2. Kenneth K. O et al., "On-Chip Antennas in Silicon ICs and Their Application", *IEEE Trans. On Electron Devices*, vol. 52, no. 7, pp 1312-1323, Jul. 2005.
3. B. Warneke, M. Last, B. Liebowitz, K.S.J. Pister, "Smart Dust: Communicating with a Cubic-Millimeter Computer", *IEEE Computer*, Jan. 2001.
4. M. Usami, A. Sato, K. Sameshima, K. Watanabe, H. Yoshigi, R. Imura, "Powder LSI - An Ultra Small RF Identification", *IEEE International Solid-State Circuit Conference*, Feb. 2003.
5. S. Selberherr, "Analysis and Simulation of Semiconductor Devices", *Springer-Verlag*, 1984.
6. K. Sundaresan, P. E. Allen, F. Ayazi, "Process and Temperature Compensation in a 7-MHz CMOS Clock Oscillator", *IEEE Journal of Solid-State Circuits*, vol. 41, no. 2, pp. 433–442, Feb. 2006.
7. S. R. Boyle, R. A. Heald, "A CMOS Circuit for Real-Time Chip Temperature Measurement", *IEEE Compcon Spring '94, Digest of Papers*, pp. 286–291, 1994.